Understanding Basic Statistics

Understanding Basic Statistics

Concepts and Methods

SECOND EDITION

Charles Henry Brase
Regis University

Corrinne Pellillo Brase
Arapahoe Community College

HOUGHTON MIFFLIN COMPANY
Boston   New York

Editor-in-Chief: Jack Shira
Sponsoring Editor: Paul Murphy
Development Editor: Mary Beckwith
Editorial Assistant: Marika Hoe
Senior Project Editor: Tamela Ambush
Senior Production/Design Coordinator: Jennifer Meyer Dare
Senior Manufacturing Coordinator: Sally Culler

Cover Design: Minko T. Dimov, MinkoImages
Cover Photograph: Chris Thomaidis, Canada, Toronto, detail of balconies
of harbourfront condo. Tony Stone Images.

Viewpoint artist: Lauren Arnest

Chapter Opener photo credits: **Chapter 1:** Gamma Liaison; **Chapter 2:** The
Granger Collection, New York; **Chapter 3:** Manni Mason's Pictures; **Chapter 4:**
Brown Brothers; **Chapter 5:** National Portrait Gallery; **Chapter 6:** Keystone Press
Agency; **Chapter 7:** Ralph Steiner; **Chapter 8:** Culver Pictures; **Chapter 9:** Brown
Brothers; **Chapter 10:** printed by permission of the Norman Rockwell Family
Trust, copyright © 1953 the Norman Rockwell Family Trust; **Chapter 11:** Glenn
Asakawa/Rocky Mountain News.

Minitab is a registered trademark of Minitab, Inc.; SAS is a registered trademark of
SAS Institute, Inc.; SPSS is a trademark of SPSS, Inc. TI is a registered trademark of
Texas Instruments, Inc.

Screen shots reprinted by permission of Microsoft Corporation.

Printed in the U.S.A.

Library of Congress Catalog Card Number: 00 - 133802

ISBN
Student Text: 0-618-05470-7

123456789-DOC-04 03 02 01 00

*This book is dedicated to the memory of
a great teacher, mathematician, and friend*

Burton W. Jones

Professor Emeritus, University of Colorado

CONTENTS

1 ORGANIZING DATA   1

2 AVERAGES AND VARIATION   69

3 REGRESSION AND CORRELATION   120

7 INTRODUCTION TO SAMPLING DISTRIBUTIONS   319

8 ESTIMATION   347

9 HYPOTHESIS TESTING INVOLVING ONE POPULATION   399

APPENDIXES

Table of Prerequisite Material

Chapter		Prerequisite Sections
1	Organizing Data	None
2	Averages and Variation	1.1, 1.2, 1.4
3	Regression and Correlation	1.1, 1.2, 2.1, 2.2
4	Elementary Probability Theory	1.1, 1.2
5	The Binomial Probability Distribution	1.1, 1.2, 1.4, 2.1, 2.2, 4.1, 4.2
6	Normal Distributions (omit 6.4) (include 6.4)	 1.1, 1.2, 1.4, 2.1, 2.2, 4.1, 4.2, 5.1 5.2, 5.3 also
7	Introduction to Sampling Distributions	1.1, 21.2, 1.4, 2.1, 2.2, 4.1, 4.2, 5.1
8	Introduction to Estimation (omit 8.3 and parts of 8.4) (include 8.3 and all of 8.4)	 1.1, 1.2, 1.4, 2.1, 2.2, 4.1, 4.2, 5.1, 6.1, 6.2, 6.3, 7.1, 7.2 5.2, 5.3, 6.4 also
9	Hypothesis Testing Involving One Population (omit 9.5) (include 9.5)	 1.1, 1.2, 1.4, 2.1, 2.2, 4.1, 4.2, 5.1, 6.1, 6.2, 6.3, 7.1, 7.2 5.2, 5.3, 6.4 also
10	Inferences About Differences (omit 10.4) (include 10.4)	 1.1, 1.2, 1.4, 2.1, 2.2, 4.1, 4.2, 5.1, 6.1, 6.2, 6.3, 7.1, 7.2, 8.1, 8.2, 9.1, 9.2, 9.3, 9.4, 5.2, 5.3, 6.4, 9.5 also
11	Additional Topics Using Inference 11.1, 11.2, 11.3 11.4, 11.5	 1.1, 1.2, 1.4, 2.1, 2.2, 4.1, 4.2, 5.1, 6.1, 6.2, 6.3, 7.1, 7.2, 9.1, 9.2, 9.3 Chapter 3, 8.1, 8.2 also

P R E F A C E

This text is designed to enable students to grasp important concepts in statistics. For a *one-semester* course in statistics, the text contains basic and essential topics such as descriptive statistics, probability, estimation, hypothesis testing, and linear regression. Two years of high school algebra serve as a prerequisite.

Many students are apprehensive about their first course in statistics. *Understanding Basic Statistics* is written to help students learn and enjoy statistics. The writing style is friendly and informal with concepts demonstrated through well-chosen and carefully presented examples. Guided exercises within the reading material lead students through additional examples. These guided exercises ask students to respond to leading questions so that the students can begin to ask the right questions themselves as they analyze and solve problems.

Each section is followed by problems selected from a wide range of fields. Most of these problems are from referenced sources such as newspapers, journals, and reference books. Problems incorporating computer displays from software packages such as MINITAB®, Microsoft® Excel®, and ComputerStat (the complimentary software package custom designed to accompany the text) are new to this edition. These problems ask the student to interpret and comment on the displayed results.

Finally, each chapter ends with a chapter summary, list of important words and symbols, comprehensive chapter problems, linking concepts questions that ask students to *write* about the relation among concepts from several chapters, and a Using Technology section. All of these features are designed to help students master the concepts, techniques, and applications of statistics appropriate to an introductory statistics course.

New Features in the Second Edition

- The extensive *revision of Section and Chapter Problems* incorporates many new problems, most based on real-world data or on situations with identified references, including Internet Web sites. As in the previous edition, these problems are from a wide range of fields, including natural science, business, economics, medicine, social science, archaeology, and consumer interest. Additional problems asking students to extend applications of the theory presented in the section have also been included.

- Labels beside each problem now specify the topic of the application. These labels show the wide variety of fields in which statistics is used. In addition, students can readily find applications of interest to them.

- Problems incorporating computer displays from software packages such as MINITAB, Microsoft Excel, and ComputerStat have been added to provide greater opportunities for students to interpret results.

- *Viewpoints* are brief essays that present diverse situations in which statistics can be used. They appear immediately before each section problem set, and relate to the content of the section. In many cases students are directed to an Internet Web site for more information or data about the topic of the essay. (Although every effort was made to verify the accuracy of Web site addresses provided, some addresses may have changed or Web sites may have closed after this book was published.) Visit the Brase and Brase Web site for links to all sites referenced in the text. Viewpoints give a humanistic touch to the study of statistics.

Content Changes in the Second Edition

- *Inferences About the Difference of Two Means (Small, Independent Samples)* is a new section included in Chapter 10. This section includes small sample techniques for hypothesis tests of the difference of two means and confidence intervals.

- Inferences about the difference of two means (Large, Independent Samples) and inferences regarding the difference of two proportions are now in two separate sections.

- *Calculator displays for the Texas Instruments TI-83 graphing calculator* have been included. Because this calculator supports confidence intervals and hypothesis testing, key instructions and displays showing these features have been included.

- *Using Technology* sections now include displays and comments regarding statistics using the software package Microsoft Excel as well as displays and comments about MINITAB, ComputerStat, and the Texas Instruments TI-83 graphing calculator.

- Other minor revisions have been made throughout the text to clarify explanations of important concepts.

Continuing Key Features

- *Detailed Examples* show students how to select and use appropriate statistical procedures.

- *Guided Exercises* immediately follow an example. Each guided exercise gives the student a chance to *work* with a new concept before another is presented. The student must examine and analyze characteristic features of a problem similar to the preceding example. Complete worked out solutions appear beside each exercise to give immediate reinforcement to the learning process.

- *Calculator Notes* give a general discussion regarding appropriate calculator use or give sample screens from the Texas Instruments TI-83 graphics calculator, showing how this calculator performs designated operations. The calculator notes are general enough to apply to a wide variety of calculators.

- *Section Problems* require the student to use all the new concepts mastered in the preceding section. The section problems reinforce the material of the examples and exercises by providing additional work in practical applications. The applications in the section problems are drawn from a variety of actual data resources such as popular newspapers, research journals, archaeological field reports, reference books, sports encyclopedias, U.S. census data, and so forth. The applications illustrate connections of statistics to a wide range of fields including health, consumer issues, environmental issues, business applications, sports, weather, education, student life, natural and physical sciences, archaeology, and human interest. *Key steps* to solutions of odd-numbered problems are contained in the solutions section at the end of the text.

- *Chapter Summaries* and *Important Words and Symbols* occur at the end of each chapter to help students review the concepts of the chapter.

- *Chapter Review Problems,* comprehensive problems that appear at the end of each chapter, cover each topic introduced in the chapter. Many chapter problems require material and concepts from several sections. Most important, the student must decide what technique to apply to a problem. As in actual applications, the position of the problem does not indicate which section the student should refer to for the method of solution.

- *Data Highlights: Group Projects* are problems based on general consumer questions or on real data from newspapers, magazines, and journals. Appearing at the end of each chapter, they ask the student to apply appropriate statistical methods from the chapter.

- *Linking Concepts: Writing Projects* provide students the opportunity to extend their thinking and look at statistical concepts from a broader perspective. This feature includes problems that ask the student to discuss and write brief essays about main concepts from the chapter and related topics from prior chapters.

- *Using Technology sections* occur at the end of each chapter. In many institutions it is possible to introduce the beginning statistics student to statistical software packages. The Using Technology sections feature published raw data and situations from a variety of fields. The problems in these sections can be solved using almost any appropriate statistical computer software package, and in many cases, by using a graphing calculator. We have tailored the exercises so that they can be completed easily by utilizing the software supplement *ComputerStat* that accompanies this text. In addition, instructions for the popular commercial software package MINITAB, Microsoft Excel, and for the TI-83 graphing calculator are also included. Displays from MINITAB, Excel, the TI-83, and ComputerStat are also included so that students who do not have direct access to these computing tools can see how such tools can be used.

- *Data Sets* from a variety of subject fields are included in Appendix II. The data sets are organized according to statistical methods appropriate for analyzing the data. These data sets are available as portable MINITAB files, TI-83 files (ASCII format), and Excel files on CDs. They are also included as classroom demonstrations in the software package ComputerStat.

- A *four-color design* highlights the text's features, enhances its visual appeal, and provides additional strength to the text's pedagogy.

Using Tools from Technology Support

Understanding Basic Statistics supports student access to a wide range of activities for computers or graphing calculators. Students can simply look at displays of computer output and screens from the TI-83 graphing calculator contained in Using Technology sections. They can solve problems or complete projects utilizing ComputerStat, MINITAB, Microsoft Excel, or other statistical software packages, or a graphing calculator such as the TI-83. Technology-based supplement support includes the following:

- *ComputerStat* is a computer software package designed to accompany this text. Institutions adopting the text may have a complimentary license to the software. ComputerStat is an interactive computer package, and it is designed to be very user friendly. The output is compatible with the results that the students obtain when they do a problem with pencil, paper, and calculator. However, the computer handles larger data sets with relative ease. ComputerStat is available in Windows and Macintosh platforms.

- *Technology Guide* is a supplement that has activities and projects for students to explore utilizing the technology of computer software or graphing calculators. Specific instructions are given for using ComputerStat, MINITAB, and the TI-83 graphing calculator.

- *Excel Guide* is a supplement that provides activities and projects utilizing the software package Excel. Instructions are given for using the statistics features of Excel.

- *Data CD* containing data files and the computer software package ComputerStat. The data files are in portable MINITAB format, Excel format, TI-83 (ASCII) format. They are also built into ComputerStat. The Data CD is packaged with the *Technology Guide*.

- *Student Version of MINITAB (Release 12) CD* can be packaged with this text. Contact Houghton Mifflin for details regarding price and platform options.

Supplements to Accompany the Second Edition

- *Instructor's Resource Guide.* This guide contains key steps and answers to even-numbered problems, two (revised) forms of chapter tests with solutions, transparency masters for use with overhead projectors, copy masters for all the statistical tables included in the sixth edition, correlation chart program in the video series *Against All Odds: Inside Statistics* (Annenberg/Corporation for Public Broadcasting Project), and hints for using this text in an Advanced Placement (AP) statistics course.

- *ComputerStat software,* a package of statistical programs designed and organized to accompany the text. This package is available without charge to institutions adopting the text. ComputerStat is available in Windows and Macintosh formats.

- *Videotapes* produced by Dana Mosely are a new supplement to the second edition. Tapes are keyed to sections of the text. They provide students additional learning support outside the classroom.

- Internet Web Page maintained by Houghton Mifflin Company. The Web page will provide resources such as: links to all Web sites referenced in the text as well as additional links to sites offering data or statistical information; Data Sets, PowerPoint slides containing definitions, formulas, and basic concepts from the text; order information for supplements.

- The *Excel Guide* contains computer activities coordinated with this text. Specific instructions for using Excel are included.

- *Data CD* provides a useful bank of large data sets for additional experimentation with MINITAB, the TI-83, or Excel. The software package ComputerStat is also included on the CD with the same data files available. Some of the data sets on the CD that are coordinated specifically with this text are also included in Appendix II of this text.

- *Technology Guide* containing computer and graphing calculator activities coordinated with the text. Specific instructions for using ComputerStat, MINITAB (both commands and menu choices), and the TI-83 graphing calculator are included. ComputerStat software is available with the guide.

- *Student Study and Solutions Guide* by Deann Christianson, Sarah Merz, and Larry Langley, University of the Pacific. This guide offers comprehensive review per text section and additional practice. Thinking About Statistics questions with their solutions and detailed solutions to selected odd-numbered problems in the text are also included.

- *Computerized Testing* in Windows and Macintosh formats. Instructors can custom-design tests from items in the test bank, add their own test items to the bank, or edit existing items. The program offers pull-down menus, dialogue boxes, pop-up windows, function keys, mouse support, and a graphical user interface.

- *Test Item File* by Diane Wagner, Regis University. This supplement serves as a convenient printed guide of all items appearing in the computerized testing software package.

Alternate Routes Through the Text

In many introductory statistics courses, most of the topics in this text will be presented. However, not every topic is required. A *Table of Prerequisite Material* precedes this preface to aid in topic selection.

Topics may also be rearranged. For instance, Chapter 3, Regression and Correlation, may be delayed until after Chapter 9. Then the descriptive topics of linear regression may be followed immediately by the inferential topics of linear regression presented in Chapter 11.

Custom Publishing Options

This text, *Understanding Basic Statistics,* contains the core topics for a one-semester course in introductory statistics. The descriptive components of the linear regression model are presented early with the inferential components discussed after estimation and hypothesis testing.

Our other statistics text, *Understandable Statistics: Concepts and Methods,* Sixth Edition, is written in the same style, with detailed examples, guided exercises, section problems, and similar chapter features. The Sixth Edition is more comprehensive and contains additional topics such as ANOVA, multiple regression, the Poisson distribution, the Geometric distribution, the Hypergeometric distribution, Bayes's Theorem, and testing two variances. Both the descriptive and inferential aspects of linear regression models are presented after hypothesis testing.

Houghton Mifflin Publishing Company supports custom publishing. If there are one or two topics contained in *Understandable Statistics,* Sixth Edition, that you wish to package with *Understanding Basic Statistics,* please contact the publisher or talk with your book representative to determine feasibility and cost.

Acknowledgments

It is our pleasure to acknowledge the prepublication reviewers of this text. All of their insights and comments have been very valuable to us. Reviewers of this text include

Charles Laws, Cleveland State Community College
Vijay Joshi, Virginia Intermont College
Jennifer Dollar, Grand Rapids Community College
Lawrence Mark Lesser, University of Northern Colorado
Geetha Ramachandran, California State University—Sacramento
Raja Khoury, Collin County Community College
C. Wayne Ehler, Anne Arundel Community College
Barbara Burrows, Santa Fe Community College
David Gurney, Southeastern Louisiana University
Rita Kolb, Community College of Baltimore County—Cantonsville
Douglas Frank, Indiana University of Pennsylvania
Michael Russo, Suffolk County Community College
Mark Ecker, University of Northern Iowa
Michael Ecker, Pennsylvania State University
Darcy Mays, Virginia Commonwealth University
Jonathan Baker, Columbus State Community College
Aileen Solomon, Trident Technical College
Tom Fox, Cleveland State Community College
Brian Bradie, Christopher Newport University
Randall Boan, Aims Community College
Don Campbell, Western Illinois University
Dennis Kimzey, Rogue Community College
Bradford Crain, Portland State University
Richard Stockbridge, University of Kentucky
Delmar Tunnell, Illinois Valley Community College
Susan Hoy, Bristol Community College
Reid Davis, University of Tennessee at Knoxville

Much of the material in this text has been drawn from our other text, *Understandable Statistics: Concepts and Methods,* Sixth Edition. We appreciate reviewer and user comments and suggestions we received for that project. We especially want to thank Dr. Diane Wagner of Regis University, who reviewed the original manuscript for accuracy.

We also acknowledge the cooperation of Minitab, Inc., and Texas Instruments. The MINITAB computer displays are from MINITAB® Release 12. The graphing calculator displays are from the TI-83.

Charles Henry Brase
Corrinne Pellillo Brase

This new edition of *Understanding Basic Statistics* includes a variety of features designed to enhance a student's understanding by providing summaries of concepts and methods, interesting real-world applications using real data sets, and information on using technology. In addition, the text has a **four-color design** that highlights important features and provides visual interest.

Key features of the text are described on the following pages. New features have been added to the second edition, and those found useful by instructors and students have been retained.

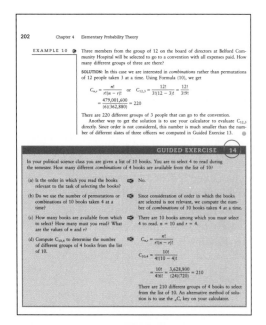

Guided Exercises

The guided exercises following selected examples are a special continuing feature of *Understanding Basic Statistics*. Students examine and analyze a problem similar to the preceding example so that they have a chance to *work* with a new concept before another is presented. Completely worked out solutions occur alongside each exercise to give immediate reinforcement in the learning process.

Real-World Applications

Many interesting real-world problems lend themselves to the study of statistics. In this edition, great emphasis is placed on including interesting problems utilizing real data and real situations using identifiable sources including some Internet Web sites. These problems come from a wide range of fields, including natural science, business, economics, medicine, social science, archaeology, and consumer interest. The applications are now titled by field so students see the connection to the real world. They can also readily locate problems from their areas of interest.

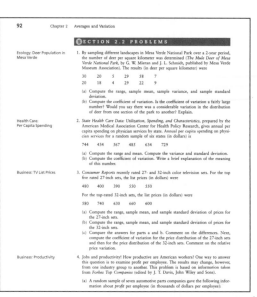

Calculator Notes

These notes are located throughout the book. They provide information regarding appropriate calculator use. Sample screens from the TI-83 graphing calculator are new to this edition. The notes show how the TI-83 performs certain operations, but are general enough to apply to most calculators.

Viewpoints

These brief illustrated essays have been added to the second edition to show the broad scope of statistical applications to a variety of human experiences and endeavors. In many cases, Internet Web site addresses are provided for students interested in further exploration of topics. The Viewpoint feature immediately precedes most section problems Sets and Chapter Review Problems.

End of chapter material includes a **Summary**, a listing of **Important Words and Symbols** with section references so that students who need additional review will be encouraged to read it in context, and **Chapter Review Problems.**

The following features are also included at the end of chapters:

Data Highlights: Group Projects

These problems can be solved using appropriate methods from the chapter. The exercises were designed to provide students with additional opportunity to put their skills into action. Newspapers, magazines, and journals are the sources for these problems.

Linking Concepts: Writing Projects

These questions help students extend and integrate their thinking, and develop a broader conceptual understanding of statistics. Students are asked to discuss and write about key concepts from the chapter and related topics from prior chapters.

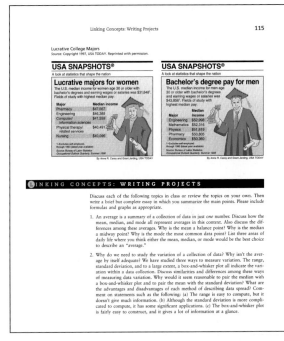

Using Technology

All these sections have been revised for this edition to include information on using MINITAB, the TI-83 graphing calculator, Excel, and ComputerStat to solve statistical problems.

Organizing Data

To guess is cheap,
To guess wrongly is expensive.

Tell me, I'll forget.
Show me, I may remember.
But involve me and I'll understand.

—Old Chinese Proverbs

Dragon Gate,
Chinatown
San Francisco

Dragon Gate is one of the many landmarks in San Francisco that exemplifies the city's ethnic diversity.

Most of the important decisions in life involve incomplete information. Such decisions often involve so many complicated factors that a complete analysis is not practical or even possible. We are often forced into the position of making a guess based on limited information. However, as the first proverb implies, a blind guess is not the best solution. Statistical methods, such as you will learn in this book, can help you make the best "educated guess."

The authors of this book want you to understand and enjoy statistics. The reading material will *tell you* about the subject. The examples will *show you* how it works. To understand, however, the second proverb says you must *get involved*. Guided exercises, calculator and computer applications, section and chapter problems, and writing exercises are all designed to get you involved in the subject. As you grow in your understanding of statistics, we believe you will enjoy learning a subject that has a world full of good applications. In fact, the applications of statistics are so numerous that in a sense you are limited only by your own imagination as to what areas in your life you may want to make use of the subject.

Populations, Samples, and Data

All of us have a built-in system of inference that helps us make decisions; without it, we would be lost. Of course, we also have a built-in set of prejudices that affects our decisions. A definite advantage of statistical methods is that they can help us make decisions without prejudice. Moreover, statistics can be used for making decisions when we are faced with uncertainties. For instance, if we wish to estimate the proportion of people who will have a severe reaction to a new flu shot without giving the shot to everyone who wants it, statistics can provide appropriate methods.

The general prerequisite for statistical decision making is the gathering of data. These data may be *quantitative* or *qualitative* in nature. Quantitative data are numerical measurements such as the number of hours a person spends driving each day, whereas qualitative data involve nonnumerical observations such as the color of a person's eyes or the yes/no response to a question. Sometimes qualitative data are referred to as *categorical* data.

Procedures for analyzing data, together with rules of inference, are central topics in the study of statistics. In short, we may say that *statistics is the study of how to collect, organize, analyze, and interpret numerical information from data.*

The statistical procedures you will learn in this book should supplement your built-in system of inference—that is, the results from statistical procedures and good sense should dovetail. Of course, statistical methods in themselves have no power to work miracles. These methods can help us make some decisions, but not all conceivable decisions. Remember, a properly applied statistical procedure is no more accurate than the data, or facts, on which it is based. Finally, statistical results should be interpreted by one who understands not only the methods but also the subject matter to which the methods have been applied.

In any investigation it is important to ask the right kinds of questions. What are the key points of the problem and its solution? In this section we will examine a variety of problems. Then, thinking as statisticians do, we will investigate some important concepts. As a start, let's look at a few important terms.

The term *population* refers to all measurements or observations of interest. For example, if we want to know the average height of people who have climbed Mt. Everest, then the population consists of the heights of all those people. Now suppose we want to know the average height of 20-year-old females in the United States. In this case, the population consists of the heights of *all* 20-year-old females in the United States.

It is interesting to note that the population is defined in terms of our *desire for knowledge*. The population can be thought of as measurements or observations for the entire group of objects or people about which information is desired. In this sense, a population can be an existing set of measurements, or it can be a set of measurements that is clear in our understanding but is not yet complete. The heights (in inches) of all U.S. presidents from George Washington to Bill Clinton can be thought of as an existing set of measurements. However, the set of heights of all

Quantitative or qualitative data?

Population

presidents from George Washington on into the future is not yet complete. Nevertheless, we have a clear understanding of how this population is to be constructed. In a way, we can think of such an (incomplete) population as being open ended.

Often it is not feasible to study an entire population. How would you measure the heights of all 20-year-old females in the United States? In such cases we look at samples. A *sample* is simply a part of the population. But not every sample is useful. The sample must represent the population. A *random sample* is such a representative sample; in the next section we will study these samples in much more detail. In the meantime, we can think of a random sample as a sample determined completely by chance.

Sample

Generally speaking, statistical problems are those that require us to draw a random sample of observations from a larger population. Then statistical methods are used to form conclusions about the population based on the information in the sample. Let's look at some examples of populations and samples.

EXAMPLE 1 ▷ The department of tropical agriculture is doing a study of pineapples in an experimental field. In this case the data under consideration are the individual weights of all pineapples in the field.

(a) The *population* is the weights of all the pineapples in the field.

(b) A random collection of 100 pineapples is taken from the field. Each pineapple is weighed. The 100 weights form a *random sample* from the population of all weights.

COMMENT When referring to a population or sample, be sure to give the *quantity* being measured or counted or the *quality* being observed. For instance, in Example 1 it is not sufficient to say that the population consists of all pineapples in the field. We also must state the quantity to be measured. Unless we do, we won't know whether to consider weight, diameter, length, sugar content, acidity level, time to mature, or any other of the many possible measurements that could be made on pineapples. On the other hand, we might be interested in a *quality* of the pineapple, such as color, taste, or ripeness.

Throughout this text you will encounter exercises embedded in the reading material. These guided exercises are included to give you an opportunity to work immediately with new ideas. Cover the answers on the right side (an index card will fit this purpose). After you have thought about or written down *your own responses,* check the answers. If there are several parts to an exercise, check each part before you continue. You will be able to answer most of the exercise questions, but don't skip them— they are important. (See Guided Exercise 1 on the following page.)

Television station QUE wants to know the proportion of TV owners in Virginia who watch the station's new program at least once a week. The station asked a random group of 1000 TV owners in Virginia if they watch the program at least once a week.

(a) What is the population?   ⮕   The population consists of the response (does, or does not, watch the new program) from each TV owner in Virginia.

(b) What is the sample?   ⮕   The sample consists of the responses of the 1000 TV owners in Virginia who were questioned.

Sometimes we do not have access to an entire population, and at other times the difficulties of working with an entire population are prohibitive. The benefit of statistical methods is that they allow us to draw conclusions about populations based only on information from samples. The handicap of these methods is that our conclusions are uncertain—probabilistic, if you will. Probabilism is one of the aspects of statistical thinking that may be unfamiliar to you, since you dealt with certainties in your previous mathematical experience. For instance, if you are asked to solve the equation $x + 2 = 5$, you know that the solution is $x = 3$. There is no uncertainty; you can check the solution in the equation. We do not demand absolute certainty from statistical methods. That is generally too much to ask. However, we do have a measurable degree of probabilistic confidence that the conclusions of a method are valid.

In future work the probability attached to a conclusion will represent the amount of confidence we should have in the conclusion. Probability will be one of our basic tools in the study and application of statistical methods. In Chapter 4 we will study some elementary probability theory, but in the meantime, let's take an intuitive point of view. Thus, if the probability that some statement is true is 0.97, then the statement is true about 97% of the time. The statement is false about 3% of the time.

Now let's look at some examples and see where these new terms fit into place. These examples do not begin to cover the complete range of statistical applications, but they are indications of the types of problems that can be solved using the methods presented in this book.

EXAMPLE 2 ❯   If you travel a long distance by air, chances are that you will not have a direct flight. You usually need to change airplanes, and you will often change airlines in the process of making connections. Normally, an agent for your first airline makes arrangements for the connecting flights, and you pay this agent a lump sum for your ticket. If more than one airline company is involved, the companies decide among themselves how much money each should receive.

In the past, a great deal of clerical work was needed to apportion the monies. Then three airlines decided to use statistical methods to determine how total revenue should be split. During a 4-month trial period, they took random samples from the overall population of all interairline tickets. From that sample they determined the proportion of total revenues to be distributed to each airline. They were able to estimate that the degree of error in their statistical process was not more than 0.07%—that is, $700 in $1,000,000. On the basis of this work, more airlines have used statistical methods in settling interairline accounts. Some of the larger airlines estimate a clerical savings of more than $250,000 each year over the old methods.

EXAMPLE 3 ➤   In 1778 Captain James Cook discovered what we now call the Hawaiian Islands. He gave the islanders a present of several goats, and over the years these animals multiplied into wild herds totaling several thousand. They eat almost anything, including the famous silver sword plant, which was once unique to Hawaii.

At one time the silver sword grew abundantly on the island of Maui (in Haleakala, a national park on that island, the silver sword can still be found), but each year there seemed to be fewer and fewer plants. The disappearance of these plants could have been due to many things (e.g., tourists picking them illegally), but a biologist hypothesized that the goats were mainly to blame.

To determine the effect of goats on the vegetation, the rangers set up stations in remote areas of Haleakala. Very few tourists came to these areas, but they were home to many goats. At each station the rangers selected two plots with about the same area, plant count, soil, and climatic conditions. One plot was carefully fenced; the other was not. At regular intervals a plant count was made in each plot.

In this example, the population can be thought of as the plant count for the entire park. The samples are plant counts from the experiment stations. The claim is that goats are in fact detrimental to plant life in the park because they reduce the plant count by one-fourth or more. Using statistical methods, the claim was confirmed with a high degree of confidence.

GUIDED EXERCISE   2

Mountain Joy Company produces ice axes, which are used by mountain climbers to catch themselves in case of a fall on a steep glacier. Most ice axes have steel heads and fiberglass shafts. A problem with this piece of equipment is that the shaft sometimes breaks after a hard fall.

The specifications for manufacturing ice axes require that 99% of all ice axes made should hold at least 400 lb on the shaft. The production manager wants to be as certain as possible that under present production methods the shafts can support at least 400 lb. To test an ice ax, a force of 400 lb is applied to the shaft. If the shaft does not break, it passes the test. It is possible for the shaft to weaken, and yet not break, when the force is applied. The manager does not wish to sell weakened ice axes, so she does not test them all.

Exercise continues

Exercise continued

(a) What does the manager wish to test?   ⇨   The manager wishes to test if 99% of ice axes made under present production methods can hold at least 400 lb.

(b) What is the population?   ⇨   The population is the strength (in pounds) of all ice ax shafts produced under present methods.

(c) Since the manager does not wish to test *each* ice ax, what should she do?   ⇨   The manager should select a random sample of ice axes and test the axes in the sample.

Producing Data

Sampling

One very popular way to produce data is to *sample* an existing population. That is, we draw subsets from an existing population. There are many popular ways to sample a population. We will delay our discussion of sampling techniques until Section 1.2. However, we want to remember that the goal of sampling techniques is to gain an accurate picture of the population and yet disturb the population as little as possible by the act of sampling. Example 2, on the apportioning of monies to the airlines, is an example of sampling.

Experiments

Another very popular way to produce data is to observe the outcome of an *experiment*. We use the word *experiment* here in a very broad sense. In our context, the term *experiment* means a study in which some treatment is deliberately imposed on units or subjects in order to observe a given response. In Example 3, the rangers of Haleakala set up two vegetation plots. One received the treatment of being fenced. An experiment might consist of exerting a given weight on a steel beam to measure how much it bends. An experiment might be done at the Colorado Shakespeare Festival that consists of changing the costumes from Elizabethan to modern twenty first-century style and then observing the effect on ticket sales at performances.

Remember, the goal of an experiment is to measure the effect of an intervention. To understand how nature (or a population) responds to a change, we must actually *impose* the change and measure the results. This is in contrast to sampling, in which we try to describe or represent the population while making every effort *not* to change the population.

Simulation

Simulation is a numerical facsimile of real-world phenomena. Sometimes simulation is called a "dry lab" approach, in the sense that it is an arithmetic imitation of a real situation. Advantages of simulation are that arithmetic and statistical simulations can fit real-world problems extremely well. The researcher can explore procedures in simulation that might be very dangerous in real life. In the real world you might not want to introduce a high level of a drug into a diabetic person's bloodstream. However, you might want to simulate the injection statistically and study the results. You will harm no one, and the information gained may be of real medical value. Similarly, you might test the effect of wind sheer on an airplane wing in a simulated environment rather than with an actual airplane in flight.

Simulations usually require many, many calculations. Computers are a very practical tool for simulation. In this text we will see several uses of computer simulation, such as computer-simulated random samples (next section), a computer simulation for tossing two dice (end of Chapter 4), and computer simulations of confidence intervals for proportions (Chapter 8) and the central limit theorem (Chapter 7).

Census

In a *census*, measurements from the *entire* population are used. The U.S. Department of Commerce, Bureau of the Census, conducts a census every 10 years. An attempt is made to reach every resident of the United States. However, even in a census, some members of the population may be missed. For instance, it is difficult to contact homeless people. In other cases, some people may not respond. Statistical estimates of missing responses are often supplied.

GUIDED EXERCISE   3

Which technique (sampling, experiment, simulation, or census) for gathering data do you think might be the most appropriate for each of the following studies?

(a) A study of the effect of stopping the cooling process in a nuclear reactor

⇨ Probably simulation, since you may not want to risk a nuclear meltdown.

(b) A study of the amount of time college students taking a full course load spend watching television

⇨ Sampling would work well. Notice that obtaining the information from a student will probably not change the amount of time the student spends watching television.

(c) A study of the effect of a calcium supplement on bone mass in young girls

⇨ Experimentation. A study by Tom Lloyd reported in the *Journal of the American Medical Association* tested 94 young girls. Half were given a placebo, and half were given calcium supplements to bring their daily calcium intake up to about 1400 milligrams per day. The group getting the experimental treatment of calcium gained 1.3% more bone mass in a year than the girls getting the placebo.

(d) A study of the credit hour load of *each* student enrolled at your college at the end of the drop/add period this semester

⇨ Census. The registrar can obtain records for *every* student.

Surveys

Once you decide you are going to use sampling or experimentation, a common means of gathering data about people is to ask them questions. This process is the essence of *surveying*. You are asked to participate in surveys in many ways. Teacher evaluations, market research conducted in shopping malls, telephone surveys, and consumer questions asked on product warranties are all familiar surveys. In the design of a survey or questionnaire, the researcher must decide how the responses will be converted into numbers. Sometimes, as in some polls, the responses might be "agree," "disagree," or "no opinion." Then the researcher looks at the percentage of people who responded in each fashion.

A number of issues can arise in using a survey. Are the questions asked in a neutral way, or is conscious or unconscious bias built into the wording? How can you be sure the respondents are answering truthfully? Is your sample representative of the population? For instance, when conducting election polls, some studies use only registered voters because these are the only people eligible to vote. Other polls use only "likely" voters. They first inquire if the respondent is planning to vote. Other problems arise if the selected respondent refuses to participate.

Voluntary response

Voluntary response samples often overrepresent people with strong opinions. A Denver newspaper that has wide circulation in Colorado decided to conduct a survey regarding the question of whether grazing fees for use of public lands should be increased. Of the many people who responded, about 85% said, "No!" However, the results were misleading as an indicator of the opinions held by the population of *all* people in Colorado. The sample who responded were self-selected people. The sample consisted mainly of ranchers who felt strongly enough to write to the newspaper. These people did not want their grazing fees increased on public land.

Many times negative opinions are especially overrepresented through voluntary response. A statistically designed opinion poll using a random sample from all (adult) Colorado residents later showed that only 34% responded no to the survey question about grazing fees. The random sample eliminated bias by sampling from all adult respondents in a way that allowed each an equal chance to be chosen.

Data from voluntary responses can be useful and interesting. However, the information collected is anecdotal in nature. It would be questionable to generalize the results to the entire population of interest. Surveys must be designed and administered carefully for the results to generalize.

Hidden bias

Whenever you gather data, whether by sampling a population, by results of experiment, or by simulation, you should view the data with a critical eye. The way the data are gathered may produce a *hidden bias*. This means that in reality you are not actually measuring what you hoped to measure. For instance, if a uniformed police officer conducts a survey regarding opinions about and use of illegal drugs, the responses might be different than if a casually dressed person asks the same questions. Asking students if having parking lots that are large enough and safe before asking if they would approve an increase in parking fees might bias responses. Care must be taken to deal with all units or subjects in the exact same way so that no (conscious or unconscious) preferential treatment or selection can occur.

Other variables

Sometimes our goal is to understand the cause-and-effect relationships between two variables (as might occur in regression and correlation of two variables presented in Chapters 3 and 10). However, the effect of one variable on another can be hidden by other variables for which no data have been obtained. For instance, a study of ticket price and attendance at a sporting event might show that higher ticket prices and higher attendance are related, since events with higher ticket prices seem to have greater attendance. One might be led to conclude that if you want to increase attendance at an event, you should raise the price of the tickets. What is missing from this analysis is the variable of *event popularity*. A Super Bowl football game is so special that people are willing to pay higher prices just to be there. A preseason football game, however, does not have the same drawing power and may need lower ticket prices to ensure a reasonable crowd at the game. Other variables, such as the location of the game, weather, and the records of the teams involved, might also influence attendance.

The problem of other variables that influence one or both of the original variables often can be overcome if the researcher is not only familiar with statistics but also well versed in the field of investigation.

Generalizing results

Some researchers want to generalize their findings to a situation wider than that of the actual data setting. The true scope of a new discovery must be determined by repeated studies in various real-world settings. Just because statistical experiments showed that a drug had a certain effect on a collection of laboratory rats does not guarantee that the drug will have a similar effect on a herd of wild horses in Montana.

GUIDED EXERCISE   4

Comment on the usefulness of the data collected as described.

(a) A uniformed police officer interviews a group of 20 college freshmen. She asks each one his or her name and then if he or she has used an illegal drug in the last month.

⮕ Respondents may not answer truthfully. Some may refuse to participate.

(b) Jessica saw some data showing that cities with more low-income housing have more homeless people. Does building low-income housing cause homelessness?

⮕ There may be some other variables such as the size of the city. Larger cities may have more low-income housing and more homeless.

(c) A survey about food in the student cafeteria was conducted by having forms available for customers to pick up at the cash register. A drop box for completed forms was available outside the cafeteria.

⮕ The voluntary response will likely produce more negative comments.

(d) Extensive studies on coronary problems were conducted using men over age 50 as the subjects.

⮕ Conclusions for men over age 50 may or may not generalize to other age and gender groups. These results may be useful for women or younger people, but studies specifically involving these groups may need to be performed.

The process of gathering data is an essential component of good statistical practice. Sampling, experimental design, simulation, and survey design are all extensive studies. Treatises are written on these topics. In a serious statistical study, a great deal of attention is devoted to the process of gathering the data. Then appropriate statistical methods need to be applied. In this course we will explore some of these statistical methods, and you will be able to carry out many of them and use the terminology associated with the methods appropriately. Even with the best statistical methodology, the results need to be interpreted by someone who is an expert in the field. Good statistics and expertise in a given field must work together to give reliable results.

Levels of Measurement

When we collect data, it is common to classify the information obtained according to one of the following four *levels of measurement:*

Nominal level	Ordinal level	Interval level	Ratio level

Among these four levels of measurement, the *nominal* level is considered to be the *lowest.* This is followed by the *ordinal* level, the *interval* level, and finally the *ratio* level, which is the highest level of measurement.

Nominal level

Since the nominal level is the lowest level, let's examine it first. A dictionary meaning of the word *nominal* is "in name only." This is an easy way to remember the meaning of the nominal level of measurement. Data at this level of measurement consist of "names only," or qualities, with no implied criteria by which the data can be identified as greater than or less than other data items.

EXAMPLE 4 ▶

The following are examples of data at the nominal level of measurement.

(a) Aspen, Vail, and Breckenridge are names of three ski resorts from the population of names of all ski resorts in Colorado.

(b) Taos, Acoma, Zuni, and Cochiti are names of four Native American pueblos from the population of all names of Native American pueblos in Arizona and New Mexico.

(c) Smith Auto Dealers has a large supply of new T50 trucks on the lot. The colors of the trucks on the lot are red, white, silver, blue, and black. ●

It is clear that the nominal data in Example 4 are not intended for numerical calculation. The specific names or qualities do not contain any implied ordering or numerical significance.

Ordinal level

The next level of measurement is the *ordinal level*. Data at the ordinal level may be arranged in some order, but actual differences between data values either cannot be determined or are meaningless.

EXAMPLE 5 ❯

The following are examples of data at the ordinal level of measurement.

(a) In a fishing tackle catalogue, 17 fishing reels are advertised. Of these reels, 6 were rated as good quality, 4 were rated as better quality, and 7 were rated as best quality.

(b) In a travel guide to California, 416 bed-and-breakfast accommodations are listed. Each bed-and-breakfast facility is given a rating from one to four stars, with four stars being the highest rating. Of the accommodations listed, 93 got one star, 115 got two stars, 172 got three stars, and 36 got four stars.

(c) In a high school graduating class of 319 students, Jim ranked 25th, June ranked 19th, Walter ranked 4th, and Julia ranked 10th.

(d) At one point in a recent baseball season, the Baltimore Orioles were ranked first, the Boston Red Sox were ranked second, and the Milwaukee Brewers were ranked third in the American League.   ●

In Example 5 we should not try to determine a specific quantitative difference between "good," "better," and "best." Nor should we try to determine an exact difference between a rating of two stars over three stars in the bed-and-breakfast rating system. The difference between June's and Jim's rank was 6, and this is the same difference that exists between Walter's and Julia's ranks. However, this difference doesn't really mean anything significant. For instance, if you look at grade point average, Walter and Julia may have had a big gap between them, whereas June and Jim may have been closer together. In any ranking system, it is only the relative standing that matters. Differences in ranks can be meaningless.

In general, the ordinal level of measurement provides information about relative comparisons, but exact differences are not computed.

Interval level

The *interval level of measurement* is like the ordinal level, but it has the additional property that meaningful differences between data values can be computed. However, interval-level data may not have an intrinsic zero or starting point. Consequently, differences are meaningful, but ratios of data values are not.

EXAMPLE 6 ❯

The following are examples of data at the interval level of measurement.

(a) Years in which Democrats won presidential elections

(b) Body temperatures (in degrees Celsius) of trout in the Yellowstone River

Temperature readings in Celsius (or Fahrenheit) are examples of data at the interval level of measurement. Such values are certainly ordered, and we can compute meaningful differences. However, for Celsius-scale temperatures, there is not an inherent starting point. The value 0°C may seem to be a starting point, but the value of 0°C does not indicate the state of "no heat." Furthermore, it is not correct to say that 20°C is twice as hot as 10°C. Calendar times are also interval measurements, since the date 0 A.D. does not signify "no time." However, a time lapse is at a higher level of measurement and is in fact an example of our top level of measurement, the *ratio level*.

Ratio level

The *ratio level of measurement* is the highest level. The ratio level is similar to the interval level, but it includes an inherent zero as a starting point for all measurements. Consequently, at this level, both differences *and* ratios are meaningful.

EXAMPLE 7 ➤   The following are examples of data at the ratio level of measurement.

(a) The core temperatures of stars in the Milky Way when measured in degrees Kelvin. Notice that in the Kelvin scale of measurement 0°K means "no heat." This is a special temperature scale used primarily by scientists.

(b) Time lapse between the deposit of a check into a bank account and the clearance of that check. An out-of-state check that clears in 2 days takes twice as long as a local check that clears in 1 day.

(c) Length of trout swimming in the Yellowstone River. A trout 18 inches long is three times as long as a 6-inch trout. Observe that we can divide 6 into 18 to determine the *ratio* of the trout lengths.   ●

In summary, there are four levels of measurement. The nominal level is considered the lowest, followed by, in ascending order, the ordinal, interval, and ratio levels. In general, calculations based on one level of measurement should not be used for a lower level.

Level of Measurement	Suitable Calculation
Nominal	We can put the data in categories.
Ordinal	We can order the data from smallest to largest or "worst" to "best." Each data value can be *compared* with another data value.
Interval	We can order the data and also take the differences between data values. At this level, it makes sense to compare the differences between data values. For instance, we can say that one data value is 5 more or 12 more than another data value.
Ratio	We can order the data, take differences, and also find the ratio between data values. For instance, it makes sense to say that one data value is twice as large as another.

GUIDED EXERCISE 5

The following items describe different types of data associated with a state senator. For each data entry, indicate the corresponding *level of measurement*.

(a) The senator's name is Sam Wilson.

⇨ Nominal level

(b) The senator is 58 years old.

⇨ Ratio level

(c) The years in which the senator was elected to the senate are 1980, 1986, 1992, and 1998.

⇨ Interval level

(d) His total taxable income last year was $878,314.19.

⇨ Ratio level

(e) The senator sponsored a bill to protect water rights. Out of 1100 voters in his district, 400 said they strongly favored the bill, 300 said they favored the bill, 200 said they were neutral, 150 said they did not favor the bill, and 50 said they strongly did not favor the bill.

⇨ The opinions about the bill are at the ordinal level.

(f) The senator is married.

⇨ Nominal level

(g) However, the senator was previously divorced in 1982 and again in 1986.

⇨ Interval level

(h) A leading news magazine claims the senator is ranked seventh for his voting record on bills regarding public education.

⇨ Ordinal level

In the next section we will examine random samples more carefully. The topics in the remainder of this chapter and in Chapter 2 discuss ways in which to organize and summarize information from samples and populations. The term *descriptive statistics* refers to this organization of data.

In Chapters 4, 5, 6, and 7 we cover elementary probability theory and some basic probability distributions. By Chapter 8 we will use the previous groundwork to begin *inferential statistics*—that is, methods of using a sample to obtain information about a population.

SECTION 1.1 PROBLEMS

Marketing: Fast-food

1. *USA Today* reported that 44.9% of those surveyed (1261 adults) ate in fast-food restaurants from one to three times each week.
   (a) What is the implied population? (b) What is the sample?

Advertising: Auto Mileage

2. What is the average gas mileage (in miles per gallon, or mpg) for all new 2000 cars? Using *Consumer Reports*, a random sample of 35 new 2000 cars gave an average of 22.5 mpg.
   (a) What is the implied population? (b) What is the sample?

Insurance: Payment Checks

3. An insurance company wants to determine the time interval between the arrival of an insurance payment check and the time that the check clears. A central payment office processes the payments for a five-state region. A random sample of 32 payment checks from this five-state region was received and processed. The time interval between receipt and check clearance was determined for each check. From this information the company estimated the time interval necessary for all checks sent to this office to clear.
   (a) What is the implied population?   (b) What is the sample?

Academic:
Teacher Evaluation

4. If you were going to apply *statistical methods* to analyze teacher evaluations, which question form, A or B, would be better?
   Form A: In your own words, tell how this teacher compares with other teachers you have had.
   Form B: Use the following scale to rank your teacher as compared with other teachers you have had.

1	2	3	4	5
worst	below average	average	above average	best

Student Life:
Levels of Measurements

5. Categorize the following measurements associated with student life according to level: nominal, ordinal, interval, or ratio.
   (a) Length of time to complete an exam
   (b) Time of first class
   (c) Class category: freshman, sophomore, junior, senior
   (d) Course evaluation scale: poor, acceptable, good
   (e) Score on last exam (based on 100 possible points)
   (f) Age of student

Business:
Levels of Measurements

6. Categorize the following measurements associated with a robotics company according to level: nominal, ordinal, interval, or ratio.
   (a) Salesperson's performance: below average, average, above average
   (b) Price of company's stock
   (c) Names of new products
   (d) Room temperature (°F) in CEO's private office
   (e) Gross income for each of past 5 years
   (f) Color of packaging

Sampling: Gathering Data

7. Which technique (sampling, experiment, simulation, or census) for gathering data do you think was used in each of the following studies?
   (a) One way to find information on the Super Bowl football game is to look at the NFL Web site <http://www.nfl.com/> and link to the history index or Super Bowl results. There, the winning scores for all the Super Bowl games played to date are given. Using the data for all the games, we find that as of Super Bowl XXXI, the average score for the winning teams was 31.
   (b) A sample of 82 healthy female and male subjects was recruited to participate in a study on pain (*Physical Therapy*, Vol. 70, No. 1). The subjects were divided into two groups. The experimental group received laser stimulation, and the control group received sham stimulation. Tests of pain tolerance were then conducted on each group.
   (c) Computer imaging of runners shows the effect of stride length on running efficiency.
   (d) Do the Chinese like chocolate? Gallup Chinese is conducting surveys in China to answer this question for the U.S. Chocolate Manufacturers Association.

Gallup Chinese is surveying a portion of the Chinese population to determine whether there is a market for chocolate in China (*Wall Street Journal*).

Sampling: Gathering Data

8. Which technique (sampling, experiment, simulation, or census) for gathering data do you think was used in each of the following studies?
   (a) An analysis of a sample of 31,000 patients from New York hospitals suggests that the poor and the elderly sue for malpractice at one-fifth the rate of wealthier patients (*Journal of the American Medical Association*).
   (b) The effects of wind sheer on airplanes during both landing and takeoff are studied by using complex computer programs that mimic actual flight.
   (c) A study of football scores attained through touchdowns and field goals was conducted by the National Football League to determine whether field goals account for more scoring events than touchdowns. Data from all regular season NFL football games from 1983 to 1992 were used (*USA Today*).
   (d) An Australian study included 588 men and women who already had some precancerous skin lesions. Half got a skin cream containing a sunscreen with a sun protection factor of 17; half got an inactive cream. After 7 months, those using the sunscreen with the sun protection had fewer new precancerous skin lesions (*New England Journal of Medicine*).

Survey: Response Manipulation

9. The *New York Times* did a special report on polling that was carried in papers across the nation. The article points out how readily the results of a survey can be manipulated. Some features that can influence the results of a poll include the following: the number of possible responses, the phrasing of the question, the sampling techniques used (voluntary response or sample designed to be representative), the fact that words may mean different things to different people, the questions that precede the question of interest, and finally, the fact that respondents can offer opinions on issues that they know nothing about.
   (a) Consider the expression "over the last few years." Do you think that this expression means the same time span to everyone? What would be a more precise phrase?
   (b) Consider this question: "Do you think fines for running stop signs should be doubled?" Do you think the response would be different if the question "Have you ever run a stop sign?" preceded the question about fines?
   (c) Consider this question: "Do you watch too much television?" What do you think the responses would be if the only responses possible were yes or no? What do you think the responses would be if the possible responses were rarely, sometimes, or frequently?

Class Data Project

10. Make a statistical profile of your own statistics class. Items of interest might be
    (a) Height, age, gender, pulse, number of siblings, marital status
    (b) Number of college credit hours completed (as of beginning of term); grade point average
    (c) Major; number of credit hours enrolled in this term
    (d) Number of scheduled hours working per week
    (e) Distance from residence to first class; time it takes to travel from residence to first class
    (f) Year, model, and color of car usually driven

What directions would you give to people answering these questions? For instance, how accurate should the measurements be? Should age be recorded as of last birthday?

Random Samples

> Eat lamb—20,000 coyotes can't be wrong!

This slogan is sometimes found on bumper stickers in the western United States. The slogan indicates the trouble that ranchers have experienced in protecting their flocks from predators. Modern methods of predator control have changed considerably from those of earlier days, and to a certain extent the changes have come about through a closer examination of the sampling techniques used.

If we are to use information obtained from a sample to draw conclusions about a population, the sample must be representative of the *entire* population. A type of sample that is representative of the entire population is a *random sample*. One of the properties of a random sample is that *every* member of the population has an equal chance of being included in the sample. This is not enough, however. Another necessary condition is that *every sample of the same size* is equally likely to be selected.

Simple random samples

> **Definition**
>
> A *simple random sample* of n measurements from a population is one selected in such a manner that every sample of size *n* from the population has an equal probability of being selected, and every member of the population has an equal probability of being included in the sample.

Consider the population of all coyotes in the western United States. The sample of the population that the ranchers observe is largely the coyotes that prefer to live near ranches. It seems that many coyotes who choose to live near ranches also like to eat lamb. In fact, most of the coyotes the ranchers observed appeared to be existing solely on sheep.

Based on their experience with this sample of the coyote population, the ranchers concluded that *all* coyotes are dangerous to their flocks and coyotes should be eliminated! The ranchers used a special poison bait to get rid of the coyotes. Not only was this poison distributed on ranch land, but with government cooperation it also was distributed widely on public lands.

GUIDED EXERCISE   6

(a) Do you think that the sample of coyotes the ranchers observed could be thought of as a random sample of the *entire* coyote population? Explain.

No; the ranchers observed only coyotes near their ranches.

Exercise continues

Exercise continued

(b) If a sample is not chosen at random from the population, is it safe to use the results of that sample to describe the *entire* population?

⇨ No.

(c) Do you think the idea of reducing the *entire* coyote population could be statistically justified on the basis of the ranchers' experience with the coyotes?

⇨ We don't think it could be statistically justified one way or another because the sample of coyotes was not a true random sample.

(d) Biological field reports indicate that coyotes who eat sheep are fairly consistent in their preference for sheep, whereas the majority of coyotes who live in the wilderness stick to foods they find in the wild. When the poison was widespread, which group of coyotes would tend to get poisoned: those that ate sheep or those that ate what they could find in the wild?

⇨ When there is poison bait scattered throughout the wilderness, it is the wilderness coyote that eats it; the sheep-eating coyote doesn't bother with it.

The ranchers found that the results were not very satisfactory. The overall coyote population dropped where the poison was applied, but the ranchers had almost as much trouble as ever. They were losing almost as many sheep to coyotes as before.

Today there is an effort to hunt selectively those specific coyotes that are known sheep killers. The important thing to learn from the preceding discussion is that if statistical methods are to be reliable, we must have a *true random sample*.

GUIDED EXERCISE   7

Why don't the following procedures give random samples for the entire population of New York City?

(a) Select every third woman who enters a hair styling salon and inquire about income.

⇨ The choice of sample subjects is biased. Only women are included. In addition, the sample is biased toward women who can afford to and who like to go to hair styling salons.

Exercise continues

Exercise continued

(b) Select every third man who places an order for a custom-tailored suit and inquire about age.   ⇨   Again, the choice of sample subjects is biased. Only men are included, and only men who can afford and want custom-tailored suits.

(c) Select every third person who leaves a performance of the Metropolitan Opera and inquire if they support funding for the National Endowment for the Arts.   ⇨   The choice of sample subjects is biased toward people who like opera and who probably support funding for the arts. Also, there may be a number of people from out of town.

We've seen several sampling procedures that do not produce random samples. How do we get random samples? Suppose you need to know if the emission system of the latest shipment of Toyotas satisfies pollution-control standards. You want to pick a random sample of 30 cars from this shipment of 500 cars and test them. One way to pick a random sample is to number the cars 1 through 500. Write these numbers on cards, mix up the cards, and then draw 30 numbers. The sample will consist of the cars with the chosen numbers. If you mix the cards sufficiently, this procedure produces a random sample.

Random-number table

An easier way to select the numbers is to use a random-number table. You can make one yourself by writing the digits 0 through 9 on separate cards and thoroughly mixing up these cards in a hat. Then draw a card, record the digit, return the card, and mix up the cards again. Draw another card, record the digit, and so on. However, Table 1 in Appendix I is a ready-made random-number table (adapted from Rand Corporation, *A Million Random Digits with 100,000 Normal Deviates*). Let's see how to pick our random sample of 30 Toyotas by using this random-number table.

EXAMPLE 8 ⊳   Use a random-number table to pick a random sample of 30 cars from a population of 500 cars.

SOLUTION: Again, we assign each car a different number between 1 and 500. Then we use the random-number table to choose the sample. Table 1 in Appendix I has 30 rows and 10 blocks of five digits each; it can be thought of as a solid mass of digits that has been broken up into rows and blocks for user convenience.

You read the digits by beginning anywhere in the table. We dropped a pin on the table, and the head of the pin landed in row 15, block 5. We'll begin there and list all the digits in that row. If we need more digits, we'll move on to row 16, and so on. The digits we begin with are

99281   59640   15221   96079   09961   05371

Since the highest number assigned to a car is 500, and this number has three digits, we regroup our digits into blocks of 3:

992   815   964   015   221   960   790   996   105   371

To construct our random sample, we use the first 30 car numbers we encounter in the random-number table when we start at row 15, block 5. We skip the first three groups—992, 815, and 964—because these numbers are all too large. The next group of three digits is 015, which corresponds to 15. Car number 15 is the first car included in our sample, and the next is car number 221. We skip the next three groups and then include car numbers 105 and 371. To get the rest of the cars in the sample, we continue to the next line and use the random-number table in this fashion. If we encounter a number we've used before, we'll skip it.   ●

COMMENT When we use the term *(simple) random sample,* we have very specific criteria in mind for selecting the sample. One proper method for selecting a simple random sample is to use a computer-based or calculator-based random-number generator or to use a table of random numbers as we did in Example 8. The term *random* should not be confused with *haphazard.*   ○

Simulation

Another important use of random-number tables is in *simulation.* We use the word *simulation* to refer to the process of providing arithmetic imitations of "real" phenomena. Simulation methods have been productive in studies of nuclear reactors, cloud formation, cardiology (and medical science in general), highway design, production control, shipbuilding, airplane design, war games, economics, electronics, and in countless other studies. A complete list would probably include something from every aspect of modern life. In Example 9 and Guided Exercise 8 we'll perform a brief simulation.

EXAMPLE 9   A well-known theory in stock market analysis is the "random walk" theory (see *A Random Walk Down Wall Street,* 6th Edition, Burton Malkiel, W. W. Norton & Co.). The term *random walk,* as applied to stock prices, means that short-term changes in stock prices cannot be predicted but rather are random. In particular, according to the random walk theory, the next move in price of a stock (up or down) is completely unpredictable on the basis of what price changes have happened before.

Let's use a very simplified model to *simulate* the stock price changes of a hypothetical company, Fun Boards (maker of skateboards, surfboards, skiboards, and in-line skates). Suppose the initial price of Fun Boards stock is $50 per share. Use the random-number table to simulate daily price changes for the next 15 trading days in the following way. Notice that we are interested in price *changes.* Days during which the stock does not change price will be ignored.

SOLUTION: The daily stock price change will be dictated by a number from the random-number table. When you encounter an even digit (0, 2, 4, 6, 8), increase the stock price by $1. When you encounter an odd digit (1, 3, 5, 7, 9), decrease the stock price by $1. Do this for a sequence of 15 trading days during which the price changes.

Beginning with line 6, block 2 in Table 1 in Appendix I, we see the 15 random digits

51709   94456   48396

Since the first random digit is odd, we will decrease the price by $1 on the first day. In fact, the next two digits are also odd, so we will decrease the price again by $1 on day 2 and then again on day 3. The next digit is even, so we will increase the price by $1 on day 4. Table 1-1 shows the simulated price changes for the 15-trading-day period.

Table 1-1   Simulated Price Moves of Fun Boards Stock

Price	50	49	48	47	48	47	46	47	48	47	48	49	50	49	48	49
Day	Initial	1	2	3	4	5	6	7	8	9	10	11	12	13	14	15
Digit		5	1	7	0	9	9	4	4	5	6	4	8	3	9	6

GUIDED EXERCISE   8

Use a random-number table to simulate the outcomes of tossing a balanced (that is, fair) penny 10 times.

(a) How many outcomes are possible when you toss a coin once?

⇨ Two; heads or tails

(b) There are several ways to assign numbers to the two outcomes. Because we assume a fair coin, assign an even digit to the outcome heads and an odd digit to the outcome tails. Then, starting at block 3 of row 2 of Table 1 in Appendix I, list the first 10 single digits.

⇨ 7 1 5 4 9 4 4 8 4 3

(c) What are the outcomes associated with the 10 digits?

⇨ T T T H T H H H H T

(d) If you start in a different block and row of Table 1 in Appendix I, will you get the same sequence of outcomes?

⇨ It is possible, but not very likely. (In Section 4.3 you will learn how to determine that there are 1024 possible sequences of outcomes for 10 tosses of a coin.)

COMMENT Recall that the random samples we have been constructing so far are called *simple random samples*. Throughout this text we will use the term *random sample* to mean simple random sample. All the statistical methods in this text assume that a simple random sampling has been used to collect the data.   ○

Other sampling techniques

Although we will always assume that (simple) random samples are used throughout this text, there are other methods of sampling that are also widely used. Appropriate statistical techniques exist for these sampling methods, but they are beyond the scope of this text.

Stratified sampling

One of these sampling methods is called *stratified sampling*. Groups or classes inside a population that share a common characteristic are called *strata* (plural of *stratum*). For example, in the population of all undergraduate college students, some strata might be freshmen, sophomores, juniors, and seniors. Other strata might be men and women, in-state students and out-of-state students, and so on. In the method of stratified sampling, the population is divided into at least two distinct strata. Then a (simple) random sample of a certain size is drawn from each stratum, and the information obtained is carefully adjusted or weighted in all resulting calculations.

The groups or strata are often sampled in proportion to their actual percentages of occurrence in the overall population. However, there are other (more sophisticated) ways to determine the optimal sample size in each stratum for best results. In general, statistical analyses and tests based on data obtained from stratified samples are somewhat different from techniques discussed in an introductory course in statistics. Methods of stratified sampling will not be emphasized in this text.

Systematic sampling

Another popular method of sampling is called *systematic sampling*. In this method, it is assumed that the elements of the population are arranged in some natural sequential order. Then we select a (random) starting point and select every *k*th element for our sample. For instance, people lining up to buy rock concert tickets are "in order." To generate a systematic sample of these people (and ask questions regarding topics such as age, smoking habits, income level, etc.), we could include every 5th person in line. The "starting" person could be selected at random from the first five.

The advantage of a systematic sample is that it is easy to get. However, there are dangers in using systematic sampling. When the population is repetitive or cyclic in nature, systematic sampling should not be used. For example, consider a fabric mill that produces dress material. Suppose the loom that produces the material makes a mistake every 17th yard, but we check only every 16th yard with an automated electronic scanner. In this case, a random starting point may or may not result in detection of fabric flaws before a large amount of fabric is produced.

Cluster sampling

Cluster sampling is a method used extensively by government agencies and certain private research organizations. In cluster sampling we begin by dividing the demographic area into sections. Then we randomly select the sections or clusters. Every member of the cluster is included in the sample. For instance, in conducting a survey of school children in a large city, we could first randomly select 5 schools and then include all the children from each selected school.

Convenience sampling

Convenience sampling simply uses results or data that are conveniently and readily obtained. In some cases this may be all that is available, and in many cases, it is better than no information at all. However, convenience sampling runs the risk of being severely biased. For instance, consider a newsperson who wishes to get the "opinions of the people" about a proposed seat tax to be imposed on tickets to all sporting events. The revenues from the seat tax will then be used to support the local symphony. The newsperson stands in front of a classical music store at noon and surveys the first five people coming out of the store who will cooperate. This method of choosing a sample will produce some opinions, and perhaps some human interest stories, but it certainly has bias. It is hoped that the city council will not use these opinions as the sole basis for a decision about the proposed tax. It is good advice to be very cautious indeed when the data come from the method of convenience sampling.

Our discussion of sampling methods is intended to be brief and general. This is not the place for an extensive treatment of theory and practice of sampling. However, the interested reader is referred to the book *A Sampler on Sampling*, by Bill Williams, John Wiley and Sons, Inc.

VIEWPOINT

Extraterrestrial Life?

Do you believe in intelligent life on other planets? Using methods of random sampling, a Fox News opinion poll found that about 54% of all U.S. men do believe in intelligent life on other planets, whereas only 47% of women believe there is such life. How could you conduct a random survey of students on your campus regarding extraterrestrial life?

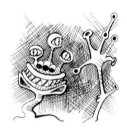

SECTION 1.2 PROBLEMS

Sampling: Random

1. (a) In your own words, explain the meaning of the terms *random numbers* and *random samples*.
   (b) Why are random samples so important in statistics?

Sampling: Random

2. Use a random-number table to get a list of eight random numbers from 1 to 976. Explain your work.

Sampling: Random

3. Use a random-number table to get a list of six random numbers from 1 to 99. Explain your work.

Sampling: Random

4. Use a random-number table to get a list of five random numbers from 1 to 900. Explain your work.

Sampling: Students in Class

5. Suppose you are given the number 1 and each of the other students in your statistics class calls out consecutive numbers until each person in class has his or her own number. Explain how you could get a random sample of four students from your statistics class.
   (a) Explain why the first four students walking into the classroom would not necessarily form a random sample.
   (b) Explain why four students coming in late would not necessarily form a random sample.
   (c) Explain why four students sitting in the back row would not necessarily form a random sample.
   (d) Explain why the four tallest students would not necessarily form a random sample.

Sampling: Quality Control

6. Quality control is an important component in manufacturing processes. Products are inspected during production, and equipment is adjusted to correct defects. It is usually not possible to examine every product, so a random sample is examined. For each of the following, give a detailed explanation of how you could get the requested random sample. Be sure to include the random numbers you use to get the random sample.

    (a) How would you draw a random sample of six of the next 500 stereo headsets coming off an assembly line?

    (b) How would you obtain a random sample of 10 men's dress shirts coming off an assembly line from 8 A.M. to 12 noon?

    (c) Serial numbers are placed on radios as they come off an assembly line. How could you get a random sample of nine radios with serial numbers from 21942 to 98756?

    (d) A truck has just delivered 800 cartons of eggs to a supermarket. How would you get a random sample of 12 cartons to check for broken eggs?

Leisure: Colorado Lotto

7. Lotto is the name of the Colorado lottery. The Lotto boards consist of 42 numbers (from 1 to 42). To play you select six distinct numbers. Every week a drawing machine randomly selects six numbered ping-pong balls. If one of your boards contains all six winning numbers, in any order, you've hit the jackpot! You can pick your numbers any way you wish. However, suppose you want to use a random-number table to pick your six numbers. Describe how you would do so, and list your selected numbers. (To play you must pay $1 to have your selections entered in a computer for a specified week's drawing.)

Psychology:
Random Selection

8. How do colds affect analytical thinking performance? Results of a study conducted by McGraw and Schleser were reported in *Psychology Today*. The study showed that under certain conditions, persons with colds do better than their healthy colleagues. The study considered 62 subjects: 40 healthy men and women and 22 suffering from colds or flu. These subjects were divided into two groups, with each group containing both healthy and sick participants. The groups were told to solve anagrams. The first group was told to "focus on trying out systematically different combinations of letters," while the second group was told to "focus on remaining loose and relaxed as you complete the task." In the first group, after the second try, the sick participants who concentrated on their anagrams scored significantly better than the healthy ones in that group. In the second group, the sick people did slightly worse than the healthy ones on the anagram solving task. A key component in this study was the formation of two groups from the 62 participants. Describe how you could take these 62 participants and, using a random-number table, divide them into two groups of equal size.

Simulation: Coin Toss

9. Use a random-number table to simulate the outcomes of tossing a quarter 25 times. Assume that the quarter is balanced (i.e., fair).

Simulation:
Birthday Problem

10. Suppose there are 30 people at a party. Do you think any two share the same birthday? Let's use the random-number table to simulate the birthdays of the 30 people at the party. Ignoring leap year, let's assume that the year has 365 days. Number the days, with 1 representing January 1, 2 representing January 2, and so forth, with 365 representing December 31. Draw a random sample of 30 days (with replacement). These days represent the birthdays of the people at the party. Were any two of the birthdays the same? Compare your results with those obtained by other students in the class. Would you expect the results to be the same or different?

Business: Benefits Package

11. An important part of employee compensation is a benefits package that might include health insurance, life insurance, child care, vacation days, retirement plan, parental leave, bonuses, etc. Suppose you want to conduct a survey of benefit packages available in private businesses in Hawaii. You want a sample size of 100. Some sampling techniques are described below. Categorize each technique as *simple random sampling, stratified sampling, systematic sampling, cluster sampling,* or *convenience sampling.*
    (a) Assign each business in the Island Business Directory a number, and then use a random-number table to select the businesses to be included in the sample.
    (b) Use the postal ZIP codes to divide the state into regions. Pick a random sample of 10 ZIP code areas and then include all the businesses in each selected ZIP code area.
    (c) Send a team of five research assistants to Bishop Street in downtown Honolulu. Let each assistant select a block or building and interview an employee from each business found. Each researcher can have the rest of the day off after getting responses from 20 different businesses.
    (d) Use the Island Business Directory. Number all the businesses. Select a starting place at random, and then use every 50th business listed until you have 100 businesses.
    (e) Group the businesses according to type: medical, shipping, retail, manufacturing, financial, construction, restaurant, hotel, tourism, other. Then select a random sample of 10 businesses from each business type.

Health Care: Sampling

12. Modern Managed Hospitals (MMH) is a national for-profit chain of hospitals. Management wants to survey patients discharged this past year to obtain patient satisfaction profiles. They wish to use a sample of such patients. Several sampling techniques are described below. Categorize each technique as *simple random sampling, stratified sampling, systematic sampling, cluster sampling,* or *convenience sampling.*
    (a) Obtain a list of patients discharged from all MMH facilities. Divide the patients according to length of hospital stay (3 days or less, 3–7 days, 8–14 days, more than 14 days). Draw simple random samples from each group.
    (b) Obtain lists of patients discharged from all MMH facilities. Number these patients, and then use a random-number table to obtain the sample.
    (c) Randomly select some MMH facilities from each of five geographic regions, and then take all patients on the discharge lists of the selected hospitals.
    (d) At the beginning of the year, instruct each MMH facility to survey every 500th patient discharged.
    (e) Instruct each MMH facility to survey 10 discharged patients this week and send in the results.

Computer Simulation:
Roll of a Die

13. A die is a cube with dots on each face. The faces have 1, 2, 3, 4, 5, or 6 dots. The table on the top of page 25 is a computer simulation (from the software package Minitab) of the results of rolling a fair die 20 times. In Minitab, version 12, the menu selections ➤**Calc** ➤**Random Data** ➤**Integer** produce a dialogue box. In the dialogue box we specified two rows for columns C1 through C10. We set the minimum value of the integers to be selected at 1 and the maximum value at 6.
    (a) Assume that each number corresponds to the number of dots on the face of the die. Is it appropriate that the same number appear more than once? Why? What is the outcome of the 4th roll?
    (b) Scan the data. What percentage of the outcomes represented "snake eyes," or 2 dots?

DATA DISPLAY										
ROW	C1	C2	C3	C4	C5	C6	C7	C8	C9	C10
1	5	2	2	2	5	3	2	3	1	4
2	3	2	4	5	4	5	3	5	3	4

(c) If we simulate more "rolls of the die," do you expect to get the same sequence of outcomes? Why or why not?

Section **1.3**

Graphs

Popular newspapers, magazines, and the Web frequently display graphs such as those in Figure 1-1. The graphs shown in the figure represent some of the basic types of graphs you will study in this section. Notice how well each graph is labeled. There is a title, there are scales on the axes, and a source is given for the data. In some cases the sample size is included as well.

Bar graphs

Let's first consider *bar graphs*. The bars can be vertical or horizontal, but they should be of *uniform width* and be *uniformly spaced*. The lengths of the bars represent values of the quantity we wish to compare under various conditions. In Figure 1-2 on page 27 we are comparing the numbers of days a person can survive at a temperature of 110°F with various amounts of water. The length of each bar represents the number of days one can survive.

Example 10 on pages 26 and 27 shows you how to construct a bar graph.

FIGURE 1-1

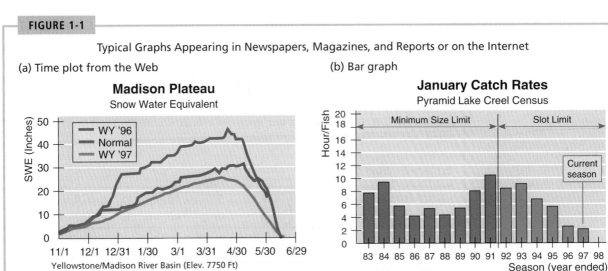

Typical Graphs Appearing in Newspapers, Magazines, and Reports or on the Internet

(a) Time plot from the Web

Madison Plateau
Snow Water Equivalent

Yellowstone/Madison River Basin (Elev. 7750 Ft)

Source: Web site <http://www.wwrc.uwyo.edu/>. Chart through courtesy of NRCS and NOAA, National Weather Service.

(b) Bar graph

January Catch Rates
Pyramid Lake Creel Census

Season (year ended)

Source: Pyramid Lake Fisheries, Resource Management Dept. (Pyramid Lake Paiute Tribe), Pyramid Lake, Nevada. Reprinted with permission.

continued

FIGURE 1-1 continued

(c) Circle graph

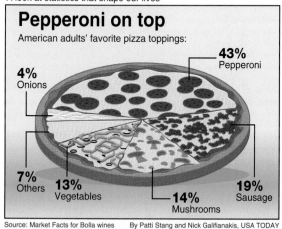

USA SNAPSHOTS®
A look at statistics that shape our lives

Pepperoni on top
American adults' favorite pizza toppings:

43% Pepperoni

4% Onions

7% Others

13% Vegetables

14% Mushrooms

19% Sausage

Source: Market Facts for Bolla wines   By Patti Stang and Nick Galifianakis, USA TODAY

Source: Copyright 1996, USA TODAY. Reprinted with permission.

(d) Pareto chart

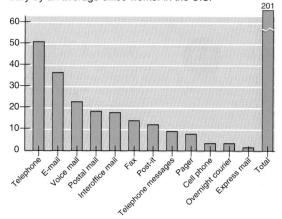

Why Things Pile Up
Staying in Touch Number of messages sent and received daily by an average office worker in the U.S.

Source: Wall Street Journal (6/21/99)

EXAMPLE 10 ➤ Hikers are often cautioned to carry extra jackets so that they will be prepared for the effect of wind chill. Suppose you are hiking in 50°F temperature and a wind comes up. A breeze of 5 miles per hour (mph) makes the effective temperature 48°F. If the wind picks up to 10 mph, then the temperature is equivalent to 40°F. A 15-mph wind drops the effective temperature to 36°F, and a 20-mph wind drops the temperature to freezing. A 25-mph wind makes the effective temperature only 30°F. Make a bar graph to display this information.

SOLUTION: Before we make a bar graph of this information, we'll make a table of wind and effective temperature (Table 1-2).

Table 1-2 Wind Chill at 50°F

Wind Speed (mph)	Calm	5	10	15	20	25
Equivalent Temperature (°F)	50	48	40	36	32	30

Source: Data from *Surviving the Unexpected Wilderness Emergency,* by Gene Fear, published by Survival Education Association, Tacoma, Washington.

To make the bar graph (Figure 1-3 on page 27), we'll list the wind speed on the horizontal axis and the effective temperature on the vertical axis. Each bar will be centered over its wind speed, and the height of the bar will represent the effective temperature. Note again that the bars have the same width and that the spacing between all the bars is the same. Both axes are labeled and have scale markings. ●

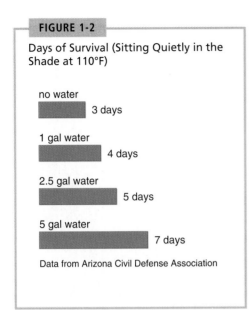

FIGURE 1-2

Days of Survival (Sitting Quietly in the Shade at 110°F)

no water
3 days

1 gal water
4 days

2.5 gal water
5 days

5 gal water
7 days

Data from Arizona Civil Defense Association

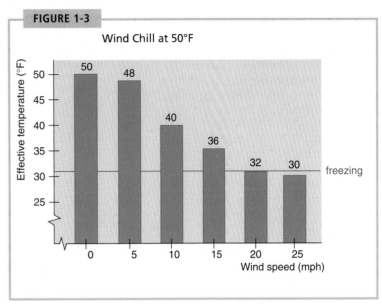

FIGURE 1-3

Wind Chill at 50°F

GUIDED EXERCISE   9

Sunshine Travel Agency offers a rain insurance policy on its Hawaiian tours. It costs an extra $100. If you buy the optional policy and it rains during more than 15% of the days of your trip, you will be reimbursed for meals and lodging during the extra rainy days (beyond the 15% and up to 5 days). You are planning a trip to Hawaii and you are debating taking the insurance. You obtain the rainfall information in Table 1-3 from the U.S. National Oceanic and Atmospheric Administration.

Table 1-3   Average Monthly Rainfall in Honolulu, Hawaii

Month	Jan.	Feb.	Mar.	Apr.	May	June	July	Aug.	Sept.	Oct.	Nov.	Dec.
Rainfall (inches)	4.40	2.46	3.18	1.36	0.96	0.32	0.60	0.76	0.67	1.51	2.99	3.64

(a) Make a bar graph of this information with the months on the horizontal axis and rainfall on the vertical axis.

⇨ See Figure 1-4 on the following page.

(b) There is the rainy season and there is the dry season. From the graph, which 6 months would you say make up the rainy season?

⇨ October, November, December, January, February, and March

Exercise continues

Exercise continued

(c) Without the rain insurance, which winter month (November, December, January, or February) would be best for your trip?

⇨ February, since it has the least average rainfall

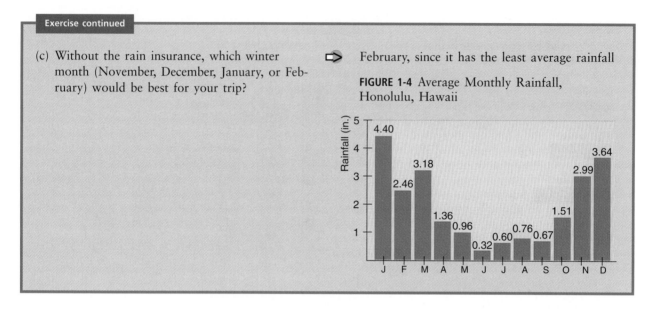

FIGURE 1-4 Average Monthly Rainfall, Honolulu, Hawaii

When you read bar graphs, be careful of changing scales. For instance, in Figure 1-5 you see two graphs showing the life expectancies of a man born in 1920 and one born in 1996 (*Statistical Abstract of the United States*, Bureau of Census). The change in life expectancy over the 76-year period illustrated in Figure 1-5 is large: from 54 years to 72.7 years, an increase of 18.7 years. But part *a* makes it seem that the life expectancy has about tripled. Notice the squiggle ⌇ at the bottom of the vertical axis. This is to inform you that the years 0 to 49 have been omitted. In part *b*, no years were skipped, and the picture immediately gives an accurate impression. Be on the alert—many magazine articles use changing scales. You also should watch for omitted values. If you omit values, be sure to give the reader fair warning with a squiggle ⌇ at the beginning of the axis.

Pareto charts

Bar graphs in which the bars are ordered according to height are used in quality-control programs. Dr. W. Edwards Deming was one of the developers of the concept of total quality management (TQM). In his book *Out of the Crisis* (MIT Center for Advanced Engineering Study), Dr. Deming outlines many strategies for improving quality in production and service industries. He recommends the use of some statistical methods to organize and analyze data from the industries so that sources of problems can be identified and then corrected. *Pareto* (pronounced "Pah-rāy-tōe") *charts* are among the many techniques used in quality-control programs. We will see another quality-control technique in Chapter 6.

The Deming Management Method, by Mary Walton (Putman Publishing Group), lists Pareto charts among seven of the most helpful charts. Pareto charts are bar graphs in which the height represents frequency. In addition, the bars are arranged according to height, with the tallest or longest bar placed on the left or on top. Pareto charts are often used to organize data about causes of problems so as to highlight major causes from left to right or from top to bottom.

FIGURE 1-5

Male Life Expectancy
from Birth

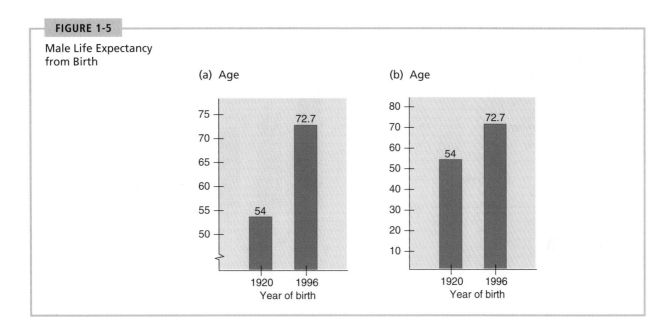

EXAMPLE 11 ➤ This is an example adapted from *The Deming Management Method*. Suppose you want to arrive at the college 15 minutes before your first class so that you can feel relaxed when you walk into class. An early arrival time also allows room for unexpected delays. However, you always find yourself arriving "just in time" or slightly late. What causes you to be late? One student made a list of possible causes and then kept a checklist for 2 months (Table 1-4 on the following page). On some days more than one item was checked because several events occurred that caused the student to be late. Make a Pareto chart showing causes for lateness.

SOLUTION: To make a Pareto chart, we make a bar chart. The bars represent the causes, and the heights of the bars represent the frequencies of occurrence. We place the specific cause that occurs most frequently first and continue with the specific cause that occurs next most frequently, and so on.

From the chart in Figure 1-6 on the following page we quickly see that last-minute studying is the most frequent cause of lateness. Perhaps a more realistic study plan is in order. Another option is to get up sufficiently early to allow for adequate morning study time. The next most frequent cause is snoozing after the alarm goes off. Here a deliberate effort to get up immediately might help. Breakfast is the next problem. Has a realistic assumption been made about the time required for breakfast? The second cup of coffee might not fit the schedule. The Pareto chart gives us a vehicle for organizing our data and then attacking the most frequent problems. Spending a lot of time and money on the car will not have much impact on arrival time at the college (assuming that the car does not get worse).   ●

Table 1-4 Causes for Lateness
(September–October)

Cause	Frequency
Snoozing after alarm goes off	15
Car trouble	5
Too long over breakfast	13
Last-minute studying	20
Finding something to wear	8
Talking too long with roommate	9
Other	3

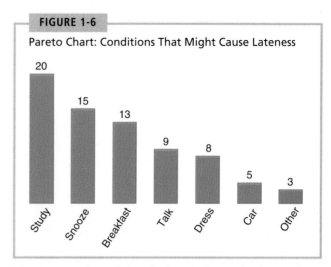

FIGURE 1-6

Pareto Chart: Conditions That Might Cause Lateness

Circle graphs
or
pie charts

Another popular pictorial representation of data is the *circle graph,* or *pie chart.* It is relatively safe from misinterpretation and is especially useful for showing the division of a total quantity into its component parts. The total quantity, or 100%, is represented by the entire circle. Each wedge of the circle represents a component part of the total. These proportional segments are usually labeled with corresponding percentages of the total. Guided Exercise 10 shows how to make a circle graph.

GUIDED EXERCISE   10

How long do we spend talking on the telephone after hours (at home after 5 P.M.)? The results from a recent survey of 500 people (as reported in *USA Today*) are shown in Table 1-5. We'll make a circle graph to display these data.

Table 1-5 Time Spent on Home Telephone After 5 P.M.

Time	Number	Fractional Part	Percentage	Number of Degrees
Less than $\frac{1}{2}$ hour	296	296/500	59.2	59.2% × 360° ≈ 213°
$\frac{1}{2}$ hour to 1 hour	83	83/500	16.6	16.6% × 360° ≈ 60°
More than 1 hour	121	____	____	____
Total	____		____	____

(a) Fill in the missing parts in Table 1-5 for "More than 1 hour." Remember that the central angle of a circle is 360°. Round to the nearest degree.

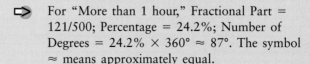 For "More than 1 hour," Fractional Part = 121/500; Percentage = 24.2%; Number of Degrees = 24.2% × 360° ≈ 87°. The symbol ≈ means approximately equal.

Exercise continues

Exercise continued

(b) Fill in the totals. What is the total number of responses? Do the percentages total 100% (within rounding error)? Do the numbers of degrees total 360° (within rounding error)?

⇨ The total number of responses is 500. The percentages total 100%. You must have such a total in order to create a circle graph. The numbers of degrees total 360°.

(c) Draw a circle graph. Divide the circle into pieces with the designated numbers of degrees. Label each piece, and show the percentage corresponding to each piece. The numbers of degrees are usually omitted from pie charts shown in newspapers, magazines, journals, and reports.

⇨ 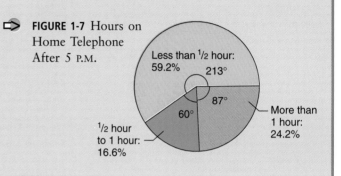 **FIGURE 1-7** Hours on Home Telephone After 5 P.M.

Less than ½ hour: 59.2%   213°

87°

60°

More than 1 hour: 24.2%

½ hour to 1 hour: 16.6%

Time plot

Suppose you begin an exercise program that involves walking or jogging for 30 minutes. You exercise several times a week but monitor yourself by logging the distance you cover in 30 minutes each Saturday. How do you display these data in a meaningful way? Making a bar chart showing the frequency of distances you cover might be interesting, but it does not really show how the distance you cover in 30 minutes has changed over time. A graph showing the distance covered on each date will let you track your performance over time.

We will use a *time plot*. A time plot is a graph showing data measurements in chronological order. To make a time plot, we put time on the horizontal scale and the variable being measured on the vertical scale. In a basic time plot, we connect the data points by lines.

EXAMPLE 12 ◈ Suppose you have been in the walking/jogging exercise program for 20 weeks, and for each week you have recorded the distance you covered in 30 minutes. Your data log is shown in Table 1-6.

Table 1-6   Distance in Miles Walked/Jogged in 30 Minutes

Week	1	2	3	4	5	6	7	8	9	10
Distance	1.5	1.4	1.7	1.6	1.9	2.0	1.8	2.0	1.9	2.0
Week	11	12	13	14	15	16	17	18	19	20
Distance	2.1	2.1	2.3	2.3	2.2	2.4	2.5	2.6	2.4	2.7

(a) Make a time plot.

SOLUTION: The data are appropriate for a time plot because they represent the same measurement (distance covered in a 30-minute period) taken at different

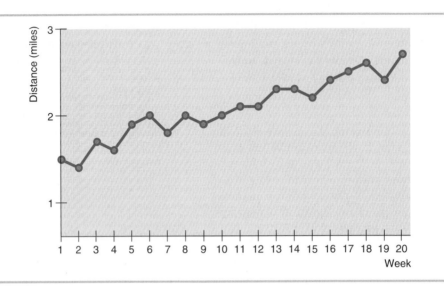

FIGURE 1-8

Time Plot of Distance (in miles) Jogged in 30 Minutes

times. The measurements are also recorded at equal time intervals (every week). To make our time plot, we list the weeks in order on the horizontal scale. Above each week, plot the distance covered that week on the vertical scale. Then connect the dots. Figure 1-8 shows the time plot. Be sure the scales are labeled.

(b) From looking at Figure 1-8, can you detect any patterns?

SOLUTION: There seems to be an upward trend in distance covered. The distances covered in the last few weeks are about a mile farther than those for the first few weeks. However, we cannot conclude that this trend will continue. Perhaps you have reached your goal for this training activity and now wish to maintain a distance of about 2.5 miles in 30 minutes.

Time series

Data sets composed of similar measurements taken at regular intervals over time are called *time series*. Time series are often used in economics, finance, sociology, medicine, and any situation where we want to study or monitor a similar measure over a period of time. A time plot can reveal some of the main features of time series.

VIEWPOINT

Do Ethical Standards Vary with the Situation?

Lutheran Brotherhood did a national survey and found that nearly 60% of all U.S. adults claim that ethical standards vary with the situation, 33% claim that there is only one ethical standard, and 7% are not sure. How could you draw a circle graph to make a visual impression of Americans' views on ethical standards?

**Education:
Does College Pay Off?**

1. It is costly in both time and money to go to college. Does it pay off? According to the *Conference Board Survey* by the Bureau of Census, the answer is yes. The average annual incomes (in thousands of dollars) of households headed by a person with the stated education level is as follows: 14.2 if grade school is the top level of education achieved, 30.1 for high school graduates, 57.4 for college graduates, and 61.0 for completion of one or more years of postgraduate studies. Make a bar graph showing household income for each education level.

**Technology:
Computer Privacy**

2. According to The Shell Poll (reported in *USA Today*), eight in 10 adults are worried that computers and the Internet are reducing privacy. Approximately 80% are concerned about credit bureaus selling personal financial information, 77% about companies selling personal information to other companies, 76% about states selling lists of licensed drivers, 65% about health companies sharing personal medical records, 53% about Social Security numbers being used for personal identification, 43% about hidden cameras at work, and 37% about employers monitoring personal phone calls. Make a bar graph that displays the percentages of adults who are concerned about these issues. Create a Pareto chart showing the same information.

Airlines: Fear of Flying

3. In an article entitled "How to Cure the Fear of Flying," *Fortune Magazine* gave the following information: The number of people who died in a calendar year while on commercial flights was 33; in the bathtub, 318; by poisoning, 5900; as pedestrians, 6500; murdered by spouses, 7000; from falls, 12,400; and in motor vehicle accidents, 33,800.
(a) Make a bar graph showing activities causing death.
(b) Make a Pareto chart showing activities causing death.

Business: Company Failures

4. Suppose you are interested in investing in a business corporation. What are some pitfalls that might cause a business corporation to fail? To answer this question, we can look at history. Buccino & Associates surveyed more than 1300 business professionals and asked them about leading causes of business failure. According to the report in *USA Today*, 88% of those surveyed cited internal problems, not external factors, as the leading causes of business failure. Of the internal problems listed, insufficient management experience was cited as the leading cause of business failure by 13% of those interviewed, poor business planning was cited by 13%, inadequate leadership was cited by 18%, and excessive corporate debt was cited by 29%.
(a) Make a Pareto chart showing the causes of business failure.
(b) Does your chart show *all* the causes of business failure cited by business professionals interviewed? Which internal cause was cited most frequently?

Lifestyle: Hide the Mess!

5. A survey of 1000 adults (reported in *USA Today*) uncovered some interesting housekeeping secrets. When unexpected company comes, where do we hide the mess? The survey showed that 68% of the respondents toss their mess in the closet, 23% shove things under the bed, 6% put things in the bathtub, and 3% hide the mess in the freezer. Make a circle graph to display this information.

Lifestyle: Fast-food

6. What meal are we most likely to eat at a fast-food restaurant? A survey of 1261 adults (reported in *USA Today*) revealed that 48.9% of the respondents are most likely to

eat lunch at a fast-food restaurant; 7.7%, breakfast; 31.6%, dinner; 10%, a snack; and 1.8% answered "don't know." Display this information in a circle graph.

Merchandise:
Consumer Electronics

7. Telephone gadgets fill our homes. The Consumer Electronics Manufacturers Association reported that 96% of U.S. households have corded phones, 66% have cordless phones, 65% have telephone answering devices, 34% have cellular phones, 28% have pagers, 18% have caller-ID equipment, 19% have modems, and 9% have fax machines. Display this information in a bar chart showing the percentage of households owning each gadget. Could this information, as reported, be put into a circle graph? Explain.

Driving: Bad Habits

8. Driving would be more pleasant if we didn't have to put up with the bad habits of other drivers. *USA Today* reported the results of a Valvoline Oil Company survey of 500 drivers in which the drivers marked their complaints about other drivers. The top complaints turned out to be tailgating, marked by 22% of the respondents; not using turn signals, marked by 19%; being cut off, 16%; driving too slowly, 11%; and being inconsiderate, 8%. Make a Pareto chart showing the percentage of drivers who listed each stated complaint. Could this information, as reported, be put in a circle graph? Why or why not?

Geography:
Pyramid Lake Elevation

9. Pyramid Lake, Nevada, is described as the pride of the Paiute Indian Nation. It is a beautiful desert lake famous for very large trout. The elevation of the lake's surface (feet above sea level) varies according to the annual flow of the Truckee River from Lake Tahoe. The U.S. Geological Survey provided the following information:

Year	Elevation	Year	Elevation
1986	3817	1992	3798
1987	3815	1993	3797
1988	3810	1994	3795
1989	3812	1995	3797
1990	3808	1996	3802
1991	3803	1997	3807

Make a time plot displaying these data.

Vital Statistics:
Height of Boys

10. How does the average height of boys change as the boys get older? According to *Physician's Handbook*, the heights at different ages are as follows:

Age (years)	0.5	1	2	3	4	5	6	7
Height (inches)	26	29	33	36	39	42	45	47

Age (years)	8	9	10	11	12	13	14
Height (inches)	50	52	54	56	58	60	62

Make a time plot of the average heights of boys aged 0.5 through 14 years.

Section **1.4**   ## Histograms and Frequency Distributions

Healthy Crunch Cereal is about to take over sponsorship of the TV program "Space Voyage." The advertising manager has requested a report on the age distribution of the viewers so that the spot ads can be tailored to appeal to the age groups with the most viewers. The viewer age report contains the graph in Figure 1-9, which was made from a random sample of viewers.

GUIDED EXERCISE   **11**

Review the graph of the viewer age distribution for the program "Space Voyage" (Figure 1-9) before answering the following questions.

(a) What does the height of each bar represent? How many viewers are represented in this graph?

⇨ The height of each bar represents the number of viewers of that age group. To get the total number of viewers, add the heights of the bars. There are 500 viewers represented.

(b) What does the width of a bar represent?

⇨ The width represents the age group.

(c) What ages are included in the group with the most viewers? Is this graph detailed enough to tell you exactly how many viewers are 21 years old?

⇨ The age group 14.5–24.5 has the most viewers. We cannot tell how many viewers are 21 years old. All we can say is that there are 190 viewers between the ages of 14.5 and 24.5 years.

(d) From the information in this graph about the ages of the viewers, which of the following ads do you think the manager might choose for "Space Voyage"?

⇨ Since the largest age group is between 14.5 and 24.5, the ad about the campers would probably be the best. This age group would not necessarily be interested in food that makes a first grader grow or food that is eaten before one reads the stock market report.

Scene 1: A grandmother and first grader at the breakfast table. The grandmother says to the child, "Eat Healthy Crunch Cereal because it will make you grow."

Scene 2: A middle-aged man reading the stock report. An empty bowl is on the table with an open box of Healthy Crunch Cereal beside it. The man puts down the paper and puts his hands on the box of Healthy Crunch as he says, "I eat Healthy Crunch even *before* I read the stock report."

Scene 3: Two young campers eating breakfast in front of their tent. A box of Healthy Crunch Cereal is clearly visible in the foreground. One camper says to the other, "Healthy Crunch will help us climb that mountain."

FIGURE 1-9   Viewer Ages for "Space Voyage"

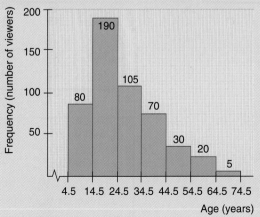

The graph in Figure 1-9 is a little different from the other bar graphs we looked at in the last section. A graph like the one in Figure 1-9 on the previous page is called a *histogram*. It differs from a bar graph in two important ways: the bars always touch, and the width of a bar represents a quantitative value, such as age. In a bar graph we could make the bars as wide as we wished, according to the visual impression we wanted to convey, but in a histogram the width of the bar has a meaning. For instance, in Figure 1-9 the width of each bar represents 10 years.

Information is presented in condensed form in a histogram. The original data for the viewer age report included the *exact* number of 21-year-olds in the sample. In the histogram this number was grouped with the ages 14.5–24.5. We can tell how many viewers are in that age group, but we cannot tell exactly how many are 21 years old. However, the condensed information in the histogram can be assimilated more quickly than the same information in more detailed form.

If you are given many pieces of data, how do you condense the information to make a histogram? The Task Force to Encourage Car Pools did a study of one-way commuting distances for workers in the downtown Dallas area. A random sample of 60 of these workers was taken. The commuting distances of the workers in the sample are given in Table 1-7.

The first thing to do is to decide how many bars or classes you want in the histogram. Five to 15 classes are usually used. If you use fewer than 5 classes, you risk losing too much information, but if you use more than 15 classes, the clarity of the diagram might be sacrificed for detail. Let the spread of the data and the purpose of the histogram be your guides when selecting the number of classes.

Next, find a convenient class width. To do this, find the difference between the largest and smallest data values and divide by the number of classes. If you want the class width to be a whole number, always *increase* the result to the next whole number so that the classes cover the data.

Table 1-7   One-Way Commuting Distances in Miles for 60 Workers in Downtown Dallas

13	47	10	3	16	20	17	40	4	2
7	25	8	21	19	15	3	17	14	6
12	45	1	8	4	16	11	18	23	12
6	2	14	13	7	15	46	12	9	18
34	13	41	28	36	17	24	27	29	9
14	26	10	24	37	31	8	16	12	16

Class Width

1. Compute

$$\frac{\text{largest data value} - \text{smallest data value}}{\text{desired number of classes}}$$

2. Increase the value computed to the next highest whole number.

Note: To ensure that the classes cover the data, we need to increase the result of step 1 to the *next* whole number, even if step 1 produces a whole number. For instance, if the calculation in step 1 produces the value 5, we make the class width 6.

In this case, let's use 6 classes. The largest distance commuted is 47 miles and the smallest is 1 mile.

$$\frac{47 - 1}{6} \approx 7.7; \quad \text{increase to 8}$$

The lowest and highest values that can fit in a class are called the *lower class limit* and *upper class limit,* respectively. The *class width* is the difference between the lower class limit of one class and the lower class limit of the next class. Each class should have the same width, although it is not uncommon to see either the first or the last class width a little longer or shorter than the others. The center of the class is called the *midpoint* (or *class mark*). This is found by adding the lower and upper class limits of one class and dividing by 2.

$$\text{Midpoint} = \frac{\text{lower class limit} + \text{upper class limit}}{2}$$

The midpoint is often used as a representative value of the entire class.

Now we can organize the commuting distance data into a *frequency table* (Table 1-8). Such a table shows the limits of each class, the frequency with which the data fall in each class, and the class midpoints. A tally will help us find the frequencies.

Now we're almost ready to make a histogram. But in a histogram we want the bars to touch. There is a space between the upper limit of one class and the lower limit of the next class. The halfway points of these intervals are called *class boundaries.* We use the class boundaries as the endpoints of the bars in the histogram. Then there is no space between the bars. Figure 1-10 on page 38 is the histogram of commuter distances. The class boundaries are shown. (Sometimes only the class midpoints are labeled.)

Table 1-8   Frequency Table of One-Way Commuting Distances for 60 Downtown Dallas Workers (Data in Miles)

Class Limits Lower–Upper	Class Boundaries Lower–Upper	Tally*	Frequency	Class Midpoint
1–8	0.5–8.5	︴︴ ︴︴ ︴︴	14	4.5
9–16	8.5–16.5	︴︴ ︴︴ ︴︴ ︴︴ ︴	21	12.5
17–24	16.5–24.5	︴︴ ︴︴ ︴	11	20.5
25–32	24.5–32.5	︴︴ ︴	6	28.5
33–40	32.5–40.5	︴︴︴︴	4	36.5
41–48	40.5–48.5	︴︴︴︴	4	44.5

*The tally column is included simply as an aid for determining the frequencies. It is not a necessary part of a frequency table.

FIGURE 1-10

One-Way Commuting Distances in Miles Driven by Downtown Dallas Workers

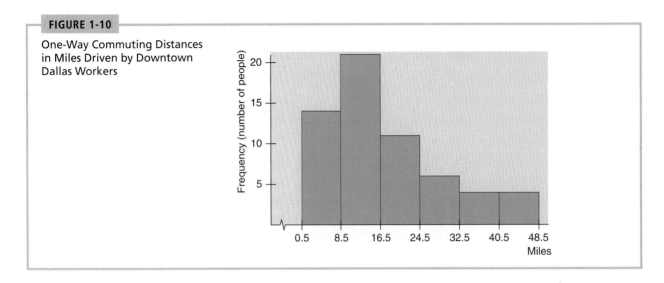

Relative-frequency tables and histograms

Other useful tools for organizing data are *relative-frequency tables* and *relative-frequency histograms*. Once we have made a frequency table, it is easy to construct a relative-frequency table. The relative frequency for a particular class is found by dividing the class frequency by the total of all frequencies (sample size).

$$\text{Relative frequency} = \frac{f}{n} = \frac{\text{class frequency}}{\text{total of all frequencies}}$$

Table 1-9 on the next page shows the relative frequencies for the commuter data in Table 1-7. Since we already have the frequency table (Table 1-8), the relative-frequency table is obtained easily. The sample size is $n = 60$. Notice that the sample size is the total of all the frequencies. Therefore, the relative frequency for the first class (the class from 1 to 8) is

$$\text{Relative frequency} = \frac{f}{n} = \frac{14}{60} \approx 0.23$$

The symbol $\approx$ means "is approximately equal to." We use this symbol because we rounded the relative frequency. Relative frequencies for the other classes are computed in a similar way.

The total of the relative frequencies should be 1. However, rounded results may make the total slightly higher or lower than 1.

Using Table 1-9 and Figure 1-10, we can quickly make a relative-frequency histogram (Figure 1-11). The horizontal scale will be the same, but the vertical scale will be marked with *relative frequencies f/n* instead of the actual frequencies *f*. The basic shape of the two graphs otherwise will be the same.

In Guided Exercise 12 we will ask you to make a frequency table, a histogram, a relative-frequency table, and a relative-frequency histogram. There are several steps to follow. The individual steps are not difficult, but you need to keep them all in mind. The steps are listed below for your convenience.

Frequency Table

1. Determine the class width.
2. Create the distinct classes. We use the convention that the *lower class limit* of the first class is the smallest data value. Add the class width to *this* number to get the *lower class limit* of the next class.
3. Tally the data into classes. Each data value should fall into exactly one class. Total the tallies to obtain each *class frequency*.
4. Compute the *midpoint* (class mark) for each class.
5. Determine the *class boundaries*.

Relative-Frequency Table

6. For each class, compute the *relative frequency f/n*, where *f* is the class frequency and *n* is the total sample size.

COMMENT The use of class boundaries in histograms assures us that the bars of the histogram touch and that no data fall on the boundaries. Both these features

Table 1-9   Relative Frequencies of One-Way Commuting Distances

Class	Frequency f	Relative Frequency f/n
1–8	14	14/60 ≈ 0.23
9–16	21	21/60 ≈ 0.35
17–24	11	11/60 ≈ 0.18
25–32	6	6/60 ≈ 0.10
33–40	4	4/60 ≈ 0.07
41–48	4	4/60 ≈ 0.07

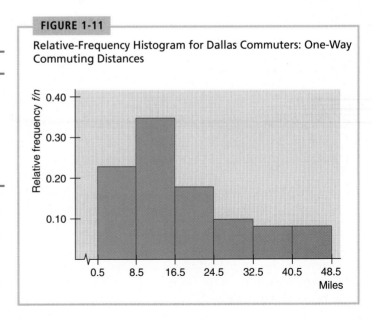

FIGURE 1-11

Relative-Frequency Histogram for Dallas Commuters: One-Way Commuting Distances

are important. But a histogram displaying class boundaries may look awkward. For instance, the age range of 14.5 to 24.5 years shown in Figure 1-9 isn't as natural a choice as an age range of 15 to 25 years. For this reason, many magazines and newspapers do not use class boundaries as labels on histograms. Instead, some use lower class limits as labels, with the convention *that a data value falling on the class limit is included in the next higher class (the class to the right of the limit)*. Another convention is to label midpoints instead of class boundaries. Determine the convention being used before creating a frequency table or a histogram on a computer.

GUIDED EXERCISE   12

One irate customer called Dollar Day Mail Order Company 40 times during the last two weeks to see why his order had not arrived. Each time he called, he recorded the length of time he was put "on hold" before being allowed to talk to a customer service representative.

Table 1-10   Length of Time on Hold, in Minutes

1	5	5	6	7	4	8	7	6	5
5	6	7	6	6	5	8	9	9	10
7	8	11	2	4	6	5	12	13	6
3	7	8	8	9	9	10	9	8	9

(a) What are the largest and smallest values in Table 1-10? If we want five classes, what should the class width be?

⇨ The largest value is 13; the smallest value is 1. The class width is

$$\frac{13 - 1}{5} = 2.4 \approx 3$$

Note that we *increase* the value to 3.

(b) Complete the following frequency table.

Table 1-11   Time on Hold

Class Limits		Tally	Frequency	Midpoint
Lower	Upper			
1	– 3	___	___	___
4	–___	___	___	___
___	– 9	___	___	___
___	–___	___	___	___
___	–___	___	___	___

⇨ **Table 1-12   Completion of Table 1-11**

Class Limits	Tally	Frequency	Midpoint
Lower–Upper			
1–3	|||	3	2
4–6	‖‖ ‖‖ ‖‖	15	5
7–9	‖‖ ‖‖ ‖‖ ||	17	8
10–12	||||	4	11
13–15			1	14

(c) Recall that the class boundary is halfway between the upper limit of one class and the lower limit of the next. Use this fact to find the class boundaries.

Exercise continues

Exercise continued

Table 1-13   Class Boundaries

Class Limits	Class Boundaries
1–3	0.5–3.5
4–6	3.5–6.5
7–9	6.5–____
10–12	____–____
13–15	____–____

Table 1-14   Completion of Table 1-13

Class Limits	Class Boundaries
1–3	0.5–3.5
4–6	3.5–6.5
7–9	6.5–9.5
10–12	9.5–12.5
13–15	12.5–15.5

(d) Finish the histogram in Figure 1-12

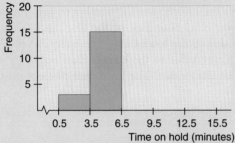

FIGURE 1-13 Completion of Figure 1-12.

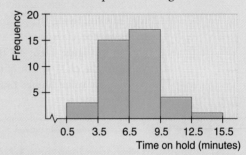

(e) Compute the relative class frequency f/n for each class.

Table 1-15   Relative Class Frequency

Class	f/n
1–3	3/40 = 0.075
4–6	15/40 = 0.375
7–9	_____
10–12	_____
13–15	_____

Table 1-16   Completion of Table 1-15

Class	f/n
1–3	0.075
4–6	0.375
7–9	0.425
10–12	0.100
13–15	0.025

(f) Finish the relative-frequency histogram in Figure 1-14.

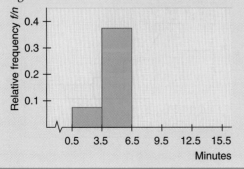

FIGURE 1-15 Completion of Figure 1-14.

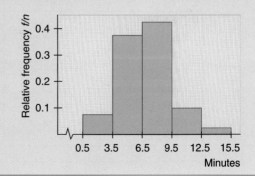

We will see relative-frequency distributions again when we study probability in Chapter 4. There we will see that if a random sample is large enough, we can estimate the probability of an event by the relative frequency of the event. The relative-frequency distribution then can be interpreted as a *probability distribution*. Such distributions will form the basis of our work in inferential statistics.

Calculator Note Many graphing calculators produce histograms. However, the procedure the calculator follows may be slightly different from our procedure. For instance, on the TI-83, the user enters the data into one list and selects the histogram plot. If the ZoomStat option is used, then the calculator determines how many classes to use and follows the convention that data falling on a lower class boundary of a class is included in that class. It is easy to have the TI-83 make histograms following the procedure of this text. Determine the number of bars you wish to have, the class width, and the lower class boundary of the first class. Then, as shown in Figure 1-16, set the Xmin value to the lowest class boundary, set the Xscl value to the class width, and set the Xmax value to the highest class boundary or a convenient value higher than the largest data value. The histogram plot option then

FIGURE 1-16

TI-83 Display of Histogram from Table 1-7, One-Way Commuting Distances

(a) Enter data in list L1

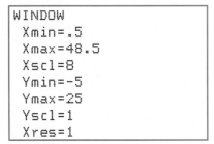

(b) Select histogram plot

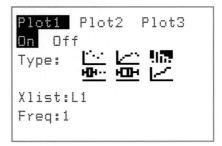

(c) Set window with Xmin = lowest class boundary, Xscl = class width

```
WINDOW
 Xmin=.5
 Xmax=48.5
 Xscl=8
 Ymin=-5
 Ymax=25
 Yscl=1
 Xres=1
```

(d) Graph with trace positioned on class 3

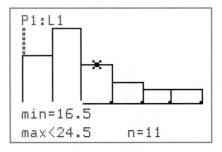

produces a histogram as shown in this text. The trace function places a blinking box at the top center of a bar and gives the left and right boundaries of the designated class (min, max) as well as the frequency *n* of the data in the class. Figure 1-16 shows how to create the histogram with six classes from the data in the preceding discussion regarding one-way commuting distances of workers in Dallas (Table 1-7).

Distribution shapes

Histograms are valuable and useful tools. If the raw data comes from a random sample of population values, the histogram constructed from your sample values should have a distribution shape that is reasonably similar to that of the population.

There are several terms that are commonly used to describe histograms and their associated population distributions.

(a) *Symmetrical:* This term refers to a histogram in which both sides are (more or less) the same when the graph is folded vertically down the middle. Figure 1-17(a) shows a typical histogram with a symmetrical shape.

(b) *Uniform or rectangular:* These terms refer to a histogram in which every class has equal frequency. From one point of view, a uniform distribution is symmetrical with the added property that the bars are of the same height. Figure 1-17(b) illustrates a typical histogram with a uniform shape.

(c) *Skewed left or skewed right:* These terms refer to a histogram in which one tail is stretched out longer than the other. The direction of skewness is on the side of the *longer* tail. So if the longer tail is on the left, we say the histogram is skewed to the left. Figure 1-17(c) shows a typical histogram skewed to the left and another skewed to the right.

(d) *Bimodal:* This term refers to a histogram in which the two classes with the largest frequencies are separated by at least one class. The top two frequencies of these classes may have slightly different values. This type of situation sometimes indicates that we are sampling from two different populations. Figure 1-17(d) on the following page illustrates a typical histogram with a bimodal shape.

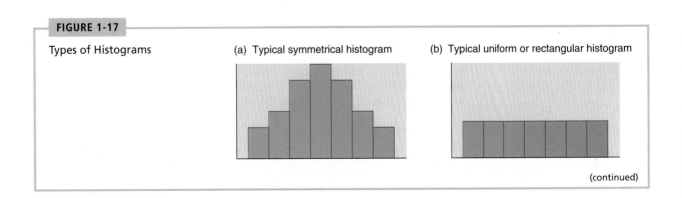

FIGURE 1-17

Types of Histograms   (a) Typical symmetrical histogram   (b) Typical uniform or rectangular histogram

(continued)

FIGURE 1-17 continued

(c) Typical skewed histograms

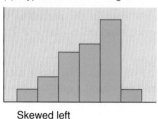

Skewed left

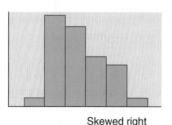

Skewed right

(d) Typical bimodal histogram

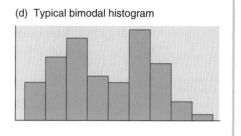

V I E W P O I N T

Mush You Huskies!

In 1925 the village of Nome, Alaska, had a terrible diphtheria epidemic. Serum was available in Anchorage but had to be brought to Nome by dogsled over the 1161-mile Iditarod Trail. Since 1973, the Iditarod Race from Anchorage to Nome has been an annual dogsled sporting event with a current winner's purse of more than $350,000. Winning times range from more than 20 days to a little over 10 days.

Use Web site <http://www.iditarod.com/> to collect data on winning times, and make a frequency distribution for these times.

S ECTION 1.4 PROBLEMS

For Problems 1–6, use the specified number of classes to do the following:
(a) Find the class width.
(b) Make a frequency table showing class limits, class boundaries, midpoints, frequencies, and relative frequencies.
(c) Draw a histogram.
(d) Draw a relative-frequency histogram.

Climate: Fairbanks, Alaska

1. How hot does it get in Fairbanks, Alaska, in January? The record maximum temperature (in °F) for each day in January can be found on the Web at <http://www.gi.alaska.edu/>. The results (in °F) for January 1 to 31 are as follows:

41	35	29	40	33	34	33	40	40	40	37
39	34	43	50	42	43	38	42	45	47	37
39	46	45	39	39	42	35	38	42		

Use six classes.

Advertising: Readability

2. "Readability Levels of Magazine Ads," by F. K. Shuptrine and D. D. McVicker, is an article in the *Journal of Advertising Research* (see Web site <http://lib.stat.cmu.edu/DASL/>. Look in Data Subjects under Consumer and then Magazine Ads Readability file). The following is a list of the numbers of three-syllable (or longer) words in advertising copy of randomly selected magazine advertisements:

34	21	37	31	10	24	39	10	17	18	32
17	3	10	6	5	6	6	13	22	25	3
5	2	9	3	0	4	29	26	5	5	24
15	3	8	16	9	10	3	12	10	10	10
11	12	13	1	9	43	13	14	32	24	15

Use eight classes.

Football: Super Bowl Games

3. How many people attend NFL Super Bowl games? For Super Bowls I to XXXI, the answer can be found on the Web at <http://www.nfl.com/history/>. Rounded to the nearest thousand, the attendance numbers are (in thousands of people) as follows:

62	76	75	81	79	81	90	72	81	80	103
76	79	104	76	81	104	73	84	74	101	73
75	73	74	63	98	73	74	76	72		

Use six classes.

Business:
Fast-food Franchises

4. Franchise and Business Opportunities *Annual Report* contains information about franchise opportunities in the United States and Canada. A franchise fee is one of the expenses associated with owning a franchise. There are other expenses, such as startup, advertising, royalties, and so on. A large category of franchises is the fast-food business, which includes franchises such as baked goods, donuts, hamburgers, chicken, and hot dogs. Franchise fees (in thousands of dollars) for the fast-food category are as follows:

21	25	25	18	44	20	25	15	19	24	10
28	30	25	25	25	25	10	25	25	20	20
20	15	10	20	25	25	13	30	15	28	15
15	35	24	40	15	20	35	5	50	30	15
25	40	15	25	15	75	33	23	30	10	15
8	25	10	8	20	25	20	30			

Use five classes.

Nursing: Workload

5. Nurses on the eighth floor of Community Hospital believe they need extra staffing at night. To estimate the night workload, a random sample of 35 nights was used. For each night the total number of room calls to the nurses' station on the eighth floor was recorded as follows:

68	60	69	70	83	58	90	86	71	71	92
95	70	74	46	18	84	82	75	63	101	77
102	80	86	85	73	86	62	100	90	37	88
70	87									

Use five classes.

Baseball:
Years in Major League

6. *The Baseball Encyclopedia,* 9th Edition (Macmillan Publishing Co., New York) gives extensive data on baseball. The Player Register section gives statistics on all the players in the major leagues from 1876 to the present. One of the statistics given is the number of years a player has been on a major league team. A sample of 46 players taken from the Player Register (not including pitchers) shows that the lengths of time (in years) that these players have been on a major league team are as follows:

2	4	18	3	12	3	5	7	19	6	11
16	10	19	3	2	8	3	1	9	1	10
11	4	8	6	12	10	21	15	16	4	11
2	9	6	9	12	13	6	8	6	9	14
4	5									

Use five classes.

Football: Weights of Players

7. To compare the distributions of two data sets, we can make histograms using the same classes. For instance, we might want to compare two football teams to see if they stack up "weightwise." *Lindy's Pro Football* guide gave preseason player statistics for all the NFL teams. Weights for 70 players with the Miami Dolphins and for 72 players with the San Diego Chargers were used to create the following histograms. The computer software package Excel generated the histograms. We entered the data, using the menu selections ►**Tools** ►**Data Analysis** ►**Histogram.** The "bin" values are the weights listed in the tables. These weights represent the *upper* class limit for each class. (For more details on making the bars touch and eliminating a "more" category, see the *Excel Guide* supplement for this text.)

Miami Dolphins

Weight	Frequency
150	0
182	3
214	17
246	24
278	14
310	12

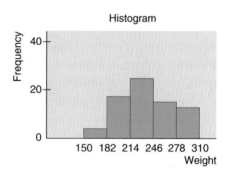

San Diego Chargers

Weight	Frequency
150	1
182	4
214	27
246	15
278	14
310	11

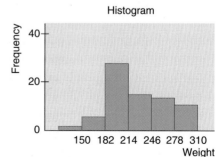

(a) From simply looking at the histograms for the weights of the two teams, how do you think the teams compare weightwise?

(b) The San Diego Chargers have two more players than the Miami Dolphins. How do you think the weights of the two additional players influenced your answer to part a?

Health Insurance: Hospital Costs

8. Hospital costs vary from state to state. The Health Insurance Association of America's *Source Book of Health Insurance Data* gave information about the average cost per day per patient in hospitals by state, including the District of Columbia. The following figure shows a histogram of these data. How many states (including the District of Columbia) have average costs per day of less than $690.50?

Average Cost per Day for Hospital Stay in Each State (Including the District of Columbia)

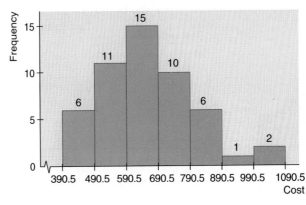

9. Professor Silva teaches anatomy and physiology. He has developed five different versions of a test on the same material. On giving each version to a different sample of 60 students, he discovered that the test score distributions looked like those shown in the histograms in the upper box on page 48.

Academic: Testing

(a) Categorize the distribution shapes as uniform, symmetric, bimodal, skewed left, or skewed right.

(b) Comment on some advantages or disadvantages of each test version. As a student, which version might you prefer? Which version would you like the least?

Consumer: Warranty Cards

10. Many products come with owner registration or warranty cards. Usually the consumer is asked a few questions about his or her family and household income. Information from random samples of warranty or registration cards for several products yielded the household income distributions shown in the histograms in the lower box on page 48.

(a) Categorize the distribution shapes as uniform, symmetric, bimodal, skewed left, or skewed right.

(b) If you were in charge of advertising, how would you use income distribution information on present customers to target ads for the indicated products?

(c) How valid do you think income information on warranty cards is?

Agriculture: Wheat Harvest

11. The following data represent tons of wheat harvested in the years 1894 to 1925 from Plot 19 at the Rothamsted Agricultural Experiment Station in England.

2.71	1.62	2.60	1.64	2.20	2.02	1.67	1.99	2.34	1.26	1.31
1.80	2.82	2.15	2.07	1.62	1.47	2.19	0.59	1.48	0.77	2.04
1.32	0.89	1.35	0.95	0.94	1.39	1.19	1.18	0.46	0.70	

Figure for Problem 9

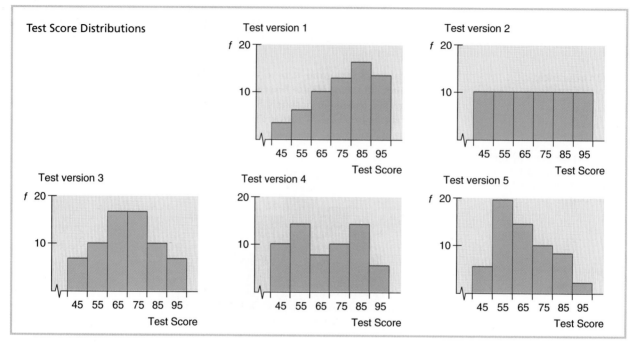

Figure for Problem 10

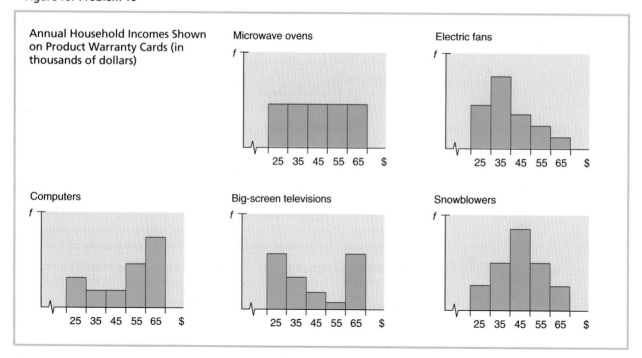

(a) Multiply each data value by 100 to "clear" the decimals.

(b) Use the standard procedures of this section to make a frequency table and histogram with your whole-number data. Use six classes.

(c) Divide class limits, class boundaries, and class midpoints by 100 to get back to your original data values.

Data Sets: Minitab, Excel, ComputerStat

12. A variety of data sets from diverse fields, such as the stock market, health care, sports, sociology, and biology, can be found in Appendix II of this text. These data sets are available in the formats: Minitab portable files, Excel files, TI-83 files, ASCII files, and class demonstrations in ComputerStat. You may use these data sets directly in the software packages or you may enter the data in your calculator or other software.

One of the advantages of using computer or calculator software to make histograms is that you can easily change the number of classes and see a new histogram without reentering data or retallying the data. You will notice that packages such as Minitab, Excel, and the TI-83 produce "default" histograms, for which the technology determines the number of classes. If you want to control the number of classes, you must do some of the work of determining class boundaries by hand and then enter the boundaries as described in the Using Technology section at the end of this chapter or in the Technology Guides that are supplements to this text. ComputerStat lets you dictate the number of classes directly.

Select several of the single variable data sets and try the histogram functions. Use different numbers of classes and decide which selection gives the best "picture" of the data.

Dotplots: Minitab

13. Another display technique that is somewhat similar to a histogram is a *dotplot*. In a dotplot, the data values are displayed along the horizontal axis. A dot is then plotted over each data value in the data set. The next display shows a dotplot generated by Minitab (▶Graph ▶Dotplot) for the number of licensed drivers per 1000 residents by state (including the District of Columbia). This dotplot was obtained from the U.S. Department of Transportation.

Dotplot for Licensed Drivers per 1000 Residents

Licensed drivers

(a) From the dotplot, how many states have 600 or fewer licensed drivers per 1000 residents?

(b) About what percentage of the states (out of 51) seem to have close to 800 licensed drivers per 1000 residents?

(c) Consider the intervals 550 to 650, 650 to 750, and 750 to 850 licensed drivers per 1000 residents. In which interval do most of the states fall?

Stem-and-Leaf Displays

Exploratory data analysis

Together with histograms and other graphics techniques, the stem-and-leaf display is one of many useful ways of studying data in a field called *exploratory data analysis* (often abbreviated as EDA). John W. Tukey wrote one of the definitive books on the subject, *Exploratory Data Analysis* (Addison-Wesley). Another very useful reference for EDA techniques is the book *Applications, Basics, and Computing of Exploratory Data Analysis*, by Paul F. Velleman and David C. Hoaglin (Duxbury Press). Exploratory data analysis techniques are particularly useful for detecting patterns and extreme data values. They are designed to help us explore a data set, to ask questions we had not thought of before, or to pursue leads in many directions.

EDA techniques are similar to those of an explorer. An explorer has a general idea of destination but is always alert to the unexpected. An explorer needs to assess situations quickly and often simplify and clarify them. An explorer makes pictures—that is, maps showing the relationships of landscape features. The aspects of rapid implementation, visual displays such as graphs and charts, data simplification, and robustness (that is, analysis that is not influenced much by extreme data values) are key ingredients of EDA techniques. In addition, these techniques are good for exploration because they require very few prior assumptions about the data.

EDA methods are especially useful when our data have been gathered for general interest and observation of subjects. For example, we may have data regarding the ages of applicants for graduate programs. We don't have a specific question in mind. We want to see what the data reveal. Are the ages fairly uniform or spread out? Are there exceptionally young or old applicants? If there are, we might look at other characteristics of these applicants, such as field of study. EDA methods help us quickly absorb some aspects of the data and then may lead us to ask specific questions for which we might apply methods of traditional statistics.

In contrast, when we design an experiment to produce data to answer a specific question, we focus on particular aspects of the data that are useful to us. If we want to determine the average highway gas mileage of a specific sports car, we use that model car in well-designed tests. We don't need to worry about unexpected road conditions, poorly trained drivers, different fuel grades, sudden stops and starts, and so on. Our experiment is designed to control outside factors. Consequently, we do not need to "explore" our data as much. We can often make valid assumptions about the data. Methods of traditional statistics will be very useful in analyzing such data and answering our specific questions.

Stem-and-leaf display

In this text we will introduce two EDA techniques: stem-and-leaf displays and, in Section 2.3, box-and-whisker plots. Let's first look at a stem-and-leaf display. We know that frequency distributions and histograms provide useful organization and summary of data. However, in a histogram, we lose most of the specific data values. A stem-and-leaf display is a device that organizes and groups data but allows us to recover the original data if desired. In the next example we will make a stem-and-leaf display.

EXAMPLE 13 ▶ Many airline passengers seem weighted down with carry-on luggage. Just how much weight are they carrying? The carry-on luggage weights for a random sample of 40 passengers returning from a vacation in Hawaii were recorded in pounds below.

Table 1-17   Weights of Carry-On Luggage in Pounds

30	27	12	42	35	47	38	36	27	35
22	17	29	3	21	0	38	32	41	33
26	45	18	43	18	32	31	32	19	21
33	31	28	29	51	12	32	18	21	26

To make a stem-and-leaf display, we break the digits of each data value into *two* groups. The left group is called a *stem* and the right group is called a *leaf*. We are free to choose the number of digits to be included in the stem.

The weights in our example consist of two-digit numbers. For a two-digit number, the stem selection is obviously the left digit. In our case, the tens digits will form the stems, and the units digits will form the leaves. For example, for the weight 12, the stem is 1 and the leaf is 2. For the weight 18, the stem is again 1 but the leaf is 8. In the stem-and-leaf display, we list each possible stem once on the left and all its leaves in the same row on the right.

Figure 1-18 shows a stem-and-leaf display for the weights of carry-on luggage. From the stem-and-leaf display in Figure 1-18, we see that two bags weighed 27 lb, one weighed 3 lb, one weighed 51 lb, and so on. We see that most of the weights were in the 30-lb range, only two were less than 10 lb, and six were more than 40 lb. Note that the length of line containing the leaves gives the visual impression that a sideways histogram would present.

As a final step, we need to indicate the scale. This is usually done by indicating the value represented by one stem value and one leaf value.   ●

FIGURE 1-18

Stem-and-Leaf Display of Carry-On Luggage Weights

```
   3 | 2   represents 32 lb
Stem | Leaves
   0 | 3 0
   1 | 2 7 8 8 9 2 8
   2 | 7 7 2 9 1 6 1 8 9 1 6
   3 | 0 5 8 6 5 8 2 3 2 1 2 3 1 2
   4 | 2 7 1 5 3
   5 | 1
```

Figure 1-18 on the previous page shows a basic stem-and-leaf display. Sometimes you will see the leaves ordered from smallest to largest, but this is not necessary in a basic display. There are no firm rules for selecting the group of digits for the stem. But whichever group you select, you must list all the possible stems from smallest to largest in the data collection.

In the example that follows, we show a stem-and-leaf display for data with three digits.

EXAMPLE 14 ➤ What does it take to win at sports? If you're talking about basketball, one sportswriter gave the answer. He listed the winning scores of the conference championship games over the last 35 years. The scores for those games are shown below. To make a stem-and-leaf display, we'll use the first *two*-digits as the stems (see Figure 1-19). We will also order the leaves.

132	118	124	109	104	101	125	83	99
131	98	125	97	106	112	92	120	103
111	117	135	143	112	112	116	106	117
119	110	105	128	112	126	105	102	

Notice in Figure 1-19 that the distribution is fairly symmetrical.

FIGURE 1-19

Winning Scores of Conference Basketball Championship Games

```
08 | 3  represents 083 or 83 points
08 | 3
09 | 2  7  8  9
10 | 1  2  3  4  5  5  6  6  9
11 | 0  1  2  2  2  2  6  7  7  8  9
12 | 0  4  5  5  6  8
13 | 1  2  5
14 | 3
```

GUIDED EXERCISE   13

Tel-a-Message is experimenting with computer-delivered telephone advertisements. Of primary concern is how much of the 4-minute advertisement is heard. A study was done to see how long the advertisement ran before the listeners hung up. A random sample of 30 calls gave the information in Table 1-18.

Exercise continues

Exercise continued

Table 1-18   Time Spent Listening to Advertisement (in Minutes)

1.3	0.7	2.1	0.5	0.2	0.9	1.1	3.2	4.0	3.8
1.4	3.1	2.5	0.6	0.5	2.1	4.0	4.0	0.3	1.2
1.0	1.5	0.4	4.0	2.3	2.7	4.0	0.7	0.5	4.0

(a) We'll make a stem-and-leaf display using the first digits as the stems and the second digits as the leaves. What is the leaf unit?

⇨ The trailing digit is in the tenths position, so

$$1 \text{ unit} = 0.1 \text{ min}$$

(b) List all the stem values.

⇨ Stems: 0
            1
            2
            3
            4

(c) Complete the stem-and-leaf display including the unit designation. Order the leaves.

⇨ **FIGURE 1-20** Time Before Hang-Up

1 | 3 represents 1.3 min

0	2 3 4 5 5 5 6 7 7 9
1	0 1 2 3 4 5
2	1 1 3 5 7
3	1 2 8
4	0 0 0 0 0 0

(d) Looking at the stem-and-leaf display, what could you say about the time intervals before people hung up?

⇨ Most people hung up before the end of the advertisement. Of those people, more hung up within the first minute than within any other 1-minute interval. There were six people who listened to the entire advertisement.

COMMENT Stem-and-leaf displays organize the data, let the data analyst spot extreme values, and are easy to create. In fact, they can be used to organize data so that frequency tables are easier to make. However, at this time, histograms are used more frequently in formal data presentations, whereas stem-and-leaf displays are used by data analysts to gain initial insights about the data. ○

VIEWPOINT

What Does It Take to Win?

Scores for NFL Super Bowl games can be found at Web site <http://www.nfl.com/> by following the links to history and Super Bowl. Of special interest in football statistics is the spread, or difference, between the scores of the winning and losing teams. If the spread is too large, the game can appear to be lopsided, and TV viewers become less interested in the game (and the accompanying commercial ads). Make a stem-and-leaf display of the spread for the NFL Super Bowl games and analyze the results.

SECTION 1.5 PROBLEMS

Consumer:
Price of Walking Shoes

1. More and more people take fitness walks before or after work or during their lunch hours. They want comfortable walking shoes. Periodically, *Consumer Reports* rates walking shoes and includes the prices as well. In one issue they rated men's and women's walking shoes and gave the prices. For a random sample of 19 of the rated walking shoes, the prices were

59	109	70	76	55	50	55	69	58	59
40	46	62	52	55	65	70	60	110	

Seven years later an issue of *Consumer Reports* again rated men's and women's walking shoes. The prices were

90	70	70	70	75	70	65	68	60	74
70	95	75	70	68	65	40	65	70	

(a) Make a stem-and-leaf diagram for the prices of walking shoes 7 years ago.
(b) Make a stem-and-leaf diagram for recent prices of walking shoes.
(c) Compare the two distributions, and comment on any differences you see. Do you think inflation might have anything to do with the distribution differences? Explain.

Ecology: Wetlands

2. Wetlands offer a diversity of benefits. They provide habitats for wildlife, spawning grounds for U.S. commercial fish, and renewable timber resources. In the last 200 years the United States has lost more than half its wetlands. *Environmental Almanac* gives the percentage of wetlands lost in each state in the last 200 years. For the lower 48 states, the percentage loss of wetlands per state is as follows:

46	37	36	42	81	20	73	59	35	50
87	52	24	27	38	56	39	74	56	31
27	91	46	9	54	52	30	33	28	35
35	23	90	72	85	42	59	50	49	
48	38	60	46	87	50	89	49	67	

Make a basic stem-and-leaf display of these data. Be sure to indicate the scale. How are the percentages distributed? Is the distribution skewed? Are there any gaps?

Health Care: Patient Visits

3. The American Medical Association Center for Health Policy Research included data, by state, on the number of community hospitals and the average patient stay (in days) in its publication *State Health Care Data: Utilization, Spending, and Characteristics*. The data (by state) are shown as follows:

State	No. of Hospitals	Average Length of Stay (days)	State	No. of Hospitals	Average Length of Stay (days)	State	No. of Hospitals	Average Length of Stay (days)
Maine	38	7.2	N. Dakota	47	11.1	Arkansas	88	7.0
New Hampshire	27	7.0	S. Dakota	52	10.3	Louisiana	136	6.7
Vermont	15	7.6	Nebraska	90	9.6	Oklahoma	113	6.7
Massachusetts	101	7.0	Kansas	133	7.8	Texas	421	6.2
Rhode Island	12	6.9	Delaware	8	6.8	Montana	53	10.0
Connecticut	35	7.4	Maryland	51	6.8	Idaho	41	7.1
New York	231	9.9	Dist. of Columbia	11	7.5	Wyoming	27	8.5
New Jersey	96	7.6	Virginia	98	7.0	Colorado	71	6.8
Pennsylvania	236	7.5	W. Virginia	59	7.1	New Mexico	37	5.5
Ohio	193	6.6	N. Carolina	117	7.3	Arizona	61	5.5
Indiana	113	6.6	S. Carolina	68	7.1	Utah	42	5.2
Illinois	209	7.3	Georgia	162	7.2	Nevada	21	6.4
Michigan	175	7.3	Florida	227	7.0	Washington	92	5.6
Wisconsin	129	7.3	Kentucky	107	6.9	Oregon	66	5.3
Minnesota	148	8.7	Tennessee	122	6.8	California	440	6.0
Iowa	123	8.4	Alabama	119	7.0	Alaska	16	5.7
Missouri	133	7.4	Mississippi	102	7.2	Hawaii	19	9.4

Make a stem-and-leaf display of the data for the average length of stay in days. Comment about the general shape of the distribution.

Health Care: Hospitals

4. Using the number of hospitals per state listed in Problem 3, make a stem-and-leaf display for the number of community hospitals per state. Which states have unusually high numbers of hospitals?

Boston Marathon: Winning Times

5. The Boston Marathon is the oldest and best known U.S. marathon. It covers a route from Hopkinton, Massachusetts, to downtown Boston. The distance is approximately 26 miles. The Web site <http://www.boston.com> with search for marathon provides a wealth of information about the history of the race. In particular, it gives the winning times for the Boston Marathon. They are all over 2 hours. The following data are the minutes over 2 hours for the winning male runners:

1958–1977

25	22	20	23	23	18	19	16	17	15
22	13	10	18	15	16	13	9	20	14

1978–1997

10	9	12	9	8	9	10	14	7	11
8	9	8	11	8	9	7	9	9	10

(a) Make a stem-and-leaf diagram for the minutes over 2 hours of the winning times for the years 1958 to 1977. Use two lines per stem. For instance, to use two lines for the stem value 0, we designate the first stem as 0* and put leaf digits 0 through 4 on this line. The second line for stem 0 is designated with stem 0•, and we put leaf digits 5 through 9 on this line.

(b) Make a stem-and-leaf diagram for the minutes over 2 hours of the winning times for the years 1978 to 1997. Use two lines per stem.

(c) Compare the two distributions. How many times under 15 minutes are in each distribution?

Golf: U.S. Open Tournament Scores

6. The 1997 U.S. Open Golf Tournament was played at Congressional Country Club, Bethesda, Maryland, with prizes ranging from $465,000 for first place to $5000. Par for the course is 70. The tournament consists of four rounds played on different days. The scores for each round of the 32 players who placed in the money (more than $17,000) were given on the Web site <http://majors.golfweb.com/>. The scores for the first round were as follows:

71	65	67	73	74	73	71	71	74	73	71
70	75	71	72	71	75	75	71	71	74	75
66	75	75	75	71	72	72	73	71	67	

The scores for the fourth round for these players were as follows:

69	69	73	74	72	72	70	71	71	70	72
73	73	72	71	71	71	69	70	71	72	73
74	72	71	68	69	70	69	71	73	74	

(a) Make a stem-and-leaf diagram for the first-round scores. Use two lines per stem. (See Problem 5.)

(b) Make a stem-and-leaf diagram for the fourth-round scores. Use two lines per stem.

(c) Compare the two distributions. How do the highest scores compare? How do the lowest scores compare?

Are cigarettes bad for people? Cigarette smoking involves tar, carbon monoxide, and nicotine. The first two are definitely not good for a person's health, and the last ingredient can cause addiction. Problems 7 and 8 on page 57 refer to the following table,

Tar and Carbon Monoxide (CO) Contents (in milligrams) for One Cigarette

Brand	Tar	CO	Brand	Tar	CO
Alpine	14.1	13.6	Merit	7.8	10.0
Benson & Hedges	16.0	16.6	MultiFilter	11.4	10.2
Bull Durham	29.8	23.5	Newport Lights	9.0	9.5
Camel Lights	8.0	10.2	Now	1.0	1.5
Carlton	4.1	5.4	Old Gold	17.0	18.5
Chesterfield	15.0	15.0	Pall Mall Lights	12.8	12.6
Golden Lights	8.8	9.0	Raleigh	15.8	17.5
Kent	12.4	12.3	Salem Ultra	4.5	4.9
Kool	16.6	16.3	Tareyton	14.5	15.9
L&M	14.9	15.4	True	7.3	8.5
Lark Lights	13.7	13.0	Viceroy Rich Light	8.6	10.0
Marlboro	15.1	14.4	Virginia Slim	15.2	13.9
			Winston Lights	12.0	14.9

Source: Journal of Statistics Education Web site <http://www.stat.ncsu.edu/info/jse/>

which was taken from the Web site <http://www.stat.ncsu.edu/info/jse/> maintained by the *Journal of Statistics Education.* Follow the links to the cigarette data.

Health: Cigarette Smoke

7. Use the data in the table to make a stem-and-leaf display for milligrams of tar per cigarette smoked.

Health: Cigarette Smoke

8. Use the data in the table to make a stem-and-leaf display for milligrams of carbon monoxide per cigarette smoked.

SUMMARY

Statistics is the study of how to collect, organize, analyze, and interpret information. In this chapter we studied ways to gather, organize, and present data. Random samples are of great importance because they are representative of the population. In later chapters we will see how we can draw conclusions about a population based on information from a random sample.

We looked at several graphical methods of presenting data, such as bar graphs, Pareto charts, circle graphs, time plots, histograms, relative-frequency histograms, and stem-and-leaf plots. These methods are all important because they can give us insights into the nature of the data and how they are distributed. From the viewpoint of future applications, histograms and relative-frequency histograms are very important because the area under a bar can represent the likelihood of data values falling in that class. Histograms reveal distribution properties such as uniformity, symmetry, and skewness.

IMPORTANT WORDS & SYMBOLS

Section 1.1
Quantitative data
Qualitative data
Statistics
Population
Sample
Descriptive statistics
Inferential statistics
Sampling
Experiment
Simulation
Census
Survey
Voluntary response
Hidden bias
Levels of measurement

nominal
ordinal
interval
ratio

Section 1.2
Random sample
Random-number table
Types of sampling
  random
  simulation
  stratified
  systematic
  cluster
  convenience

Section 1.3
Bar graph
Pareto chart
Circle graph or pie chart
Time plot
Time series

Section 1.4
Frequency
Frequency distribution
Relative frequency
Relative-frequency
  distribution
Class width
Class mark or midpoint
Histogram

Distribution shapes
  symmetrical
  uniform
  skewed left or right
  bimodal

Section 1.5
Exploratory data
  analysis (EDA)
Stem-and-leaf display

VIEWPOINT

"This land is your land, This land is my land"*
—Woody Guthrie

But who actually owns the forest? On many maps, forest land (including national forests) is colored green. Such maps give the impression that vast areas of the western United States are public land. This is not the case! *USA Today* gave the following information about ownership of U.S. timber lands: state/local, 17%; federal, 10%; forest industry, 14%; and private nonindustry, 59%. Organize these data for better visual presentation using a Pareto chart and a circle graph.

*Words and Music by Woody Guthrie TRO © Copyright 1956 (Renewed) 1958 (Renewed) Ludlow Music, Inc. New York, New York. Used by permission.

CHAPTER REVIEW PROBLEMS

Student Life:
Levels of Measurement

1. Give the highest level of measurement (ratio, interval, ordinal, nominal) appropriate for the following data about your college.
   (a) Year the college was founded
   (b) Number of students enrolled this term
   (c) Name of your favorite professor this term
   (d) Rating of college cafeteria food on a scale from poor to excellent

Ecology: CO₂ Concentration

2. The following figure appears in the book *Earth in the Balance: Ecology and the Human Spirit*, by Al Gore. This time plot gives the carbon dioxide (CO₂) concentration in the atmosphere. The time plot is based on data gathered at the Mauna Loa Observatory in Hawaii.

Concentration of CO₂

Source: American Geophysical Union.

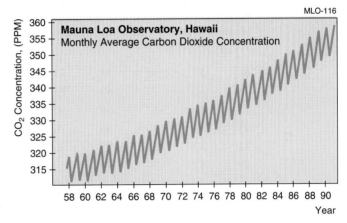

The concentration of CO_2 in the atmosphere from April 1958 until June 1991. In summer, the line goes down as vegetation in the Northern Hemisphere (with most of the earth's land area) breathes in vast quantities of CO_2. In winter, when the leaves have fallen, the line climbs again. The peak concentration has grown steadily higher because of such human activities as the burning of fossil fuels and the destruction of forests.

(a) Estimate the range of CO_2 concentration for 1970.
(b) Estimate the range of CO_2 concentration for 1988.
(c) Estimate the change in the highest levels of CO_2 concentration from 1970 to 1988.

IRS: Filing Tax Returns

3. Almost everyone files (or will sometime file) a federal income tax return. A research poll for Turbo Tax (a computer software package to aid in tax-return preparation) asked what aspect of filing a return people thought to be the most difficult. The results showed that 43% of the people said understanding the IRS jargon, 28% said knowing deductions, 10% said getting the right form, 8% said calculating the numbers, and 10% didn't know. Make a circle graph to display this information. *Note:* Percentages will not total 100% because of rounding.

Driving: DUI Arrests

4. Driving under the influence of alcohol (DUI) is a serious offense. The following data give the ages of a random sample of 50 drivers arrested while driving under the influence of alcohol. This distribution is based on the age distribution of DUI arrests given in the *Statistical Abstract of the United States* (112th Edition).

46	16	41	26	22	33	30	22	36	34
63	21	26	18	27	24	31	38	26	55
31	47	27	43	35	22	64	40	58	20
49	37	53	25	29	32	23	49	39	40
24	56	30	51	21	45	27	34	47	35

(a) Make a stem-and-leaf display of the age distribution.
(b) Make a frequency table using seven classes.
(c) Make a histogram showing class boundaries.

Health: Glucose Blood Test

5. Fasting glucose blood tests were given to 53 (nonpregnant) women at Boston City Hospital. The results of the blood tests (milligrams of glucose per 100 milliliters of blood) are shown below. (Reference: *American Journal of Clinical Nutrition*, Vol. 19, pp. 345–351.)

66	59	69	84	85	96	85	82	80	90
75	80	73	69	85	82	91	84	77	86
75	83	87	74	80	83	77	69	78	93
93	87	70	85	76	87	87	79	91	86
82	82	84	85	90	79	83	65	87	70
89	81	74							

(a) Make a frequency table with seven classes showing class limits, class boundaries, midpoints, frequencies, and relative frequencies.
(b) Draw a histogram.
(c) Draw a relative-frequency histogram.

Corporations: Lawsuits

6. Many people say that the civil justice system is overburdened. Many cases center on suits involving businesses. The following data are based on a *Wall Street Journal* report. Researchers conducted a study of lawsuits involving 1908 businesses ranked in the Fortune 1000 over a 20-year period. They found the following distribution of civil justice caseloads brought before the federal courts involving these businesses:

Case Type	Number of Filings (in thousands)
Contracts	107
General torts (personal injury)	191
Asbestos liability	49
Other product liability	38
All other	21

Note: Contracts involve disputes over contracts between businesses.
(a) Make a Pareto chart of the caseloads. Which type of case occurs most frequently?
(b) Make a circle chart showing the percentage of cases in each type.

Archaeology: Tree Ring Data

7. *The Sand Canyon Archaeological Project,* edited by W. D. Lipe and published by Crow Canyon Archaeological Center, contains the stem-and-leaf diagram shown in the following figure. The study uses tree rings to determine accurately the year in which a tree was cut. The figure gives the tree-ring cutting dates for samples of timbers found in the architectural units at Sand Canyon Pueblo. The text referring to the figure says, "The three-digit numbers in the left column represent centuries and decades A.D. The numbers to the right represent individual years, with each number derived from an individual sample. Thus, **124 2 2 2** represents three samples dated A.D. 1242."

Tree-Ring Cutting Dates from Architectural Units at Sand Canyon Pueblo: *The Sand Canyon Archaeological Project*

```
119 | 5 6
120 | 0 0 1 2 3 3 3 3 3 3 3 3 3 3 3 3 3 3 3 3 3 3 3 3 3 3 3 3 3 3 3 3
120 |
121 | 2
121 | 5 5
122 | 0 0 1 1 1 1 2 2 3 4 4 4 4 4 4 4
122 | 5 8 9
123 | 0 1 2 3 3 4
123 | 5 5 5 5 5 5 5 5 5 5 5 5 5 5 6 8 8 9
124 | 1 2 2 2 2 2 2 2 2 2 2 2 2 2 2 2 2 2 2 2 2 2 3 4 4
124 | 5 6 8 9 9 9 9 9 9 9 9 9
125 | 0 0 0 0 0 0 0 0 0 0 0 0 0 0 0 1 1 1 1 1 1 1 2 2 2
125 |
126 | 0 0 0 1 2 2 2 2 2 2 2 2 2 2 2 2 2 2 4 4 4 4 4 4 4
126 | 5 5 5 6 6 7
127 | 0 1 1 1 1 4 4
```

Use the figure and the verbal description to answer the following questions.

(a) Which decade contained the most samples?

(b) How many samples had a tree-ring cutting date between 1200 A.D. and 1239 A.D., inclusive?

(c) What are the dates of the longest interval in which no tree cutting samples occurred? What might this indicate about new construction or renovation of the pueblo structures during this period?

Health Care: Age of Patients

8. The data in the following figure are based on information from the *Statistical Abstract of the United States* (112th Edition). The horizontal axis represents the ages of hospital patients (from 5 years of age on). The vertical axis gives the relative frequencies.

Age of Hospital Patients (5 years of age or older)

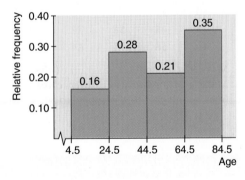

(a) Which age group has the greatest frequency?

(b) Convert the relative frequencies into percentages. Then find the percentage of patients older than age 44. What percentage of patients are 44 years old or younger?

Football: Most Valuable Player

9. At each Super Bowl, a most valuable player award is given (in Super Bowl XII, two awards were given). Since Super Bowl XXV, the award has been named the Pete Rozelle

Trophy. The Web site <http://www.nfl.com/history/> gives a record of all the awards. As of Super Bowl XXX, 16 awards had gone to quarterbacks, six to running backs, three to wide receivers, two to defensive ends, and one each to a defensive tackle, a center, a linebacker, and a safety. Make a bar graph representing this information.

Sampling: Type of Sample

10. Categorize the type (simple random, stratified, systematic, cluster, or convenience) of sampling used in each of the following situations.
    (a) To conduct a preelection opinion poll on a proposed amendment to the state constitution, a random sample of 10 telephone prefixes (first three digits of the phone number) was selected, and all households from the phone prefixes selected were called.
    (b) To conduct a study on depression among the elderly, a sample of 30 patients in one nursing home was used.
    (c) To maintain quality control in a brewery, every 20th bottle of beer coming off the production line was opened and tested.
    (d) Subscribers to the magazine *Sound Alive* were assigned numbers. Then a sample of 30 subscribers was selected by using a random-number table. The subscribers in the sample were invited to rate new compact disc players for a "What the Subscribers Think" column.
    (e) To judge the appeal of a proposed television sitcom, a random sample of 10 people from each of three different age categories was selected and those chosen were asked to rate a pilot show.

Lifestyle:
World's Richest People

11. *Forbes Richest People* gives profiles of the world's wealthiest men and women. Do you have to be old to be worth at least $2 billion? You can answer this question yourself by studying the following data—ages of men and women worth at least $2 billion:

40	66	43	82	52	58	77	52	50	48	47
68	66	73	76	53	67	88	40	79	73	66
65	70	72	77	48	75	82	54	76	41	93
65	60	57	74	70	83	67	68	77	66	34
66	59	48	56	71	40	53	63	52	57	83
52	60	56	71	64	61	53	53	73	70	

    (a) Make a stem-and-leaf diagram of this information.
    (b) Make a histogram using seven classes. Describe the shape of the distribution (that is, indicate if it is symmetrical, skewed, or bimodal).

DATA HIGHLIGHTS: GROUP PROJECTS

Break into small groups and discuss the following topics. Organize a brief outline in which you summarize the main points of your group discussion.

1. Examine the following figure, entitled "Slobs make worst roommates." This is a double bar graph because two percentages are given for each response category: response from men and responses from women. Comment about how the artistic rendition has slightly changed the format of this bar graph. Do the bars seem to have lengths that accurately reflect the relative percentages of the responses? In your own opinion, does the artistic

USA SNAPSHOTS®

A look at statistics that shape our lives

Genders agree:
Slobs make worst roommates

What men and women say bothers them the most when living with another person:

Sloppiness — 41% / 35%

Uneven sharing of chores — 15% / 32%

Irritating personal habits — 22% / 24%

Invasions of privacy — 22% / 9%

☐ Men
☐ Women

Source: Advantage Business Research for Mattel *Compatibility*

By Anne R. Carey and Marcia Staimer, USA TODAY

Source: Copyright 1996, USA TODAY. Reprinted with permission.

rendition garble the information? Explain. Which characteristic of the "worst roommates" does the graphic seem to illustrate? Can this graph be considered a Pareto chart for men? For women? Why or why not? From the information given in the figure, do you think the survey just listed the four given annoying characteristics? Do you think a respondent could choose more than one characteristic? Explain your answer in terms of the percentages given and in terms of the explanation given in the graphic. Could this information also be displayed in one circle graph for men and another for women? Explain.

2. Examine the following figure, entitled "Global teen worries." How many countries were contained in the sample? The graph contains bars and a circle. Which bar is the longest? Which bar represents the greatest percentage? Is this a bar graph or not? If not, what changes would need to be made to put the information into a bar graph? Could it be made into a Pareto chart? Could it be made into a circle graph? Explain.

USA SNAPSHOTS®

A look at statistics that shape our lives

Global teen worries

Leading sources of major concern for teenagers, ages 15–18, surveyed in 41 countries:

Getting a good job **70%**

Parents' health **63%**

Losing someone they love **57%**

Finishing education **54%**

Personal health **54%**

Getting university degree **52%**

Note: May name more than one

Source: BrainWaves Group's New World Teen Study

By Anne R. Carey and Genevieve Lynn, USA TODAY

Source: Copyright 1996, USA TODAY. Reprinted with permission.

Discuss each of the following topics in class or review the topics on your own. Then write a brief but complete essay in which you summarize the main points. Please include formulas and graphs as appropriate.

1. What does it mean to say that we are going to use a sample to draw an inference about a population? Why is a random sample so important for this process? If we wanted a random sample of students in the cafeteria, why couldn't we just take the students who order Diet Pepsi with their lunch? Comment on the statement, "A random sample is kind of a miniature population, whereas samples that are not random are likely to be biased." Why would the students who order Diet Pepsi with lunch not be a random sample of students in the cafeteria?

2. In your own words, explain the differences among the following sampling techniques: simple random sampling, stratified sampling, systematic sampling, cluster sampling, and convenience sampling. Describe situations in which each type might be useful.

3. In your own words, explain the differences among bar graphs, circle graphs, time plots, Pareto charts, relative-frequency histograms, and histograms. If you have nominal data, which graphic displays might be useful? What if you have ordinal, interval, or ratio data?

4. What do we mean when we say that a histogram is skewed to the left? To the right? What is a bimodal histogram? Discuss the following statement: "A bimodal histogram usually results if we draw a sample from two populations at once." Suppose you took a sample of weights of college football players and with this sample you included weights of cheerleaders. Do you think a histogram made from the combined weights would be bimodal? Explain.

5. Discuss the statement that stem-and-leaf displays are quick and easy to construct. How can we use a stem-and-leaf display to make the construction of a frequency table easier? How does a stem-and-leaf display help you spot extreme values fast?

6. Go to the library and pick up a current issue of the *Wall Street Journal, Newsweek, Time, USA Today,* or some other periodical. Examine each newspaper or magazine for graphs of the type discussed in this chapter. List the variables used, the method of data collection, and the general type of conclusion drawn from the graphs. Another source for information is the Internet. Explore several Web sites, and categorize the graphs you find as you did for the print media. Recommended Web sites are the Social Statistics Briefing Room at <http://www.whitehouse.gov/fsbr/ssbr.html>, a law enforcement Web site at <http://www.copnet.org>, and a golf Web site at <http://majors.golfweb.com/>.

USING TECHNOLOGY

The problems in this section may be done using software or calculators with statistical functions. Displays and suggestions are given for Minitab (Release 12), the TI-83, ComputerStat, and Excel.

Professional statisticians in industry and research use computers to help them analyze and process statistical data. There are many computer programs available for statistics. Some commonly used statistical packages include Minitab®, SAS®, and SPSS.* Spreadsheet packages such as Excel support many statistical functions.

In addition, graphing calculators such as those available from Casio, Hewlett-Packard, Sharp, and Texas Instruments have built-in statistics routines, data editing, data sorting, and graphical support. These calculators can be programmed to perform statistical functions not included in the basic statistics modes.

ComputerStat is a software package specifically written to go with this text (it is available without charge to institutions using this text). The package is designed for the beginning statistics student with no previous computer experience. Users need only enter data and select appropriate programs. Screen prompts guide the user through the software.

Technology guides that are supplements to this text give detailed instructions for the TI-83, Minitab, ComputerStat, and Excel.

Random-Number Generators

Statistical software packages and graphing calculators all have random-number generators. Suggestions for using Minitab, the TI-83, and ComputerStat to generate random numbers follow.

*Minitab is a registered trademark of Minitab, Inc. SAS is a registered trademark of SAS Institute, Inc. SPSS is a trademark of SPSS, Inc.

Minitab

In Minitab, you can draw a random sample of integers between two values a and b and store the results in specified columns. Sampling is done with replacement. The menu selections ➤Calc ➤Random Data ➤Integer open the dialogue box where you specify the values of a and b, the number of rows of data desired in designated columns.

TI-83

On the TI-83, the PRB menu under MATH contains the selection *rand*. This selection generates a random number between 0 and 1. To generate a random number that is a whole number with up to three digits, multiply the random number by 1000 and take the integer part (*ipart* under MATH NUM). Then, each time you press ENTER, a random number with up to three digits is displayed.

```
iPart(rand*1000)
                339
                995
                200
                798
                951
```

ComputerStat

In ComputerStat, select the program Random Samples found in the Descriptive Statistics menu. This program lets you select the range of integers from which you wish to draw random samples. It also gives you the option of sampling with or without replacement.

65

Excel

In Excel, select the paste function button f_x. Select the function category All, and then use the command RANDBETWEEN to select a random number between two specified values. Again, see the *Excel Guide* supplement to this text for more details.

Using the appropriate software or calculator that is available to you, do the following:

1. In a large condominium complex there are 473 condominiums numbered 1 to 473. You want to check the thermostats in a random sample of 50 different condominiums. Find the numbers of your 50 condominiums.

2. A theater is showing a new movie. After the movie, you want to ask a random sample of 30 people for their opinions on the movie. As people buy their tickets, you write a number on the back. Because 278 tickets are purchased, you will use the numbers from 1 to 278. Just before the movie begins, you announce that a small prize will be given to the people whose numbers are called and who respond to a questionnaire after the movie. Find the 30 different numbers you will call.

3. The phone book indicates that there are 83 sporting goods stores in Kansas City. Using the alphabetical order of their appearance in the phone book, you number the stores from 1 to 83. Find a random sample of 10 different sporting goods stores you will call.

4. You have a group of 18 people and you wish to split it into two groups for baseball teams. Use random numbers to assign players to the teams.

Histograms

Suggestions for creating histograms in Minitab (professional graphics), on the TI-83, and in ComputerStat follow.

Minitab

In Minitab, after you enter the data, you can make a histogram. You can control the number of classes by designating values for the class boundaries (called *cutpoints*). The menu selections ➤**Graph** ➤**Histogram** open a dialogue box. Press the Options button to specify that you want a histogram. Select cutpoints, and list the class boundaries as cutpoints. More details and examples can be found in the *Technology Guide* supplement for this text.

TI-83

On the TI-83, enter the data in a list and select *histogram* in the STAT PLOT menu. In the WINDOW screen, make the following settings:

$$Xmin = \text{lowest class boundary}$$
$$Xmax = \text{highest class boundary}$$
$$Xscl = \text{class width}$$

ComputerStat

In ComputerStat, select Frequency Distribution under the Descriptive Statistics menu. Then follow the instructions given on the screen.

Excel

The menu selections ➤**Tools** ➤**Data Analysis** ➤**Histogram** open the histogram dialogue box. Note that you may have to load the Analysis ToolPak as an Add-In to your installation of Excel. Put upper class boundaries in the Bin Range column. Select New Workbook and check the chart output. Right click on a bar in the histogram, select format, select the options tab, and set the gap width to 0 to make the bars touch. Again, see the *Excel Guide* supplement for more information.

Each business day the Dow Jones Information Retrieval Service gives closing prices and volumes of sales for all major stocks on the New York Stock Exchange. The phrase "volume leads price" is often heard in discussions about market activity. In fact, history has shown that an unusually high volume of sales of a stock generally indicates that an imminent change (either up or down) in stock price is about to take place. What is a high or low volume for a particular stock? What is an everyday or ordinary volume? How frequently do these volumes occur?

Perhaps the best way to answer these questions is to track the market activity of your stock over a period of time. In particular, a frequency table and histogram of volumes over, say, a 10-week period would help answer some of these questions about volume. The following table lists the volumes of IBM (International Business Machines) stock (in hundreds of shares sold) for a 10-week period from March 17, 1997 through May 27, 1997.

Volume of IBM Stock Trades

Date	Volume (in hundreds)	Date	Volume (in hundreds)
03/17/97	50944	04/22/97	30222
03/18/97	36361	04/23/97	39171
03/19/97	41194	04/24/97	106591
03/20/97	37262	04/25/97	30050
03/21/97	65373	04/28/97	24960
03/24/97	45042	04/29/97	46250
03/25/97	34593	04/30/97	45794
03/26/97	35149	05/01/97	29933
03/27/97	32080	05/02/97	30329
03/31/97	38770	05/05/97	31041
04/01/97	29294	05/06/97	32973
04/02/97	35380	05/07/97	28405
04/03/97	53845	05/08/97	47924
04/04/97	62518	05/09/97	29369
04/07/97	26676	05/12/97	37451
04/08/97	27080	05/13/97	56621
04/09/97	34459	05/14/97	34481
04/10/97	24831	05/15/97	27361
04/11/97	41386	05/16/97	35243
04/14/97	26344	05/19/97	24237
04/15/97	39319	05/20/97	31326
04/16/97	18453	05/21/97	31703
04/17/97	26551	05/22/97	18074
04/18/97	23432	05/23/97	17290
04/21/97	23567	05/27/97	32377

Using the appropriate available software or calculator, do the following:

1. Enter the given IBM volumes. Then make histograms with three classes; five classes; eight classes, and 10 classes.

2. Compare the effect of lumping all the data together in only three classes with the opposite effect of thinning the data out into as many as 10 classes. Both these extremes have drawbacks. Comment on them.

3. From the histograms, what would you say is a high volume of sales? What is an everyday or ordinary volume? What is a low volume? What are the frequencies of days on which high, low, and ordinary volumes occur? In the given period of observation, which volumes (or volume ranges) seem to occur most frequently? Which occur least frequently? Notice that 4/24/97 had an unusually high volume. This was a day on which the price of one share of IBM stock rose almost 10%.

4. On 5/28/97, IBM announced a two-for-one stock split. This means that if an investor held one share worth $180, the investor would hold two shares each worth $90 after the split. After the split, how

do you suppose the volume will change if investors want to trade shares so that the dollar amount of the shares traded is the same as before the split?

Computer Displays

Different computer software packages follow slightly different procedures for generating and displaying histograms, as shown in the following two figures.

FIGURE 1-21

ComputerStat Histogram (for Data in Table 1-7, Commuting Distance), Windows Version. Reprinted with permission of Houghton Mifflin Company.

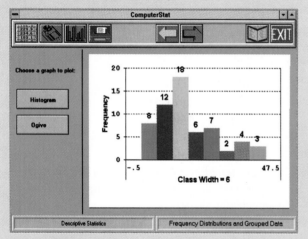

FIGURE 1-22

Minitab Release 12 Histogram (for Data in Table 1-7), Commuting Distance

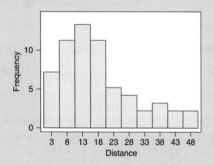

Windows Pull-Down Menu Selection: ➤**Graph** ➤**Histogram**

In the HISTOGRAM dialogue box, SELECT the column containing data, select BAR for the display, and press the OPTIONS button. In the Options dialogue box, select FREQUENCY, MIDPOINTS, and specify MIDPOINT positions as the class midpoints from lowest to highest. If you use cutpoints instead, you can show the class boundaries.

Minitab also generates stem-and-leaf displays, as shown in the figure below. Notice that it automatically orders the leaves.

FIGURE 1-23

Minitab Release 12 Stem-and-Leaf Display (for Data in Table 1-18)

```
Stem-and-Leaf of C1          N=30
LeafUnit=0.10

  10                0    2345556779
 (6)                1    012345
  14                2    11357
   9                3    128
   6                4    000000
```

Windows Pull-Down Menu Selection: ➤Graph ➤Character Graphs ➤Stem-and-Leaf

In the STEM-AND-LEAF dialogue box, SELECT the column containing the data. Specify the desired increment.

2 Averages and Variation

*While the individual man is an insoluble puzzle,
in the aggregate he becomes
a mathematical certainty. You can,
for example, never foretell what any one man
will do, but you can say
with precision what an average number will be up to.*

—Arthur Conan Doyle,
The Sign of Four

Sherlock Holmes

Sherlock Holmes spoke these words to his colleague Dr. Watson as the two were unraveling a mystery. The detective was commenting that if a single member is drawn at random from a population, we cannot predict *exactly* what that member will look like. However, there are some "average" features of the entire population that an individual is likely to exhibit. The degree of certainty with which we would expect to observe such average features in any individual depends on our knowledge of the variation among individuals in the population. Sherlock Holmes has led us to two of the most important statistical concepts: average and variation.

Section 2.1

Measures of Central Tendency: Mode, Median, and Mean

The average salary of a Major League Baseball player is $2.1 million. The Zippy car averages 39 miles per gallon on the highway. A survey shows that the average shoe size for women is size 8.

In each of the preceding statements, *one* number is used to describe an entire sample or population. Such a number is called an *average*. There are many ways to compute averages, but we will study only three of the major ones.

Mode

The easiest average to compute is the mode. The *mode* is the value or property that occurs most frequently in the data. For instance, if you count the number of letters in each word in the preceding paragraph, you will see that the mode is two letters. In other words, there are more words with exactly two letters than with any other number of letters (see Table 2-1). Sometimes a distribution will not have a mode. If Professor Fair gives an *equal* number of A's, B's, C's, D's, and F's, then there is no modal grade. The data in Table 2-2 have no mode because no data value occurs more frequently than all the others.

GUIDED EXERCISE   1

On the first day of finals, 20 students at La Platta College were selected at random. They were asked how many hours they had slept the night before (rounded to the nearest hour). The results in hours were 8, 6, 5, 6, 4, 3, 5, 8, 7, 7, 5, 6, 2, 0, 5, 7, 6, 6, 7, and 8.

(a) Complete Table 2-3.

Table 2-3 Hours Slept Before Finals

Hours Slept	Number of Students
0	1
1	0
2	1
3	1
4	____
5	____
6	____
7	____
8	____

⇨

Table 2-4 Completion of Table 2-3

Hours Slept	Number of Students
0	1
1	0
2	1
3	1
4	1
5	4
6	5
7	4
8	3

(b) Is there a mode?   ⇨   Yes, one quantity has the greatest frequency.

(c) What is the modal number of hours slept?   ⇨   The modal number of hours slept is 6 hours.

The mode is an easy average to compute, but it is not very stable. For example, if one of the students in Guided Exercise 1 had slept 5 hours instead of 6 on the night before the first finals, the mode would change to 5. However, if you are interested in the *most common* value in a distribution, the mode is appropriate to use.

Median

Another average is the *median*, or central value, of an ordered distribution. When you are given the median, you know there are an equal number of values above and below it. To obtain the median, we *order* the data from the smallest value to the largest. Then we pick or construct the middle value.

Table 2-1 Word Length of Second Paragraph in This Section

Number of Letters in Word	Number of Words
1	1
Mode → 2	11 ← Greatest number of words
3	5
4	8
5	4
6	5
7	2
8	2
9	1
10	2

Table 2-2 Number of Minutes Spent by Professor Adams's Students Using the School Computer Terminals Last Wednesday Afternoon

27	30	42	42
36	36	50	

The median of the following set of test scores for English literature is 75.

50 51 60 64 65 70	median	80 81 85 90 95 97
6 below	↓ 75	6 above

There are as many test scores above as below the median.

For an even number of test scores, the median must be constructed. It is not necessarily one of the given test scores. For instance, the following list has an even number of scores.

median
↓

51   60   64   69   70   75   [?]   78   80   85   90   91   95

middle values

To construct the median of a set of data with an *even* number of entries, add the two middle values and divide by 2.

$$\text{Median} = \frac{\text{sum of two middle scores}}{2}$$

$$= \frac{75 + 78}{2}$$

$$= 76.5$$

Belleview College must make a report to the budget committee about the average credit hour load a full-time student takes. (A 12-credit-hour load is the minimum requirement for full-time status. For the same tuition, students may take up to 20 credit hours.) A random sample of 40 students yielded the following information (in credit hours):

17	12	14	17	13	16	18	20	13	12
12	17	16	15	14	12	12	13	17	14
15	12	15	16	12	18	20	19	12	15
18	14	16	17	15	19	12	13	12	15

(a) Organize the data from smallest to largest number of credit hours.

⇨   12 12 12 12 12 12 12 12 12 12
13 13 13 13 14 14 14 14 15 ⑮
⑮ 15 15 15 16 16 16 16 17 17
17 17 17 18 18 18 19 19 20 20

(b) Since there is an _____ (odd, even) number of values, we add the two middle values and divide by 2 to get the median. What is the median credit hour load?

⇨   There is an even number of entries. The two middle values are circled in part a.

$$\text{Median} = \frac{15 + 15}{2} = 15$$

(c) What is the mode of this distribution? Is it different from the median? If the budget committee is going to fund the school according to the average student credit hour load (more money for higher loads), which of these two averages do you think the college will use?

⇨   The mode is 12. It is different from the median. Since the median is higher, the school will probably use it and indicate that the average being used is the median.

The median is a more stable average than the mode, but it does not indicate the range of values above or below it. For instance, the median is 20 for both of the following groups of scores:

(a) 10   15   20   25   30

(b) 1   10   20   40   100

In the first group, all scores are within 10 points of the median; in the second group, one score is 80 points above the median. The median is based on the *position* rather than the specific value of each data entry.

Mean

An average that uses the exact value of each entry is the *mean* (sometimes called the *arithmetic mean*). To compute the mean, we add the values of all the entries and then divide by the number of entries.

$$\text{Mean} = \frac{\text{sum of all the entries}}{\text{number of entries}}$$

The mean is the average usually used to compute a test average.

EXAMPLE 1 ❯   To graduate, Linda needs at least a B in biology. She did not do very well on her first three tests; however, she did well on the last four. Here are her scores:

58   67   60   84   93   98   100

Compute the mean and determine if Linda's grade will be a B (80 to 89 average) or a C (70 to 79 average).

SOLUTION:

$$\text{Mean} = \frac{\text{sum of scores}}{\text{number of scores}} = \frac{58 + 67 + 60 + 84 + 93 + 98 + 100}{7} = \frac{560}{7} = 80$$

Since the average is 80, Linda will get the needed B.   ●

COMMENT When we compute the mean, we sum the given data. There is a convenient notation for indicating the sum. Let x represent any value in the data set. Then the notation

Σx (read "the sum of all given x values")

means that we are to sum all the data values. In other words, we are to sum all the entries in the distribution. The symbol Σ means *sum the following* and is a capital sigma, the S of the Greek alphabet.   ○

Formulas for the mean

The symbol for the mean of a *sample* distribution of x values is denoted by $\bar{x}$ (read "x bar"). If your data comprises the entire *population*, we use the symbol μ (lowercase Greek letter mu, pronounced "mew") to represent the mean. The procedure for computing the mean is the same regardless of whether we have population or sample data. If we let n represent the number of entries in a *sample* data set and let N represent the number of entries in a *population* data set, the formulas are

$$\text{Sample mean} = \bar{x} = \frac{\Sigma x}{n} \qquad\qquad \text{Population mean} = \mu = \frac{\Sigma x}{N} \qquad (1)$$

GUIDED EXERCISE   3

A fabric store manager is eager to see if the latest patterns for size 12 dresses show a longer hemline than last year's. If so, she can expect to sell more fabric because each pattern will call for more material. She takes a random sample of 10 dress patterns and finds the finished lengths from back of neck to bottom of hem (in inches) to be

| 41.5 | 42 | 39 | 44 | 43.5 | 45 | 43 | 45 | 42 | 46 |

(a) What is the value of n?

⇨ Since there are 10 data entries, $n = 10$.

(b) How do you find Σx? What is the value of Σx?

⇨ To find Σx, we add all the data entries together.

$$\Sigma x = 41.5 + 42 + 39 + 44 + 43.5 + 45$$
$$+ 43 + 45 + 42 + 46$$
$$= 431$$

(c) Compute the mean, $\bar{x}$.

⇨ $\bar{x} = \dfrac{\Sigma x}{n} = \dfrac{431}{10} = 43.1$

(d) Last year the mean length of size 12 dresses was 36 in. How much longer is the mean length now? Can the manager expect to sell more material per dress?

⇨ The difference is $43.1 - 36 = 7.1$ in. The manager can expect to sell more fabric for each dress.

Calculator Note It is very easy to compute the mean on *any* calculator: simply add the data values and divide the total by the number of data. However, on calculators with a statistics mode, you place the calculator in that mode, *enter* the data, and then press the key for the mean, often labeled $\bar{x}$. Once the data have been entered under the statistics mode, other statistical measures are readily available. For instance, if you forget how many data values there are, you can press a key usually labeled n. The number of data values appears. A Σx key tells you the sum of the data values.

Most *graphing* calculators will provide all the basic statistical measurements, and many will order the data for you through a *sort list* command. From the ordered list it is then easy to find or compute the median. Some graphing calculators, such as the TI-83, actually give you the value of the median under the one-variable statistics calculation menu.

All these calculators are wonderful aids in analyzing data. *However, a measurement has no meaning if you do not know what it represents or how a change in data values might affect it.* The defining formulas and procedures for computing the measures tell you a great deal about the measures. Even if you use a calculator to evaluate all the statistical measures, pay attention to the information the formulas give you about the components or features of the measurement.

Figure 2-1 shows the TI-83 displays for the data in Guided Exercise 2 on the average credit hour load of full-time students. The data are entered as list 1 (L1). The command 1-Var Stats gives the summary statistics. Notice that $\bar{x}$ is given as well as the sum Σx. On the second screen, the median Med is given. The SortA command on the Edit menu sorts the data in ascending order. You can scan the sorted data to find the mode.

FIGURE 2-1

TI-83 Screens: Credit Hour Load (see data in Guided Exercise 2)

```
1-Var Stats
 x̄=14.975
 Σx=599
 Σx²=9213
 Sx=2.496022477
 σx=2.464624718
↓n=40
█
```

```
1-Var Stats
↑n=40
 minX=12
 Q1=12.5
 Med=15
 Q3=17
 maxX=20
```

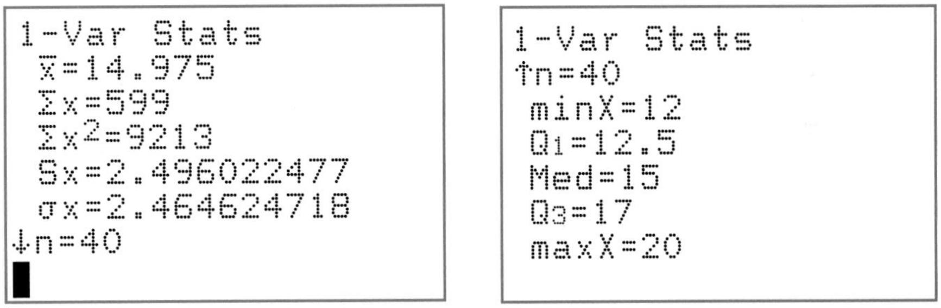

GUIDED EXERCISE   4

Rowdy Rho fraternity is in danger of losing campus approval if it does not raise the mean grade point average of the entire group to at least 2.2 on a four-point scale. This term, the averages of the members were

    1.8   2.0   2.0   2.0   2.0   1.9   1.8   2.3   2.5   2.3   1.9   2.2   2.0   2.3

(a) What is the mean of the grade point averages?

⇨   Mean $= \dfrac{\Sigma x}{n} = \dfrac{29.0}{14} \approx 2.07$

(b) Rod made a 2.0 average this term because he was in the hospital 6 weeks. He believes he would have made a 3.9 average if he had been well. Recompute the mean with the first 2.0 replaced by 3.9. Would Rod have saved the fraternity if he had made a 3.9 average?

⇨   If we replace the first 2.0 by 3.9, the new mean is

$$\text{Mean} = \dfrac{\Sigma x}{n} = \dfrac{30.9}{14} \approx 2.21$$

If Rod had made a 3.9 instead of a 2.0, the fraternity would have been saved.

Exercise continues

Exercise continued

(c) Suppose the college had required the fraternity to raise the *median* grade point average to 2.2. Would Rod's potential 3.9 have saved the fraternity? What can you say about the effect of the exceptional value 3.9 on the median and mean?

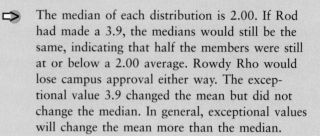

The median of each distribution is 2.00. If Rod had made a 3.9, the medians would still be the same, indicating that half the members were still at or below a 2.00 average. Rowdy Rho would lose campus approval either way. The exceptional value 3.9 changed the mean but did not change the median. In general, exceptional values will change the mean more than the median.

We have seen three averages: the mode, the median, and the mean. For later work, the mean is the most important. A disadvantage of the mean, however, is that it can be affected by exceptional values, as shown in Guided Exercise 4. In such cases, the median is a better representation of the general level of the distribution.

Resistant measures

A *resistant measure* is one that is not influenced by extremely high or low data values. The mean is not a resistant measure of center because we can make the mean as large as we want by increasing the size of only one data value. The median, on the other hand, is more resistant. However, a disadvantage of the median is that it is not sensitive to the specific size of a data value.

Trimmed mean

A measure of center that is more resistant than the mean but still sensitive to specific data values is the *trimmed mean*. To compute a 5% trimmed mean, order the data from smallest to largest, delete the bottom 5% of the data, and then delete the top 5% of the data. Finally, compute the mean of the remaining 90% of the data. Trimming eliminates the influence of unusually small or large data values. Both 5% and 10% trimmed means are presented in the outputs of many statistical computer packages. The package Minitab gives a 5% trimmed mean along with the mean and median of a data set.

If the calculation of 5% of the number of data values does not produce a whole number, *round* to the nearest integer to determine how many data values to remove from each end of the distribution (the convention Minitab uses). For instance, if 5% of the number of data values gives 2.3, you would round down and drop the two largest and the two smallest data values. If 5% of the number of data values gives 2.7, you would round up and drop the three largest and the three smallest data values.

EXAMPLE 2 ➤

Barron's Profiles of American Colleges, 19th Edition, lists average class size for introductory lecture courses at each of the profiled institutions. A sample of 20 colleges and universities in California showed class sizes for introductory lecture courses to be

| ⑭ | 20 | 20 | 20 | 20 | 23 | 25 | 30 | 30 | 30 |
| 35 | 35 | 35 | 40 | 40 | 42 | 50 | 50 | 80 | ⑧⓪ |

Compute a 5% trimmed mean for this data set.

SOLUTION: The data are already arranged in order. Since 5% of 20 = 1, we delete *one* data value from the bottom of the ordered data list and *one* data value from the top. The data values to be deleted are circled. Now we take the mean of the remaining 18 data values.

$$5\% \text{ trimmed mean} = \frac{\Sigma x}{n} = \frac{625}{18} \approx 34.7$$

The average you use depends on what you want to do with that average. If you want to know which value occurs most frequently in a distribution, use the mode. If a store wants to know which shirt size is requested most frequently, the mode is the proper average to use, and the store will know to carry more shirts of that size than any other. If you want to cut a distribution in half, use the median. A report showing the average salary of workers at Gator Tire Factory should show the median because the top-level administrative salaries will pull up the mean salary and make the production-line salaries look higher than they are. If you want each entry value in the data to enter into the average, use the mean. As we shall see in later chapters, we use the mean if we want to estimate population average from a sample average. The mean uses all the data entries, and the mean can be analyzed more conveniently by statistical methods.

COMMENT In Chapter 1 we examined four levels of data: nominal, ordinal, interval, and ratio. The mode (if it exists) can be used with all four levels, including nominal. For instance, the modal color of all passenger cars sold last year might be blue. The median may be used with data at the ordinal level or above. If we ranked the passenger cars in order of customer satisfaction level, we could identify the median satisfaction level. For the mean, our data need to be at the interval or ratio level (although there are exceptions in which the mean of ordinal-level data is computed). We can certainly find the mean model year of used passenger cars sold or the mean price of new passenger cars.

V I E W P O I N T

Scholastic Aptitude Test Is Flawed!

The SAT given on October 12, 1996 had a defective mathematics question involving the *median* of a sequence of numbers. Of course, it is a very rare event for the SAT to have a flawed question. Nevertheless, it can happen. SAT mathematics scores can be very influential for a student's college admission and possible scholarships. Because of the flawed question, SAT tests containing the question were rescored. See Web site <http://www.collegeboard.org/> and look under news.

SECTION 2.1 PROBLEMS

Agriculture: Growing Season

1. The average length of the growing season is often measured in average number of frost-free days. The front range of Colorado (Fort Collins, Boulder, Denver, Colorado Springs, Pueblo) was studied by J. F. Benci and T. B. McKee, from the Department of Atmospheric Science at Colorado State University. Based on data from their Climatology Report No. 77-3, different locations in the Colorado front range had the following average numbers of frost-free days per year:

| 156 | 161 | 152 | 162 | 144 | 153 |
| 148 | 157 | 168 | 157 | 161 | 157 |

Compute the mean, median, and mode. Write a brief description of the meanings of these numbers from the point of view of a gardener.

Baseball: Babe Ruth

2. Babe Ruth was the American League home run champion 12 times (during the period from 1918 to 1931). The number of home runs he hit to earn the 12 titles were

| 11 | 29 | 54 | 59 | 41 | 46 |
| 47 | 60 | 54 | 46 | 49 | 46 |

Find the mean, median, and mode of the number of home runs.

Climate: Death Valley

3. How hot does it get in Death Valley? The following data are taken from a study conducted by the National Park System, of which Death Valley National Monument is a unit. The ground temperatures (°F) were taken from May to November in the vicinity of Furnace Creek.

| 146 | 152 | 168 | 174 | 180 | 178 | 179 |
| 180 | 178 | 178 | 168 | 165 | 152 | 144 |

Compute the mean, median, and mode for these ground temperatures.

Ecology: Wolf Packs

4. How large is a wolf pack? The following information is from a random sample of winter wolf packs in regions of Alaska, Minnesota, Michigan, Wisconsin, Canada, and Finland. (Source: *The Wolf*, by L. D. Mech, University of Minnesota Press.) Winter pack size:

| 13 | 10 | 7 | 5 | 7 | 7 | 2 | 4 | 3 |
| 2 | 3 | 15 | 4 | 4 | 2 | 8 | 7 | 8 |

Compute the mean, median, and mode for the size of winter wolf packs.

Leisure: Vacation Rentals

5. How expensive is Maui? If you want a vacation rental condominium (for up to four people), then *The Maui News* (Web site <http://www.maui.net/>, then follow the

links to accommodations) gave the following cost in dollars per day for a random sample of condominiums located throughout the island of Maui.

89	50	68	60	375	55	500	71	40	350
60	50	250	45	45	125	235	65	60	130

(a) Compute the mean, median, and mode for these data.
(b) Compute a 5% trimmed mean for the data, and compare it with the mean computed in part a. Does the trimmed mean more accurately reflect the general level of the daily rental costs?
(c) If you were a travel agent and a client asked about the daily cost of renting a condominium on Maui, what average would you use? Explain. Is there any other information about the costs that you think might be useful, such as the spread of the costs?

Pro-football: Age of Players

6. How old are professional football players? The 11th Edition of *The Pro Football Encyclopedia* gave the following information.

Random sample of pro football player ages in years:

24	23	25	23	30	29	28	26	33	29
24	37	25	23	22	27	28	25	31	29
25	22	31	29	22	28	27	26	23	21
25	21	25	24	22	26	25	32	26	29

(a) Compute the mean, median, and mode of these ages.
(b) Compute a 5% trimmed mean.
(c) Compare the averages. Does one seem to represent the ages of the pro football players most accurately? Explain.

Business: Salaries

7. Brookridge National Bank is a small bank in a rural Iowa town. The 12 people who work at the bank are the president, the vice president (the president's son-in-law), eight tellers, and two secretaries. The annual salaries for these people in thousands of dollars are:

President: 93

Vice President: 80

Tellers: 15, 25, 14, 18, 21, 16, 19, 20

Secretaries: 12, 13

(a) Compute the mean of all 12 salaries.
(b) Compute the median of all 12 salaries, and compare your answer with the mean in part a. Which average better describes the salaries of the *majority* of employees?
(c) Omit the salaries of the president and vice president. Calculate the mean and median for the remaining 10 people.
(d) Compare your answers for part c with those for parts a and b. Comment on the effects of extreme values on the mean and median.

Business: Shipping Time

8. *Thriving on Chaos,* by Tom Peters, has some excellent cautions about utilizing measurements and averages. In discussing ways to provide superior service, he says, "The use of averages is downright dangerous." He illustrates with examples. Suppose that a manufacturing company claims, "On average, we ship parts within 37 hours of order entry" but that a careful look at data shows that for the "worst-off 10 percent of customers," the shipping time was within 89 hours of order entry. Peters' advice is to "focus attention on the worst-off 1, 5, 10, or 25 percent of customers" instead of on the average. Comment on this advice.

General: Mean, Median

9. Consider a data set of 15 distinct measurements with mean *A* and median *B*.
   (a) If the highest number were increased, what would be the effects on the median and mean? Explain.
   (b) If the highest number were decreased to a value that was still larger than *B*, what would be the effects on the median and mean?
   (c) If the highest number were decreased to a value smaller than *B*, what would be the effects on the median and mean?

Student Data Project

10. For the next week, record the length of time you spend on each telephone call you make or receive. Round each time to the nearest minute. Compute the mean, median, and mode. What accounts for some of your shorter phone calls (leaving messages, responding to marketing calls, etc.)?

Section **2.2**

Measures of Variation

An average is a summary of a set of data in just one number. We have studied several averages. As some of our examples have shown, an average taken by itself may not always be very meaningful. We need a statistical cross-reference. This cross-reference should be a measure of the *variance,* or spread, of the data.

Range

The *range* is one such measure of variance. The range is the difference between the largest and smallest values of a distribution. For example, the distance between rows (in inches) in the various sections of Flicker Auditorium are

14	15	18	20	35

The range of these distances is

$$\text{Range} = \text{largest value} - \text{smallest value}$$

$$= 35 - 14 = 21 \text{ in.}$$

The range indicates the variation between the smallest and largest entries, but it does not tell us how much other values vary from one another. We need a different measure of variation, as the next example shows.

EXAMPLE 3 ▷

You are trying to decide which compact disc club to join: Discount Disks or Selecta Disc. Both have the same bonuses for new members, and both require members to buy one compact disc a month from the monthly selections for at least 1 year. They both charge the same price for compact discs, and they both advertise that the number of compact discs in the monthly selections has mean 31 and range 79. The only advertised difference is that Discount Disks has no membership fee and Selecta Disc costs $25 to join. Which club would you join? Before you make up your mind, look at the additional information in Table 2-5.

Table 2-5   Compact Disc Selections per Month

Month	Selecta Disc	Discount Disks
January	100	80
February	30	1
March	30	70
April	30	2
May	21	70
June	23	1
July	21	1
August	21	70
September	24	3
October	21	70
November	30	2
December	21	2
	Mean = 31	Mean = 31
	Range = 79	Range = 79

SOLUTION: The mean and range are not enough to tell you how much the number of monthly disc selections varies from the advertised mean. Discount Disks could be a disappointment because some months it gives you only one selection. Selecta Disc is more consistent in its offerings; you always have at least 21 compact discs from which to choose.

Sample standard deviation

A measurement that will give you a better idea of how the data entries differ from the mean is the *standard deviation*. The formula for the standard deviation differs slightly depending on whether you are using an entire population or just a sample. At the moment, we will compute the standard deviation for sample data only. When we have sample data, we use the letter s to denote the standard deviation. The formula for the sample standard deviation is

$$\text{Sample standard deviation} = s = \sqrt{\frac{\Sigma(x - \bar{x})^2}{n - 1}} \qquad (2)$$

where x is any entry in the distribution, $\bar{x}$ is the mean, and n is the number of entries.

Notice that the standard deviation uses the difference between each entry x and the mean $\bar{x}$. This quantity $(x - \bar{x})$ will be negative if the mean $\bar{x}$ is greater than the entry x. If you take the sum

$$\Sigma(x - \bar{x})$$

then the negative values will cancel the positive values, leaving you with a variation measure of 0 even if some entries vary greatly from the mean.

In the formula for the standard deviation, the quantities $(x - \bar{x})$ are *squared* before they are summed. This device eliminates the possibility of having negative values in the sum. So, in the formula, we have the quantity

$$\Sigma(x - \bar{x})^2$$

Then we divide this sum by $n - 1$ to get the quantity under the square root sign:

$$\frac{\Sigma(x - \bar{x})^2}{n - 1}$$

Sample variance

These three steps have given us a quantity called the *variance* of a sample, denoted by s^2:

$$\text{Sample variance} = s^2 = \frac{\Sigma(x - \bar{x})^2}{n - 1} \tag{3}$$

The compact disc data in Example 3 initially described the number of selections per month, but the variance s^2 of these data would be in *square number of selections*. Square number of selections—what's that? We need to take the square root of the variance to return to number of selections per month. This brings us to the standard deviation of a sample.

$$\text{Sample standard deviation} = s = \sqrt{\frac{\Sigma(x - \bar{x})^2}{n - 1}}$$

The next example shows how to use this formula.

EXAMPLE 4 ➤ Big Blossom Greenhouse was commissioned to develop an extra large rose for the Rose Bowl Parade. A random sample of blossoms from hybrid A bushes yielded these diameters (in inches) for mature peak blossoms:

$$2 \quad 3 \quad 4 \quad 5 \quad 6 \quad 8 \quad 10 \quad 10$$

Find the standard deviation.

SOLUTION: There are several steps involved in computing the standard deviation, and a table will be helpful (see Table 2-6 on the next page). Since $n = 8$, we take the total sum of the entries in the first column of Table 2-6 and divide by 8 to find the mean $\bar{x}$.

$$\bar{x} = \frac{\Sigma x}{n} = \frac{48}{8} = 6.0$$

Table 2-6   Diameter of Rose Blossoms (in inches)

Column I x	Column II $x - \bar{x}$	Column III $(x - \bar{x})^2$
2	$2 - 6 = -4$	$(-4)^2 = 16$
3	$3 - 6 = -3$	$(-3)^2 = 9$
4	$4 - 6 = -2$	$(-2)^2 = 4$
5	$5 - 6 = -1$	$(-1)^2 = 1$
6	$6 - 6 = 0$	$(0)^2 = 0$
8	$8 - 6 = 2$	$(2)^2 = 4$
10	$10 - 6 = 4$	$(4)^2 = 16$
10	$10 - 6 = 4$	$(4)^2 = 16$
$\Sigma x = 48$		$\Sigma(x - \bar{x})^2 = 66$

Using this value for $\bar{x}$, we obtain the values in column II. Square each value in column II to obtain the values in column III, and then add the values in column III. To get the variance, divide the sum of the values in column III by $n - 1$. Since $n = 8$, $n - 1 = 7$.

$$s^2 = \frac{\Sigma(x - \bar{x})^2}{n - 1} = \frac{66}{7} \approx 9.43$$

Now obtain the standard deviation by taking the square root of the variance.

$$s = \sqrt{s^2} \approx \sqrt{9.43} \approx 3.07$$

(Use a calculator to compute the square root.)

GUIDED EXERCISE   5

Big Blossom Greenhouse gathered another random sample of mature peak blooms from hybrid B bushes. The eight blossoms had these widths (in inches):

    5   5   5   6   6   6   7   8

(a) Again, we will construct a table so we can find the mean, variance, and standard deviation more easily. In this case, what is the value of n? Find the sum of the values in column I in Table 2-7, and compute the mean.

 $n = 8$. The sum of the values in column I is $\Sigma x = 48$, so the mean is

$$\bar{x} = \frac{48}{8} = 6 \text{ in.}$$

Exercise continues

Exercise continued

Table 2-7 Complete Columns II and III

Column I x	Column II $x - \bar{x}$	Column III $(x - \bar{x})^2$
5	___	___
5	___	___
5	___	___
6	___	___
6	___	___
6	___	___
7	___	___
8	___	___
$\Sigma x = $ ___		$\Sigma(x - \bar{x})^2 = $ ___

Table 2-8 Completion of Table 2-7

Column I x	Column II $x - \bar{x}$	Column III $(x - \bar{x})^2$
5	−1	1
5	−1	1
5	−1	1
6	0	0
6	0	0
6	0	0
7	1	1
8	2	4
$\Sigma x = 48$		$\Sigma(x - \bar{x})^2 = 8$

(b) What is the value of $n - 1$? Divide the total sum of the values in column III by $n - 1$ to find the variance.

⟹ $n - 1 = 7$

Variance $= s^2 = \dfrac{\Sigma(x - \bar{x})^2}{n - 1} = \dfrac{8}{7} \approx 1.14$

(c) Use a calculator to find the square root of the variance. Is this the standard deviation?

⟹ $\sqrt{\text{variance}} = \sqrt{s^2} = \sqrt{1.14} \approx 1.07$ in. The square root of the variance is the standard deviation. (*Note:* We say $\sqrt{1.14} \approx 1.07$. The symbol $\approx$ means "is approximately equal to." We use $\approx$ because 1.07 is not exactly equal to $\sqrt{1.14}$.)

Let's summarize and compare the results of Example 4 and Guided Exercise 5. The greenhouse found the following blossom diameters for hybrid A and hybrid B:

Hybrid A: mean, 6.0 in.; standard deviation, 3.07 in.

Hybrid B: mean, 6.0 in.; standard deviation, 1.07 in.

In both cases, the means are the same: 6 in. But the first hybrid has a larger standard deviation. This means that the blossoms of hybrid A are less consistent than those of hybrid B. If you want a rosebush that occasionally has 10-in. blooms and 2-in. blooms, use the first hybrid. But if you want a bush that consistently produces roses close to 6 in. across, use hybrid B.

Computation formula for *s*

There is another formula for the standard deviation that gives the same results as those of Formula (2). It is easier to use with a calculator, since there are fewer subtractions involved.

The computation formula depends on the fact that

$$\Sigma(x - \bar{x})^2 = \Sigma x^2 - \frac{(\Sigma x)^2}{n}$$

which can be proved with the aid of some algebra. The expression $\Sigma(x - \bar{x})^2$ is a sum of squares. Using the notation SS_x to indicate this sum of squares, we get the relation

$$SS_x = \Sigma(x - \bar{x})^2 = \Sigma x^2 - \frac{(\Sigma x)^2}{n}$$

Then the computation formula for the standard deviation s is

Computation Formula for the Sample Standard Deviation s

$$s = \sqrt{\frac{SS_x}{n-1}}$$

(4)

where $SS_x = \Sigma x^2 - \frac{(\Sigma x)^2}{n}$

To compute Σx^2, we *first square* all the x values and then take the sum. To compute $(\Sigma x)^2$, we *first sum* the x values and then square the total.

The next example shows you how to use the computation formula. The expression SS_x will be used in Chapter 3 on regression and correlation.

EXAMPLE 5 ▶

Table 2-9 Number of Days Books Are Overdue

x	x^2
5	25
5	25
6	36
6	36
6	36
7	49
7	49
8	64
9	81
10	100
$\Sigma x = 69$	$\Sigma x^2 = 501$

Rockwood Library was having difficulty because some books were being kept out long after the due dates. The original late fine was 25¢ per day. The mean overdue time was found to be 10.8 days, with a standard deviation of 5.02 days. The librarian decided to change the late-fine rate to $1.00 per day. Table 2-9 contains data from a random sample of overdue books under the new fine system. Compute the mean and standard deviation using the computation formula. Have overdue times been reduced?

SOLUTION: To use the computation formula, we need the sum of x values and the sum of x^2 values. Table 2-9 arranges our work for us.

The mean is

$$\bar{x} = \Sigma x/n = 69/10 = 6.9$$

To use the computation formula for the standard deviation, we first evaluate the sum of squares SS_x.

$$SS_x = \Sigma x^2 - \frac{(\Sigma x)^2}{n} = 501 - \frac{69^2}{10} = 24.9$$

Next, we compute s by dividing SS_x by $n - 1$ and taking the square root of the result.

$$s = \sqrt{\frac{SS_x}{n-1}} = \sqrt{\frac{24.9}{9}} \approx 1.66$$

Notice that the new fine system has reduced the average overdue time and, moreover, that the smaller standard deviation indicates that the distribution is more closely clustered about the mean.

Rounding Note Rounding errors cannot be completely eliminated, even if a computer or calculator does all the computations. However, software and calculator routines are designed to minimize the error. If the mean is rounded, the value of the standard deviation will change slightly depending on how much the mean is rounded. If you do your calculations "by hand" or reenter intermediate values into a calculator, try to carry one or two more digits than occur in the original data. If your resulting answers vary slightly from those in this text, do not be overly concerned. The text answers are computer- or calculator-generated.

Calculator Note The computation formula for standard deviation (4) using Σx and Σx^2 is fairly easy to implement on any calculator. You compute SS_x, divide by $n - 1$, and then take the square root of the result. If your calculator has a statistics mode, you set the mode and enter the data. Pressing a key usually labeled s_x or σ_{n-1} provides the sample standard deviation.

EXAMPLE 6 ▷ Let's return to Example 3. In Example 3 we were trying to decide which compact disc club to join, Selecta Disc or Discount Disks. We noted that the mean and range for the number of monthly disc selections were the same for both clubs. Let's see how the sample standard deviations compare. The data showing the numbers of selections offered each month by each club are repeated.

Selecta Disc

100	30	30	30	21	23
21	21	24	21	30	21

Discount Disks

80	1	70	2	70	1
1	70	3	70	2	2

SOLUTION: Using a calculator that has a statistics mode, set the calculator in that mode. Then enter the 12 data values from Selecta Disc. Press the appropriate keys to find the sample standard deviation s. Clear the calculator and then enter the data from Discount Disks. Press appropriate keys to find s. Figure 2-2 on the next page shows the TI-83 displays for the two data sets.

FIGURE 2-2

TI-83 Displays for
(a) Selecta Disc and
(b) Discount Disks

(a)
```
1-Var Stats
 x̄=31
 Σx=372
 Σx²=16910
 Sx=22.1128877
 σx=21.16994725
↓n=12
█
```

(b)
```
1-Var Stats
 x̄=31
 Σx=372
 Σx²=26024
 Sx=36.29675668
 σx=34.75149877
↓n=12
█
```

We see from that displays that Selecta Disc has standard deviation $s = 22.11$, whereas Discount Disks has standard deviation $s = 36.30$. The larger standard deviation of Discount Disks reflects the greater variability of the data around the mean $\bar{x} = 31$. ●

Population mean and
standard deviation

In almost all applications of statistics we work with a random sample of data rather than the entire population of *all* possible data values. However, if we do in fact have data for the entire population, we can compute the *population mean μ* (lowercase Greek letter mu, pronounced "mew") *and population standard deviation σ* (lowercase Greek letter sigma) using the following formulas:

$$\mu = \frac{\Sigma x}{N} \quad \text{population mean}$$

$$\sigma = \sqrt{\frac{\Sigma (x - \mu)^2}{N}} \quad \text{population standard deviation}$$

where N is the number of data values in the population, and x represents the individual data values of the population. We note that the formula for μ is the same as the formula for $\bar{x}$ (the sample mean) and the formula for σ is the same as the formula for s (the sample standard deviation), except that N is used instead of $n - 1$ and μ is used instead of $\bar{x}$ in the formula for σ.

In the formulas for s and σ we use $n - 1$ to compute s and N to compute σ. Why? The reason is that N (capital letter) represents the population size, while n (lowercase letter) represents the sample size. Since a random sample usually will not contain extreme data values (large or small), we divide by $n - 1$ in the formula for s to make s a little larger than it would have been had we divided by n. Courses in advanced theoretical statistics show that this procedure gives us the best possible estimate for the standard deviation σ. In fact, s is called the *unbiased estimate* for σ. If we have the population of all data values, then extreme data values are, of course, present, so we divide by N instead of $N - 1$.

We've seen that the standard deviation (sample or population) is a measure of data spread. We will use the standard deviation extensively in later chapters. In Chapter 6 we will use it to study standard z values and areas under normal curves. In Chapters 8 and 9 we will use it to study the inferential statistics topics of estimation and testing. The standard deviation will appear again in our study of regression and correlation.

For now, though, we will show you two immediate applications of the standard deviation. The first is the coefficient of variation, and the second is Chebyshev's theorem.

Coefficient of Variation

A disadvantage of the standard deviation as a comparative measure of variation is that it depends on the units of measurement. This means that it is difficult to use the standard deviation to compare measurements from different populations. For this reason, statisticians have defined the *coefficient of variation,* which expresses the standard deviation as a percentage of the sample or population mean.

> If $\bar{x}$ and s represent the sample mean and sample standard deviation, then the *coefficient of variation* CV is defined to be
>
> $$CV = \frac{s}{\bar{x}} \cdot 100$$
>
> If μ and σ represent the population mean and standard deviation, then the coefficient of variation CV is defined to be
>
> $$CV = \frac{\sigma}{\mu} \cdot 100$$

Notice that the numerator and denominator in the definition of CV have the same units, so CV itself has no units of measurement. This gives us the advantage of being able to compare directly the variabilities of two different populations using the coefficient of variation.

In the next example and guided exercise we will compute the CV values of a population and of a sample and then compare the results.

EXAMPLE 7 ➤ The Trading Post on Grand Mesa is a small, family-run store in a remote part of Colorado. The Grand Mesa region contains many good fishing lakes, so the Trading Post sells spinners (a type of fishing lure). The store has a very limited selection of spinners, however. In fact, the Trading Post has only eight different types of spinners for sale. The prices (in dollars) are

2.10   1.95   2.60   2.00   1.85   2.25   2.15   2.25

Since the Trading Post has only eight different kinds of spinners for sale, we consider the eight data values to be the *population.*

(a) Use a calculator with appropriate statistics keys to verify that for the Trading Post data, $\mu = \$2.14$ and $\sigma = \$0.22$.

SOLUTION: Since the computation formulas for $\bar{x}$ and μ are identical, most calculators provide the value of $\bar{x}$ only. Use the output of this key for μ. The computation formulas for the sample standard deviation s and the population standard deviation σ are slightly different. Be sure that you use the key for σ (sometimes designated as σ_n or σ_x).

(b) Compute the CV of the prices of spinners at the Trading Post and comment on the meaning of the result.

SOLUTION:

$$CV = \frac{\sigma}{\mu} \cdot 100 = \frac{0.22}{2.14} \cdot 100 = 10.28\%$$

The coefficient of variation can be thought of as a measure of the spread of the data relative to the average of the data. Since the Trading Post is very small, it carries a small selection of spinners that are all priced similarly. The CV tells us that the standard deviation of the spinner prices is only 10.28% of the mean.

GUIDED EXERCISE   6

Cabela's in Sidney, Nebraska, is a very large outfitter that carries a broad selection of fishing tackle. It markets its products nationwide through a catalogue service. A random sample of 10 spinners from Cabela's extensive spring catalogue gave the following prices (in dollars):

1.69	1.49	3.09	1.79	1.39
2.89	1.49	1.39	1.49	1.99

(a) Use a calculator with sample mean and sample standard deviation keys to compute $\bar{x}$ and s.

⇨ $\bar{x} = \$1.87$ and $s = \$0.62$

(b) Compute the CV for the spinner prices at Cabela's.

⇨ $CV = \frac{s}{\bar{x}} \cdot 100 = \frac{0.62}{1.87} \cdot 100 = 33.16\%$

(c) Compare the mean, standard deviation, and CV for the spinner prices at the Grand Mesa Trading Post (Example 7) and Cabela's. Comment on the differences.

⇨   The CV's for Cabela's and the Trading Post are pure numbers (no units), so a direct comparison is possible. The CV for Cabela's is more than three times the CV for the Trading Post. Why? First, because of its remote location, the Trading Post tends to have somewhat higher prices (larger μ). Second, the Trading Post is very small, so it has a rather limited selection of spinners with a smaller variation in price. For Cabela's, however, the average price $\bar{x}$ is lower and the variety larger (larger s). It makes sense that the CV for Cabela's is larger than the CV for the Trading Post.

Chebyshev's Theorem

From our earlier discussion of standard deviation, recall that the spread or dispersion of a set of data about the mean will be small if the standard deviation is small, and it will be large if the standard deviation is large. If we are dealing with a symmetrical bell-shaped distribution, then we can make very definite statements about the proportion of the data that must lie within a certain number of standard deviations on either side of the mean. This will be discussed in detail in Chapter 6 when we talk about normal distributions.

However, the concept of data spread about the mean can be expressed quite generally for *all data distributions* (skewed, symmetric, or other shape) by using the remarkable theorem of Chebyshev.

Chebyshev's Theorem

For *any* set of data (either population or sample) and for any constant k greater than 1, the proportion of the data that must lie within k standard deviations on either side of the mean is *at least*

$$1 - \frac{1}{k^2}$$

Results of Chebyshev's Theorem

For *any* set of data:

- at *least* 75% of the data fall in the interval from $\mu - 2\sigma$ to $\mu + 2\sigma$
- at *least* 88.9% of the data fall in the interval from $\mu - 3\sigma$ to $\mu + 3\sigma$
- at *least* 93.8% of the data fall in the interval from $\mu - 4\sigma$ to $\mu + 4\sigma$

The results of Chebyshev's theorem can be derived by using the theorem and a little arithmetic. For instance, if we create an interval $k = 2$ standard deviations on either side of the mean, Chebyshev's theorem tells us that

$$1 - \frac{1}{2^2} = 1 - \frac{1}{4} = \frac{3}{4} \text{ or } 75\%$$

is the minimum percentage of data in the $\mu - 2\sigma$ to $\mu + 2\sigma$ interval.

Notice that Chebyshev's theorem refers to the *minimum* percentage of data that must fall within the specified number of standard deviations of the mean. If the distribution is mound-shaped, an even *greater* percentage of data will fall in the specified intervals (see the Empirical Rule in Section 6.1).

EXAMPLE 8 ➤   Students Who Care is a student volunteer program in which college students donate work time in community centers for homeless people. Professor Gill is the faculty sponsor for this student volunteer program. For several years Dr. Gill has kept a careful record of x = total number of work hours volunteered by a student in the program each semester. For students in the program, the mean number of hours is $\bar{x} = 29.1$ hours each semester, with a standard deviation $s = 1.7$ hours each semester. Find an interval A to B for the number of hours volunteered in which at least 75% of the students in this program would fit.

SOLUTION: According to results of Chebyshev's theorem, at least 75% of the data must fall within two standard deviations of the mean. Because the mean $\bar{x} = 29.1$ and the standard deviation $s = 1.7$, the interval is

$$\bar{x} - 2s \text{ to } \bar{x} + 2s$$

$$29.1 - 2(1.7) \text{ to } 29.1 + 2(1.7)$$

$$25.7 \text{ to } 32.5$$

At least 75% of the students would fit in the group that volunteered from 25.7 to 32.5 hours each semester.

GUIDED EXERCISE   7

The *East Coast Independent News* periodically runs ads in its own classified section offering a month's free subscription to those who respond. In this way, management can get a sense about the number of subscribers who read the classified section each day. Over a period of 2 years, careful records have been kept. The mean number of responses per ad is $\bar{x} = 525$ with standard deviation $s = 30$.

Determine the interval about the mean in which at least 88.9% of the data fall.

 By Chebyshev's theorem, at least 88.9% of the data fall in the interval

$$\bar{x} - 3s \text{ to } \bar{x} + 3s$$

Because $\bar{x} = 525$ and $s = 30$, the interval is

$$525 - 3(30) \text{ to } 525 + 3(30)$$

or from 435 to 615 responses per ad.

VIEWPOINT

Earth in the Balance

Vice President Al Gore's book *Earth in the Balance* has received recognition and praise from the general news media as well as from outstanding scientists of our time. In his book, Gore discusses droughts in Africa that have killed tens of thousands and dramatic increases in rain and floods in mirroring northern latitudes that include Europe. Is all this only a normal variation? Scientific measurements of atmospheric variances s^2 seem to indicate otherwise.

SECTION 2.2 PROBLEMS

Ecology: Deer Population in Mesa Verde

1. By sampling different landscapes in Mesa Verde National Park over a 2-year period, the number of deer per square kilometer was determined (*The Mule Deer of Mesa Verde National Park*, by G. W. Mieran and J. L. Schmidt, published by Mesa Verde Museum Association). The results (in deer per square kilometer) were

30	20	5	29	58	7
20	18	4	29	22	9

(a) Compute the range, sample mean, sample variance, and sample standard deviation.
(b) Compute the coefficient of variation. Is the coefficient of variation a fairly large number? Would you say there was a considerable variation in the distribution of deer from one section of the park to another? Explain.

Health Care: Per Capita Spending

2. *State Health Care Data: Utilization, Spending, and Characteristics*, prepared by the American Medical Association Center for Health Policy Research, gives annual per capita spending on physician services by state. Annual per capita spending on physician services for a random sample of six states (in dollars) is

744   434   567   485   634   729

(a) Compute the range and mean. Compute the variance and standard deviation.
(b) Compute the coefficient of variation. Write a brief explanation of the meaning of this number.

Business: TV List Prices

3. *Consumer Reports* recently rated 27- and 32-inch color television sets. For the top five rated 27-inch sets, the list prices (in dollars) were

480   400   390   550   550

For the top-rated 32-inch sets, the list prices (in dollars) were

580   740   630   660   600

(a) Compute the range, sample mean, and sample standard deviation of prices for the 27-inch sets.
(b) Compute the range, sample mean, and sample standard deviation of prices for the 32-inch sets.
(c) Compare the answers for parts a and b. Comment on the differences. Next, compute the coefficient of variation for the price distribution of the 27-inch sets and then for the price distribution of the 32-inch sets. Comment on the relative price variation.

Business: Productivity

4. Jobs and productivity! How productive are American workers? One way to answer this question is to examine profit per employee. The results may change, however, from one industry group to another. This problem is based on information taken from *Forbes Top Companies* (edited by J. T. Davis, John Wiley and Sons).

(a) A random sample of seven automotive parts companies gave the following information about *profit* per employee (in thousands of dollars per employee):

| 14.1 | 12.4 | 7.7 | 6.9 | 9.0 | 6.8 | 6.8 |

Compute the range, mean, sample variance, sample standard deviation, and coefficient of variation.

(b) For a random sample of seven computer companies, the values of profit per employee (in the same units) were

| 29.0 | 24.5 | 31.0 | 29.8 | 21.8 | 27.7 | 19.1 |

Compute the range, mean, sample variance, sample standard deviation, and coefficient of variation.

(c) Examine your answers for parts a and b. Although the standard deviation for part a is the smaller of the two, the mean for part b is the larger of the two. The CV for part b is almost half that of part a. In this case, why would a small CV and high mean indicate more reliable productivity from the point of view of profit per employee? Explain.

Airlines: Cost per Mile

5. What does it cost to fly? One way to answer this question is to compute the cents per mile on your next flight. The following data are based on information taken from *Consumer Reports* (Vol. 62, No. 7).

(a) Average cost of flying (cents per mile) for a random sample of six airlines for short flights (less than 300 miles):

| 47 | 17 | 23 | 53 | 20 | 41 |

Compute the range, mean, sample variance, sample standard deviation, and coefficient of variation.

(b) Average cost of flying (cents per mile) for a random sample of six airlines for long flights (1500 miles or more):

| 9 | 11 | 12 | 7 | 8 | 10 |

Compute the range, mean, sample variance, sample standard deviation, and coefficient of variation.

(c) Does it pay to compare fares on different airlines? Look at the information from parts a and b. Compare the means and standard deviations of the data in parts a and b. Which flight lengths have higher average costs? Greater variability? Compare the CVs. Which flight lengths have more consistent costs relative to the average? How do you think this information might be reflected in airfares from different airlines for longer flights? For shorter flights? Are there other factors that might influence airfares between two cities, such as competition?

Driving: Gas Mileage

6. Ralph and Gloria did a 4-H project to demonstrate ways to get better gasoline mileage. They kept the car windows rolled up to prevent air drag, used only moderate acceleration from a standstill, and kept their speed down in general. Ralph recorded the mean miles per gallon for 5 days selected at random from the period in which he drove the car. Gloria did the same. The results are shown below.

Ralph	22.3	21.2	20.8	19.8	23.8
Gloria	25.2	19.1	18.0	24.4	20.3

(a) Find the ranges for Ralph and for Gloria.

(b) Find the mean, sample standard deviation, and coefficient of variation for each.

(c) Who consistently seems to have gotten better mileage: Ralph or Gloria? Who had the smaller coefficient of variation?

Astronomy: Sun Spot Cycles

7. The National Aeronautics and Space Administration (NASA) has studied data on sun spot cycles collected for the years 1745 to the present. During this time, the mean length of a cycle (maximum to maximum) was 11.01 years, with a standard deviation of 2.17 years.

(a) Use Chebyshev's theorem to find an interval centered about the mean for the cycle length in which you would expect at least 75% of the cycles to fall.

(b) Use Chebyshev's theorem to find an interval centered about the mean for the cycle length in which you would expect at least 93.8% of the cycles to fall.

Climate: Tornados

8. The U.S. Weather Bureau has provided the following information about the total annual numbers of reported tornados in the United States for the years 1956 to 1975:

504	856	564	604	616	697	657	464	704	906
585	926	660	608	653	888	741	1102	947	918

(a) Use a calculator with mean and standard deviation keys to verify that the mean numbers of tornados per year is 730, with a sample standard deviation of 172 tornados.

(b) Use Chebyshev's theorem to find an interval centered about the mean for the annual number of tornados in which you would expect at least 75% of the years to fall.

(c) Use Chebyshev's theorem to find an interval centered about the mean in which you would expect at least 88.9% of the years to fall.

History: Billy the Kid

9. In 1881, Billy the Kid killed two deputies and escaped from a jail cell in the Lincoln County Courthouse, New Mexico Territory. Famous frontier personalities such as Kit Carson, Jesse Chisum, and Sheriff Pat Garrett (who eventually shot Billy the Kid) also were involved in the notorious Lincoln County Cattle Baron Wars. Before the 1985 renovation of the now famous courthouse, the Museum of New Mexico commissioned anthropologist/historian Yvonne Oakes to do a thorough analysis of the area. The distribution of artifacts (of the 1880s period) for seven excavation sites were

851	596	444	956	576	219	326

(Source: *Museum of New Mexico: Laboratory of Anthropology Notes No. 357.*)

(a) Compute the range and mean.

(b) Compute the sample variance and standard deviation.

(c) Compute the coefficient of variation. Write a brief explanation of the meaning of this number in the context of this problem.

(d) Use Chebyshev's theorem to find an interval centered about the mean in which at least 75% of the artifact counts for all such excavation sites would fall.

Wildlife Management: Game Birds

10. Hatching success of game birds is a topic discussed in the book *Wildlife Management Techniques,* edited by R. H. Giles (The Wildlife Society Press, Washington,

D.C.). What percentage of Canada goose nests are successful (at least one gosling survives)? Studies in regions of Montana, Illinois, Wyoming, Utah, and California recorded percentages of successful nests. In the same studies, hatching success rates for mallard duck nests were also recorded. The following Minitab display (menu selection: ➤Stat ➤Basic Statistics ➤Display Descriptive Statistics) gives summary information for goose nests and mallard duck nests.

Minitab Display:
Hatching success rates for goose nests and mallard duck nests

VARIABLE	N	MEAN	MEDIAN	TRMEAN	STDEV	SE MEAN
Goose	10	49.08	55.00	50.25	18.87	5.97
Duck	10	55.67	53.60	56.71	23.17	7.33

VARIABLE	MINIMUM	MAXIMUM	Q_1	Q_3
Goose	17.80	71.00	27.35	62.15
Duck	12.70	90.30	39.47	73.80

The sample size for each variable is given under N, the mean is given under Mean, and the sample standard deviation is given under StDev. The header TrMean gives a 5% trimmed mean. We will make extensive use of the Median, and of Q_1 (first quartile) and Q_3 (third quartile) in the next section. In Chapter 7 we will use SE Mean (the standard error of the mean), which is simply the standard deviation divided by the square root of the sample size.

(a) Compare the mean hatching success rates for Canada goose and mallard duck nests. Compare the median hatching success rates. Briefly comment on these comparisons.
(b) Compute the ranges of the hatching success rates for Canada goose and mallard duck nests. Compare these ranges.
(c) Compare the standard deviations of the success rates for the two types of nests. What does this tell you about the relative data spread?
(d) Compute the coefficient of variation of the success rate for the two types of nests. Does the hatching success rate of one species seem more volatile than the other? Explain your answer.
(e) Based on the information in the display of the hatching success rates for the two species, write a brief paragraph in which you compare the hatching success rates.

Grouped Data

When data are grouped, such as in a frequency table or histogram, we can estimate the mean and standard deviation using the following formulas. Notice that all data values in a given class are treated as though each of them equals the midpoint x of the class.

Sample mean for a frequency distribution:

$$\bar{x} = \frac{\Sigma xf}{n} \tag{5}$$

where x is the midpoint of a class,
  f is the number of entries in that class,
  n is the total number of entries in the distribution, and
  the summation Σ is over all classes in the distribution.   *(continued)*

Sample standard deviation for a frequency distribution:

$$s = \sqrt{\frac{\Sigma(x - \bar{x})^2 f}{n - 1}} \qquad (6)$$

where x is the midpoint of a class,
f is the number of entries in that class,
n is the total number of entries in the distribution, and
the summation Σ is over all classes in the distribution.

Use these formulas to solve Problems 11–15. *Hint:* Make a computation table using the column heads.

Midpoint x	Frequency f	xf	$(x - \bar{x})$	$(x - \bar{x})^2$	$(x - \bar{x})^2 f$

and sum the entries in the third column and the last column.
Note: On the TI-83 calculator, enter the midpoints in column L_1 and the frequencies in column L_2. Then use 1-VarStats L_1, L_2.

Vital Statistics:
Life Expectancy

11. In the United States, the life expectancy of a male child born between 1979 and 1981 varies by state (including the District of Columbia). Information given in the *Statistical Abstract of the United States*, 109th edition, shows that the life expectancies range from 64.55 to 74.08 years. In the following table, life expectancies are grouped by years.

Life expectancy for men, x (in years)	64–67	68–71	72–75
Number of states, f	1	38	12

(a) Estimate the mean life expectancy in years for all the states.
(b) Estimate the sample standard deviation of life expectancy in years.

Vital Statistics:
Life Expectancy

12. The life expectancy of a female child born between 1979 and 1981 also varies by state (including the District of Columbia), from 73.70 to 80.33 years. In the following table, life expectancies are grouped by years.

Life expectancy for women, x (in years)	73–75	76–78	79–81
Number of states, f	1	36	14

(a) Estimate the mean life expectancy in years for all the states.
(b) Estimate the sample standard deviation of life expectancy in years.

Vital Statistics:
Life Expectancy

13. For both men and women, the lowest life expectancy occurs in the District of Columbia. That is the only entry in each of the first classes in Problems 11 and 12. Remove those entries and recalculate estimates for the mean and sample standard deviation of life expectancies
(a) for men.
(b) for women.

Business:
Catalogue Shoppers

14. Based on data from *USA Today*, the ages of a random sample of 300 adults who shop by catalogue are

Age	18–24	25–34	35–44	45–54	55–64	65–74*
Number	78	75	48	33	33	33

*Ages over 74 are counted as 74.

Estimate the mean age of the adults who shop by catalogue. Estimate the standard deviation of the ages of the shoppers and the coefficient of variation.

Medical: Sleep

15. Alexander Borbely is a professor at the University of Zurich Medical School, where he is director of the sleep laboratory. The histogram in the following figure is based on information in his book *Secrets of Sleep*. The histogram displays results from a random sample of 200 subjects. Estimate the mean hours of sleep, standard deviation of hours of sleep, and coefficient of variation.

Hours of Sleep
Each Day
(24-Hour Period)

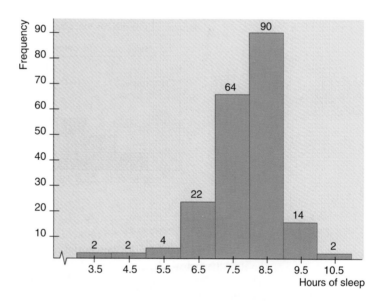

Section 2.3 — Percentiles and Box-and-Whisker Plots

Percentiles

We've seen measures of central tendency and spread for a set of data. The arithmetic mean $\bar{x}$ and the standard deviation s will be very useful in later work. However, because they each utilize every data value, they can be heavily influenced by one or two extreme data values. In cases where our data distributions are heavily skewed or even bimodal, we often get a better summary of the distribution by utilizing relative positions of data rather than exact values.

Recall that the median is an average computed by using relative positions of the data. If we are told that 81 is the median score on a biology test, we know that after the data have been ordered, 50% of the data fall at or below the median value of 81. The median is an example of a percentile; in fact, it is the 50th percentile. The general definition of the *P*th percentile follows.

For whole numbers *P* (where $1 \leq P \leq 99$), the *P*th *percentile* of a distribution is a value such that *P*% of the data fall at or below it and $(100 - P)$% of the data fall at or above it.

FIGURE 2-3

A Histogram with the 60th Percentile Shown

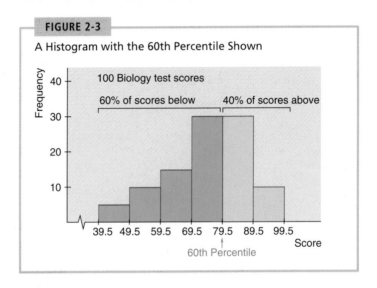

In Figure 2-3, we see the 60th percentile marked on a histogram. We see that 60% of the data lie below the mark and 40% lie above it.

There are 99 percentiles, and in an ideal situation the 99 percentiles divide the data set into 100 equal parts. (See Figure 2-4.)

However, if the number of data elements is not exactly divisible by 100, the percentiles will not divide the data into equal parts.

FIGURE 2-4

Percentiles

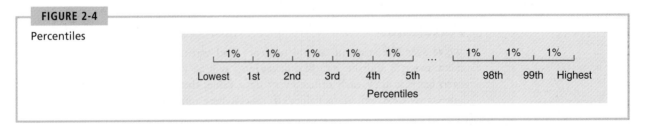

Suppose you took the English achievement test to obtain college credit in freshman English by examination.

(a) If your score was at the 89th percentile, what percentage of scores were at or below yours?

⇨ The 89th percentile means that 89% of the scores were at or below yours.

(b) If the scores ranged from 1 to 100 and your raw score was 95, does this necessarily mean that your score was at the 95th percentile?

⇨ No, the percentile gives an indication of the relative positions of the scores. The determination of your percentile has to do with the number of scores at or below yours. If everyone did very well and only 80% of the scores fell at or below yours, you would be at the 80th percentile even though you got 95 out of 100 points on the exam.

There are several widely used conventions for finding percentiles. They lead to slightly different values for different situations, but these values are close together. For all conventions, the data are first *ranked* or ordered from smallest to largest. A natural way to find the Pth percentile is then to find a value at or below which P% of the data fall. This will not always be possible, so we take the nearest value satisfying the criterion. It is at this point that there are a variety of processes to determine the exact value of the percentile.

We will not be very concerned about exact procedures for evaluating percentiles in general. However, *quartiles* are special percentiles used so frequently that we want to adopt a specific procedure for their computation.

Quartiles

Quartiles are those percentiles that divide the data into fourths. The *first quartile* Q_1 is the 25th percentile, the *second quartile* Q_2 is the median, and the *third quartile* Q_3 is the 75th percentile. (See Figure 2-5.)

FIGURE 2-5

Quartiles

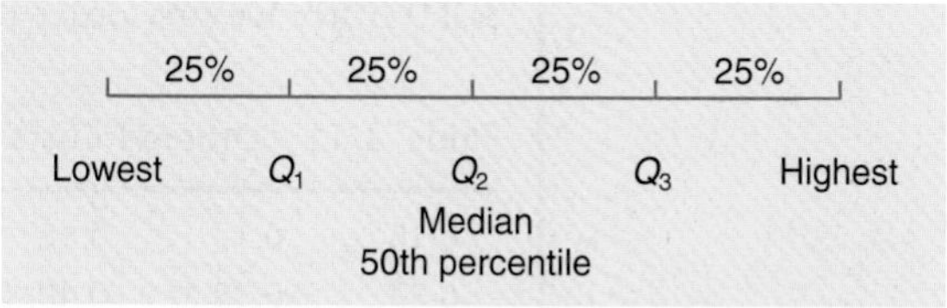

Again, there are many conventions used for computing quartiles, but the following one utilizes the median and is widely adopted.

> **Procedure for Computing Quartiles**
> 1. Order the data from smallest to largest.
> 2. Find the median. This is the second quartile.
> 3. The first quartile Q_1 is then the median of the lower half of the data—that is, it is the median of the data falling *below* the Q_2 position (and not including Q_2).
> 4. The third quartile Q_3 is the median of the upper half of the data—that is, it is the median of the data falling *above* the Q_2 position (and not including Q_2).

In short, all we do to find the quartiles is to find three medians.

Interquartile range

The median, or second quartile, is a popular measure of the center utilizing relative position. A useful measure of data spread utilizing relative position is the *interquartile range (IQR)*. It is simply the difference between the third and first quartiles.

$$\text{Interquartile range} = Q_3 - Q_1$$

The interquartile range tells us the spread of the middle half of the data. Now let's look at an example to see how to compute all these quantities.

EXAMPLE 9 ➤ In a hurry? On the run? Hungry as well? How about an ice cream bar as a snack? Ice cream bars are popular among all age groups. *Consumer Reports* did a study of ice cream bars. Twenty-seven bars with taste ratings of at least "fair" were listed, and cost per bar was included in the report. Just how much will an ice cream bar cost? The data, expressed in dollars, appear in Table 2-10. As you can see, the costs vary quite a bit, partly because the bars are not of uniform size.

Table 2-10   Costs of Ice Cream Bars (in dollars)

0.99	1.07	1.00	0.50	0.37	1.03	1.07	1.07
0.97	0.63	0.33	0.50	0.97	1.08	0.47	0.84
1.23	0.25	0.50	0.40	0.33	0.35	0.17	0.38
0.20	0.18	0.16					

(a) Find the quartiles.

SOLUTION: We first order the data from smallest to largest, as shown in Table 2-11.

Table 2-11   Ordered Costs of Ice Cream Bars (in dollars)

0.16	0.17	0.18	0.20	0.25	0.33	0.33	0.35
0.37	0.38	0.40	0.47	0.50	0.50	0.50	0.63
0.84	0.97	0.97	0.99	1.00	1.03	1.07	1.07
1.07	1.08	1.23					

Next, we find the median. Since the number of data values is 27, there is an odd number of values, and the median is simply the center or 14th value. The value is shown boxed in Table 2-11.

$$\text{Median} = Q_2 = 0.50$$

There are 13 values below the median position, and Q_1 is the median of these values. It is the middle or 7th value and is boxed in Table 2-11.

$$\text{First quartile} = Q_1 = 0.33$$

There are also 13 values above the median position. The median of these is the 7th value from the right end. This value is also boxed in Table 2-11.

$$\text{Third quartile} = Q_3 = 1.00$$

(b) Find the interquartile range.

SOLUTION:

$$IQR = Q_3 - Q_1$$
$$= 1.00 - 0.33$$
$$= 0.67$$

This means that the middle half of the data has a cost spread of 67¢.

GUIDED EXERCISE   **9**

Many people consider the number of calories in an ice cream bar as important as, if not more important than, the cost. The *Consumer Reports* article cited in Example 9 also included the calorie counts of the rated ice cream bars (Table 2-12). There were 22 vanilla-flavored bars rated. Again, the bars varied in size, and some of the smaller bars had fewer calories.

Table 2-12   Calories in Vanilla-Flavored Ice Cream Bars

342	377	319	353	295
234	294	286	377	182
310	439	111	201	182
197	209	147	190	151
131	151			

(a) Our first step is to order the data. Do so.

Table 2-13   Ordered Data

111	131	147	151	151	182
182	190	197	201	209	234
286	294	295	310	319	342
353	377	377	439		

Exercise continues

(b) There are 22 data values. Find the median.   ⇨   Average the 11th and 12th data values boxed together in Table 2-13.

$$\text{Median} = \frac{209 + 234}{2} = 221.5$$

(c) How many values are below the median position? Find Q_1.   ⇨   Since the median lies halfway between the 11th and 12th values, there are 11 values below the median position. Q_1 is the median of these values.

$Q_1 = 182$

(d) The same number of data lie above the median as below it. Use this fact to find Q_3.   ⇨ Q_3 is the median of the upper half of the data. There are 11 values in the upper portion.

$Q_3 = 319$

(e) Find the interquartile range and comment on its meaning.   ⇨ $IQR = Q_3 - Q_1$

$= 319 - 182$

$= 137$

The middle portion of the data has a spread of 137 calories.

Box-and-Whisker Plots

The quartiles together with the low and high data values give us a very useful *five-number summary* of the data and their spread.

> **Five-Number Summary**
> Lowest value, Q_1, median, Q_3, highest value

We will use these five numbers to create a graphic sketch of the data called a *box-and-whisker plot* (see Figure 2-6). Box-and-whisker plots provide another useful technique for describing data from exploratory data analysis (EDA).

Procedure for Making a Box-and-Whisker Plot

1. Draw a vertical scale to include the lowest and highest data values.

2. To the right of the scale, draw a box from Q_1 to Q_3.

3. Include a solid line through the box at the median level.

4. Draw solid lines, called *whiskers*, from Q_1 to the lowest value and from Q_3 to the highest value.

FIGURE 2-6   Box-and-Whisker Plot

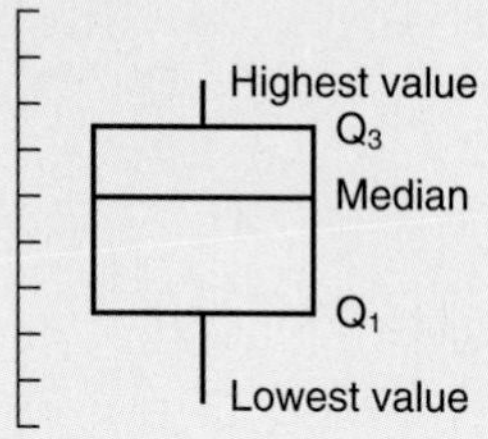

The next example demonstrates the process of making a box-and-whisker plot.

EXAMPLE 10 ❯

Renata College is a small college offering baccalaureate degrees in liberal arts and business. The development office (fund raising) did a salary survey of alumni who graduated 2 years ago and have jobs. Sixteen alumni responded to the survey the first week. Table 2-14 shows their annual salaries (in thousands of dollars). Make a box-and-whisker plot for these data.

Table 2-14   Annual Salaries (in thousands of dollars)

38.5	39.5	32.0	30.5	36.8	29.2	23.7	34.1
28.3	27.9	33.6	37.0	43.5	34.6	33.8	36.1

SOLUTION: We order the data (Table 2-15).

Table 2-15   Annual Salaries Ordered

23.7	27.9	28.3	29.2	30.5	32.0	33.6	33.8
34.1	34.6	36.1	36.8	37.0	38.5	39.5	43.5

Since there are 16 data values, we see that the median is the mean of the 8th and 9th values.

Median = (33.8 + 34.1)/2 = 33.95

Because there are eight values below the median position, Q_1 is the mean of the 4th and 5th values.

Q_1 = (29.2 + 30.5)/2 = 29.85

Q_3 is the mean of the 4th and 5th values from the *high* end.

Q_3 = (36.8 + 37)/2 = 36.90

The five-number summary is then

Low = 23.7 Q_1 = 29.85   Median = 33.95 Q_3 = 36.90   High = 43.5

Figure 2-7 shows the box-and-whisker plot.

FIGURE 2-7

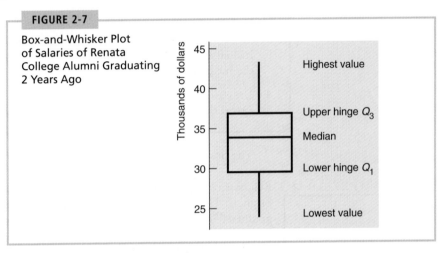

Box-and-Whisker Plot of Salaries of Renata College Alumni Graduating 2 Years Ago

A quick glance at the box-and-whisker plot reveals the following:

(a) The box tells us where the middle half of the data lie, so 50% of the salaries lie in the box part and range from about $30,000 to about $37,000. The length of the box equals the interquartile range.

(b) The median is about $34,000 and is slightly closer to the top of the box. This means that the higher salaries are more concentrated and closer together. The distribution is skewed slightly to the low salaries.

(c) The whiskers are of about the same length, which says that the ranges from Q_1 and from Q_3 to the corresponding extreme values are about the same.

Calculator Note Box-and-whisker plots are so useful that some graphing calculators such as the TI-83 produce them. On the TI-83, the quartiles Q_1 and Q_3 are calculated as we calculate them in this text. Once the box plot is graphed on the TI-83, the trace key gives the values for minX, Q1, Med (for median), Q3, and maxX (see Figure 2-8 on the next page).

COMMENT In exploratory data analysis, *hinges* rather than quartiles are used to create the box. Hinges are computed in a manner similar to the way we compute quartiles. However, in the case of an odd number of data values, include the median itself in both the lower and upper halves of the data (see *Applications, Basics, and Computing of Exploratory Data Analysis,* by Paul Velleman and David Hoaglin, Duxbury Press). This has the effect of shrinking the box and moving the ends of the box slightly toward the median. For an even number of data, the quartiles as we computed them equal the hinges.

FIGURE 2-8

Box Plot Produced on the TI-83 for the Data in Table 2-14, Annual Salaries (Window settings: X from 20 to 45, Y from 0 to 10)

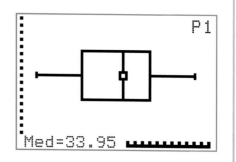

```
1-Var Stats
↑n=16
 minX=23.7
 Q1=29.85
 Med=33.95
 Q3=36.9
 maxX=43.5
■
```

Med=33.95

The Renata College development office also sent a salary survey to alumni who graduated 5 years ago. Again, questions about annual salary were asked. The responses received the first week are summarized in the box-and-whisker plot in Figure 2-9. The plot for the alumni graduating 2 years ago is repeated for comparison.

FIGURE 2-9 Box-and-Whisker Plots for Alumni Salaries (in thousands of dollars)

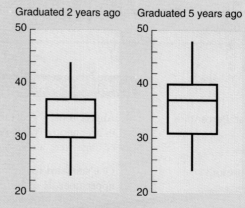

(a) From Figure 2-9, estimate the median and the extreme values of salaries of alumni graduating 5 years ago. What is the location of the middle half of the salaries?

➡ The median seems to be $37,000. The extremes are $24,000 and $48,000. The middle half of the data is enclosed by the box with low side at $31,000 and high side at $40,000.

(b) Compare the two box plots and make comments about the salaries of alumni graduating 2 and 5 years ago.

➡ The salaries of the alumni graduating 5 years ago have a larger spread but begin slightly higher and extend to levels about $5000 above those of alumni graduating 2 years ago. The middle half of the data is also more spread out, with higher boundaries and a higher median.

We have developed the skeletal box-and-whisker plot. Other variations may include *fences*, which are marks placed on either side of the box to represent various portions of the data. Values that lie outside the fences are called *outliers* (see Problem 11 in this section). These values seem to stand by themselves, away from most of the data. They might be exceptional values and deserve closer study. Or they may be the result of data entry error. For a more complete discussion of outliers and variations of box plots, see *Applications, Basics, and Computing of Exploratory Data Analysis,* by Velleman and Hoaglin.

 V I E W P O I N T

Is Shorter Higher?

Can you estimate a person's *height* from the *pitch* of his or her voice? Is a soprano shorter than an alto? Is a bass taller than a tenor? A statistical study of singers in the New York Choral Society provided information. See Web site <http://lib.stat.cmu.edu/DASL/>, and from Data Subjects select music, and then singers. The methods presented in this chapter can be used with new methods we will learn in Chapters 8 and 9 for examining such questions from a statistical point of view.

SECTION 2.3 PROBLEMS

Aptitude Test: Percentiles

1. Angela took a general aptitude test and scored in the 82nd percentile for aptitude in accounting. What percentage of the scores were at or below her score? What percentage were above her score?

College Admissions: Quartiles

2. One standard for admission to Redfield College is that the student must rank in the upper quartile of his or her graduating high school class. What is the minimal percentile rank of a successful applicant?

Education: Percentiles

3. The town of Butler, Nebraska, decided to give a teacher competency exam and defined the passing scores to be those in the 70th percentile or higher. The raw test scores ranged from 0 to 100. Was a raw score of 82 necessarily a passing score? Explain.

Education: Percentile Ranks

4. Clayton and Timothy took different sections of Introduction to Economics. Each section had a different final exam. Timothy scored 83 out of 100 and had a percentile rank in his class of 72. Clayton scored 85 out of 100 but his percentile rank in his class was 70. Who performed better with respect to the rest of the students in the class: Clayton or Timothy? Explain your answer.

Careers: Income

5. The Medical Group Management Association provided *The New York Times* with data on incomes for medical doctors in group practice by specialty. The specialty with the highest annual income is cardiovascular surgeons. The median income is

$574,769, and the 90th percentile is $887,057.
(a) What percentage of the cardiovascular surgeons make more than $574,769?
(b) What percentage make $887,057 or more?
(c) What percentage make between $574,769 and $887,057?

Careers: Income

6. The Medical Group Management Association data show that the median income for anesthesiologists in group practice is $253,511, while the 90th percentile is 378,261.
(a) What percentage of the anesthesiologists make less than $253,511?
(b) What percentage of the anesthesiologists make less than $378,261?
(c) What percentage make between $253,511 and $378,261?

Careers: Salary Increase

7. The following data are percentage increases in annual salary for faculty at the professor rank at Oregon colleges and universities. Source: *Academe: Bulletin of the American Association of University Professors*, (Vol 85, No. 2).

5.8	6.4	6.4	10.2	6.0	6.2	6.0	6.1
6.1	1.3	7.2	6.7	10.1	6.4	4.0	

(b) Compute the five-number summary and the interquartile range.
(c) Make a box-and-whisker plot.

Business: Cell Phones

8. Cellular phones are growing in popularity. How much does it cost to make a call from a cellular phone? *Consumer Reports* (Vol. 62, No. 2) gave the lowest costs of cellular phone service for 20 regions. Among the charges listed were the per-minute charge when a call is made during peak time from the "home" area (that is, the basic service area). These costs (in cents per minute) are

75	49	35	31	35	30	38	47	35	45	39
25	34	38	35	49	40	49	27	48	32	49
48	42	39	26	50	50	50	45	25	25	49
79	43	49	38	25	45	39	25	50	27	45
49	35	49	60	69	65	37	36	35	35	49
39	39	40	85	39	45	49	50	40	39	50

Make a box-and-whisker plot and find the interquartile range.

Business: Cell Phones

9. Does it cost more to use your cellular phone when you are roaming outside your home district (that is, when you are outside your basic service area)? In the *Consumer Reports* article referred to in Problem 8, the per-minute costs for "roaming" calls from a cellular phone also were listed. These costs (in cents per minute) are

35	65	59	50	44	59	59	44	59	59	50
25	59	35	99	75	59	49	27	50	59	49
48	59	50	99	59	50	49	49	25	59	79
65	59	30	38	25	59	50	25	60	59	45
49	59	69	60	69	59	59	99	59	39	99
49	5	59	75	49	59	49	50	59	50	3

Make a box-and-whisker plot and find the interquartile range. If you did Problem 8, compare the "roaming" cost per minute for cellular phone use with the "home" cost.

Business: Auto Insurance

10. *Consumer Reports* rated automobile insurance companies and gave annual premiums for top-rated companies in several states. The figure below shows box plots for annual premiums for urban customers (married couples with one 17-year-old son) in three states. The box plots are all drawn on the same scale on a TI-83 calculator.

Insurance Premiums (Annual, Urban)

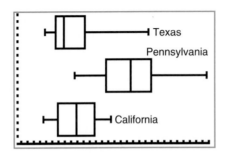

(a) Which state has the lowest premium? The highest?
(b) Which state has the highest median premium?
(c) Which state has the smallest range of premiums? The smallest interquartile range?
(d) The following displays give the five-number summaries generated on the TI-83 for the box plots in the figure above. Match the five-number summaries with the appropriate box plots.

Five-Number Summaries for Insurance Premiums

(a)
```
1-Var Stats
↑n=10
  minX=2382
  Q₁=2758
  Med=2991
  Q₃=3652
  maxX=5715
```

(b)
```
1-Var Stats
↑n=10
  minX=3314
  Q₁=4326
  Med=5116.5
  Q₃=5801
  maxX=7527
```

(c)
```
1-Var Stats
↑n=10
  minX=2323
  Q₁=2801
  Med=3377.5
  Q₃=3966
  maxX=4482
```

General: Outliers

11. Some data sets include values so high or so low that they seem to stand apart from the rest of the data. These data are called *outliers*. Outliers may be from data collection errors, data entry errors, or simply valid but unusual data values. Regardless of the reason, it is important to identify outliers in the data set and examine the outliers carefully to determine if they are in error. One way to detect outliers is to use a box-and-whisker plot. Data values that fall beyond the limits

Lower Limit: $Q_1 - 1.5 \times (IQR)$

Upper Limit: $Q_3 + 1.5 \times (IQR)$

where IQR is the interquartile range, are suspected outliers. In the computer software package Minitab, values beyond these limits are plotted with asterisks (*).

Students from a statistics class were asked to record their heights in inches. The heights (as recorded) were

65	72	68	64	60	55	73	71
52	63	61	74	69	67	74	50
4	75	67	62	66	80	64	65

(a) Make a box-and-whisker plot of the data.
(b) Find the value of the interquartile range (IQR).
(c) Multiply the IQR by 1.5 and find the lower and upper limits.
(d) Are there any data values below the lower limit? above the upper limit? List any suspected outliers. What might be some explanations for the outliers?

Careers: Salary Increases

12. The following Minitab display shows box-and-whisker plots for the percentage increases in annual salaries for the university and college faculty in Tennessee (by rank). Source: *Academe: Bulletin of the American Association of University Professors*, Volume 85, Number 2.

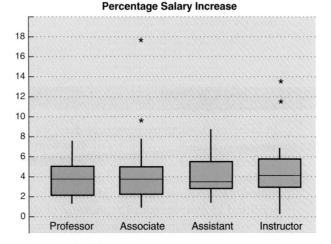

Percentage Salary Increase

(The * symbol represents an outlier.)

Minitab menu selections: ►**Graph** ►**Boxplot.**
Note: Minitab uses a procedure to find Q_1 and Q_3 that is slightly different from the method shown in this text. As a result, Minitab values for Q_1 and Q_3 will differ slightly from text or TI-83 values.

(a) Which faculty rank had the smallest median percentage salary increase? Which faculty rank had the single highest percentage salary increase?
(b) Which faculty rank had the largest spread between the first and third quartiles?
(c) Which faculty rank had the smallest spread for the lower 50% of the percentage salary increases?
(d) Which faculty rank had the most symmetric percentage salary increases? If the outliers for the associate professors were omitted, would that distribution appear to be symmetric?
(e) Look at the following descriptive statistics from Minitab.

Minitab Display: Descriptive Statistics

VARIABLE	N	MEAN	MEDIAN	TRMEAN	STDEV	SE MEAN
Professo	20	3.685	3.700	3.611	1.766	0.395
Associat	24	4.362	3.750	3.927	3.592	0.733
Assistan	23	4.330	3.500	4.262	2.007	0.419
Instruct	21	4.781	4.100	4.568	3.132	0.684

VARIABLE	MINIMUM	MAXIMUM	Q_1	Q_3
Prefesso	1.200	7.500	2.100	4.975
Associat	0.600	17.700	2.350	5.075
Assistan	1.500	8.600	2.900	5.500
Instruct	0.200	13.400	2.850	5.800

For the ranks of associate professor and instructor, compute the upper and lower limits for outliers (see Problem 11). Do these values appear to correspond to the endpoints of the basic whiskers?

SUMMARY

In order to characterize numerical data, we use both averages and variation. An average is a summary of a set of data with just one number. We studied several important averages: the mode, the median, the mean, and weighted averages. We also looked at trimmed means, which are more resistant to the effects of extreme values. However, an average alone can be misleading; we really need another statistical cross-reference. The measure of data spread, or variation, satisfies this purpose. The variations that we looked at most carefully were the range, the variance, the standard deviation, and the interquartile range. Chebyshev's theorem enables us to estimate the data spread. The coefficient of variation lets us compare relative spreads of different data sets. A box-and-whisker plot gives a good visual impression of the range of the data and the location of the middle half of the data.

In later work, the average we will use most is the mean, and the measure of variation we will use most is the standard deviation.

IMPORTANT WORDS & SYMBOLS

Section 2.1
Average
Mode
Median
Sample mean, $\bar{x}$
Population mean, μ
Trimmed mean
Summation symbol, Σ
Resistant measure

Section 2.2
Variation
Range
Sample variance, s^2
Sample standard deviation, s

Sum of the squares, SS_x
Square of the sum $(\Sigma x)^2$
Population size, N
Population standard deviation, σ
Coefficient of variation
Chebyshev's theorem

Section 2.3
Percentile
Quartile
Interquartile range
Whisker
Box-and-whisker plot
Outlier

VIEWPOINT

Are Students Ready for Work?

Of all high school seniors, 62% say they have "very good" mathematics skills. Only 8% of all employers agree with this claim. Employers appear to be telling us they need people with *more training* in mathematics, statistics, and general quantitative reasoning skills.

CHAPTER REVIEW PROBLEMS

Criminal Justice:
Armed Robbery

1. The following set of numbers consists of the ages (rounded to the nearest year) of 10 people selected at random who were convicted of armed robbery last year in Brooks County, Mississippi.

   | 19 | 24 | 26 | 23 | 19 | 27 | 46 | 52 | 27 | 27 |

   (a) Find the mean, median, and mode.
   (b) Find the range and sample standard deviation.

Legal:
Medical Malpractice Claims

2. A random sample of medical malpractice claims that had to be paid last year showed the following amounts (in millions of dollars):

   | 3.2 | 1.4 | 0.7 | 0.2 | 0.5 | 2.1 |
   | 0.8 | 1.7 | 1.6 | 0.5 | 0.6 | 0.2 |

(a) Make a box-and-whisker plot for these data.
(b) Compute the sample mean and median. Which is larger?
(c) Compute the range, sample standard deviation, and coefficient of variation.

Business: Auto Insurance

3. Is there a difference in auto insurance cost for people who reside in suburban and urban areas? *Consumer Reports* (Vol. 62, No. 1) gave suburban and urban costs of auto insurance for several states. In Illinois, the lowest costs for similar policies (in dollars) were

Suburban					**Urban**				
1170	1216	1211	1282	1292	2356	2584	2674	2840	2910
808	874	972	986	992	1768	1968	1968	2083	2107

For each group, compute the five-number summary and interquartile range. Then make a box-and-whisker plot for each group and comment on the differences between urban and suburban auto insurance rates.

Health: Radon

4. "Radon: The Problem No One Wants to Face" is the title of an article that appeared in *Consumer Reports*. Radon is a gas emitted from the ground that can collect in houses and buildings. At certain levels it can cause lung cancer. Radon concentrations are measured in picocuries per liter (pCi/L). A radon level of 4 pCi/L is considered "acceptable." Radon levels in a house vary from week to week. In one house, radon levels (in pCi/L) measured during 8 successive weeks were as follows.

1.9   2.8   5.7   4.2   1.9   8.6   3.9   7.2

(a) Find the mean, median, and the mode.
(b) Find the sample standard deviation, coefficient of variation, and range.

General: Mean Weight

5. An elevator is loaded with 16 people and is at its load limit of 2500 lb. What is the mean weight of these people?

Driving: Rush Hour Traffic

6. Radar was used to check speeds on a random sample of 20 cars in rush-hour traffic on an Atlanta expressway at 7:30 A.M. with the following results (in miles per hour):

50	60	48	60	56	55	60	58	50	45
55	40	45	66	50	60	55	38	55	60

(a) Make a box-and-whisker plot of the data. Find the interquartile range.
(b) Find the mean, median, and mode.
(c) Comment on the statement: "On the average, cars do not exceed the speed limit (55) during the early morning rush hour." For which averages is this statement *not* true?

Health Care: Glucose Blood Level

7. The data in the following table represent glucose blood level (mg/100 ml) after a 12-hour fast for a random sample of 70 women (based on information from the *American Journal of Clinical Nutrition,* Vol. 19, 345–351).

Glucose Blood Level (mg/100 ml)

45	66	83	71	76	64	59	59	76	82	80
81	85	77	82	90	87	72	79	69	83	71
87	69	81	76	96	83	67	94	101	94	89
94	73	99	93	85	83	80	78	80	85	83
84	74	81	70	65	89	70	80	84	77	65
46	80	70	75	45	101	71	109	73	73	80
72	81	63	74							

(a) Compute the five-number summary and the interquartile range, and make a box-and-whisker plot.

(b) Make a frequency table using 3 classes. Then estimate the mean and sample standard deviation. Use the formulas for grouped data found on pages 95 and 96.

Business:
Consumer Price Index

8. Inflation is a stealth enemy of family budgets. One measure of inflation is the Consumer Price Index (CPI). The following data give percentage changes in the CPI from 1970 to 1996. (Data source: Web site <http://www.census.gov/>, then use the index to find Consumer Price Index.)

5.9	4.3	3.3	6.2	11.0	9.1	5.8	6.5
7.7	11.3	13.5	10.4	6.1	3.2	4.3	3.6
1.9	4.0	4.4	4.6	5.4	4.2	3.0	3.0
2.6	2.8	2.9					

(a) Use a calculator with mean and standard deviation keys to verify that the mean CPI percent change from 1970 to 1996 was 5.59, with a sample standard deviation of 3.05.

(b) Use Chebyshev's theorem to find an interval of CPI rates centered about the mean in which you would expect at least 75% of the years to fall.

DATA HIGHLIGHTS: GROUP PROJECTS

Break into small groups and discuss the following topics. Organize a brief outline in which you summarize the main points of your group discussion.

1. *The Story of Old Faithful* is a short book written by George Marler and published by the Yellowstone Association. Chapter 7 of this interesting book talks about the effect of the 1959 earthquake on eruption intervals for Old Faithful Geyser. Dr. John Rinehart, a senior research scientist with the National Oceanic and Atmospheric Administration, has done extensive studies of the eruption intervals before and after the 1959 earthquake. Examine the figure at the top of page 114 for the period before

Typical Behavior of Old Faithful Geyser
Before 1959 Quake

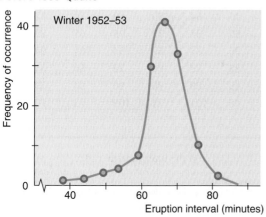

Typical Behavior of Old Faithful Geyser
After 1959 Quake

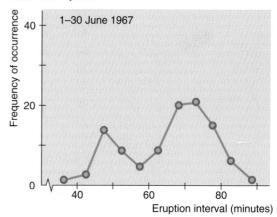

the 1959 earthquake. Notice the general shape. Is the graph more or less symmetrical? Does it have a single mode frequency? The mean interval between eruptions has remained steady at about 65 minutes for the past 100 years. Therefore, the 1959 earthquake did not significantly change the mean, but it did change the distribution of eruption intervals. Examine the figure above for the period after the 1959 earthquake. Would you say there are really two frequency modes? One shorter and the other longer? Explain. The overall mean is about the same for both graphs, but one graph has a much larger standard deviation (for eruption intervals) than the other. Do no calculations, just look at both graphs, and then explain which graph has the smaller and which has the larger standard deviation. Which distribution will have the larger coefficient of variation? In everyday terms, what would this mean if you were actually at Yellowstone waiting to see the next eruption of Old Faithful? Explain your answer.

2. Most academic advisors tell students to major in a field the student really loves. After all, it is true that money can't buy happiness. Nevertheless, it is interesting at least to look at some of the higher-paying fields of study. After all, a field such as mathematics can be a lot of fun once you get into it. We see that women's salaries tend to be lower than men's salaries. However, women's salaries are rapidly catching up, and this benefits the entire work force in different ways. The figure at the top of page 115 shows the median incomes for college graduates with different majors. The employees in the sample are all at least 30 years old. Does it seem reasonable to assume that many of the employees are in jobs beyond the entry level? Explain. Compare the median incomes shown for all women aged 30 or older holding bachelor's degrees with the median income for men of similar age holding bachelor's degrees. Look at the particular majors listed. What percentage of men holding bachelor's degrees in mathematics make $52,316 or more? What percentage of women holding computer/information science degrees make $41,559 or more? How do median incomes for men and women holding engineering degrees compare? What about pharmacy degrees?

Lucrative College Majors
Source: Copyright 1997, USA TODAY. Reprinted with permission.

USA SNAPSHOTS®

A look at statistics that shape the nation

Lucrative majors for women

The U.S. median income for women age 30 or older with bachelor's degrees and earning wages or salaries was $31,848[1]. Fields of study with highest median pay:

Major	Median income
Pharmacy	$47,567
Engineering	$46,389
Computer/ information sciences	$41,559
Physical therapy/ related services	$40,491
Nursing	$40,096

1–Excludes self-employed, through 1993 (latest year available)
Source: Bureau of Labor Statistics
Occupational Outlook Quarterly, Summer 1998

By Anne R. Carey and Grant Jerding, USA TODAY

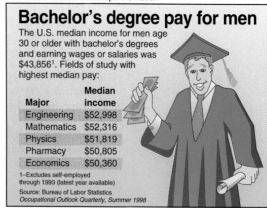

USA SNAPSHOTS®

A look at statistics that shape the nation

Bachelor's degree pay for men

The U.S. median income for men age 30 or older with bachelor's degrees and earning wages or salaries was $43,856[1]. Fields of study with highest median pay:

Major	Median income
Engineering	$52,998
Mathematics	$52,316
Physics	$51,819
Pharmacy	$50,805
Economics	$50,360

1–Excludes self-employed, through 1993 (latest year available)
Source: Bureau of Labor Statistics
Occupational Outlook Quarterly, Summer 1998

By Anne R. Carey and Grant Jerding, USA TODAY

ⓁINKING CONCEPTS: WRITING PROJECTS

Discuss each of the following topics in class or review the topics on your own. Then write a brief but complete essay in which you summarize the main points. Please include formulas and graphs as appropriate.

1. An average is a summary of a collection of data in just *one* number. Discuss how the mean, median, and mode all represent averages in this context. Also discuss the differences among these averages. Why is the mean a balance point? Why is the median a midway point? Why is the mode the most common data point? List three areas of daily life where you think either the mean, median, or mode would be the best choice to describe an "average."

2. Why do we need to study the variation of a collection of data? Why isn't the average by itself adequate? We have studied three ways to measure variation. The range, standard deviation, and to a large extent, a box-and-whisker plot all indicate the variation within a data collection. Discuss similarities and differences among these ways of measuring data variation. Why would it seem reasonable to pair the median with a box-and-whisker plot and to pair the mean with the standard deviation? What are the advantages and disadvantages of each method of describing data spread? Comment on statements such as the following: (a) The range is easy to compute, but it doesn't give much information. (b) Although the standard deviation is more complicated to compute, it has some significant applications. (c) The box-and-whisker plot is fairly easy to construct, and it gives a lot of information at a glance.

3. Why is the coefficient of variation important? What do we mean when we say that the coefficient of variation has no units? What advantage can there be in having no units? Why is *relative size* important?

   Consider robin eggs; the mean weight of a collection of robin eggs is 0.72 ounces and the standard deviation is 0.12 ounces. Now consider elephants; the mean weight of elephants in the zoo is 6.42 tons, with a standard deviation of 1.07 tons. The units of measurement are different and there is a great deal of difference between the size of an elephant and the size of a robin's egg. Yet the coefficient of variation is about the same for both. Comment on this from the viewpoint of the size of the standard deviation relative to the mean.

4. What is Chebyshev's theorem? Suppose you have a friend who knows very little about statistics. Write a paragraph or two in which you describe Chebyshev's theorem for your friend. Keep the discussion as simple as possible, but be sure to get the main ideas across to your friend. Suppose he or she asks, "What is this stuff good for?" and suppose you respond (a little sarcastically) that Chebyshev's theorem applies to everything from butterflies to the orbits of the planets. Would you be correct? Explain.

The problems in this section may be done using statistical computer software or calculators with statistical functions. Displays and suggestions are given for Minitab (Release 12), the TI-83 graphing calculator, ComputerStat, and Excel.

Descriptive Statistics

Statistical software packages and graphing calculators all provide descriptive statistics such as mean, median, and standard deviation for data. Most statistical software and graphing calculators will order the data for you so that you can find the mode. In some cases the quartile values Q_1 and Q_3 are also given.

Minitab

Minitab supports several data entry methods. However, the easiest method is to use the data worksheet. Enter the data in columns. Then, to obtain the mean, median, 5% trimmed mean, sample standard deviation, minimum, and maximum, follow the menu selections ➤Stat ➤Basic Statistics ➤Descriptive Statistics. In the dialogue box, list the columns containing the data as the variables. If you press the graphs button, you have the option of producing a boxplot.

The ➤Graph ➤Boxplot menu selections produce a dialogue box dedicated to creating a boxplot. You can control the title, control the locations of tick marks on the axes, opt to create several boxplots on the same display, and so on.

Note: Minitab uses a slightly different method to compute Q_1 and Q_3 than that shown in this text, so the values for Q_1 and Q_3 shown in Minitab will be slightly different from those you compute or those shown on the TI-83.

TI-83

Enter the data in a list such as L1. Then, in the STAT menu, select the CALC option. Select 1-Var Stats. At the 1-Var Stats prompt, type L1. The following screens show you output from a typical data set (Problem 11 of Section 2.3).

```
EDIT CALC TESTS
1:1-Var Stats
2:2-Var Stats
3:Med-Med
4:LinReg(ax+b)
5:QuadReg
6↓CubicReg
```

```
1-Var Stats L1
```

```
1-Var Stats
 x̄=63.375
 Σx=1521
 Σx²=101271
 Sx=14.56264578
 σx=14.25602943
↓n=24
```

117

```
1-Var Stats
↑n=24
 minX=4
 Q1=61.5
 Med=65.5
 Q3=71.5
 maxX=80
█
```

The TI-83 also produces a box plot. Press STAT PLOT and select box plot. Then set the WINDOW with Xmin = minimum data value and Xmax = maximum data value. The following display uses the same data as summarized in the previous displays. ZoomStat will set the window automatically.

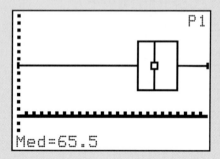

```
                              P1
 ┌──────────┐
 │    □     │
 └──────────┘

·······························
Med=65.5
```

ComputerStat

In ComputerStat, follow the instructions on the screen to enter your data. Then under the Descriptive Statistics menu, select Averages and Variation for Ungrouped Data.

Excel

In Excel, enter your data in columns in the worksheet. Then individual summary statistics commands are available through the paste function f_x. Press the paste function button, select statistical, and then choose the desired function. In particular, you will find the commands for the mean, median, mode, sample standard deviation, population standard deviation, 5% trimmed

mean, minimum, maximum, quartile, and so on. The method that Excel uses to compute quartiles is slightly different from the method shown in this text, and so you will notice that the values for Q_1 and Q_3 may differ slightly from those you compute by hand.

If the Analysis Toolpak is installed with your version of Excel, the menu choices ➤Tools ➤Data Analysis ➤Descriptive Statistics open a dialogue box. Select the columns to be included in the Input Range, select Output Range, and check Summary statistics. Then Excel provides summary statistics including the mean, median, mode, sample standard deviation, sample variance, range, minimum, maximum, and a number of data values. In addition, the standard error is given as well as some measure of skewness.

Excel does not produce a boxplot, but with the five-number statistics available under the paste function, you can easily create a boxplot by hand.

APPLICATION

Using the software or calculator available to you, do the following.

1. Trade winds are one of the beautiful features of island life in Hawaii. The following data represent total air movement in miles each day over a weather station in Hawaii as determined by a continuous anemometer recorder. The period of observation is January 1 to February 15, 1971.

26	14	18	14	113	50	13	22
27	57	28	50	72	52	105	138
16	33	18	16	32	26	11	16
17	14	57	100	35	20	21	34
18	13	18	28	21	13	25	19
11	19	22	19	15	20		

Source: United States Department of Commerce, National Oceanic and Atmospheric Administration, Environmental Data Service. *Climatological Data, Annual Summary, Hawaii,* Vol. 67, No. 13. Asheville: National Climatic Center, 1971, pp. 11, 24.

(a) Enter the data.

(b) Use the computer to find the sample mean, median, and (if it exists) mode.

(c) Use the computer to find the range, sample variance, and sample standard deviation.

(d) As a topic in exploratory data analysis (EDA), we studied the box-and-whisker plot. Use the five-number summary provided by the computer to make a box-and-whisker plot of total air movement over the weather station.

(e) There are four exceptionally high data values: 113, 105, 138, and 100. The strong winds of January 5 (113 reading) brought in a cold front that dropped snow on Haleakala National Park (at the 8000-ft elevation). The residents were so excited that they drove up to see the snow and caused such a massive traffic jam that the Park Service had to close the road. The winds of January 15 and 16 (readings 105 and 138) brought in a storm that created more damaging funnel clouds than any other storm to that date in the recorded history of Hawaii. The strong winds of January 28 (reading 100) accompanied a storm with funnel clouds that did much damage. Eliminate these values (i.e., 100, 105, 113, and 138) from the data bank and redo parts a through d. Compare your results with those previously obtained. Which average is most affected? What happens to the standard deviation? How do the two box-and-whisker plots compare?

Computer Displays

Minitab-generated displays of box-and-whisker plots for percentage salary increases by rank of college faculty in Tennessee are given in Problem 12 of Section 2.3. Displays from ComputerStat and Excel for some of the same data follow.

ComputerStat Windows Display:
Percentage Salary Increases for Assistant Professors

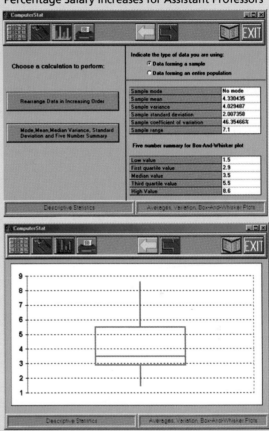

Excel Display:

Percentage Salary Increases for College Faculty

Professor		Associate	
Mean	3.685	Mean	4.3625
Standard Error	0.394986675	Standard Error	0.7331887
Median	3.7	Median	3.75
Mode	3.8	Mode	4
Standard Deviation	1.766434112	Standard Deviation	3.591876402
Sample Variance	3.120289474	Sample Variance	12.90157609
Kurtosis	−0.412408394	Kurtosis	7.838493433
Skewness	0.402782982	Skewness	2.441099713
Range	6.3	Range	17.1
Minimum	1.2	Minimum	0.6
Maximum	7.5	Maximum	17.7
Sum	73.7	Sum	104.7
Count	20	Count	24

Regression and Correlation

It seems, therefore, that even if we find a unified theory, we may be able to make only statistical predictions.

—Stephen Hawking

We live in the most probable of all possible worlds.

—Voltaire, paraphrased by Stephen Hawking

Stephen Hawking

Hawking is the current Lucasian Professor of Mathematics at Cambridge University, a post once held by Isaac Newton. In spite of having a severe motor neurone disease, he has made brilliant contributions to mathematics and science. His courage and creative spirit inspire many people facing their own limitations.

In his book *Black Holes and Baby Universes*, Professor Stephen Hawking (Cambridge University) discusses what it would mean if we had a complete and unified theory of physics. Then he quickly adds that except in any but the simplest situations, we would have to use *statistical methods* and abandon any pretense of solving the equations (in the unified theory) exactly.

Professor Hawking goes on to say that in a unified theory of physics we may have to adopt a picture in which there is an ensemble of many possible universes, each with its own probability distribution. Perhaps this is what the eighteenth-century French philosopher Voltaire refers to in his novel *Candide*.

In this chapter we study regression, correlation, and forecasting. This material may be used to study finance, economics, business administration, social science, natural science, and (if you wish) black holes and baby universes.

Section **3.1** **Introduction to Paired Data and Scatter Diagrams**

The study of correlation and regression of two variables usually begins with a table and/or a graph of *paired data values*. Example 1 shows what we mean.

EXAMPLE 1 ❯ For her botany class project, Jan wonders if there is a connection between the average weight of watermelons a vine produces and the root depth of the vine. Jan suspects that vines with deeper roots have a better water supply and produce larger melons. From a large watermelon field, 30 vines are chosen at random. At the end of 8 weeks, the watermelons are removed from each vine, and the average weight of watermelons from each vine is determined. Each plant is carefully dug up and its root depth (or length) is measured. Table 3-1 shows the results.

(a) For each plant there is an ordered pair (x, y) of data values. If we plot the pairs (x, y) as points on a coordinate system, we obtain the graph shown in Figure 3-1. Figure 3-1 is called a *scatter diagram* for the paired data values in Table 3-1.

Scatter diagrams

(b) By inspecting Figure 3-1, we see that to some extent larger values of x tend to be associated with larger y values and smaller x values tend to be associated with smaller y values. Roughly speaking, the general trend seems to be reasonably well represented by the line segment shown in the figure.   ●

Introduction to linear correlation

Of course, it is possible to draw many curves in Figure 3-1, but the straight line is the simplest and most widely used curve for elementary studies of paired data. We can draw many lines in Figure 3-1, but in some sense the "best" line should be the one that comes closest to each of the points of the scatter diagram.

Table 3-1 Results of a Botany Experiment

x = Root Depth (nearest inch)	y = Mean Weight of Watermelon (nearest pound)
26	20
14	10
18	13
10	9
26	19
21	17
7	8
26	15
13	9
19	13
17	12
13	7
16	9
28	17
23	14

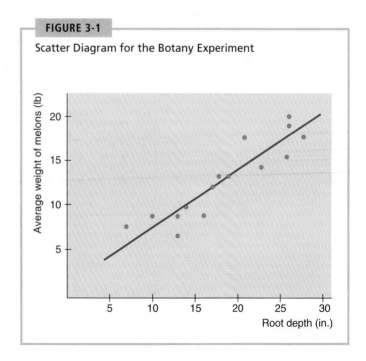

FIGURE 3-1

Scatter Diagram for the Botany Experiment

To single out one line as the "best-fitting line," we must find a mathematical criterion for this line and a formula representing the line. This will be done in Section 3.2 by the *method of least squares.*

Another problem precedes that of finding the "best-fitting line"—that is, the problem of determining how well the points of the scatter diagram are suited for fitting *any* line. Certainly if the points are a very poor fit to *any* line, there is little use in trying to find the "best" line. This problem will be dealt with in Section 3.3 by use of the Pearson product-moment correlation coefficient.

If the points of a scatter diagram are located so that *no* line is realistically a "good" fit, we then say that the points have *no linear correlation.* We see two examples of scatter diagrams for which there is no linear correlation in Figure 3-2.

On the other hand, if all the points do in fact lie on a line, then we have *perfect linear correlation.* In Figure 3-3 on page 123, we see diagrams with perfect linear correlation. In statistical applications, perfect linear correlation almost never occurs.

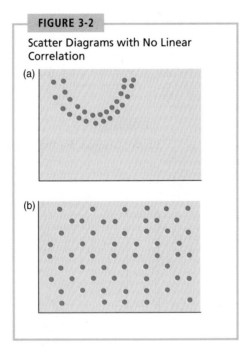

FIGURE 3-2

Scatter Diagrams with No Linear Correlation

(a)

(b)

We say a scatter diagram has *high* linear correlation if the points lie close to a straight line. If the points are not close to a straight line, we say the correlation is *moderate* or *low.* If the points fit no straight-line pattern, we say there is *no* linear correlation.

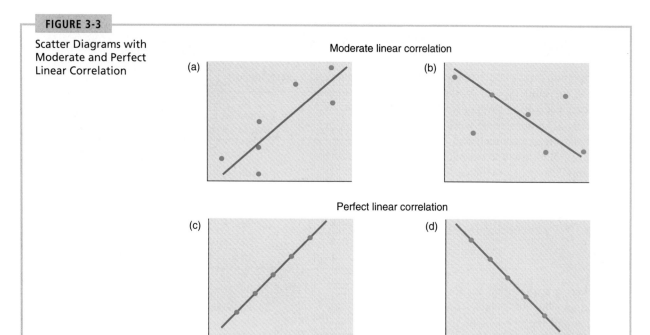

FIGURE 3-3

Scatter Diagrams with Moderate and Perfect Linear Correlation

Moderate linear correlation

(a)   (b)

Perfect linear correlation

(c)   (d)

GUIDED EXERCISE   1

Examine the scatter diagrams in Figure 3-4 and then answer the following questions.

FIGURE 3-4   Scatter Diagrams

(a)   (b)   (c)

(a) Which diagram has no linear correlation?   ⟹   Figure 3-4c has no linear correlation. No straight-line fit should be attempted.

(b) Which has perfect linear correlation?   ⟹   Figure 3-4a has perfect linear correlation and can be fitted exactly by a straight line.

(c) Which can be reasonably fitted by a straight line?   ⟹   Figure 3-4b can be reasonably fitted by a straight line.

GUIDED EXERCISE   2

A large industrial plant has seven divisions that do the same type of work. A safety inspector visits each division of 20 workers quarterly. The number of work-hours devoted to safety training and the number of work-hours lost due to industry-related accidents are recorded for each separate division in Table 3-2.

Table 3-2 Safety Report

Division	x (number of work-hours in safety training)	y (number of work-hours lost due to accidents)
1	10.0	80
2	19.5	65
3	30.0	68
4	45.0	55
5	50.0	35
6	65.0	10
7	80.0	12

(a) Make a scatter diagram for these pairs of values. Plot the x values on the horizontal axis and the y values on the vertical axis.

⇨ **FIGURE 3-5** Scatter Diagram for Safety Report

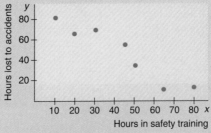

(b) As the number of hours spent on safety training increases, what happens in general to the number of hours lost due to industry-related accidents?

⇨ In general, as the number of hours of safety training goes up, the number of hours lost due to accidents goes down.

(c) Does a line fit the data reasonably well?

⇨ A line fits reasonably well.

(d) Draw a line that you think "fits best."

⇨ Any line that seems to be the best fit to you is correct. Later, you will see the equation of a line that is a "best fit."

Calculator Note Many graphing calculators produce scatter plots. On the TI-83, put x values in list L_1 and y values in list L_2. Then press the **STAT PLOT** key. The displays show the graph for Guided Exercise 2 regarding the safety training and hours lost because of accidents.

Notice that the scales on the *x* and *y* axes are not shown. A different choice of scales can change the look of the diagram. To check the scales, look at the settings displayed when you press **WINDOW**.

Use **STAT PLOT**, enter 1,
and use options shown.

Press **ZOOM**, and choose option
9: ZoomStat.

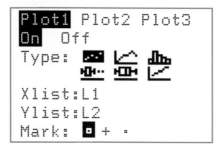

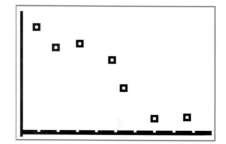

 I E W P O I N T

Hawaii Island Hopping

Suppose you want to go camping in Hawaii. Hawaii has both state and federal parks where you can enjoy camping on the beach or in the mountains. However, you will probably need to rent a car to get to the different campgrounds. How much will the car rental cost? That depends on the islands you visit. For car rental data and regression statistics you can compute regarding costs on different Hawaiian Islands, see Web site <http://www.garden-isle.com/>.

SECTION 3.1 PROBLEMS

For Problems 1–6, look at the scatter diagrams and state which of the following conditions you think is true for each diagram.
(a) High linear correlation
(b) Moderate or low linear correlation
(c) No linear correlation

1.
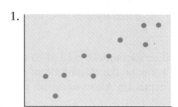

2.

3.

4.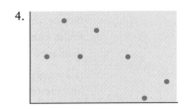

5.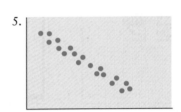

6.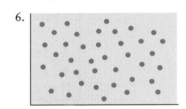

Veterinary Science:
Shetland Ponies

7. How much should a healthy Shetland pony weigh? Let x be the age of the pony (in months), and let y be the average weight of the pony (in kilograms). The following information is based on data taken from *The Merck Veterinary Manual* (a reference used in most veterinary colleges).

x	3	6	12	18	24
y	60	95	140	170	185

(a) Draw a scatter diagram for the given data.
(b) Draw a straight line that you think best fits the data.
(c) Would you say the correlation is low, moderate, or high?

Health Care: Insurance Plans

8. The following data are based on information from *Domestic Affairs*. Let x be the average number of employees in a group health insurance plan, and let y be the average administrative cost as a percentage of claims.

x	3	7	15	35	75
y	40	35	30	25	18

(a) Draw a scatter diagram for the given data.
(b) Draw a straight line that you think best fits the data.
(c) Would you say the correlation is low, moderate, or high?

Science: Earthquakes

9. Is the magnitude of an earthquake related to the depth below the surface at which the quake occurs? Let x be the magnitude of an earthquake (on the Richter scale), and let y be the depth (in kilometers) of the quake below the surface at the

epicenter. The following data are based on information taken from the National Earthquake Information Service of the U.S. Geological Survey. Additional data may be found at Web site <http://www.seismo.usbr.gov/quake/>.

x	2.9	4.2	3.3	4.5	2.6	3.2	3.4
y	5.0	10.0	11.2	10.0	7.9	3.9	5.5

(a) Draw a scatter diagram for the given data.
(b) Draw a straight line that you think best fits the data.
(c) Would you say the correlation is low, moderate, or high?

Archaeology: Pottery

10. Wind Mountain archaeological site is located in southwest New Mexico. Ancient, prehistoric pottery vessels are usually found as sherds (broken pieces) and carefully reconstructed if enough sherds can be found. For reconstructed (or even rare unbroken) pottery vessels, let x be the body diameter (in centimeters), and let y be the height (in centimeters) of the vessel. The following data are based on information taken from *Mimbres Mogollon Archaeology*, by A. I. Woosley and A. J. McIntyre (University of New Mexico Press).

x	7.3	31.0	18.4	6.5	4.9	2.6	19.5	9.2	23.7
y	5.5	28.5	19.7	5.0	5.7	2.1	11.5	5.0	11.6

(a) Draw a scatter diagram for the given data.
(b) Draw a straight line that you think best fits the data.
(c) Would you say the correlation is low, moderate, or high?

General:
Graphical Interpretation

11. The initial visual impact of a scatter diagram depends on the scales used on the x and y axes. Consider the following data.

x	1	2	3	4	5	6
y	1	4	6	3	6	7

(a) Make a scatter diagram using the same scale on both the x and y axes (i.e., make sure the unit lengths on the two axes are equal).
(b) Make a scatter diagram using a scale on the y axis that is twice as long as that on the x axis.
(c) Make a scatter diagram using a scale on the y axis that is half as long as that on the x axis.
(d) On each of the three graphs, draw the straight line that you think best fits the data points. How do the slopes (or directions) of the three lines appear to change? (*Note:* The actual slopes will be the same; they just appear different because of the choice of scale factors.)

Business: Hotels

12. Hotels often use computer models to set prices of rooms. One factor is the occupancy rate. *Consumer Reports*, Volume 63, No. 6, listed hotel occupancy rates in several popular destination cities (New York City, San Francisco, Boston, Honolulu, Chicago, Tampa, Detroit, Houston, St. Louis, and Norfolk-Virginia Beach) and the average room rate charged in the city. The first figure on page 128 is a scatter diagram produced by Excel (➤**Chart Wizard**, ➤**Scatter Diagram**).

The second figure shows a scatter diagram for the same data produced by Minitab (➤**Stat** ➤**Regression** ➤**Fitted Line Plot**). Minitab also shows the best fit line.

(a) Compare the scatter diagrams produced by the two different software packages. Besides the line shown in the Minitab plot, what is a key difference between the plots? (*Hint:* Look at the "origin.")

(b) Does a linear model seem appropriate for these data?

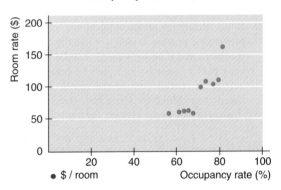

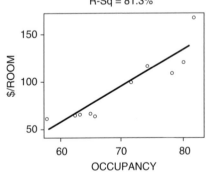

Linear Regression

Anyone who has been outdoors on a summer evening has probably heard crickets. Did you know that it is possible to use the cricket as a thermometer? Crickets tend to chirp more frequently as temperature increases. A Harvard physics professor made a detailed study of this phenomenon. Using sophisticated equipment, Professor George W. Pierce studied the striped ground cricket and compiled the data in Table 3-3.

Do the data indicate a linear relation between chirping frequency and temperature? Is there a way we can use the data to predict the temperature that corresponds to a chirp frequency that is not listed in the table? For instance, how can we use the data to predict the temperature for $x = 19$ chirps per second? Let us first make a scatter diagram (Figure 3-6) for the data in Table 3-3.

Looking at the scatter diagram in Figure 3-6, we ask two questions:

1. Can we find a relationship between x and y?

2. If so, how strong is the relationship?

The first step in answering these questions is to try to express the relationship as a mathematical equation. There are many possible equations, but the simplest and most widely used is the linear equation, or the equation of a straight line. Since we will be using this line to predict the y values (temperature) from the x values (chirps per second), we call x the *explanatory variable* and y the *response variable*.

Our job is to find the "best" linear equation representing the points in the scatter diagram. For our criterion of best-fitting line, we use the *least-squares criterion*, which says that the line we fit to the data points must be such that *the*

Explanatory variable
Response variable

Least-squares criterion

Table 3-3   Chirping Frequency and Temperature for the Striped Ground Cricket

x (chirps/s)	y (temperature, °F)
20.0	88.6
16.0	71.6
19.8	93.3
18.4	84.3
17.1	80.6
15.5	75.2
14.7	69.7
17.1	82.0
15.4	69.4
16.2	83.3
15.0	79.6
17.2	82.6
16.0	80.6
17.0	83.5
14.4	76.3

Source: Reprinted by permission of the publisher from *The Song of Insects* by George W. Pierce, Cambridge, Mass.: Harvard University Press, Copyright © 1948 by the President and Fellows of Harvard College.

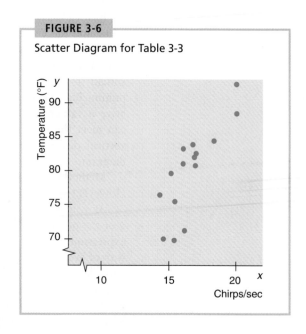

FIGURE 3-6

Scatter Diagram for Table 3-3

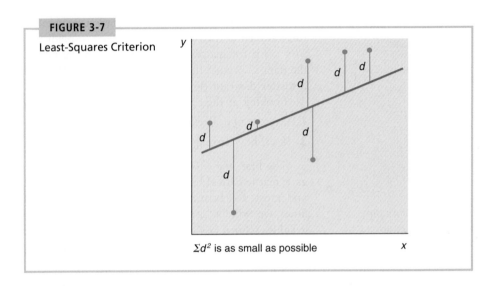

FIGURE 3-7

Least-Squares Criterion

Σd^2 is as small as possible

sum of the squares of the vertical distances from the points to the line be made as small as possible. The least-squares criterion is illustrated in Figure 3-7.

In Figure 3-7, d represents the difference between the y coordinate of the data point and the corresponding y coordinate on the line. Thus, if the data point lies above the line, d is positive, but if the data point lies below the line, d is negative. As a result, the sum of the d values can be small even if the points are widely spread in the scatter diagram. However, the squares d^2 cannot be negative. By minimizing the sum of the squares, we are in effect not allowing positive and negative d values to "cancel out" one another in the sum. It is in this way that we can meet the least-squares criterion of minimizing the sum of the squares of the vertical distances between the points and the line over *all* points in the scatter diagram.

Least-squares line

Methods of calculus are required to derive the formula for the least-squares line. However, we will simply use the formula rather than derive it.

As you study Formula (1) for the least-squares line, notice that the means $\bar{x}$ and $\bar{y}$ of the points in the scatter diagram enter into the computations. In addition, the expression SS_x and SS_{xy} are used. Recall from Section 2.2 that SS_x is just a faster way of evaluating $\Sigma(x - \bar{x})^2$. We used SS_x in the computation formula for the standard deviation of x values. Likewise, the formula for SS_{xy} is a more efficient way to compute $\Sigma(x - \bar{x})(y - \bar{y})$.

The simplest way to find a and b is to organize your work into a table. In Table 3-4, we use the data relating rate of cricket chirps to temperature to obtain the sums needed for Formulas (4) and (5). We will then use these sums to compute the formula for the least-squares line.

Least-Squares Line

$$y = a + bx \tag{1}$$

where $b = \dfrac{SS_{xy}}{SS_x}$ b is the slope $\tag{2}$

$a = \bar{y} - b\bar{x}$ a is the y-intercept $\tag{3}$

and $\bar{y}$ = mean of y values in scatter diagram
$\bar{x}$ = mean of x values in scatter diagram

$$SS_{xy} = \Sigma xy - \frac{(\Sigma x)(\Sigma y)}{n} \tag{4}$$

$$SS_x = \Sigma x^2 - \frac{(\Sigma x)^2}{n} \tag{5}$$

n = number of points in a scatter diagram
In Formulas (4) and (5), the sums are taken over all x or y values in the scatter diagram.

Table 3-4   Sums for Computing $\bar{x}$, $\bar{y}$, SS_x and SS_{xy}

x (chirps/s)	y (°F)	x^2	xy
20.0	88.6	400.0	1,772.0
16.0	71.6	256.0	1,145.6
19.8	93.3	392.0	1,847.3
18.4	84.3	338.6	1,551.1
17.1	80.6	292.4	1,378.3
15.5	75.2	240.3	1,165.6
14.7	69.7	216.1	1,024.6
17.1	82.0	292.4	1,402.2
15.4	69.4	237.2	1,068.8
16.2	83.3	262.4	1,349.5
15.0	79.6	225.0	1,194.0
17.2	82.6	295.8	1,420.7
16.0	80.6	256.0	1,289.6
17.0	83.5	289.0	1,419.5
14.4	76.3	207.4	1,098.7
Σx = 249.8	Σy = 1200.6	Σx^2 = 4200.6	Σxy = 20,127.5

COMPUTATION NOTES

1. The formulas used to compute the slope and y intercept of the least-squares line are sensitive to rounding. Answers to the problems at the end of each section and at the end of the chapter are computer-generated, so you may expect slight differences in your answers depending on how you round intermediate steps.

2. Do not confuse Σx^2 and $(\Sigma x)^2$. For Σx^2, we *first square* each x value and then find the total sum of these values. For $(\Sigma x)^2$, we *first sum* the x values and then square the total.

3. On a calculator with a statistics mode, you can compute the standard deviation s and then the variance s^2 of the x values. Then

$$SS_x = s^2(n - 1)$$

Using the formulas to find the equation of the least-squares line

From Table 3-4, we have

$$SS_x = \Sigma x^2 - \frac{(\Sigma x)^2}{n} = 4200.6 - \frac{(249.8)^2}{15} = 40.6$$

and we also have

$$SS_{xy} = \Sigma xy - \frac{(\Sigma x)(\Sigma y)}{n} = 20{,}127.5 - \frac{(249.8)(1200.6)}{15} = 133.5$$

We also find

$$\bar{x} = \frac{\Sigma x}{n} = \frac{249.8}{15} = 16.7$$

$$\bar{y} = \frac{\Sigma y}{n} = \frac{1200.6}{15} = 80.0$$

Therefore, using Formulas (2) and (3), we find a and b in the equation of the least-squares line.

Slope: $b = \dfrac{SS_{xy}}{SS_x} = \dfrac{133.5}{40.6} = 3.3$

y intercept: $a = \bar{y} - b\bar{x} = (80.0) - (3.3)(16.7) = 24.9$

We conclude that the least-squares line for the data in Table 3-3 is

$$y = a + bx \qquad\qquad (6)$$

$$y = 24.9 + 3.3x$$

Graphing the least-squares line

To graph the least-squares line (6), we have several options. The slope-intercept method of college algebra is probably the quickest. The slope is $b = 3.3$, and the y intercept is $a = 24.9$. However, if you don't remember this method, it is almost as easy to plot two points and connect them with a straight line. For x values, we usually use any two values in the range of x values. Corresponding y values are computed from the equation of the least-squares line.

The value of $\bar{x}$ will always be in the range of x values. When we use $x = \bar{x}$ in Formulas (1) and (3), we see that the corresponding y value is $\bar{y}$.

Therefore, the point $(\bar{x}, \bar{y})$ will always be on the least-squares line.

Since we have already computed these values, the point $(\bar{x}, \bar{y})$ is a convenient choice for one of the two points we use to graph the least-squares line. From Table 3-3 we see that $x = 20$ is also in the range of x values. We compute the corresponding y value by using the equation of the least-squares line.

x	$y = 24.9 + 3.3x$
When we choose $x = \bar{x} = 16.7$	$y = 24.9 + 3.3(16.7) = 80.0 = \bar{y}$
When we choose $x = 20.0$	$y = 24.9 + 3.3(20.0) = 90.9$

The line going through the points (16.7, 80.0) and (20.0, 90.9) is the least-squares line for the scatter diagram in Figure 3-6. This line is shown in Figure 3-8 on page 134.

Now suppose we find a striped ground cricket and, with a listening device, discover that it chirps at the rate of 19.0 chirps per second. What should we predict for the temperature? We could read the y value above $x = 19.0$ from the least-squares line graphed in Figure 3-8. But a more accurate estimate can be obtained by using the value $x = 19.0$ in the equation of the least-squares line and computing the corresponding y-value.

$y = 24.9 + 3.3x$	Equation of least-squares line
$y = 24.9 + 3.3(19.0)$	Using 19.0 in place of x
$y = 87.6°F$	Evaluating y

Rounded to the nearest whole number, we should predict the temperature to be 88°F. Of course, this is just a prediction, and we would be quite happy if the temperature turned out to be relatively close to our prediction. This brings up the natural question: How *good* are predictions based on the least-squares line? This is a fairly difficult question, and much of the answer requires advanced mathematics; however, a partial answer will be given in Section 3.3.

Calculator Note Calculators supporting two-variable statistics directly produce the values for a and b in the least-squares line $y = a + bx$. Read your calculator manual for instructions on entering the data and on finding the values for a and b. Many calculators also directly evaluate the predicted y value for a given x value. Again, see the manual for your calculator.

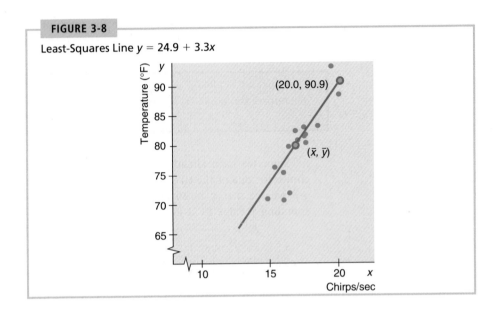

FIGURE 3-8

Least-Squares Line $y = 24.9 + 3.3x$

The TI-83 screens below show the output for the linear regression line of the following Guided Exercise 3. Enter data pairs with x in L_1 and y in L_2.

Press **STAT,** choose **Calculate,**
use option **8:LinReg(a + bx).**
Press **ENTER.**

Use **STAT PLOT** for scatter diagram;
Y1 = least-squares line for line. Use
value command under **CALC** to get y
when $x = 12$.

```
LinReg
y=a+bx
a=6.54065941
b=1.010989011
r²=.8449399438
r=.9192061487
```

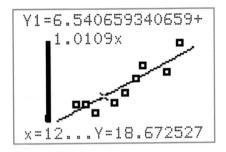

Interpolation

We have used the least-squares line to predict y values for x values that were *between* x values observed in the experiment. Predicting y values for x values that are between x values of points in the scatter diagram is called *interpolation*. The least-squares line can be used for interpolation. Predicting y values for x values beyond the range of observed x values is a complex problem that is not treated in this book. Prediction beyond the range of observations is called *extrapolation*.

Extrapolation

The Quick Sell car dealership has been using 1-minute spot ads on a local TV station. The ads always run during the evening hours and advertise the different models and price ranges of cars on the lot that week. During a 10-week period, the Quick Sell dealer kept a weekly record of the number of TV ads versus the number of cars sold. The results are given in Table 3-5.

The manager decided that Quick Sell can afford only 12 ads per week. At that level of advertisement, how many cars can Quick Sell expect to sell each week? We'll answer this question in several steps.

Table 3-5

x (number of ads in a week)	y (number of cars sold that week)
6	15
20	31
0	10
14	16
25	28
16	20
28	40
18	25
10	12
8	15

(a) Draw a scatter diagram of the data.

⟹ The scatter diagram is shown in Figure 3-9 on page 136.

(b) Look at Formulas (1) to (5) pertaining to the least-squares line. Two of the quantities we need to find b are (Σx) and (Σxy). List the others.

⟹ We also need n, (Σy), (Σx^2), and $(\Sigma x)^2$.

(c) Complete Table 3-6(a).

⟹ The missing table entries are shown in Table 3-6(b).

Table 3-6(a)

x	y	x^2	xy
6	15	36	90
20	31	400	620
0	10	0	0
14	16	196	224
25	28	625	700
16	20	256	320
28	40	___	___
18	25	___	___
10	12	___	___
8	15	64	120
$\Sigma x = 145$	$\Sigma y = 212$	$\Sigma x^2 =$ ___	$\Sigma xy =$ ___

Table 3-6(b)

x^2	xy
$(28)^2 = 784$	$28(40) = 1120$
$(18)^2 = 324$	$18(25) = 450$
$(10)^2 = 100$	$10(12) = 120$
$\Sigma x^2 = 2785$	$\Sigma xy = 3764$

Exercise continues

(d) Use Table 3-6(a) to compute SS_x, SS_{xy}, $\bar{x}$, and $\bar{y}$.

$\Rightarrow$ $SS_x = \Sigma x^2 - \dfrac{(\Sigma x)^2}{n} = 2785 - \dfrac{(145)^2}{10} = 682.5$

$SS_{xy} = \Sigma xy - \dfrac{(\Sigma x)(\Sigma y)}{n} = 3764 - \dfrac{(145)(212)}{10} = 690.0$

$\bar{x} = \dfrac{\Sigma x}{n} = \dfrac{145}{10} = 14.5$

$\bar{y} = \dfrac{\Sigma y}{n} = \dfrac{212}{10} = 21.2$

(e) Compute a and b in the formula

$y = a + bx$

for the least-squares line. What is the equation of the least-squares line?

$\Rightarrow$ $b = \dfrac{SS_{xy}}{SS_x} = \dfrac{690.0}{682.5} \approx 1.01$

$a = \bar{y} - b\bar{x}$

$= 21.2 - 1.01(14.5)$

$= 6.56$

The equation for the least-squares line is

$y = 6.56 + 1.01x$

(f) Plot the least-squares line on your scatter diagram.

$\Rightarrow$ The least-squares line goes through the point $(\bar{x}, \bar{y}) = (14.5, 21.2)$. To get another point on the line, select a value for x and compute the corresponding y value using the equation $y = 6.56 + 1.01x$. For $x = 20$, we get $y = 6.56 + 1.01(20) = 26.8$, so the point $(20, 26.8)$ is also on the line. The least-squares line is shown in Figure 3-9.

(g) Read the y value for $x = 12$ from your graph. Then use the equation of the least-squares line to calculate y when $x = 12$. How many cars can the manager expect to sell if 12 ads per week are aired on TV?

$\Rightarrow$ The graph gives $y \approx 19$. From the equation we get

$y = 6.56 + 1.01x$

$= 6.56 + 1.01(12)$ using 12 in place of x

$= 18.68$

To the nearest whole number, the manager can expect to sell 19 cars when 12 ads are aired on TV each week.

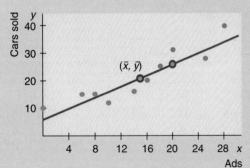

◀ **FIGURE 3-9** Scatter Diagram and Least-Squares Line for Table 3-5

The least-squares line

$$y = a + bx$$

was developed with y as the response variable and x as the explanatory variable. This model can be used only to predict y values from specified x values. If you wish to begin with y values and predict corresponding x values, you must use a different formula. Such a formula would be developed using a model with x as the response variable and y as the explanatory variable (see problem 10 at the end of this section). For our purposes, we'll always arrange to predict y values from given x values.

Effect of extreme data
points

The least-squares line can be affected greatly by extreme data points. This fact is illustrated in Example 2.

EXAMPLE 2 ➤ Consider the two data sets shown below. They are identical except for the y coordinate of the last point.

Set 1	3	5	7	9	15	20	25	27	33	35
	60	55	57	50	47	44	48	40	35	30
Set 2	3	5	7	9	15	20	25	27	33	35
	60	55	57	50	47	44	48	40	35	55

Figure 3-10 shows the scatter plots for these two data sets as well as the equation of the least-squares line for each set. Compare the slopes of the lines. Notice that the line for data set 2 has a slope closer to zero. This line is not as steep.

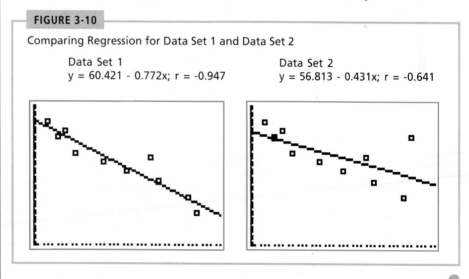

FIGURE 3-10

Comparing Regression for Data Set 1 and Data Set 2

Data Set 1
$y = 60.421 - 0.772x$; $r = -0.947$

Data Set 2
$y = 56.813 - 0.431x$; $r = -0.641$

Sometimes a scatter diagram clearly indicates the existence of a linear relationship between x and y, but it can happen that the points are widely scattered around the least-squares line. We need a method (besides just looking) for measuring the spread of points about the least-squares line. The next section gives us such a measure. It is the value of r shown in Figure 3-10.

V I E W P O I N T

It's Freezing!

Can we use average temperatures in January to predict how bad the rest of the winter will be? Can you predict the number of days with freezing temperatures for the entire calendar year using conditions in January? How good would such a forecast be for predicting growing season or number of frost-free days? The methods presented in this section can help you answer such questions. For more information, see Web site <http://www.ugems.psu.edu/~owens/min32.html>.

SECTION 3.2 PROBLEMS

For Problems 1–8, do the following:
(a) Draw a scatter diagram for the data.
(b) Find $\bar{x}$, $\bar{y}$, and b. Then find the equation of the least-squares line.
(c) Graph the least-squares line on your scatter diagram. Be sure to use the point $(\bar{x}, \bar{y})$ as one of the points on the line.

Basketball: Number of Fouls

1. Data for this problem are based on information from *STATS Basketball Scoreboard*. It is thought that basketball teams that make too many fouls in a game tend to lose the game even if they play well otherwise. Let x be the number of fouls more than (i.e., over and above) the opposing team. Let y be the percentage of times the team with the larger number of fouls wins the game.

x	0	2	5	6
y	50	45	33	26

Complete parts a through c.
(d) If a team had $x = 4$ fouls over and above the opposing team, what does the least-squares equation forecast for y?

Veterinary Science: Calves

2. You are the foreman of the Bar-S cattle ranch in Colorado. A neighboring ranch has calves for sale, and you are going to buy some calves to add to the Bar-S herd. How much should a healthy calf weigh? Let x be the age of the calf (in weeks), and let y be the weight of the calf (in kilograms). The following data are based on information taken from *The Merck Veterinary Manual* (a reference used by many ranchers).

x	1	3	10	16	26	36
y	42	50	75	100	150	200

Complete parts a through c.

(d) The calves you want to buy are 12 weeks old. What does the least-squares line predict for a healthy weight?

Environment:
Gas Consumption

3. Do heavier cars really use more gasoline? Suppose that a car is chosen at random. Let x be the weight of the car (in hundreds of pounds), and let y be the miles per gallon (mpg). The following data are based on information taken from *Consumer Reports* (Vol. 62, No. 4).

x	27	44	32	47	23	40	34	52
y	30	19	24	13	29	17	21	14

Complete parts a through c.
(d) Suppose that a car weighs $x = 38$ (hundred pounds). What does the least-squares line forecast for $y =$ miles per gallon?

Psychology:
Irrelevant Responses

4. A child psychiatrist is studying the mental development of children. A random sample of nine children were given a standard set of questions appropriate to the age of each child. The number of irrelevant responses to the questions was recorded for each child. In the following data, $x =$ age of child in years and $y =$ number of irrelevant responses.

x	2	3	4	5	7	9	10	11	12
y	15	15	12	13	11	10	8	6	5

Complete parts a through c.
(d) If a child is 9.5 years old, what does the least-squares line predict for the number of irrelevant responses?

Sociology: College vs Income

5. The following data are based on information from the book *Life in America's Small Cities* (by G. S. Thomas, Prometheus Books). Let x be the percentage of those 25 years or older with 4 or more years of college. Let y be the per capita income in thousands of dollars. Five small cities in South Carolina (Greenwood, Hilton Head Island, Myrtle Beach, Orangeburg, and Sumpter) reported the following information regarding the x and y variables.

x	13.8	21.9	12.5	12.7	11.5
y	9.0	10.8	8.8	6.9	7.2

Complete parts a through c.
(d) In a small city in South Carolina where $x = 20$ percent of the population 25 years or older who have had 4 or more years of college, what would the least-squares equation forecast for $y =$ per capita income (in thousands of dollars) in this community?

Sociology: High School
Dropouts vs Income

6. Five small cities in California (El Centro, Eureka, Hanford, Madera, and San Luis Obispo–Atascadero) reported the following information. Let x be the percentage of 16- to 19-year-olds not in school and not high school graduates. Let y be the per capita income in thousands of dollars. The following information was obtained from the reference cited in Problem 5.

x	16.2	9.9	19.5	19.7	9.8
y	7.2	8.8	7.9	8.1	10.3

Complete parts a through c.
(d) In a small city in California where $x = 17$, what would the least-squares equation forecast for $y =$ per capita income (in thousands of dollars) in this community?

Archaeology:
Cultural Affiliation

7. Data for this problem are based on information taken from *Prehistoric New Mexico: Background for Survey* (by D. E. Stuart and R. P. Gauthier, University of New Mexico Press). It is thought that prehistoric Indians did not take their best tools, pottery, and household items when they visited higher elevations for their summer camps. It is hypothesized that archaeological sites tend to lose their cultural identity and specific cultural affiliation as the elevation of the site increases. Let x be the elevation (in thousands of feet) for an archaeological site in the southwestern United States. Let y be the percentage of unidentified artifacts (no specific cultural affiliation) at a given elevation. The following data were obtained for a collection of archaeological sites in New Mexico.

x	5.25	5.75	6.25	6.75	7.25
y	19	13	33	37	62

Complete parts a through c.
(d) At an archaeological site with elevation 6.5 (thousand feet), what does the least-squares equation forecast for $y =$ percentage of culturally unidentified artifacts?

Climate: Frost-free Days

8. Data for this problem are from *Climatology Report No. 77-3* (by J. F. Benci and T. B. McKee, Department of Atmospheric Science, Colorado State University). Let x be the elevation (in thousands of feet), and let y be the average number of frost-free days in a year. For Denver, Gunnison, Aspen, Crested Butte, and Dillon, Colorado, the following data were obtained.

x	5.3	7.7	7.9	8.9	9.8
y	162	63	73	49	21

Complete parts a through c.
(d) Colorado Springs is at an elevation of 6 (thousand feet). What does the least-squares equation forecast for the average number of frost-free days per year in Colorado Springs?

Environment:
Registered Vehicles

9. Does an increase in licensed drivers mean an increase in cars and trucks on the road? One way to answer this question is to look at data from the U.S. Department of Transportation. The Department has data by state, including the District of Columbia, giving the number of licensed drivers and the number of registered vehicles per 1000 residents. The following figure shows a Minitab display of linear regression on the data with the number of licensed drivers as the explanatory variable and the number of registered vehicles as the response variable. Notice that the regression equation is included in the diagram.

**Number of Licensed Drivers vs.
Number of Registered Vehicles per 1000 Residents**

Y = 148.047 + 0.960036X

R-Sq = 14.2%

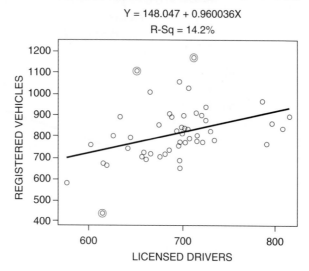

(a) (i)   Use the least-squares line equation $y = 148 + 0.96x$ to predict the number of registered vehicles for 700 licensed drivers per 1000 residents.

(ii)   If the number of licensed drivers increases to 800 per 1000 residents, what is the predicted number of registered vehicles?

(iii)   What is the increase in the number of registered vehicles (i.e., the difference between the numbers of registered vehicles at 700 licensed drivers and at 800 licensed drivers per 1000 residents)?

(b)   Look at the circled points in the figure above. They seem to be somewhat unusual and lie the greatest distance from the least-squares line. The higher points represent the sparsely populated states of Montana and Wyoming, where the number of registered vehicles per licensed driver is particularly high. The lower point represents the District of Columbia, where the number of registered vehicles per licensed driver is particularly low. What would happen if we removed these unusual data points from the model? The figure on the top of page 142 shows the result with these three unusual points removed. Now use the least-squares line $y = 233.5 + 0.83x$ and repeat parts a(i), a(ii), and a(iii).

(c) (i)   At 700 licensed drivers per 1000 residents, the model in part a predicts 5.5 more registered vehicles per 1000 residents than the model in part b. This does not seem to be a large difference. However, how many vehicles does this represent in a state such as Massachusetts, with an estimated population of 6200 thousand residents? What about Texas, with an estimated population of 20,100 thousand residents?

**Number of Licensed Drivers vs.
Number of Registered Vehicles per 1000 Residents**

Y = 233.464 + 0.826216X

R-Sq = 17.2%

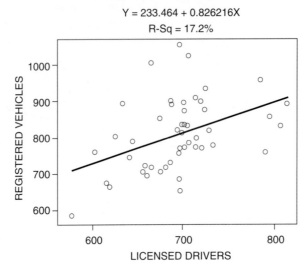

(ii) At 800 licensed drivers per 1000 residents, the model in part a predicts 18.5 more registered vehicles per 1000 residents than the model in part b. What does this difference mean for the number of registered vehicles in Massachusetts? In Texas?

General:
Least-squares Equations

10. (a) Suppose that you are given the following (x, y) data pairs.

x	1	3	4	9	15
y	2	1	6	8	12

Show that the least-squares equation for these data is $y = 1 + 0.75x$ (where we round to three digits after the decimal).

(b) Now suppose that you are given these (x, y) data pairs:

x	2	1	6	8	12
y	1	3	4	9	15

Show that the least-squares equation for these data is $y = -0.448 + 1.18x$ (where we round to three digits after the decimal).

(c) In the data for parts a and b, did we simply exchange the x and y values of each data pair?

(d) Solve $y = 1 + 0.75x$ for x. Do you get the least-squares equation in part b with the symbols x and y exchanged?

(e) In general, suppose that we have the least-squares equation $y = a + bx$ for a set of data pairs (x, y). If we solve this equation for x, will we *necessarily* get the least-squares equation for the set of data pairs (y, x) (with x and y exchanged)? Explain using results of parts a through d.

Section **3.3**

The Linear Correlation Coefficient

If we are given a set of data pairs, we know how to find the equation of the line that "best" predicts y from a given x. This equation is called the *least-squares regression line of y on x*. However, we need more information about how well the line "fits" the data.

Correlation coefficient, r

We need a unitless measurement to describe the strength of the linear association that exists between two variables regardless of which is listed first. Such a measure is the *correlation coefficient r*. The full name for r is the *Pearson product-moment correlation coefficient*, named in honor of the English statistician Karl Pearson (1857–1936), who is credited with formulating r. We'll develop the defining formula for r and then give a more convenient computation formula. Many calculators will give you the value of r directly.

Development of Formula for r

If there is a *positive* linear relation between variables x and y, then high values of x are paired with high values of y and low values of x are paired with low values of y. (See Figure 3-11a.) In the case of *negative* linear correlation, high values of x are paired with low values of y and low values of x are paired with high values of y. This relation is pictured in Figure 3-11b. On the other hand, if there is little or no linear correlation between x and y, then we will find both high and low x values sometimes paired with high y values and sometimes paired with low y values. This relation is shown in Figure 3-11c.

FIGURE 3-11

Patterns of Linear Correlation

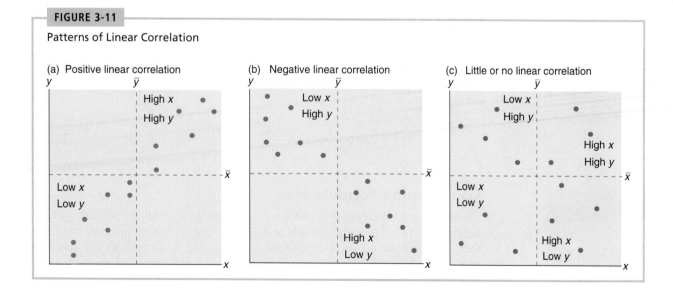

These observations lead us to the development of the formula for the correlation coefficient r. Taking *high* to mean "above the mean," we can express the relationships pictured in Figure 3-11 by considering the products

$$(x - \bar{x})(y - \bar{y})$$

If both x and y are high, both factors will be positive, and the product will be positive as well. The sign of this product will depend on the relative values of x and y compared with their respective means.

$$(x - \bar{x})(y - \bar{y}) \begin{cases} \text{is positive if } x \text{ and } y \text{ are both "high"} \\ \text{is positive if } x \text{ and } y \text{ are both "low"} \\ \\ \text{is negative if } x \text{ is "low" but } y \text{ is "high"} \\ \text{is negative if } x \text{ is "high" but } y \text{ is "low"} \end{cases}$$

In the case of positive linear correlation, most of the products $(x - \bar{x})(y - \bar{y})$ will be positive and so will the sum over all the data pairs

$$\Sigma(x - \bar{x})(y - \bar{y})$$

For negative linear correlation, the products will tend to be negative, so the sum also will be negative. On the other hand, in the case of little if any linear correlation, the sum will tend to be zero.

One trouble with the preceding sum is that it will be larger or smaller depending on the units of x and y. We want r to be unitless so we can make direct comparisons of r values for different data sets (with possibly different units). Since we want r to be unitless, we standardize both x and y of a data pair by dividing each factor $(x - \bar{x})$ by the sample standard deviation s_x and each factor $(y - \bar{y})$ by s_y. Finally, we take an average of all the products. For technical reasons, we take the average by dividing by $n - 1$ instead of by n. This process leads us to the desired measurement, r.

$$r = \frac{1}{n - 1} \Sigma \frac{(y - \bar{y})}{s_y} \cdot \frac{(x - \bar{x})}{s_x}$$

Computation Formula for r

The defining formula for r is awkward to use because of all the subtractions. As before, we can simplify the formula and produce one that is much easier to use. In fact, it should be no surprise that the sums utilized in computing the standard deviations of x and y and the slope of the least-squares line are also used in the computational formula for r.

Formula for Calculating Correlation Coefficient, r

$$r = \frac{SS_{xy}}{\sqrt{SS_x SS_y}}$$
(7)

where $SS_{xy} = \Sigma xy - \dfrac{(\Sigma x)(\Sigma y)}{n}$ *Note:* $SS_{xy} = \Sigma(x - \bar{x})(y - \bar{y})$

$SS_x = \Sigma x^2 - \dfrac{(\Sigma x)^2}{n}$

$SS_y = \Sigma y^2 - \dfrac{(\Sigma y)^2}{n}$

n = number of data pairs in scatter diagram

This formula for r is sensitive to rounding. As in other formulas involving SS_x, SS_{xy}, and SS_y, you want to carry as many digits as is reasonable for your problem until the last step. Again, depending on the rounding process used, answers will vary slightly. The answers for the end-of-section and end-of-chapter problems were computer-generated, so your answers might differ from them slightly and still be essentially correct.

Let us delay an example showing how to compute r until we know a little more about the meaning of the correlation coefficient. It can be shown mathematically that r is always a number between $+1$ and -1 ($-1 \le r \le +1$). Table 3-7 gives a quick summary of some basic facts about r.

Table 3-7 Some Facts About the Correlation Coefficient

If r Is	Then	The Scatter Diagram Might Look Something Like
0	There is no linear relation among the points in the scatter diagram.	
1 or -1	There is a perfect linear relation between x and y values; all points lie on the least-squares line.	

Table continues top of page 146

If *r* Is	Then	The Scatter Diagram Might Look Something Like	
Between 0 and 1 $(0 < r < 1)$	The *x* and *y* values have a *positive correlation.* By this we mean that *large x* values are associated with *large y* values and *small x* values are associated with *small y* values.	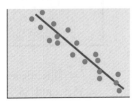	As we go from left to right, the least-squares line goes *up.*
Between −1 and 0 $(-1 < r < 0)$	The *x* and *y* values have a *negative correlation.* By this we mean *large x* values are associated with *small y* values and *small x* values are associated with *large y* values.	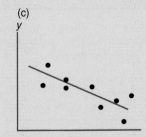	As we go from left to right, the least-squares line goes *down.*

GUIDED EXERCISE   4

Match the appropriate statement about *r* to each scatter diagram in Figure 3-12.

(1) $r = 0$. (3) $r = -1$. (5) *r* is between −1 and 0.

(2) $r = 1$. (4) *r* is between 0 and 1.

FIGURE 3-12 Scatter Diagrams

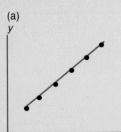

(a) $r = 1$ because all the points are on the line and the line goes up from left to right.

(b) $r = 0$ because there is no apparent linear relation among the points.

(c) *r* is between −1 and 0 because the points are fairly close to the line, and as we read from left to right the least-squares line goes down.

Now let's actually compute r for some data.

EXAMPLE 3 ➤ Most of us have heard someone say that more intelligent people tend to do better in school. Is this always true? Experienced teachers know that it is only partially true. Students with higher IQs (intelligence quotients) often do better schoolwork, but factors other than IQ can affect academic success. However, let's see if there is a correlation between IQ and grade point average (GPA). The principal of Delta High School chose 12 students from the senior class at random and compiled the data in Table 3-8.

Table 3-8   IQ versus GPA (on a Four-Point Scale) of 12 High School Seniors

IQ, x	117	92	102	115	87	76	107	108	121	91	113	98
GPA, y	3.7	2.6	3.3	2.2	2.4	1.8	2.8	3.2	3.8	3.0	4.0	3.5

(a) First, make a scatter diagram, and determine if r is positive, close to 0, or negative.

SOLUTION: The scatter diagram indicates that r is positive. See Figure 3-13.

(b) Compute r.

SOLUTION: To find r, we must compute Σx, Σy, Σx^2, Σy^2, and Σxy. The values for $(\Sigma x)^2$ and $(\Sigma y)^2$ can be obtained from Σx and Σy. It is easiest to organize our work into a table of five columns (Table 3-9). The first two columns are just a repetition of Table 3-8.

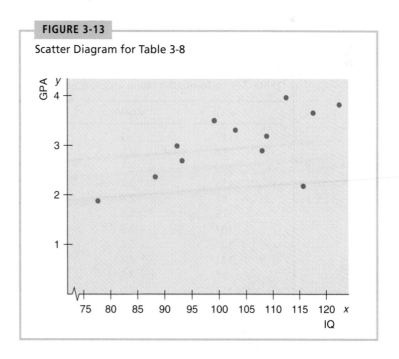

FIGURE 3-13

Scatter Diagram for Table 3-8

To use Formula (7) to calculate r, we will first compute SS_{xy}, SS_x, and SS_y.

$$SS_{xy} = \Sigma xy - \frac{(\Sigma x)(\Sigma y)}{n}$$

$$= 3780.3 - \frac{(1227)(36.3)}{12}$$

$$= 68.63$$

$$SS_x = \Sigma x^2 - \frac{(\Sigma x)^2}{n}$$

$$= 127{,}535 - \frac{(1227)^2}{12}$$

$$= 2074.25$$

$$SS_y = \Sigma y^2 - \frac{(\Sigma y)^2}{n}$$

$$= 114.9 - \frac{(36.3)^2}{12}$$

$$= 5.09$$

Therefore, the correlation coefficient is

$$r = \frac{SS_{xy}}{\sqrt{SS_x SS_y}}$$

$$= \frac{68.63}{\sqrt{(2074.25)(5.09)}}$$

$$= 0.6679 \approx 0.67$$

Table 3-9   Information Necessary to Compute r

x(IQ)	y(GPA)	x^2	y^2	xy
117	3.7	13,689	13.7	432.9
92	2.6	8,464	6.8	239.2
102	3.3	10,404	10.9	336.6
115	2.2	13,225	4.8	253.0
87	2.4	7,569	5.8	208.8
76	1.8	5,776	3.2	136.8
107	2.8	11,449	7.8	299.6
108	3.2	11,664	10.2	345.6
121	3.8	14,641	14.4	459.8
91	3.0	8,281	9.0	273.0
113	4.0	12,769	16.0	452.0
98	3.5	9,604	12.3	343.0
$\Sigma x = 1227$	$\Sigma y = 36.3$	$\Sigma x^2 = 127{,}535$	$\Sigma y^2 = 114.9$	$\Sigma xy = 3780.3$
$(\Sigma x)^2 = 1{,}505{,}529$	$(\Sigma y)^2 = 1317.7$			

Our correlation coefficient is $r \approx 0.67$. Let's make sure that this answer agrees with what we expect from a quick glance at the scatter diagram (Figure 3-13). In Figure 3-13 the general trend is upward as we read from left to right, so we would expect a positive value for r. The value 0.67 is in the expected range—that is, it is between 0 and 1.

It is quite a task to compute r for even 12 data pairs. The use of columns as in Example 3 is extremely helpful. Your value for r should always be between -1 and 1. Use a scatter diagram to get a rough idea of the value of r. If your computed value of r is outside the allowable range or if it disagrees quite a bit with the scatter diagram, recheck your calculations. Be sure you distinguish between expressions such as (Σx^2) and $(\Sigma x)^2$. Negligible rounding errors may occur depending on how you (or your calculator) round.

Calculator Note Most calculators that support two-variable statistics provide the value of the correlation coefficient r directly. The screens below show the results of Example 3 as provided by the TI-83 calculator. [On the TI-83 calculator, first use **CATALOGUE**, find **DiagnosticOn**, and press **ENTER** twice in order to show the values for r and r^2 when you use **STAT, CALC**, option **8:LinReg(a+bx)**.]

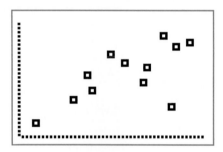

```
LinReg
 y=a+bx
 a=-.3578642883
 b=.0330842473
 r²=.4414985848
 r=.6644535987
```

GUIDED EXERCISE   5

In one of the Boston city parks there has been a problem with muggings in the summer months. A police cadet took a random sample of 10 days (out of the 90-day summer) and compiled the following data. For each day, x represents the number of police officers on duty in the park and y represents the number of reported muggings on that day.

x	10	15	16	1	4	6	18	12	14	7
y	5	2	1	9	7	8	1	5	3	6

(a) Construct a scatter diagram of x and y values.   ⟹   Figure 3-14 shows the scatter diagram.

Exercise continues

Exercise continued

FIGURE 3-14 Scatter Diagram for Number of Police Officers Versus Number of Muggings

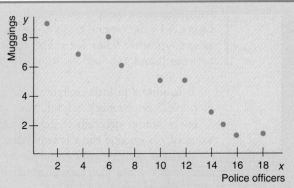

(b) From the scatter diagram, do you think the computed value of r will be positive, negative, or zero? Explain.

(c) Complete Table 3-10.

Table 3-10

x	y	x^2	y^2	xy
10	5	100	25	50
15	2	225	4	30
16	1	256	1	16
1	9	1	81	9
4	7	16	49	28
6	8	___	___	___
18	1	___	___	___
12	5	___	___	___
14	3	___	___	___
7	6	49	36	42
$\Sigma x = 103$	$\Sigma y = 47$	$\Sigma x^2 =$ __	$\Sigma y^2 =$ __	$\Sigma xy =$ __
$(\Sigma x)^2 =$ __	$(\Sigma y)^2 =$ __			

(d) Compute SS_{xy}, SS_x, SS_y, and then r.

⇨ r will be negative. The general trend is that large x values are associated with small y values and vice versa. From left to right, the least-squares line goes down.

⇨ **Table 3-11 Completion of Table 3-10**

x	y	x^2	y^2	xy
6	8	36	64	48
18	1	324	1	18
12	5	144	25	60
14	3	196	9	42
		$\Sigma x^2 = 1{,}347$	$\Sigma y^2 = 295$	$\Sigma xy = 343$
		$(\Sigma x)^2 = 10{,}609$	$(\Sigma y)^2 = 2209$	

⇨

$$SS_{xy} = \Sigma xy - \frac{(\Sigma x)(\Sigma y)}{n}$$

$$= 343 - \frac{(103)(47)}{10} = -141.1$$

$$SS_x = \Sigma x^2 - \frac{(\Sigma x)^2}{n}$$

$$= 1347 - \frac{(103)^2}{10} = 286.1$$

$$SS_y = \Sigma y^2 - \frac{(\Sigma y)^2}{n}$$

$$= 295 - \frac{(47)^2}{10} = 74.1$$

$$r = \frac{SS_{xy}}{\sqrt{SS_x SS_y}}$$

$$= \frac{-141.1}{\sqrt{(286.1)(74.1)}} = -0.9691 \approx -0.97$$

The correlation coefficient can be thought of as another measure of how "good" the least-squares line fits the data points of the scatter diagram. The closer r is to $+1$ or -1, the better the least-squares line "fits" the data. Values of r close to 0 indicate a poor "fit."

Usually our scatter diagram does not contain *all* possible data points that could be gathered. Most scatter diagrams represent only a *random sample* of data pairs taken from a very large population of possible pairs. Because r is computed by Formula (7) on the basis of a random sample of (x, y) pairs, we expect the values of r to vary from one sample to the next (much as sample means $\bar{x}$ varied from sample to sample). This brings up the question of the *significance* of r. Or put another way, what are the chances that our random sample of data pairs indicates a high correlation when in fact the population x and y values are not so strongly correlated? Right now let us just say the significance of r is a separate issue that is left to Chapter 11.

COMMENT As we use computing formulas for the slope of the least-squares line, for r, and for standard deviations s_x and s_y, we see many of the same sums used. There is, in fact, a relationship between the correlation coefficient r and the slope of the least-squares line b. In instances when we know r, s_x, and s_y, we can use the following formula to compute b.

$$b = r\left(\frac{s_y}{s_x}\right)$$

Coefficient of Determination

There is another way to answer the question "How good is the least-squares line as an instrument of regression?" The *coefficient of determination* r^2 is the square of the sample correlation coefficient r.

> Coefficient of determination $= r^2$

The coefficient of determination r^2 is a measure of the proportion of variation in y that is explained by the regression line using x as the predicting variable. If $r = 0.90$, then $r^2 = 0.81$ is the coefficient of determination, and we can say that about 81% of the (variation) behavior of the y variable can be explained by the corresponding (variation) behavior of the x variable if we use the equation of the least-squares line. The remaining 19% of the (variation) behavior of the y variable is due to random chance or to the presence of other variables besides the x that may influence y.

(a) In Example 3 we found the correlation coefficient r of the relationship between IQ and GPA. In that case, r was 0.67. How would you describe the strength of this relationship?

⮕ The correlation coefficient $r = 0.67$ is moderate, but not extremely high. It seems that other factors besides IQ are significant in determining GPA.

(b) Compute the coefficient of determination for the data in Example 3 and comment on the meaning of this number.

⮕ Since $r = 0.67$, then $r^2 = 0.449$ is the value of the coefficient of determination. This says that 44.9% of the variation of $y = $ GPA can be explained by the least-squares line and $x = $ IQ. The remaining $100 - 44.9 = 55.1\%$ of the variation of y is due to random chance or other variables besides x that influence y (possibly, amount of time spent studying).

(c) In Guided Exercise 5, dealing with the relation between the number of police officers in the park and the number of muggings, we found r to be -0.97. How would you describe the strength of this relationship? Do you think the city is justified in asking for more police officers to be assigned to park duty?

⮕ $r = -0.97$ is a high correlation. The relation between the number of police officers in the park and the number of muggings in the park is a strong and dependable negative correlation. The authors feel that the city would be wise to hire more police officers to patrol the park, but many other aspects of the situation must be considered. Perhaps more crimes would be prevented by putting those officers elsewhere.

(d) Compute the coefficient of determination for the data in Guided Exercise 5 and comment on the meaning of this number.

⮕ Since $r = -0.97$, then $r^2 = 0.941$ is the coefficient of determination. About 94.1% of the variation of y can be explained by the least-squares line and the x variable. The remaining $100 - 94.1 = 5.9\%$ of the variation of y is due to random chance or other variables besides x that influence y.

Causation

The correlation coefficient is a mathematical tool for measuring the strength of the linear relationship between two variables. As such, it makes no implication about cause or effect. Just because two variables tend to increase or decrease together does not mean a change in one is *causing* a change in the other. A strong correlation between x and y is sometimes due to other (either known or unknown) variables.

EXAMPLE 4 ➤ Over a period of years, a certain town observed that the correlation between x, the number of people attending churches, and y, the number of people in the city jail, was $r = 0.90$.

Does going to church cause people to go to jail? We hope not. During this period, there was a steady increase in population. Therefore, it is not too surprising that both the number of people attending churches and the number of people in jail increased together. The high correlation between x and y was due to the common effect of the increase in the general population.

V I E W P O I N T

The Best of Times

Many American runners have now beaten the 4-minute mile. In fact, as the years go by, it seems that the best times for running the mile get shorter and shorter. Could there be a correlation here? What would the correlation coefficient be? Could you forecast next year's best time for running the mile? For more information, see Web site: <http://www.runnersworld.com/home.html>, and look under statistics for sub-4-minute milers.

S E C T I O N   3 . 3   P R O B L E M S

Driving: Traffic Accidents

1. Over the past 10 years, there has been a high positive correlation between the number of South Dakota safety inspection stickers issued and the number of South Dakota traffic accidents.
   (a) Do safety inspection stickers cause traffic accidents?
   (b) What third factor might cause traffic accidents and the number of safety stickers to increase together?

General: Teachers' Salaries

2. There is a high positive correlation in the United States between teachers' salaries and annual consumption of liquor.
   (a) Do you think increasing teachers' salaries has caused increased liquor consumption?
   (b) As teachers' salaries have been going up, most other salaries have been going up, too. To some extent this means an upward trend in buying power for everyone. How might this explain the high correlation between teachers' salaries and liquor consumption?

General: Infant Deaths

3. Over the past 30 years in the United States there has been a strong negative correlation between the number of infant deaths at birth and the number of people over age 65.
   (a) Is the fact that people are living longer causing a decrease in infant mortalities at birth?
   (b) What third factor might be decreasing infant mortalities and at the same time increasing life span?

General: Diet Soda Pop

4. Over the past few years, there has been a strong positive correlation between the annual consumption of diet soda pop and the number of traffic accidents.
    (a) Do you think that an increasing consumption of diet pop has led to more traffic accidents?
    (b) What third factor or factors might be causing both the annual consumption of diet pop and the number of traffic accidents to increase together?

For Problems 5–9, do the following:
(a) Draw a scatter diagram for the data.
(b) From the scatter diagram, would you estimate the correlation coefficient r to be closest to 1, 0, or -1?
(c) Compute the correlation coefficient r and the coefficient of determination r^2. What percentage of the variation in y can be *explained* by the corresponding variation in x using the least-squares line? What percentage of the variation is *unexplained*?

Business: Travel Cost

5. Data for this problem are based on information taken from the *Wall Street Journal* (Dow Jones Travel Index). For randomly selected major U.S. cities, let x be the cost of daily car rental, and let y be the cost of daily room rental.

x ($)	52	51	43	65	45	39
y ($)	152	190	117	222	134	131

Complete parts a through c for these data, and comment on the meanings of r and r^2 in the context of this problem.

Leisure: Airfare

6. Data for this problem are based on information taken from the *Wall Street Journal* (Dow Jones Travel Index). For randomly selected major U.S. cities, let x be the cost of round-trip leisure airfare (overnight Saturday required) between cities, and let y be the cost of round-trip business airfare between the same cities.

x ($)	264	144	222	163	262
y ($)	804	284	510	524	994

Complete parts a through c for these data, and comment on the meanings of r and r^2 in the context of this problem.

Driving: Fatal Accidents

7. Data for this problem are based on information taken from the *Wall Street Journal*. Let x be the age in years of a licensed automobile driver. Let y be the percentage of all fatal accidents (for a given age) due to speeding. For example, the first data pair indicates that 36% of all fatal accidents of 17-year-olds are due to speeding.

x	17	27	37	47	57	67	77
y	36	25	20	12	10	7	5

Complete parts a through c for these data, and comment on the meanings of r and r^2 in the context of this problem.

Driving Fatal Accidents

8. Let x be the age of a licensed driver in years. Let y be the percentage of all fatal accidents (for a given age) due to failure to yield the right of way. For example, the first data pair says that 5% of all fatal accidents of 37-year-olds are due to failure to yield the right of way. The *Wall Street Journal* article referenced in Problem 7 reported the following data.

x	37	47	57	67	77	87
y	5	8	10	16	30	43

Complete parts a through c for these data, and comment on the meanings of r and r^2 in the context of this problem.

Climate: Tropical Cyclones

9. Can a low barometer reading be used to predict maximum wind speed of an approaching tropical cyclone? Data for this problem are based on information taken from *Weatherwise* (Vol. 46, No. 1), a publication of the American Meteorological Society. For a random sample of tropical cyclones, let x be the lowest pressure (in millibars) as a cyclone approaches, and let y be the maximum wind speed (in miles per hour) of the cyclone.

x	1004	975	992	935	985	932
y	40	100	65	145	80	150

Complete parts a through c for these data, and comment on the meanings of r and r^2 in the context of this problem.

Academic:
Faculty vs Student

10. Is there a strong correlation between the number of students matriculated at a college and the number of faculty members at that college? *Peterson's College Guide* gives a great deal of information about colleges in the United States. Among the data presented is the number of students matriculated at a college and the number of faculty members for two-year colleges.

Silva took a random sample of 18 of the colleges, and used Minitab to obtain the correlation coefficient (▶**Basic Statistics** ▶**Correlation**). Her printout was

CORRELATIONS (PEARSON)

```
Correlation of St_2yr and Fac_2yr = 0.913, P-Value = 0.000
```

George took another random sample of 18 of the colleges and used Minitab to obtain the printout

CORRELATIONS (PEARSON)

```
Correlation of St_2yr and Fac_2y = 0.864, P-Value = 0.000
```

Lucinda combined the sample data gathered by Silva and George and obtained the printout

CORRELATIONS (PEARSON)

```
Correlation of St_2yr and Fac_2yr = 0.909, P-Value = 0.000
```

(a) Compare the correlation coefficients obtained by Silva, George, and Lucinda. Are they the same or different?

(b) Based on your observations, would you say that different samples from the same population may produce different values for the sample Pearson correlation coefficient? Would you say that r is a measurement that depends on the sample?

General:
Correlation Coefficient

11. Examine the computation formula for r, the sample correlation coefficient [Formula (7) in this section].

(a) In the formula for SS_{xy}, if we exchange the symbols x and y, do we get a different result or do we get the same (i.e., equivalent) result? Explain.

(b) In the formula for r [see Formula (7) in this section], if we exchange the symbols x and y, do we get a different result or do we get the same (equivalent) result? Explain.

(c) If we have a set of x and y values and we exchange each corresponding x and y value to get a new data set, should the sample correlation coefficient be the same for both sets of data? Explain.

(d) Compute the sample correlation coefficient r for each of the following data sets and show that they are the same.

x	1	3	4	9	15	x	2	1	6	8	12
y	2	1	6	8	12	y	1	3	4	9	15

What can you say about the least-squares equation for each data set? Are the equations algebraically equivalent? (*Hint:* See Problem 10 in Section 3.2.)

SUMMARY

We can often get information about a population of y values by attempting to predict them from known values of another population x. If data pairs (x, y) are obtained from an overall population of such data pairs, we may use the equation of the least-squares line

$$y = a + bx$$

to predict y values from given x values. In this equation, a and b are computed from Formulas (2) to (5) in Section 3.2. When we use the least-squares line to predict y values from x values, we call y the *response variable* and x the *explanatory variable*.

If two sets of data are obtained from the same sample, the extent to which they are related can be measured by the Pearson product-moment correlation coefficient r given in Formula (7) in Section 3.3. The quantity r^2 is called the *coefficient of determination*. We use this value to measure the proportion of variation in y explained by the least-squares line.

Scatter diagrams of (x, y) values are helpful in determining visually if there is any linear relation between the x and y values and, if so, how strong the relation might be.

ⓘMPORTANT WORDS & SYMBOLS

Section 3.1
Paired data values
Scatter diagram
Moderate linear correlation
Perfect linear correlation
No linear correlation

Section 3.2
Explanatory variable
Response variable
Linear regression

Least-squares criterion
Least-squares line $y = a + bx$
Interpolation
Extrapolation

Section 3.3
Correlation coefficient r (later called sample
    correlation coefficient)
Linear correlation
Positive, negative, and zero correlation
Coefficient of determination r^2

ⓋIEWPOINT

Living Arrangements

Male, female, married, single, living alone, living with friends or relatives—
all these categories are of interest to the U.S. Census Bureau. In addition to
these categories, there are others, such as age, income, and health needs.
How strongly correlated are these variables? Can we use one or more of
these variables to predict the others? How good is such a prediction? The
methods discussed in this chapter can help you answer such questions. For
more information regarding such data, see Web site <http://www.census.gov>.

ⒸHAPTER REVIEW PROBLEMS

Wildlife:
Desert Bighorn Sheep

1. Bighorn sheep are beautiful wild animals found throughout the western United
   States. Data for this problem are based on information taken from *The Desert
   Bighorn,* edited by Monson and Sumner (University of Arizona Press). Let x be the
   age of a bighorn sheep (in years), and let y be the mortality rate (percent that die)
   in this age group. For example, $x = 1$, $y = 14$ means that 14% of the bighorn
   sheep between 1 and 2 years old died. A random sample of Arizona bighorn sheep
   gave the following information.

x	1	2	3	4	5
y	14	18.9	14.4	19.6	20.0

   (a) Draw a scatter diagram.
   (b) Find the equation of the least-squares line.
   (c) Find r. Find the coefficient of determination r^2.

2. A sociologist is interested in the relation between x = number of job changes and y = annual salary (in thousands of dollars) for people living in the Nashville area. A random sample of 10 people employed in Nashville provided the following information.

x (no. of job changes)	4	7	5	6	1	5	9	10	10	3
y (salary in $1000)	53	57	54	52	52	58	63	57	60	53

(a) Draw a scatter diagram for the data.
(b) Find $\bar{x}, \bar{y}$, and b and then the equation of the least-squares line.
(c) Graph the least-squares line on your scatter diagram.
(d) If someone had $x = 2$ job changes, what does the least-squares line predict for y, the annual salary?
(e) Looking at the scatter diagram for the least-squares line, do you think the correlation coefficient will be positive, negative, or zero?
(f) Find r. Find the coefficient of determination.

3. Modern medical practice tells us not to encourage babies to become too fat. Medical research indicates that there is a positive correlation between the weight x of a 1-year-old baby and the weight y of a mature adult (30 years old). A random sample of medical files produced the following information for 14 females.

x (lb)	21	25	23	24	20	15	25	21	17	24	26	22	18	19
y (lb)	125	125	120	125	130	120	145	130	130	130	130	140	110	115

(a) Draw a scatter diagram for the data.
(b) Find $\bar{x}, \bar{y}$, and b and the equation of the least-squares line.
(c) Graph the least-squares line on your scatter diagram.
(d) If a female baby weighs 20 lb at 1 year, what would you predict she would weigh at 30 years of age?
(e) Looking at the scatter diagram, do you think the correlation coefficient is positive, negative, or zero?
(f) Find r. Find the coefficient of determination.

4. Dorothy Kelly sells life insurance for the Prudence Insurance Company. She sells insurance by making visits to her clients' homes. Dorothy believes that the number of sales should depend to some degree on the number of visits made. For the past several years she has kept careful records of the number of visits (x) she made each week and the number of people (y) who bought insurance that week. For a random sample of 15 such weeks, the x and y values are as follows.

x	11	19	16	13	28	5	20	14	22	7	15	29	8	25	16
y	3	11	8	5	8	2	5	6	8	3	5	10	6	10	7

(a) Draw a scatter diagram for the data.
(b) Find $\bar{x}, \bar{y}$, and b and the equation of the least-squares line.
(c) Graph the least-squares line on your scatter diagram.
(d) On a week in which Dorothy made 18 visits, how many people would you predict would buy insurance from her?
(e) Find the correlation coefficient and the coefficient of determination.

Business: Cereal Coupons

5. Each box of Healthy Crunch breakfast cereal contains a coupon entitling you to a free package of garden seeds. At the Healthy Crunch home office they use the weight of incoming mail to determine how many of their employees are to be assigned to collecting coupons and mailing out seed packages on a given day. (Healthy Crunch has a policy of answering all its mail the day it is received.)

Let x be the weight of incoming mail and y be the number of employees required to process the mail in one working day. A random sample of 8 days gave the following data.

x (lb)	11	20	16	6	12	18	23	25
y (no. of employees)	6	10	9	5	8	14	13	16

(a) Draw a scatter diagram for the data.
(b) Find $\bar{x}$, $\bar{y}$, and b and the equation of the least-squares line.
(c) Graph the least-squares line on your scatter diagram.
(d) If Healthy Crunch receives 15 lb of mail, how many employees should be assigned mail duty?
(e) Find r. Find the coefficient of determination.

DATA HIGHLIGHTS: GROUP PROJECTS

Break into small groups and discuss the following topics. Organize a brief outline in which you summarize the main points of your group discussion.

All the problems use the following data. You may find it helpful to use a calculator with two-variable statistics or appropriate computer software to compute the specified quantities.

Plate tectonics and the spread of the ocean floor are very important in modern studies of earthquakes and earth science in general. The following data give x = age of volcanic islands in the Atlantic and Indian Oceans and y = distance of the island from the center of the midoceanic ridge. As you can see, the oldest islands are the farthest from the ridge crest. This fact is explained by the spreading of the ocean floor on which the islands stand. The following data are adapted from Cuchlaine A. M. King, *Physical Geography* (Oxford: Basil Blackwell, 1980, pp. 196–206) by permission of Basil Blackwell Limited, Oxford, England.

x (age $\times 10^6$ years)	120	120	120	83	60	50	50	50	35	35
y (distance $\times 10^3$ km)	3.0	2.2	2.0	1.6	1.55	1.45	0.6	0.2	2.2	1.6
x (age $\times 10^6$ years)	30	20	20	20	17	10	2	1	0	
y (distance $\times 10^3$ km)	1.8	1.2	0.7	0.3	0.0	0.0	0.0	0.2	0.0	

1. Make a scatter diagram of the data. Does there appear to be a linear relation between the age of the islands x and the distance y from the center of the midoceanic ridge?

2. Find the equation of the least-squares line, and graph it on your scatter plot. Find the values of $\bar{x}$ and $\bar{y}$.

3. Compute the sample correlation coefficient r.

4. If an island is 70 million years old, how far does the least-squares line predict it will be from the midoceanic ridge crest?

5. If an island is estimated to be 100 million years old, how far does the least-squares line predict it will be from the midoceanic ridge crest?

6. Consider the number b, which is the slope of the least-squares line. How can this number be used to estimate the rate at which the ocean floor is moving? In 1 year, how many centimeters will the ocean floor be expected to move? (*Hint:* Express the units of b in centimeters per year.)

⒧INKING CONCEPTS: WRITING PROJECTS

Discuss each of the following topics in class or review the topics on your own. Then write a brief but complete essay in which you summarize the main points. Please include formulas and graphs as appropriate.

1. What do we mean when we say two variables have a strong positive (or negative) linear correlation? What would a scatter diagram for these variables look like? Is it possible that two variables could be strongly related somehow, but have a low *linear* correlation? Explain and draw a scatter diagram to demonstrate your point.

2. What do we mean by the least-squares criterion? Give a very general description of how the least-squares criterion is involved in the construction of the least-squares line. Why do we say the least-squares line is the "best-fitting" line for the data set?

3. Use the Internet or go to the library and find a magazine or journal article in your field of major interest where the content of this chapter could be applied. List the variables used, the method of data collection, and the general type of information used and conclusions drawn.

The problems in this section may be done using statistical computer software or calculators that have statistical functions. Displays and suggestions are given for Minitab (Release 12), the TI-83 graphing calculator, ComputerStat, and Excel.

Simple Linear Regression (One Explanatory Variable)

Minitab

Minitab provides several options for finding the equation of the least-squares line when there is only one explanatory variable.

The menu selection ➤Stat ➤Regression ➤Fitted Line Plot produces a scatter diagram, the equation of the least-squares line, and the value of the coefficient of determination r^2. In the dialogue box, list the column containing the response variable. The predictor variable refers to the column containing the explanatory variable. Under the Storage option you may check fits to get a column of predicted values y_p for the explanatory variable data.

The menu selection ➤Stat ➤Basic Statistics ➤Correlation produces the Pearson correlation coefficient r for the explanatory and response variable columns specified in the dialogue box.

Use the menu selection ➤Stat ➤Regression ➤Regression for predictions, select ➤Options . . . , and list the value of x as a new observation. (You will also get a confidence interval for the prediction, which is discussed in Section 11.4.)

TI-83

Enter the values of the explanatory variable x in list L_1 and the corresponding values of the response variable in list L_2. Press STAT, select CALC, and use option 8: LinReg(a + bx). The coefficients for the regression line will appear as well as the value of the correlation coefficient r and r^2. [*Note:* Be sure that you have entered DiagnosticOn (located under CATALOG) first. Once DiagnosticOn is entered, it remains on until you enter DiagnosticOff.]

To draw a scatter plot, press STAT PLOT and select 1:Plot 1. Turn the plot on, use Type showing a scatter plot, select L_1 for Xlist and L_2 for Ylist, and pick the little square for the Mark. Then press ZOOM and select option 9:ZoomStat.

To draw the least-squares line on your scatter plot, press Y=. Then press VARS and select item 5:Statistics. In the menu appearing on the screen, select EQ and then select item 1:RegEQ. This sequence of choices will automatically set Y_1 = your regression equation. Press GRAPH.

To use the least-squares line to make a prediction for a specified x value, press the CALC key and select item 1:Value. A prompt X= will appear at the bottom of the graph. Enter the desired x value, and the corresponding y value will be computed and displayed. If your x value is outside the data range for x, you will need to adjust the graphing window to include it.

ComputerStat

In ComputerStat, select the program Linear Regression and Correlation, and follow the instructions on the screen.

161

Excel

In Excel, there is a group of commands under the paste function f_x button that provide the slope of the least-squares line, the intercept, the correlation coefficient, and a forecast value.

The menu choices ➤ f_x button ➤Statistical ➤Slope produce a dialogue box in which you select the cells containing the known x's (explanatory variable) and the cell containing the known y's (response variable). Then the slope b of the least-squares line $y = a + bx$ is given.

The menu choices ➤f_x button ➤Statistical ➤Intercept produce a similar dialogue box. The result gives the intercept a of the least-squares line $y = a + bx$.

The menu choices ➤f_x button ➤Statistical ➤Correl produce a similar dialogue box. The result gives the Pearson product-moment correlation coefficient r.

The menu choices ➤f_x button ➤Statistical ➤Forecast produce a dialogue box asking for the x value for which you wish a prediction and the cells containing the y values and the cells containing the x values. The result gives the predicted value y_p from the least-squares line for the designated x value.

The menu choices ➤Tools ➤Data Analysis ➤Regression produce summary statistics regarding regression, including the coefficients for the least-squares line, the value of r, and the value of r^2.

A graphical display can be designed using the chart wizard to give the scatter plot, the graph of the least-squares line, the equation of the least-squares line, and the value of r^2.

APPLICATIONS

The data in this section are taken from:

> King, Cuchlaine A. M. *Physical Geography*. Oxford: Basil Blackwell, 1980, 77–86, 196–206. Reprinted with permission of Basil Blackwell Limited, Oxford, England.

Throughout the world, natural ocean beaches are beautiful sights to see. If you have visited natural beaches, you may have noticed that when the gradient or dropoff is steep, the grains of sand tend to be larger. In fact, a manmade beach with sand granules of the "wrong" size tends to be washed away and eventually replaced when the proper grain size is selected by the action of the ocean and the gradient of the bottom. Since manmade beaches are expensive, grain size is an important consideration.

In the data that follows, x = median diameter (in millimeters) of granules of sand and y = gradient of beach slope in degrees on natural ocean beaches.

To complete these problems, use ComputerStat as described or do equivalent calculations on your software or calculator.

In the ComputerStat menu of topics, select **Linear Regression** and **Correlation**.

1. We have nine data pairs (x, y). Enter the data shown below as directed by the computer.

x	y
0.17	0.63
0.19	0.70
0.22	0.82
0.235	0.88
0.235	1.15
0.30	1.50
0.35	4.40
0.42	7.30
0.85	11.30

2. Scan the information summary. What is the value of $\bar{x}$? Of $\bar{y}$? What are the values of the slope and intercept of the least-squares line? What is the value of the sample correlation coefficient?

3. Select the graphing option. First, graph the data points. Just looking at the scatter diagram, would you expect moderately high correlation and a good fit for the least-squares line? Press return and graph the least-squares line.

4. Suppose that you have a truckload of sifted sand in which the median size of the granules is 0.38 mm. If you want to put this sand on a beach and you don't want the sand to wash away, then what does the least-squares line predict for the angle of the beach? *Note:* Heavy storms that produce abnormal waves may also wash out the sand. However, in the long run, the sizes of sand granules that remain on the beach or that are brought back to the beach by long-term wave action are determined to a large extent by the angle at which the beach drops off.

To solve the problem, select the option to predict y values from x values. Enter 0.38 for your x value.

5. Suppose we now have a truckload of sifted sand where the median size of the granules is 0.45 mm. Repeat Problem 4 for this new load of sand.

Computer Displays

Computer software support is very convenient for linear regression problems of every kind. (See Figures 3-15, 3-16, and 3-17.)

FIGURE 3-15 ComputerStat Display: Linear Regression for Cricket Chirps/S with Temperature

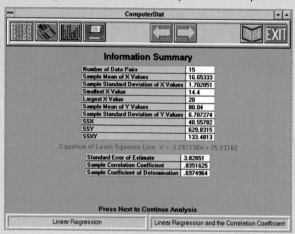

FIGURE 3-16 Minitab Display: Linear Regression for Cricket Chirps/S with Temperature

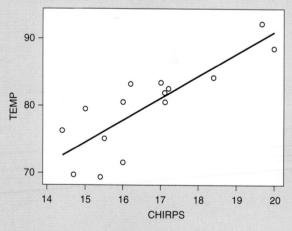

```
The regression equation is
TEMP = 25.2 + 3.29 CHIRPS

Predictor     Coef    Stdev    t-ratio      p
Constant     25.23    10.06       2.51  0.026
CHIRPS      3.2911   0.6012       5.47  0.000

s = 3.829   R-Sq = 69.7%  R-Sq(adj) = 67.4%
```

163

```
Analysis of Variance

SOURCE        DF       SS       MS       F       p
Regression     1    439.29   439.29   29.97   0.000
Error         13    190.55    14.66
Total         14    629.84

  Fit   Stdev.Fit      95% C.I.           95% P.I.
77.890     1.064   ( 75.591, 80.188)( 69.303, 86.476)
```

FIGURE 3-17 Excel Display: LInear Regression for Cricket Chirps/S with Temperature

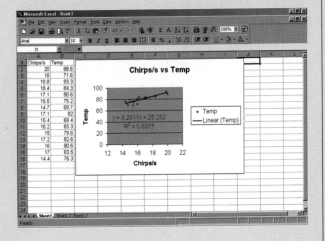

Select ➤**Chart Wizard Button** ➤**xy (scatter)** and follow instructions in the dialogue boxes. Then right click on *x* axis to set scale and right click on a data point of the scatter diagram to open the Add Trendline dialogue box. In the dialogue box select **Linear** as the type and then click on options tab under options. Check **Display Equation** on chart and check **Display R-squared Value.**

4 Elementary Probability Theory

In dwelling upon the vital importance of sound observation, it must never be lost sight of what observation is for. It is not for the sake of piling up miscellaneous information or curious facts, but for the sake of saving life and increasing health and comfort.

—Florence Nightingale,
Notes on Nursing

Florence Nightingale
(1820–1910)

Florence Nightingale has been described as a "passionate statistician" and a "relevant statistician." She viewed statistics as a science that allows one to transcend his or her narrow individual experience and aspire to the broader service of humanity. She was one of the first nurses to use graphic representation of statistics, illustrating with charts and diagrams how improved sanitation decreased the rate of mortality. Her statistical reports about the appalling sanitary conditions at Scutari (the main British hospital during the Crimean War) were taken very seriously by the English Secretary at War, Sidney Herbert. When sanitary reforms recommended by Nightingale were instituted in military hospitals, the mortality rate dropped from an incredible 42.7% to only 2.2%.

In her lifetime, Florence Nightingale used statistics to make a positive difference in health care. Perhaps you can use statistics to make a positive difference in a field that interests you. In this chapter we investigate probability as an essential tool in the study of statistics.

What Is Probability?

We encounter statements in terms of probability all the time. An excited sports announcer claims that Sheila has a 90% chance of breaking the world record in the upcoming 100-yard dash. Henry figures that if he guesses on a true–false question, the probability of getting it right is 1/2. The Right to Health Lobby claims that the probability of getting an erroneous report from a medical laboratory in one low-cost health center is 0.40. It is consequently lobbying for a federal agency to license and monitor all medical laboratories.

Basic concepts

When we use probability in a statement, we're using *a number between 0 and 1* to indicate the likelihood of an event. We'll use the notation $P(A)$ (read "P of A") to denote the probability of event A. The closer to 1 the probability assignment is, the more likely the event is to occur. If the event A is certain to occur, then $P(A)$ should be 1.

It is important to know what probability statements mean and how to compute or assign probabilities to events, because probability is the language of inferential statistics. For instance, suppose a college counselor claims that 70% of first-year students receive counseling to help plan their schedules. Because of the high percentage of students needing help, he is requesting that an additional counselor be hired. You want to test the counselor's claim. In doing so, you pick a random sample of 30 first-year students and find that only 3 of them got help from a counselor. Can you challenge the counselor's claim on the basis of this random sample in which only 10% of the students got counselor help with their schedules? To answer this question, we need to find the *probability* of picking a random sample of first-year students in which only 10% got counselor help under the assumption that the counselor's claim is correct. This is the kind of question we will consider in hypothesis testing (Chapter 9).

In the meantime, we need to learn how to find probabilities or assign them to events. There are three major methods. One is *intuition*. The sports announcer probably used Sheila's performances in past track events and his own confidence in her running ability as a basis for his prediction that she has a 90% chance of breaking the world record. In other words, the announcer feels that the probability is 0.90 that Sheila will break the world record.

Relative frequency

The Right to Health Lobby used another method to arrive at its probability statement. It took the *relative frequency* with which erroneous laboratory reports occurred. From a random sample of $n = 100$, it found $f = 40$ erroneous laboratory reports. From this it computed the relative frequency of erroneous laboratory reports by means of Formula (1).

> **Probability Formula for Relative Frequency**
>
> $$\text{Probability of an event} = \text{relative frequency} = \frac{f}{n} \tag{1}$$
>
> where f is the frequency of occurrence of an event,
> n is the sample size.

In the case of the laboratory reports, we have

$$\text{Relative frequency} = \frac{f}{n} = \frac{40}{100} = 0.40$$

The relative frequency of erroneous laboratory reports was used as the *probability* of erroneous reports.

The technique of using the relative frequency of an event as the probability of that event is a common way of assigning probabilities and will be used a great deal in later chapters. The underlying assumption we make is that if events have occurred a certain percentage of times in the past, they will occur about the same percentage of times in the future. In fact, this can be strengthened to a very general statement called the *law of large numbers*.

Law of large numbers

> **Law of Large Numbers**
>
> In the long run, as the sample size increases and increases, the relative frequencies of outcomes get closer and closer to the theoretical (or actual) probability value.

The law of large numbers is the reason such businesses as health insurance, automobile insurance, and gambling casinos can exist and make a profit. In Central City, Colorado, there are many casinos with many slot machines. The winnings of a gambler on a single play or even a few plays are uncertain (small sample size). This is one of the reasons gambling is exciting. However, over tens of thousands of plays, the theoretical or actual probability of winning favors the casino. That's why the casino and its owners regard gambling as a business. The house is guaranteed a profit in the long run.

Henry used the third method of assigning probabilities when he determined the probability of correctly guessing the answer to a true–false question. Essentially, he used the probability formula for *equally likely outcomes*.

Equally likely outcomes

> **Probability Formula When Outcomes Are Equally Likely**
>
> $$\text{Probability of an event} = \frac{\text{number of outcomes favorable to event}}{\text{total number of outcomes}} \qquad (2)$$

In Henry's case, there are two possible outcomes. A test answer will be either correct or incorrect. Since he is guessing, we assume that the outcomes are equally likely, and only one is "favorable" to being correct. So, by Formula (2),

$$P(\text{correct answer}) = \frac{\text{number of favorable outcomes}}{\text{total number of outcomes}} = \frac{1}{2}$$

We've seen three ways to assign probabilities: intuition, relative frequency, and, when outcomes are equally likely, a formula. Which do we use? Most of the time it depends on the information that is at hand or that can be feasibly obtained. Our choice of methods also depends on the particular problem. In Guided Exercise 1 you will see three different situations, and you will decide which way to assign the probabilities. *Remember, probabilities are numbers between 0 and 1, so don't assign probabilities outside this range.*

GUIDED EXERCISE   1

Assign a probability to the indicated event on the basis of the information provided. Indicate the technique you use: intuition, relative frequency, or the formula for equally likely outcomes.

(a) The director of the Readlot College Health Center wishes to open an eye clinic. To justify the expense of such a clinic, the director reports the probability that a student selected at random from the college roster needs corrective lenses. She took a random sample of 500 students to compute this probability and found that 375 of them needed corrective lenses. What is the probability that a Readlot College student selected at random needs corrective lenses?

➡ In this case we are given a sample size of 500, and we are told that 375 of these students need glasses. It is appropriate to use a relative frequency for the desired probability:

$$P(\text{student needs glasses}) = \frac{f}{n} = \frac{375}{500} = 0.75$$

(b) The Friends of the Library host a fund-raising barbecue. George is on the cleanup committee. There are four members on this committee, and they draw lots to see who will clean the grills. Assuming that each member is equally likely to be drawn, what is the probability that George will be assigned the grill cleaning job?

➡ There are four people on the committee, and each is equally likely to be drawn. It is appropriate to use the formula for equally likely events. George can be drawn in only one way, so there is only one outcome favorable to that event.

$$P(\text{George}) = \frac{\text{no. of favorable outcomes}}{\text{total no. of outcomes}}$$

$$= \frac{1}{4} = 0.25$$

(c) Joanna photographs whales for Sea Life Adventure Films. On her next expedition, she is to film blue whales feeding. Her boss asks her what she thinks the probability of success will be for this particular assignment. She gives an answer based on her knowledge of the habits of blue whales and the region she is to visit. She is almost certain she will be successful. What specific number do you suppose she gave for the probability of success, and how do you suppose she arrived at it?

➡ Since Joanna is almost certain of success, she should make the probability close to 1. We would say $P(\text{success})$ is above 0.90 but less than 1. We think the probability assignment was based on intuition.

No matter how we compute probabilities, it is useful to know what outcomes are possible in a given setting. For instance, if you are going to decide the probability that Hardscrabble will win the Kentucky Derby, you need to know which other horses will be running.

Sample space

A *statistical experiment* (or simply an *experiment*) can be thought of as any activity that results in a definite outcome. Usually the outcome is in the form of a description, count, or measurement. For example, tossing a coin can be thought of as an experiment. There are only two possible outcomes: heads or tails. The set of all possible outcomes of an experiment is the *sample space*. If you toss a coin, the sample space for that experiment consists of the two outcomes (heads or tails).

It is especially convenient to know the sample space in the case where all outcomes are equally likely because then we can compute probabilities of various events by using Formula (2).

$$P(\text{event } A) = \frac{\text{number of outcomes favorable to } A}{\text{total number of outcomes}} \qquad (2)$$

To use this formula, we need to know the sample space so that we can determine which outcomes are favorable to the event in question as well as the total number of outcomes.

EXAMPLE 1

Human eye color is controlled by a single pair of genes (one from the father and one from the mother) called a *genotype*. Brown eye color, B, is dominant over blue eye color, ℓ. Therefore, in the genotype Bℓ, consisting of one brown gene B and one blue gene ℓ, the brown gene dominates. A person with a Bℓ genotype has brown eyes.

If both parents have brown eyes and have genotype Bℓ, what is the probability that their child will have blue eyes? What is the probability the child will have brown eyes?

SOLUTION: To answer these questions we need to look at the sample space of all possible eye color genotypes for the child. They are given in Table 4-1.

According to genetics theory, the four possible genotypes for the child are equally likely. Therefore, we can use Formula (2) to compute probabilities. Blue eyes can occur only with the $\ell\ell$ genotype, so there is only one outcome favorable to blue eyes. By Formula (2),

$$P(\text{blue eyes}) = \frac{\text{number of favorable outcomes}}{\text{total number of outcomes}} = \frac{1}{4}$$

Brown eyes occur with the three remaining genotypes: BB, Bℓ, and ℓB. By Formula (2),

$$P(\text{brown eyes}) = \frac{\text{number of favorable outcomes}}{\text{total number of outcomes}} = \frac{3}{4}$$

Table 4-1 Eye Color Genotypes for Child

	Mother	
Father	B	ℓ
B	BB	Bℓ
ℓ	ℓB	$\ell\ell$

GUIDED EXERCISE   2

Professor Gutierrez is making up a final exam for a course in literature of the Southwest. He wants the last three questions to be of the true–false type. In order to guarantee that the answers do not follow his favorite pattern, he lists all possible true–false combinations for three questions on slips of paper and then picks one at random from a hat.

(a) Finish listing the outcomes in the given sample space.

$\Rightarrow$ The missing outcomes are FFT and FFF.

TTT	FTT	TFT	_____
TTF	FTF	TFF	_____

(b) What is the probability that all three items will be false? Use the formula

$$P(\text{all F}) = \frac{\text{no. of favorable outcomes}}{\text{total no. of outcomes}}$$

$\Rightarrow$ There is only one outcome, FFF, favorable to all false, so

$$P(\text{all F}) = \frac{1}{8}$$

(c) What is the probability that exactly two items will be true?

$\Rightarrow$ There are three outcomes that have exactly two true items: TTF, TFT, and FTT. Thus,

$$P(\text{two T}) = \frac{\text{no. of favorable outcomes}}{\text{total no. of outcomes}} = \frac{3}{8}$$

Complement of an event

There is one more important point about probability assignments. The sum of all the probabilities assigned to outcomes in a sample space must be 1. This makes sense. If you think the probability is 0.65 that you will win a tennis match, you assume the probability is 0.35 that your opponent will win. This fact is particularly useful, for if the probability that an event occurs is denoted by p and the probability that it *does not* occur is denoted by q, we have

$p + q = 1$ since the sum of the probabilities of the outcomes must be 1

or

$q = 1 - p$ (3)

For an event A, the event *not A* is called the *complement of A*.* To compute the probability of the complement of A, we use Formula (3) and find

$$P(\textit{not A}) = 1 - P(A)$$

---

*The complement of event A is often symbolized by the compact notation A^c or $\overline{A}$. However, in this text we will continue to use the more expanded description *not A* to refer to the complement of A. That is, *not A* refers to all the outcomes of the sample space that are not favorable to event A.

EXAMPLE 2 ❯ The probability that a college student without a flu shot will get the flu is 0.45. What is the probability that a college student will *not* get the flu if the student has not had the flu shot?

SOLUTION: In this case, we have

$P(\text{will get flu}) = p = 0.45$

$P(\text{will } \textit{not} \text{ get flu}) = q = 1 - p = 1 - 0.45 = 0.55$

GUIDED EXERCISE   3

A veterinarian tells you that if you breed two cream-colored guinea pigs, the probability that an offspring will be pure white is 0.25. What is the probability that it will not be pure white?

(a) $P(\text{pure white}) + P(\textit{not} \text{ pure white}) = $ ____   1

(b) $P(\textit{not} \text{ pure white}) = $ ____   ⇨   1 − 0.25, or 0.75

The important facts about probabilities we have seen in this section are

1. The probability of an event A is denoted by $P(A)$.

2. The probability of any event is a number between 0 and 1. The closer to 1 the probability is, the more likely the event is.

3. The sum of the probabilities of all possible outcomes in a sample space is 1.

4. Probabilities can be assigned by using any of three methods: intuition, relative frequencies, or the formula for equally likely outcomes.

5. The probability that an event occurs plus the probability that the same event does not occur is 1.

Probability Related to Statistics

We conclude this section with a few comments on the nature of statistics versus probability. Although statistics and probability are closely related fields of mathematics, they are nevertheless separate fields. It can be said that probability is the medium through which statistical work is done. In fact, if it were not for probability theory, inferential statistics would not be possible.

Put very briefly, probability is the field of study that makes statements about what will occur when samples are drawn from a *known population*. Statistics is the field of study that describes how samples are to be obtained and how inferences are to be made about *unknown populations*.

A simple but effective illustration of the difference between these two subjects can be made by considering how we treat the contents of two boxes.

● Box 1 contains three green balls, five red balls, and four white balls.

● Box 2 contains a collection of colored balls, but the exact number and colors of the balls are unknown.

The study of probability would investigate box 1, where we already know the contents of the box. Typical probability questions would be

1. If one ball is drawn from box 1, what is the probability that the ball is green?

2. If three balls are drawn from box 1, what is the probability that one is white and two are red?

3. If four balls are drawn from box 1, what is the probability that none is red?

Box 1   Probability
Given: 3 green balls, 5 red balls, 4 white balls.

Box 2   Statistics
Exact number and colors of balls are unknown.

The study of statistics would investigate box 2, where we do not know the exact contents of the box. Typical work in statistics would ask us to draw a random sample of balls from box 2 and, based on the sample results, make a conjecture about the colors and number of the population of balls in box 2.

In another sense, probability and statistics are like flip sides of the same coin. On the probability side, you know the overall description of the population (contents of box 1). The central problem is to compute the likelihood that a specific outcome will happen. On the statistics side, you know only the results of a sample drawn from box 2. The central problem is to describe the sample (descriptive statistics) and to draw conclusions about the population of box 2 based on the sample results (inferential statistics).

In statistical work, the inferences we draw about an unknown population are not claimed to be absolutely correct. Since the population remains unknown (in a theoretical sense), we must accept a "best guess" for our conclusions and act on the basis of the most probable answer rather than absolute certainty.

Probability is the topic of this chapter. However, we will not study probability just for its own sake. Probability is a wonderful field of mathematics, but we will study mainly the ideas from probability that are needed for a proper understanding of statistics.

Even though we have only a brief introduction to probability, we will be able to answer questions raised in situations such as the following.

● Commercial salmon fishing is very important in Alaska. Since the salmon are sold by weight, it is useful to know the average weight of a freshly caught salmon. How large a sample of freshly caught salmon is needed if we want to be 95% sure the mean weight of the sample is within plus or minus 1 ounce of the mean weight of all catchable salmon in Alaskan waters? For the answer to this type of question, see Example 6 in Section 8.4, in which we discuss probability as applied to sample size.

● Many practical situations call on us to choose between two competing statements. *Consumer Reports* did a study on the maintenance records of two competing brands of cellular phones. Each brand claims to have the lowest maintenance costs. Is one brand actually better than the other? If there is an apparent difference, is the difference statistically significant at, say, a 1% level (of risk)? For answers to these kinds of questions, see Sections 10.2 and 10.3, in which we discuss tests for differences.

All the listed examples require some use of probability. This is why we encourage you to study this chapter carefully. Your time in doing so will be well spent.

V I E W P O I N T

What Makes a Good Teacher?

A survey of 735 students at nine colleges in the United States was taken to determine instructor behaviors that help students succeed. Using the data provided at Web site <http://lib.stat.cmu.edu/DASL/>, where you select Data Subjects, then Psychology, then Instructor Behavior, you can estimate the probability of how a student would respond (for example, very positive, neutral, or very negative) to different instructor behaviors. For example, more than 90% of the students responded "very positive" to the instructor's use of real-world examples in the classroom.

S E C T I O N   4 . 1   P R O B L E M S

General: Concepts

1. In your own words, carefully answer the question: What is probability? List three methods of assigning probabilities.

General: Provide Examples

2. List examples of where probability might be applied in business, medicine, social science, and natural science. Why do you think probability will be useful in the study of statistics?

General: Valid Probability

3. Which of the following numbers cannot be the probability of an event?

    (a) 0.71   (b) 4.1   (c) $\dfrac{1}{8}$ (d) -0.5

    (e) 0.5   (f) 0   (g) 1   (h) 150%

General: Valid Probability

4. (a) Explain why -0.41 cannot be the probability of an event.
   (b) Explain why 1.21 cannot be the probability of an event.
   (c) Explain why 120% cannot be the probability of an event.
   (d) Can the number 0.56 be the probability of an event? Explain.

Probability Estimate:
Wiggle Your Ears

5. Can you wiggle your ears? Use the students in your statistics class (or a group of friends) to estimate the percentage of people who can wiggle their ears. How can your result be thought of as an estimate of the probability that a person chosen at random can wiggle his or her ears? Comment: National statistics indicate that about 13% of Americans can wiggle their ears (Source: Bernice Kanner, *Are You Normal?* St. Martin's Press, New York).

Probability Estimate:
Raise One Eyebrow

6. Can you raise one eyebrow at a time? Use the students in your statistics class (or a group of friends) to estimate the percentage of people who can raise one eyebrow at a time. How can your result be thought of as an estimate of the probability that a person chosen at random can raise one eyebrow at a time? Comment: National statistics indicate that about 30% of Americans can raise one eyebrow at a time (see source in Problem 5).

Myers-Briggs: Personality
Types and Marriage

7. Isabel Briggs Myers was a pioneer in the study of personality types. The personality types are broadly defined according to four main preferences. Do married couples choose similar or different personality types in their mates? The following data give an indication (Source: I. B. Myers and M. H. McCaulley, *A Guide to the Development and Use of the Myers-Briggs Type Indicators*, p. 71).

**Similarities and Differences in a Random Sample
of 375 Married Couples**

Number of Similar Preferences	Number of Married Couples
All four	34
Three	131
Two	124
One	71
None	15

Suppose that a married couple is selected at random.
(a) Use the data to estimate the probability that they will have 0, 1, 2, 3, or 4 personality preferences in common.
(b) Do the probabilities in part a add up to 1? Why should they? What is the sample space in this problem?

Sociology: Dating Couples

8. Do couples get engaged or not? If they are engaged, how long did they date before becoming engaged? A poll of 1000 couples conducted by Bruskin and Goldring

Research for Korbel Champagne Cellars gave the following information (*USA Today*).

Length of Dating Time Before Engagement

Time	Number of Couples
Never engaged	200
Less than 1 year	240
1 to 2 years	210
More than 2 years	350

(a) Use the data to estimate the probability of each event for a dating couple chosen at random: The couple is not engaged, dated less than 1 year before getting engaged, dated 1 to 2 years before getting engaged, or dated more than 2 years before getting engaged.

(b) Do the probabilities of part a add up to 1? Why should they? What is the sample space in this problem?

Psychology: Best Ideas

9. When do creative people get their *best* ideas? *USA Today* did a survey of 966 inventors (who hold U.S. patents) and obtained the following information.

Time of Day When Best Ideas Occur

Time	Number of Inventors
6 A.M.–12 noon	290
12 noon–6 P.M.	135
6 P.M.–12 midnight	319
12 midnight–6 A.M.	222

(a) Assuming that the time interval includes the left limit and all the times up to but not including the right limit, estimate the probability that an inventor has a best idea during each time interval: from 6 A.M. to 12 noon, from 12 noon to 6 P.M., from 6 P.M. to 12 midnight, and from 12 midnight to 6 A.M.

(b) Do the probabilities of part a add up to 1? Why should they? What is the sample space in this problem?

Business: Customers

10. John runs a computer software store. Yesterday he counted 127 people who walked by his store, 58 of whom came into the store. Of the 58, only 25 bought something in the store.

(a) Estimate the probability that a person who walks by the store will enter the store.

(b) Estimate the probability that a person who walks into the store will buy something.

(c) Estimate the probability that a person who walks by the store will come in *and* buy something.

(d) Estimate the probability that a person who comes into the store will buy nothing.

Some Probability Rules: Compound Events

Probability of *A* and *B*

You roll two dice. What is the probability that you will get a 5 on each die? You draw two cards from a well-shuffled, standard deck without replacing the first card before drawing the second. What is the probability that they will both be aces?

It seems that these two problems are nearly alike. They are alike in the sense that in each case you are to find the probability of two events occurring *together*. In the first problem you are to find

$$P(5 \text{ on 1st die } and \text{ 5 on 2nd die})$$

In the second you want

$$P(\text{ace on 1st card } and \text{ ace on 2nd card})$$

Independent events

The two problems differ in one important aspect, however. In the dice problem, the outcome of a 5 on the first die does not have any effect on the probability of getting a 5 on the second die. Because of this, the events are *independent*. In general, two events are independent if the occurrence or nonoccurrence of one does not change the probability that the other will occur.

In the card problem, the probability of an ace on the first card is 4/52, since there are 52 cards in the deck and 4 of them are aces. If you get an ace on the first card, then the probability of an ace on the second is changed to 3/51, because one ace has already been drawn and only 51 cards remain in the deck. Therefore, the two events in the card-draw problem are *not* independent. They are, in fact, *dependent*, since the outcome of the first draw changes the probability of getting an ace on the second draw.

Why does the *independence* or *dependence* of two events matter? The type of events determines the way we compute the probability of the two events happening together. If two events *A* and *B* are *independent*, then we use Formula (4) to compute the probability of the event *A* and *B*:

> For independent events,
> $$P(A \text{ and } B) = P(A) \cdot P(B) \tag{4}$$

Conditional probability

If the events are *dependent*, then we must take into account the changes in the probability of one event caused by the occurrence of the other event. The notation $P(A, given B)$ denotes the probability that event *A* will occur, *given* that event *B* has occurred. This is called a *conditional probability*. We read $P(A, given B)$ as "probability of *A* given *B*." If *A* and *B* are dependent events, then $P(A) \neq P(A, given B)$ because the occurrence of event *B* has changed the probability that event *A* will occur. A standard notation for $P(A, given B)$ is $P(A \mid B)$. However, we will use the more expanded notation $P(A, given B)$ to remind you that we assume that event *B* has already occurred. We use either Formula (5) or Formula (6) to compute the probability of *A* *and* *B* when the events *A* and *B* are dependent.

> For dependent events,
>
> $$P(A \text{ and } B) = P(A) \cdot P(B, \text{ given that } A \text{ has occurred}) \qquad (5)$$
>
> $$P(A \text{ and } B) = P(B) \cdot P(A, \text{ given that } B \text{ has occurred}) \qquad (6)$$

We will use either Formula (5) or Formula (6) according to the information available.

Formulas (4), (5), and (6) constitute the *multiplication rules* of probability. They help us compute the probability of events happening together when the sample space is too large for convenient reference or when it is not completely known.

Let's use the multiplication rules to complete the dice and card problems. We'll compare the results with those obtained by using the sample space directly.

EXAMPLE 3 ▷

Suppose you are going to throw two fair dice. What is the probability of getting a 5 on each die?

SOLUTION USING SAMPLE SPACE: The first task is to write down the sample space. Each die has six equally likely outcomes, and each outcome of the second die can be paired with each of the first. The sample space is shown in Figure 4-1. The total number of outcomes is 36, and only one is favorable to a 5 on the first die *and* a 5 on the second. The 36 outcomes are equally likely, so by Formula (2) for equally likely outcomes,

$$P(5 \text{ on 1st } and \text{ 5 on 2nd}) = \frac{1}{36}$$

FIGURE 4-1

Sample Space for Two Dice

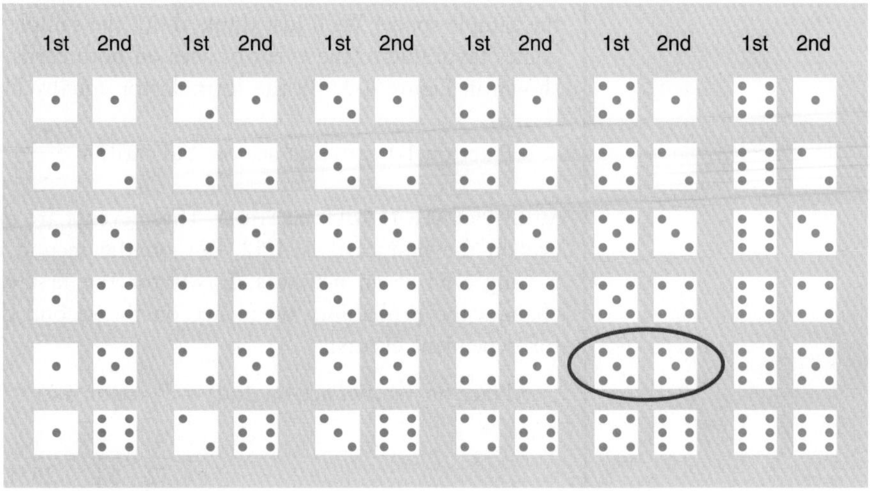

SOLUTION USING THE MULTIPLICATION RULE: The two events are independent, so we should use Formula (4). P(5 on 1st die *and* 5 on 2nd die) = P(5 on 1st) · P(5 on 2nd). To finish the problem, we need only compute the probability of getting a 5 when we throw one die.

There are six faces on a die, and on a fair die each is equally likely to come up when you throw the die. Only one face has five dots, so by Formula (2) for equally likely outcomes,

$$P(5 \text{ on die}) = \frac{1}{6}$$

Now we can complete the calculation.

$$P(5 \text{ on 1st die } and \text{ 5 on 2nd die}) = P(5 \text{ on 1st}) \cdot P(5 \text{ on 2nd})$$

$$= \frac{1}{6} \cdot \frac{1}{6}$$

$$= \frac{1}{36}$$

The two methods yield the same result. The multiplication rule was easier to use because we did not need to look at all 36 outcomes in the sample space for tossing two dice.

EXAMPLE 4 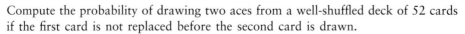 Compute the probability of drawing two aces from a well-shuffled deck of 52 cards if the first card is not replaced before the second card is drawn.

SAMPLE SPACE METHOD: We won't actually look at the sample space because each of the 51 possible outcomes for the second card must be paired with each of the 52 possible outcomes for the first card. This gives us a total of 2652 outcomes in the sample space! We'll just think about the sample space and try to list all the outcomes favorable to the event of aces on both cards. The 12 favorable outcomes are shown in Figure 4-2. By the formula for equally likely outcomes,

$$P(\text{ace on 1st card } and \text{ ace on 2nd card}) = \frac{12}{2652} \approx 0.0045$$

MULTIPLICATION RULE METHOD: These events are *dependent*. The probability of an ace on the first card is 4/52, but on the second card the probability of an ace is only 3/51 if an ace was drawn for the first card. An ace on the first draw changes the probability of an ace on the second draw. By the multiplication rule for dependent events,

$$P(\text{ace on 1st } and \text{ ace on 2nd}) = P(\text{ace on 1st}) \cdot P(\text{ace on 2nd}, given \text{ ace on 1st})$$

$$= \frac{4}{52} \cdot \frac{3}{51} = \frac{12}{2652} \approx 0.0045$$

Again, the two methods agree.

FIGURE 4-2

Outcomes Favorable to
Drawing Two Aces

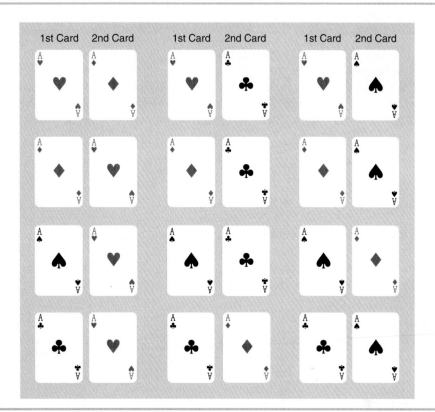

GUIDED EXERCISE   4

Andrew is 55, and the probability that he will be alive in 10 years is 0.72. Ellen is 35, and the probability that she will be alive in 10 years is 0.92. Assuming that the life span of one will have no effect on the life span of the other, what is the probability they will both be alive in 10 years?

(a) Are these events dependent or independent?

⇨ Since the life span of one does not affect the life span of the other, the events are independent.

(b) Use the appropriate multiplication rule to find

P(Andrew alive in 10 years *and* Ellen alive in 10 years)

⇨ We use the rule for independent events:

$P(A \text{ and } B) = P(A) \cdot P(B)$

P(Andrew alive *and* Ellen alive)

$= P$(Andrew alive) $\cdot P$(Ellen alive)

$= (0.72)(0.92) \approx 0.66$

GUIDED EXERCISE   5

A quality-control procedure for testing Ready-Flash disposable cameras consists of drawing two cameras at random from each lot of 100 without replacing the first camera before drawing the second. If both are defective, the entire lot is rejected. Find the probability that both cameras are defective if the lot contains 10 defective cameras. Since we are drawing the cameras at random, assume that each camera in the lot has an equal chance of being drawn.

(a) What is the probability of getting a defective camera on the first draw?

⇨ The sample space consists of all 100 cameras. Since each is equally likely to be drawn and there are 10 defective ones,

$$P(\text{defective camera}) = \frac{10}{100} = \frac{1}{10}$$

(b) The first camera drawn is not replaced, so there are only 99 cameras for the second draw. What is the probability of getting a defective camera on the second draw if the first camera was defective?

⇨ If the first camera is defective, then there are only 9 defective cameras left among the 99 remaining cameras in the lot.

$$P(\text{defective camera on 2nd draw, } given \text{ defective camera on 1st}) = \frac{9}{99} = \frac{1}{11}$$

(c) Are the probabilities computed in parts a and b different? Does drawing a defective camera on the first draw change the probability of getting a defective camera on the second draw? Are the events dependent?

⇨ The answer to all these questions is yes.

(d) Use the formula for dependent events,

$$P(A \text{ and } B) = P(A) \cdot P(B, given \text{ } A \text{ has occurred})$$

to compute P(1st camera defective *and* 2nd camera defective).

⇨ $$P(\text{1st defective } and \text{ 2nd defective}) = \frac{1}{10} \cdot \frac{1}{11}$$

$$= \frac{1}{110}$$

$$\approx 0.009$$

The multiplication rules apply whenever we wish to determine the probability of two events happening *together*. To indicate together, we use *and* between the events. But before you use a multiplication rule to compute the probability of *A and B*, you must determine if *A* and *B* are independent or dependent events.

One of the multiplication rules can be used any time we are trying to find the probability of two events happening *together*. Pictorially, we are looking for the probability of the shaded region on page 181 in Figure 4-3.

Probability of *A* or *B*

Another way to combine events is to consider the possibility of one event *or* another occurring. For instance, if a sports car saleswoman gets an extra bonus if she sells a convertible or a car with leather upholstery, she is interested in the prob-

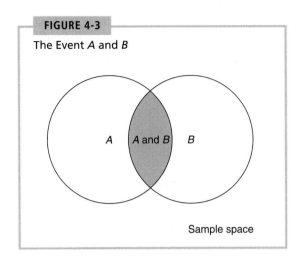

FIGURE 4-3

The Event *A* and *B*

A   *A* and *B*   *B*

Sample space

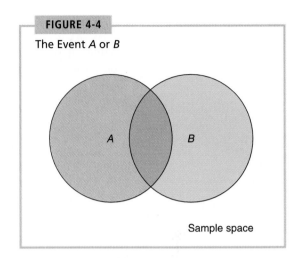

FIGURE 4-4

The Event *A* or *B*

A   *B*

Sample space

ability that you will buy a car that is a convertible *or* has leather upholstery. Of course, if you bought a convertible with leather upholstery, that would be fine, too. Pictorially, the shaded portion of Figure 4-4 represents the outcomes satisfying the *or* condition. Notice that the condition *A or B* is satisfied by any one of the following conditions:

1. Any outcome in *A* occurs.

2. Any outcome in *B* occurs.

3. Any outcome in both *A* and *B* occurs.

It is important to distinguish between the *or* combinations and the *and* combinations because we apply different rules to compute their probabilities.

GUIDED EXERCISE   6

Indicate how each of the following pairs of events are combined. Use either the *and* combination or the *or* combination.

(a) Satisfying the humanities requirement by taking a course in the history of Japan or by taking a course in classical literature

⇨   Use the *or* combination.

(b) Buying new tires and aligning the tires

⇨   Use the *and* combination.

(c) Getting an A not only in psychology but also in biology

⇨   Use the *and* combination.

(d) Having at least one of these pets: cat, dog, bird, rabbit

⇨   Use the *or* combination.

Once you decide that you are to find the probability of an *or* combination rather than an *and* combination, what formula do you use? Again, it depends on the situation. If you want to compute the probability of drawing either a jack or a king on a single draw from a well-shuffled deck of cards, the formula is simple:

$$P(\text{jack } or \text{ king}) = P(\text{jack}) + P(\text{king}) = \frac{4}{52} + \frac{4}{52} = \frac{8}{52} = \frac{2}{13}$$

since there are 4 jacks and 4 kings in a deck of 52 cards.

If you want to compute the probability of drawing a king or a diamond on a single draw, the formula is a bit more complicated. We have to take the overlap of the two events into account so that we do not count the outcomes twice. We can see the overlap of the two events in Figure 4-5 below.

$$P(\text{king}) = \frac{4}{52} \quad P(\text{diamond}) = \frac{13}{52} \quad P(\text{king } and \text{ diamond}) = \frac{1}{52}$$

If we simply add $P(\text{king})$ and $P(\text{diamond})$, we're including $P(\text{king } and \text{ diamond})$ twice in the sum. To compensate for this double summing, we simply subtract $P(\text{king } and \text{ diamond})$ from the sum. Therefore,

$$P(\text{king } or \text{ diamond}) = P(\text{king}) + P(\text{diamond}) - P(\text{king } and \text{ diamond})$$

FIGURE 4-5

Drawing a King or a Diamond from a Standard Deck

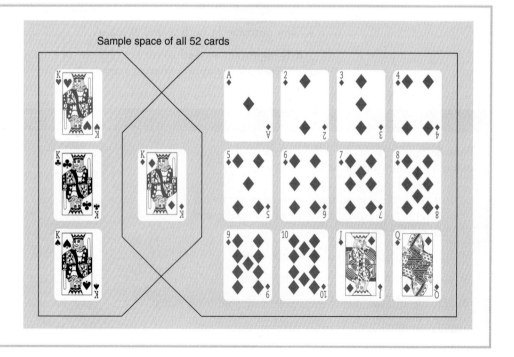

$$= \frac{4}{52} + \frac{13}{52} - \frac{1}{52}$$

$$= \frac{16}{52} = \frac{4}{13}$$

Mutually exclusive events

We say the events A and B are *mutually exclusive* or *disjoint* if they cannot occur together. This means that A and B have no outcomes in common or, put another way, that $P(A \text{ and } B) = 0$. Formula (7) is the addition rule for *mutually exclusive* events A and B.

> For *mutually exclusive* events A and B
> $$P(A \text{ or } B) = P(A) + P(B)$$ (7)

If the events are not mutually exclusive, we must use the more general Formula (8), which is the general addition rule for any events A and B.

> For any events A and B,
> $$P(A \text{ or } B) = P(A) + P(B) - P(A \text{ and } B)$$ (8)

You may ask: Which formula should we use? The answer is: Use Formula (7) only if you know that A and B are mutually exclusive (i.e., cannot occur together); if you do not know whether A and B are mutually exclusive, then use Formula (8). Formula (8) is valid either way. Notice that when A and B are mutually exclusive, then $P(A \text{ and } B) = 0$, so Formula (8) reduces to Formula (7).

GUIDED EXERCISE   7

The "Cost Less Clothing" store carries manufacturer's seconds in slacks. If you buy a pair of slacks in your regular waist size without trying them on, the probability that the waist will be too tight is 0.30 and the probability that it will be too loose is 0.10.

(a) Are the events too tight and too loose mutually exclusive?

⇨ The waist cannot be both too tight and too loose at the same time, so the events are mutually exclusive.

(b) If you choose a pair of slacks at random in your regular waist size, what is the probability that the waist will be too tight *or* too loose?

⇨ Since the events are mutually exclusive,

$P(\text{too tight } or \text{ too loose})$

$= P(\text{too tight}) + P(\text{too loose})$

$= 0.03 + 0.10$

$= 0.40$

GUIDED EXERCISE	8

Professor Jackson is in charge of a program to prepare people for a high school equivalency exam. Records show that 80% of the students need work in math, 70% need work in English, and 55% need work in both areas.

(a) Are the events need math and need English mutually exclusive?

⇨ These events are not mutually exclusive, since some students need both. In fact,

P(need math *and* need English) = 0.55

(b) Use the appropriate formula to compute the probability that a student selected at random needs math *or* needs English.

⇨ Since the events are not mutually exclusive, we use Formula (8):

P(need math *or* need English)

= P(need math) + P(need English)

– P(need math *and* English)

= 0.80 + 0.70 – 0.55

= 0.95

Combination of several events

The addition rule for mutually exclusive events can be extended so that it applies to the situation in which we have more than two events that are each mutually exclusive to all the other events.

EXAMPLE 5   Laura is playing Monopoly. On her next move she needs to throw a sum bigger than 8 on the two dice in order to land on her own property and pass Go. What is the probability that Laura will roll a sum bigger than 8?

SOLUTION: When two dice are thrown, the largest sum that can come up is 12. Consequently, the only sums larger than 8 are 9, 10, 11, and 12. These outcomes are mutually exclusive, since only one of these sums can possibly occur on one throw of the dice. The probability of throwing more than 8 is the same as

P(9 *or* 10 *or* 11 *or* 12)

Since the events are mutually exclusive,

$$P(9 \ or \ 10 \ or \ 11 \ or \ 12) = P(9) + P(10) + P(11) + P(12)$$

$$= \frac{4}{36} + \frac{3}{36} + \frac{2}{36} + \frac{1}{36}$$

$$= \frac{10}{36} = \frac{5}{18}$$

To get the specific values of $P(9)$, $P(10)$, $P(11)$, and $P(12)$, we used the sample space for throwing two dice (see Figure 4-1). There are 36 equally likely outcomes—for example, those favorable to 9 are 6, 3; 3, 6; 5, 4; and 4, 5. So $P(9) = 4/36$. The other values can be computed in a similar way.   ●

The multiplication rule for independent events also extends to more than two independent events. If you toss a fair coin, then roll a fair die, and finally draw a card from a standard deck of 52 cards, the three events are independent. To compute the probability of the outcome heads on the coin *and* five on the die *and* an ace for the card, we use the extended multiplication rule for independent events together with the facts

$$P(\text{head}) = \frac{1}{2} \quad P(5) = \frac{1}{6} \quad P(\text{ace}) = \frac{4}{52} = \frac{1}{13}$$

Then

$$P(\text{head } and \text{ five } and \text{ ace}) = \frac{1}{2} \cdot \frac{1}{6} \cdot \frac{1}{13}$$

$$= \frac{1}{156}$$

Further Examples

Most of us have been asked to participate in a survey. Schools, retail stores, churches, and government offices all conduct surveys. There are many types of surveys, and it is not our intention to give a general discussion of this topic. Let us study a very popular method called the *simple tally survey*. Such a survey consists of questions for which the responses can be recorded in rows and columns of a table. These questions are appropriate to the information you want and are designed to cover the *entire* population of interest. In addition, the questions should be designed so that we can partition the sample space of responses into distinct (that is, mutually exclusive) sectors.

If the survey includes responses from a reasonably large random sample, then the results should be representative of your population. In this case we can estimate simple probabilities, conditional probabilities, and the probabilities of some combinations of events directly from the results of the survey.

EXAMPLE 6 ⬧ At Hopewell Electronics, all 140 employees were asked about their political affiliations. The employees were grouped by type of work, as executives or production workers. The results with row and column totals are shown in Table 4-2 on the following page.

Suppose an employee is selected at random from the 140 Hopewell employees. Let us use the following notation to represent different events of choosing: E = executive; PW = production worker; D = Democrat; R = Republican; I = Independent.

Table 4-2 Employee Type and Political Affiliation

| Employee Type | Political Affiliation | | | Row Total |
	Democrat (D)	Republican (R)	Independent (I)	
Executive (E)	5	34	9	48
Production worker (PW)	63	21	8	92
Column total	68	55	17	140 Grand Total

(a) Compute $P(D)$ and $P(E)$.

SOLUTION: To find these probabilities, we look at the *entire* sample space.

$$P(D) = \frac{\text{number of Democrats}}{\text{number of employees}} = \frac{68}{140} \approx 0.486$$

$$P(E) = \frac{\text{number of executives}}{\text{number of employees}} = \frac{48}{140} \approx 0.343$$

(b) Compute $P(D, \text{ given } E)$.

SOLUTION: For the conditional probability, we *restrict* our attention to the portion of the sample space satisfying the condition of being an executive.

$$P(D, \text{ given } E) = \frac{\text{number of executives who are Democrats}}{\text{number of executives}} = \frac{5}{48} \approx 0.104$$

(c) Are the events D and E independent?

SOLUTION: One way to determine if the events D and E are independent is to see if $P(D) = P(D, \text{ given } E)$ [or equivalently, if $P(E) = P(E, \text{ given } D)$]. Since $P(D) = 0.486$ and $P(D, \text{ given } E) = 0.104$, we see that $P(D) \neq P(D, \text{ given } E)$. This means that the events D and E are *not* independent. The probability of event D "depends on" whether or not event E has occurred.

(d) Compute $P(D \text{ and } E)$.

SOLUTION: This probability is not conditional, so we must look at the entire sample space.

$$P(D \text{ and } E) = \frac{\text{number of executives who are Democrats}}{\text{total number of employees}} = \frac{5}{140} \approx 0.036$$

Let's recompute this probability using the rules of probability for dependent events.

$$P(D \text{ and } E) = P(E) \cdot P(D, \text{ given } E) = \frac{48}{140} \cdot \frac{5}{48} = \frac{5}{140} \approx 0.036$$

The results using the rules are consistent with those using the sample space.

(e) Compute $P(D\ or\ E)$.

SOLUTION: From part d we know that the events Democrat and executive are not mutually exclusive, because $P(D\ and\ E) \neq 0$. Therefore,

$$P(D\ or\ E) = P(D) + P(E) - P(D\ and\ E)$$

$$= \frac{68}{140} + \frac{48}{140} - \frac{5}{140} = \frac{111}{140} \approx 0.793$$

Using Table 4-2, let's consider other probabilities regarding the type of employees at Hopewell and their political affiliations. This time let's consider the production worker and the affiliation of Independent. Suppose an employee is selected at random from the group of 140.

(a) Compute $P(I)$ and $P(PW)$.

$\Rightarrow$ $P(I) = \dfrac{\text{no. of independents}}{\text{total no. of employees}}$

$= \dfrac{17}{140} \approx 0.121$

$P(PW) = \dfrac{\text{no. of production workers}}{\text{total no. of employees}}$

$= \dfrac{92}{140} \approx 0.657$

(b) Compute $P(I,\ given\ PW)$. This is a conditional probability. Be sure to restrict your attention to production workers since that is the condition given.

$\Rightarrow$ $P(I,\ given\ PW) = \dfrac{\text{no. of independent production workers}}{\text{no. of production workers}}$

$= \dfrac{8}{92} \approx 0.087$

(c) Compute $P(I\ and\ PW)$. In this case look at the entire sample space and at the number of employees who are both Independent and in production.

$\Rightarrow$ $P(I\ and\ PW) = \dfrac{\text{no. of independent production workers}}{\text{total no. of employees}}$

$= \dfrac{8}{140} \approx 0.057$

Exercise continues

Exercise continued

(d) Use the multiplication rule for dependent events to calculate $P(I \text{ and } PW)$. Is the result the same as that of part c?

⇨ By the multiplication rule,

$$P(I \text{ and } PW) = P(PW) \cdot P(I, \text{ given } PW)$$
$$= \frac{92}{140} \cdot \frac{8}{92} = \frac{8}{140} \approx 0.057$$

The results are the same.

(e) Compute $P(I \text{ or } PW)$. Are the events mutually exclusive?

⇨ Since the events are not mutually exclusive,

$$P(I \text{ or } PW) = P(I) + P(PW) - P(I \text{ and } PW)$$
$$= \frac{17}{140} + \frac{92}{140} - \frac{8}{140}$$
$$= \frac{101}{140} \approx 0.721$$

Basic Probability Rules

As you apply probability to a variety of settings, keep the following rules in mind.

Summary of Basic Probability Rules for Events A and B

1. For any event A: $0 \le P(A) \le 1$

2. Complement of A: $P(\text{not } A) = 1 - P(A)$

3. Events A and B are independent events if $P(A) = P(A, \text{ given } B)$

4. Multiplication Rules
   If A and B are independent events: $P(A \text{ and } B) = P(A) P(B)$
   If A and B are dependent events: $P(A \text{ and } B) = P(A) P(B, \text{ given } A)$
   or equivalently; $P(B, \text{ given } A) = P(A \text{ and } B)/P(A)$

5. Events A and B are mutually exclusive if $P(A \text{ and } B) = 0$

6. Addition Rules
   If A and B are mutually exclusive events: $P(A \text{ or } B) = P(A) + P(B)$
   If A and B are not mutually exclusive: $P(A \text{ or } B) = P(A) + P(B) -$
   $\qquad\qquad\qquad\qquad\qquad\qquad\qquad P(A \text{ and } B)$

SECTION 4.2 PROBLEMS

In Problems 1–12, use the appropriate addition or multiplication rules. When possible, verify results by considering the sample space.

General: M&M Candy

1. M&M plain candies come in a variety of colors. According to the M&M/Mars Department of Consumer Affairs (Web site <www.m-ms.com>), the distribution of colors for plain M&M candies is

Color	Brown	Yellow	Red	Orange	Green	Blue
Percentage	30%	20%	20%	10%	10%	10%

Suppose you have a large bag of plain M&M candies and you take one candy at random. Find
   (a) P(orange candy *or* blue candy). Are these outcomes mutually exclusive? Why?
   (b) P(yellow candy *or* red candy). Are these outcomes mutually exclusive? Why?
   (c) P(*not* brown candy).

General: M&M Candy

2. According to the Department of Consumer Affairs of M&M/Mars, the color distribution of peanut M&M candies is

Color	Brown	Yellow	Red	Orange	Green	Blue
Percentage	20%	20%	20%	10%	10%	20%

Suppose you have a large bag of peanut M&M candies and you take one candy at random. Compute the probabilities in parts a through c of Problem 1 for peanut M&M candies. Compare the results with those for plain M&M candies. Do you expect any differences? Why or why not?

General: M&M Candy

3. Almond M&M candies have another color distribution, utilizing only five colors (see Problem 1). The color distribution for almond M&M candies is uniform. There are only five colors: brown, red, yellow, green, and blue. Each color comprises 20% of the almond M&M mix. Compute the probabilities in parts a through c of Problem 1 for almond M&M candies. Compare the results with those for plain M&M candies. Do you expect any differences? Why or why not?

Environment:
Arches National Park

4. Arches National Park is located in southern Utah. The park is famous for its beautiful desert landscape and its many natural sandstone arches. Park Ranger Edward McCarrick started an inventory (not yet complete) of natural arches within the park that have an opening of at least 3 feet. The following table is based on information taken from the book *Canyon Country Arches and Bridges*, by F. A. Barnes. The height of the arch opening is rounded to the nearest foot.

Height of arch, feet	3–9	10–29	30–49	50–74	75 and higher
Number of arches in park	111	96	30	33	18

For an arch chosen at random in Arches National Park, use the preceding information to estimate the probability that the height of the arch opening is
(a) 3 to 9 feet tall
(b) 30 feet or taller
(c) 3 to 49 feet tall
(d) 10 to 74 feet tall
(e) 75 feet or taller

General: Roll Two Dice

5. You roll two fair dice, a green one and a red one.
(a) Are the outcomes on the dice independent?
(b) Find $P(5$ on green die *and* 3 on red die$)$.
(c) Find $P(3$ on green die *and* 5 on red die$)$.
(d) Find $P[(5$ on green die *and* 3 on red die$)$ *or* $(3$ on green die *and* 5 on red die$)]$.

General: Roll Two Dice

6. You roll two fair dice, a green one and a red one.
(a) Are the outcomes on the dice independent?
(b) Find $P(1$ on green die *and* 2 on red die$)$.
(c) Find $P(2$ on green die *and* 1 on red die$)$.
(d) Find $P[(1$ on green die *and* 2 on red die$)$ *or* $(2$ on green die *and* 1 on red die$)]$.

General: Roll Two Dice

7. You roll two fair dice, a green one and a red one.
(a) What is the probability of getting a sum of 6?
(b) What is the probability of getting a sum of 4?
(c) What is the probability of getting a sum of 6 *or* 4? Are these outcomes mutually exclusive?

General: Roll Two Dice

8. You roll two fair dice, a green one and a red one.
(a) What is the probability of getting a sum of 7?
(b) What is the probability of getting a sum of 11?
(c) What is the probability of getting a sum of 7 *or* 11? Are these outcomes mutually exclusive?

General: Deck of 52 Cards

9. You draw two cards from a standard deck of 52 cards without replacing the first one before drawing the second.
(a) Are the outcomes on the two cards independent? Why?
(b) Find $P($ace on 1st card *and* king on 2nd card$)$.
(c) Find $P($king on 1st card *and* ace on 2nd card$)$.
(d) Find the probability of drawing an ace and a king in either order.

General: Deck of 52 Cards

10. You draw two cards from a standard deck of 52 cards without replacing the first one before drawing the second.
    (a) Are the outcomes on the two cards independent? Why?
    (b) Find P(3 on 1st card *and* 10 on 2nd).
    (c) Find P(10 on 1st card *and* 3 on 2nd).
    (d) Find the probability of drawing a 10 and a 3 in either order.

General: Deck of 52 Cards

11. You draw two cards from a standard deck of 52 cards, but before you draw the second card, you put the first one back and reshuffle the deck.
    (a) Are the outcomes on the two cards independent? Why?
    (b) Find P(ace on 1st card *and* king on 2nd).
    (c) Find P(king on 1st card *and* ace on 2nd).
    (d) Find the probability of drawing an ace and a king in either order.

General: Deck of 52 Cards

12. You draw two cards from a standard deck of 52 cards, but before you draw the second card, you put the first one back and reshuffle the deck.
    (a) Are the outcomes on the two cards independent? Why?
    (b) Find P(3 on 1st card *and* 10 on 2nd).
    (c) Find P(10 on 1st card *and* 3 on 2nd).
    (d) Find the probability of drawing a 10 and a 3 in either order.

Marketing:
Age of Children and Toys

13. *USA Today* gave the following information about ages of children receiving toys. The percentages represent all toys sold.

    What is the probability that a toy is purchased for someone

Age (years)	Percentage of Toys
2 and under	15%
3–5	22%
6–9	27%
10–12	14%
13 and over	22%

    (a) 6 years or older?
    (b) 12 years or younger?
    (c) between 6 and 12 years old?
    (d) between 3 and 9 years old?

    A child between 10 and 12 years old looks at this probability distribution and asks, "Why are people more likely to buy toys for kids older than I am (13 and over) than for kids in my age group (10–12)?" How would you respond?

Health: Incidence of Flu

14. Based on data from the *Statistical Abstract of the United States* (112th Edition), only about 14% of senior citizens (65 years or older) get the flu each year. However, about 24% of people under 65 years old get the flu each year. In the general population, there are 12.5% senior citizens (65 years or older).
    (a) What is the probability that a person selected at random from the general population is a senior citizen who will get the flu this year?
    (b) What is the probability that a person selected at random from the general population is a person under age 65 who will get the flu this year?
    (c) Answer parts a and b for a community that has 95% senior citizens.
    (d) Answer parts a and b for a community that has 50% senior citizens.

Psychology:
Lie Detector Tests

15. In his book *Chances: Risk and Odds in Every Day Life,* James Burke says that there is a 72% chance a polygraph test (lie detector test) will catch a person who is in

fact lying. Furthermore, there is approximately a 7% chance that the polygraph will falsely accuse someone of lying.

(a) Suppose that a person answers 90% of a long battery of questions truthfully. What percentage of the story will the polygraph *wrongly* indicate is a lie?

(b) Suppose that a person answers 10% of a long battery of questions with lies. What percentage of the story will the polygraph *correctly* indicate is a lie?

(c) Repeat parts a and b if 50% of the questions are answered truthfully and 50% are answered with lies.

(d) Repeat parts a and b if 15% of the questions are answered truthfully and the rest are answered with lies.

Psychology: Lie Detector Tests

16. This problem continues Problem 15. The solution involves applying several basic probability rules and a little algebra to solve an equation.

(a) If the polygraph in Problem 15 said 30% of the questions were answered with lies, what would you estimate for the actual percentage of lies in the story? *Hint:* Let B = event detector indicates a lie. We are given $P(B) = 0.30$. Let A = event person is lying, so *not A* = event person is not lying. Then

$$P(B) = P(A \text{ and } B) + P(not\ A \text{ and } B)$$
$$P(B) = P(A)P(B, given\ A) + P(not\ A)P(B, given\ not\ A)$$

Replacing $P(not\ A)$ by $1 - P(A)$ gives

$$P(B) = P(A)P(B, given\ A) + [1 - P(A)]P(B, given\ not\ A)$$

Substitute known values for $P(B)$, $P(B, given\ A)$, and $P(B, given\ not\ A)$ into the last equation and solve for $P(A)$.

(b) If the polygraph in Problem 15 said 70% of questions were answered with lies, what would you estimate for the actual percentage of lies in the story?

Business: Sales Approach

17. In a sales effectiveness seminar, a group of sales representatives tried two approaches to selling a customer a new automobile: the aggressive approach and the passive approach. From 1160 customers, the following record was kept.

	Sale	No Sale	Row Total
Aggressive	270	310	580
Passive	416	164	580
Column total	686	474	1160

Suppose that a customer is selected at random from the 1160 participating customers. Let us use the following notation for events: A = aggressive approach, Pa = passive approach, S = sale, N = no sale. So $P(A)$ is the probability that an aggressive approach was used, and so on.

(a) Compute $P(S)$, $P(S, given\ A)$, and $P(S, given\ Pa)$.

(b) Are the events S = sale and Pa = passive approach independent? Explain.

(c) Compute $P(A \text{ and } S)$ and $P(Pa \text{ and } S)$.

(d) Compute $P(N)$ and $P(N, given\ A)$.

(e) Are the events N = no sale and A = aggressive approach independent? Explain.

(f) Compute $P(A \text{ or } S)$.

Medical: Sensitivity of a Test

18. Diagnostic tests of medical conditions have several results. The test result can be positive or negative, whether or not a patient has the condition (+ indicates a patient has the condition). Consider a random sample of 200 patients, some of whom have a medical condition and some of whom do not. Results of a new diagnostic test for the condition are shown.

	Condition Present	Condition Absent	Row Total
Test Result +	110	20	130
Test Result −	20	50	70
Column total	130	70	200

Assume the sample is representative of the entire population. For a person selected at random, compute the following probabilities:

(a) $P(+,$ *given* condition present); this is known as the *sensitivity* of a test.
(b) $P(-,$ *given* condition present); this is known as the false-negative rate.
(c) $P(-,$ *given* condition absent); this is known as the *specificity* of a test.
(d) $P(+,$ *given* condition absent); this is known as the false-positive rate.
(e) P(condition present *and* +); this is the predictive value of the test.
(f) P(condition present *and* −).

Medical: High Fever

19. In an article entitled "Diagnostic accuracy of fever as a measure of postoperative pulmonary complications" (*Heart Lung* 10, No. 1:61), J. Roberts and colleagues discuss using a fever of 38°C or higher as a diagnostic indicator of postoperative atelectasis (collapse of the lung) as evidenced by x-ray observation. For fever ≥ 38°C as the diagnostic test, the results for postoperative patients are

	Condition Present	Condition Absent	Row Total
Test Result +	72	37	109
Test Result −	82	79	161
Column total	154	116	270

Complete parts a–f from Problem 20.

Trees and Counting Techniques

When outcomes are equally likely, we compute the probability of an event by using the formula

$$P(A) = \frac{\text{number of outcomes favorable to the event } A}{\text{number of outcomes in the sample space}}$$

The probability formula requires that we be able to determine the number of outcomes in the sample space. In the problems we have done in previous sections, this task has not been difficult because the number of outcomes was small or the sample space consisted of fairly straightforward events. The tools we present in this section will help you count the number of possible outcomes in larger sample spaces or those formed by more complicated events.

Tree diagrams

A *tree diagram* helps us display the outcomes of an experiment consisting of a series of activities. The total number of outcomes corresponds to the total number of final branches in the tree. Perhaps the best way to learn to make a tree diagram is to see one. In the next example we will see a tree diagram and analyze its parts.

EXAMPLE 7

Jacqueline is in the nursing program and is required to take a course in psychology and one in anatomy and physiology (*A* and *P*) next semester. She also wants to take Spanish II. If there are four sections of psychology, two of *A* and *P*, and three of Spanish II, how many different class schedules can Jacqueline choose from? (Assume that the times of the sections do not conflict with each other.) Figure 4-6 below shows a tree diagram for Jacqueline's possible schedules.

SOLUTION: Let's study the tree diagram and see how it shows Jacqueline's schedule choices. There are four branches from Start. These branches indicate the four possible choices for psychology sections. No matter which section of psychology

FIGURE 4-6

Tree Diagram for Selecting Class Schedules

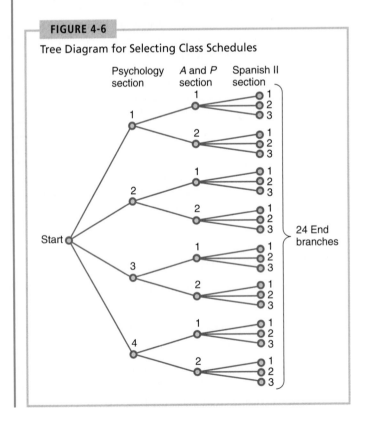

Jacqueline chooses, she can choose from the two available *A* and *P* sections. Therefore, we have two branches leading from *each* psychology branch. Finally, after the psychology and *A* and *P* sections are selected, there are three choices for Spanish II. That is why there are three branches from *each A* and *P* section.

The tree ends with a total of 24 branches. This number of end branches tells us the number of possible schedules. The outcomes themselves can be listed from the tree by following each series of branches from Start to End. For instance, the top branch from Start generates the schedules shown below in Table 4-3.

Table 4-3   Schedules Utilizing Section I of Psychology

Psychology Section	*A* and *P* Section	Spanish II Section
1	1	1
1	1	2
1	1	3
1	2	1
1	2	2
1	2	3

Following the second branch from Start, we see all the possible schedules utilizing Section 2 of psychology (see Table 4-4). The other 12 schedules can be listed in a similar manner.

Table 4-4   Schedules Utilizing Section 2 of Psychology

Psychology Section	*A* and *P* Section	Spanish II Section
2	1	1
2	1	2
2	1	3
2	2	1
2	2	2
2	2	3

We draw a tree diagram in stages, indicating the possible outcomes for the first event, the second event, and so forth. The next exercise will lead you through the process.

GUIDED EXERCISE   10

Louis plays three tennis matches. Use a tree diagram to list the possible win and loss sequences Louis can experience for the set of three matches. W = Win, L = Lose

(a) On the first match, Louis can win or lose. From Start, indicate these two branches.

⇨ **FIGURE 4-7** First Match

1st Match
W
Start
L

(b) Regardless of whether Louis wins or loses the first match, he plays the second match and can again win or lose. Attach branches representing these two outcomes to *each* of the first match results.

⇨ **FIGURE 4-8** Second Match

1st Match   2nd Match
W
W
L
Start
L
W
L

(c) Louis may win or lose the third match. Attach branches representing these two outcomes to *each* of the second match results.

⇨ **FIGURE 4-9** Third Match

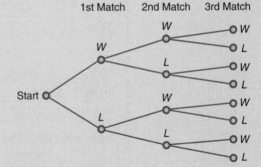

(d) How many possible win–lose sequences are there for the three matches?

⇨ Since there are eight branches at the end, there are eight sequences.

(e) Complete this list of win–lose sequences.

1st	2nd	3rd
W	W	W
W	W	L
W	L	W
W	L	L
___	___	___
___	___	___
___	___	___

⇨ The last four sequences all involve a loss in Match 1.

1st	2nd	3rd
L	W	W
L	W	L
L	L	W
L	L	L

Multiplication rule of counting

When an outcome is composed of a series of events, tree diagrams tell us how many possible outcomes there are. They also help us list the individual outcomes. However, if we are interested only in the number of outcomes created by a series of events, the multiplication rule will give us the total number of outcomes more directly. We state the multiplication rule for an outcome composed of a series of two events.

Multiplication Rule of Counting

If there are n possible outcomes for event E_1 and m possible outcomes for event E_2, then there are a total of $n \times m$ or nm possible outcomes for the series of events E_1 followed by E_2.

The rule extends to outcomes created by a series of three, four, or more events. We simply multiply the number of outcomes possible for each step in the series of events to get the total number of outcomes for the series.

EXAMPLE 8 > The Night Hawk is the new car model produced by Limited Motors, Inc. It comes with a choice of two body styles, three interior package options, and four different colors, as well as the choice of automatic or standard transmission. Select-an-Auto Car Dealership wants to carry one of each of the different types of Night Hawks. How many cars are required?

SOLUTION: There are four items to select. We take the product of the numbers of choices for each of the four items.

$$\begin{pmatrix} \text{no. of body} \\ \text{styles} \end{pmatrix} \begin{pmatrix} \text{no. of} \\ \text{interiors} \end{pmatrix} \begin{pmatrix} \text{no. of} \\ \text{colors} \end{pmatrix} \begin{pmatrix} \text{no. of transmission} \\ \text{types} \end{pmatrix}$$

$$(2)(3)(4)(2) = 48$$

Select-an-Auto must stock 48 cars in order to have one of each possible type. ●

GUIDED EXERCISE   11

The Old Sage Inn offers a special dinner menu each night. There are two appetizers to choose from, three main courses, and four desserts. A customer can select one item from each category. How many different meals can be ordered from the special dinner menu?

(a) Each special dinner consists of three items. List the item and the number of choices per item.   ⇨   Appetizer—2; main course—3; dessert—4

Exercise continues

(b) To find the number of different dinners composed of the three items, multiply the number of choices per item together. (2)(3)(4) = 24
There are 24 different dinners that can be ordered from the special dinner menu.

Sometimes when we consider *n* items, we need to know the number of different ordered *arrangements* of the *n* items that are possible. The multiplication rules can help us find the number of possible ordered arrangements. Let's consider the classic example of determining the number of different ways in which eight people can be seated at a dinner table. For the first chair, there are eight choices. For the second chair, there are seven choices, since one person is already seated. For the third chair, there are six choices, since two people are already seated. By the time we get to the last chair, there is only one person left for that seat. We can view each arrangement as an outcome of a series of eight events. Event 1 is *fill the first chair*, event 2 is *fill the second chair*, and so forth. The multiplication rule will tell us the number of different outcomes.

Choices for	1st	2nd	3rd	4th	5th	6th	7th	8th	Chair position
	↓	↓	↓	↓	↓	↓	↓	↓	
	(8)	(7)	(6)	(5)	(4)	(3)	(2)	(1)	= 40,320

In all, there are 40,320 different seating arrangements for eight people. It is no wonder that it takes a little time to seat guests at a dinner table.

The multiplication pattern shown above is not unusual. In fact, it is an example of the multiplication indicated by the factorial notation 8!.

Factorial notation

    ! is read *factorial*

    8! is read *8 factorial*

    $8! = 8 \cdot 7 \cdot 6 \cdot 5 \cdot 4 \cdot 3 \cdot 2 \cdot 1$

In general, *n*! indicates the product of *n* with each of the positive counting numbers less than *n*. By special definition 0! = 1.

Factorial Notation

For a counting number *n*,

$$n! = n(n - 1)(n - 2) \cdots 1$$
$$0! = 1$$
$$1! = 1$$

(a) Evaluate 3!.

    $3! = 3 \cdot 2 \cdot 1 = 6$

(b) How many different ways can three objects be arranged in order? How many choices do you have for the first position, for the second position, and for the third position?

    We have three choices for the first position, two for the second position, and one for the third position. By the multiplication rule, we have

$$(3)(2)(1) = 3! = 6 \text{ arrangements}$$

Permutations

We have considered the number of ordered arrangements of n objects taken as an entire group. Specifically, we considered a dinner party for eight and found the number of ordered seating arrangements for all eight people. However, suppose you have an open house and have only five chairs. How many ways can five of the eight people seat themselves in the chairs? The formula we use to compute this number is called the *permutation formula*. We will simply state the formula and show you how to use it.

Counting Rule for Permutations

The number of ways to *arrange in order* n distinct objects, taking them r at a time, is

$$P_{n,r} = \frac{n!}{(n-r)!} \tag{9}$$

where n and r are whole numbers and $n \geq r$. Another commonly used notation for permutations is nPr.

EXAMPLE 9 ▸ Let's compute the number of ordered seating arrangements we have for eight people in five chairs.

SOLUTION: In this case we are considering a total of $n = 8$ different people, and we wish to arrange $r = 5$ of these people. Substituting into Formula (9), we have

$$P_{8,5} = \frac{8!}{(8-5)!} = \frac{8!}{3!} = \frac{40,320}{6} = 6720$$

    Calculator Note Most scientific calculators have a factorial key often designated x! or n!. On TI graphing calculators, the factorial operation ! is found in the MATH menu under the PRB (for probability) option. If you have the ! function available on your calculator, use it to calculate 8! and 3! directly, and divide the first result by the second result. Many of these same calculators have the permutation function built in, often labeled nPr. Again, on the TI graphing calculators, you find the permutation operation nPr within the MATH menu under PRB. You can, of course, use the permutation function directly with the numbers 8 and 5 and obtain the result 6720. In addition to the nPr operation, these calculators also have the combinations operation discussed after the next exercise. The label for the combinations operation is usually nCr. Figure 4-10 below shows these calculations on a TI-83 screen.

FIGURE 4-10

TI-83 Display of !, $P_{8,5}$

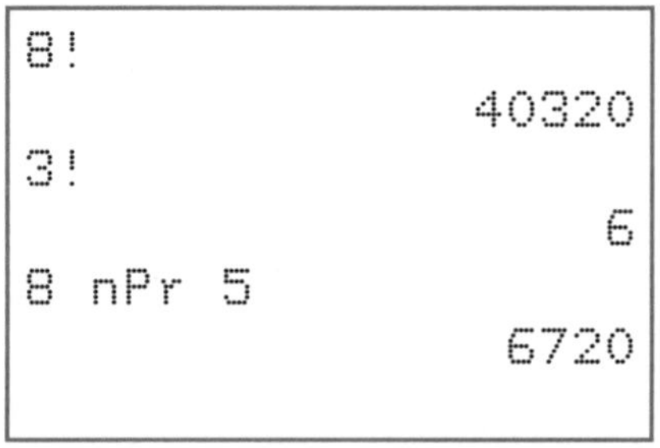

```
8!
                    40320
3!
                        6
8 nPr 5
                     6720
```

GUIDED EXERCISE   13

The board of directors of Belford Community Hospital has 12 members. Three officers—president, vice president, and treasurer—must be elected from the members. How many different possible slates of officers are there? We will view a slate of officers as a list of three people with one person for president listed first, one person for vice president listed second, and one person for treasurer listed third. For instance, if Mr. Acosta, Ms. Hill, and Mr. Smith wish to be on a slate together, there are several different slates possible, depending on which one will run for president, which for vice president, and which for treasurer. Not only are we asking for the number of different groups of three names for a slate, we are also concerned about order, since it makes a difference which name is listed in which position.

(a) What is the size of the group from which $\Rightarrow$ $n = 12$
the slates of officers will be selected? This is
the value of n.

(b) How many people will be selected for each $\Rightarrow$ $r = 3$
slate of officers? This is the value of r.

Exercise continues

(c) Each slate of officers is composed of three candidates. Different slates occur as we arrange the three candidates in the positions of president, vice president, and treasurer. For this reason, we need to consider the number of *permutations* of 12 items arranged in groups of 3. Compute $P_{n,r}$.

⟹

$$P_{n,r} = \frac{n!}{(n-r)!}$$

$$P_{12,3} = \frac{12!}{(12-3)!} = \frac{12!}{9!} = \frac{479{,}001{,}600}{362{,}880}$$

$$= 1320$$

There are 1320 different possible slates of officers. An alternative is to use your calculator to compute $P_{12,3}$ directly.

Combinations

In each of our previous counting formulas, we have taken the *order* of the objects or people into account. But what if order is not important? For instance, suppose we need to choose 3 members from the 12-member board of directors of Belford Community Hospital to go to a convention. We are interested in *different groupings* of 12 people so that each group contains 3 people. The order is of no concern, since all 3 will go to the convention. In other words, we need to consider the number of *different combinations* of 12 people taken 3 at a time. Our next formula will help us compute this number of different combinations.

Counting Rule for Combinations

The number of *combinations* of n objects taken r at a time is

$$C_{n,r} = \frac{n!}{r!(n-r)!} \tag{10}$$

where n and r are whole numbers and $n \geq r$. Other commonly used notations for combinations include $_nC_r$ and $\binom{n}{r}$.

Notice the difference between the concepts of permutations and combinations. When we consider permutations, we are considering groupings *and order*. When we consider combinations, we are considering only the number of different groupings. For combinations, order within the groupings is not considered. As a result, the number of combinations of n objects taken r at a time is generally smaller than the number of permutations of the same n objects taken r at a time. In fact, the combinations formula is simply the permutations formula with the number of permutations of each distinct group divided out. In the formula for combinations, notice the factor $r!$ in the denominator.

Now let's look at an example in which we compute the number of *combinations* of 12 people taken 3 at a time.

EXAMPLE 10 ▷ Three members from the group of 12 on the board of directors at Belford Community Hospital will be selected to go to a convention with all expenses paid. How many different groups of three are there?

SOLUTION: In this case we are interested in *combinations* rather than permutations of 12 people taken 3 at a time. Using Formula (10), we get

$$C_{n,r} = \frac{n!}{r!(n-r)!} \quad \text{or} \quad C_{12,3} = \frac{12!}{3!(12-3)!} = \frac{12!}{3!9!}$$

$$= \frac{479,001,600}{(6)(362,880)} = 220$$

There are 220 different groups of 3 people that can go to the convention.

Another way to get the solution is to use your calculator to evaluate $C_{12,3}$ directly. Since order is not considered, this number is much smaller than the number of different slates of three officers we computed in Guided Exercise 13. ●

GUIDED EXERCISE   14

In your political science class you are given a list of 10 books. You are to select 4 to read during the semester. How many different *combinations* of 4 books are available from the list of 10?

(a) Is the order in which you read the books relevant to the task of selecting the books?

▷ No.

(b) Do we use the number of permutations or combinations of 10 books taken 4 at a time?

▷ Since consideration of order in which the books are selected is not relevant, we compute the number of *combinations* of 10 books taken 4 at a time.

(c) How many books are available from which to select? How many must you read? What are the values of n and r?

▷ There are 10 books among which you must select 4 to read. $n = 10$ and $r = 4$.

(d) Compute $C_{10,4}$ to determine the number of different groups of 4 books from the list of 10.

▷ $C_{n,r} = \dfrac{n!}{r!(n-r)!}$

$C_{10,4} = \dfrac{10!}{4!(10-4)!}$

$= \dfrac{10!}{4!6!} = \dfrac{3,628,800}{(24)(720)} = 210$

There are 210 different groups of 4 books to select from the list of 10. An alternative method of solution is to use the $_nC_r$ key on your calculator.

We have different formulas for permutations and combinations of *n* objects taken *r* at a time. How do you decide which one to use? Always ask yourself if order within each group of *r* objects is relevant. If it is, use $P_{n,r}$. If order is not relevant, use $C_{n,r}$.

We have introduced you to three counting formulas: the multiplication rule, the permutations rule, and the combinations rule. There are other rules that apply when the objects are not distinct. Many counting problems are easy to state and fairly difficult to solve. Some have you combine several counting rules. However, the problems for this section are all straightforward. Some ask you to use your counting abilities to compute probabilities.

 V I E W P O I N T

Powerball

Powerball is a multistate lottery game that consists of drawing five distinct whole numbers between 1 and 49. Then one more number between 1 and 42 is selected as the Powerball number (this number could be one of the original five). Powerball numbers are drawn every Wednesday and Saturday. If you match all six numbers, you win the jackpot, which is worth at least 10 million dollars. Use the methods of this section to show that the odds of winning the jackpot are 1 in 80,089,128. For more information about the game of Powerball and the probability of winning different prizes, see Web site <http://www.qgm.com/lottoinf> and select Powerball.

S E C T I O N   4 . 3   P R O B L E M S

General: Tree Diagrams

1. (a) Draw a tree diagram to display all the possible head–tail sequences that can occur when you flip a coin three times.
   (b) How many sequences contain exactly two heads?

General: Tree Diagrams

2. (a) Draw a tree diagram to display all the possible outcomes that can occur when you flip a coin and then toss a die.
   (b) How many outcomes contain a head and a number greater than 4?

General: Tree Diagrams

3. Consider three true–false questions. There are two possible outcomes for each question: true or false. Draw a tree diagram showing all possible sequences of responses for the three questions. Does your tree diagram look similar to the one in Problem 1? Why would you expect this result?

General: Tree Diagrams

4. Make a tree diagram to show all the possible sequences of answers for three multiple-choice questions, each with four possible responses.

Industrial Production: Electronics

5. Four wires (red, green, blue, and yellow) need to be attached to a circuit board. A robotic device will attach the wires. The wires can be attached in any order, and the production manager wishes to determine which order would be fastest for the robot to use. Use the multiplication rule of counting to determine the number of all the possible sequences of assembly that must be tested. (*Hint:* There are four choices for the first wire, three for the second, two for the third, and only one for the fourth.)

Business: Air Connections

6. A sales representative must visit four cities: Omaha, Dallas, Wichita, and Oklahoma City. There are direct air connections between each of the cities. Use the multiplication rule of counting to determine the number of different choices the sales representative has for the order in which to visit the cities. How is this problem similar to Problem 5?

General: Deck of 52 Cards

7. You have two decks of cards (52 cards per deck), and you draw one card from each deck.
   (a) Use the multiplication rule of counting to determine the number of pairs of cards possible.
   (b) There are four kings in each deck. How many pairs of kings are possible?
   (c) *Probability extension:* Assuming all pairs are equally likely to be drawn, what is the probability of drawing two kings?

General: Roll Two Dice

8. You toss a pair of dice.
   (a) Use the multiplication rule of counting to determine the number of possible pairs of outcomes. (Recall that there are six possible outcomes for each die.)
   (b) There are three even numbers on each die. How many outcomes are possible with even numbers appearing on each die?
   (c) *Probability extension:* What is the probability that both dice will show even numbers?

General: Permutations and Combinations

9. Compute $P_{5,2}$.   10. Compute $P_{8,3}$.

11. Compute $P_{7,7}$.   12. Compute $P_{9,9}$.

13. Compute $C_{5,2}$.   14. Compute $C_{8,3}$.

15. Compute $C_{7,7}$.   16. Compute $C_{8,8}$.

Employment: Nursing Positions

17. There are three nursing positions to be filled at Lilly Hospital. Position one is the day nursing supervisor; position two is the night nursing supervisor; and position three is the nursing coordinator. There are 15 candidates qualified for all three of the positions. Use the permutation rule to determine the number of different ways the positions can be filled by these applicants.

Business: Combinations of Software

18. During the Computer Daze special promotion, a customer purchasing a computer and printer is given a choice of three free software packages. There are 10 different software packages from which to select. How many different combinations of software packages can be selected?

Management:
Trainee Positions

19. There are 15 qualified applicants for five trainee positions in a fast-food management program. How many different groups of trainees can be selected? (*Hint:* Is order important? If not, use the formula for combinations.)

Academic: Grading

20. One professor grades homework by randomly choosing 5 out of 12 homework problems to grade.
    (a) How many different groups of 5 problems are there from the 12 problems?
    (b) *Probability extension:* Jerry did only 5 problems of one assignment. What is the probability that the problems he did comprised the group that was selected to be graded?
    (c) Silvia did 7 problems. How many different groups of 5 did she complete? What is the probability that one of the groups of 5 she completed comprised the group selected to be graded?

Management:
Trainee Positions

21. The qualified applicant pool for six management trainee positions consists of seven women and five men.
    (a) How many different groups of applicants can be selected for the positions?
    (b) How many different groups of trainees would consist entirely of women?
    (c) *Probability extension:* If the applicants are equally qualified and the trainee positions are selected by drawing the names at random so that all groups of six are equally likely, what is the probability that the trainee class will consist entirely of women?

Leisure: Colorado Lotto

22. In the Colorado State Lotto game, there are 42 numbers. Players choose any 6. Then the state selects 6 of the numbers at random. The winning tickets (for the grand prize) are those on which the player's 6 numbers match the state's 6 numbers.
    (a) From 42 numbers, how many groups of 6 are possible?
    (b) *Probability extension:* If you buy one lottery ticket, what is the probability of winning the grand prize?
    (c) *Probability extension:* If you buy 10 lottery tickets, what is the probability of winning the grand prize?

SUMMARY

In this chapter we first examined the question: What is probability? We found that probabilities can be assigned to events by intuition, by the method of relative frequency, or by the method of equally likely outcomes. Next, we studied some probability rules. The most important rules are the multiplication rules for independent and dependent events and the addition rules for mutually exclusive and general events. We also looked at some counting techniques useful in computing probabilities. These techniques included tree diagrams, the multiplication rule for counting, combinations, and permutations.

IMPORTANT WORDS & SYMBOLS

Section 4.1
Relative frequency
Law of large numbers
Equally likely outcomes
Sample space
Probability of an event A, $P(A)$
Complement of A

Section 4.2
$P(A \text{ and } B)$
Independent events
Dependent events
$P(A, \text{ given } B)$
Multiplication rules (for independent and
  dependent events)

$P(A \text{ or } B)$
Mutually exclusive events
Addition rules (for mutually exclusive and
  general events

Section 4.3
Tree diagram
Multiplication rule of counting
Factorial notation, $n!$
Permutation, $P_{n,r}$
Permutation rule
Combination, $C_{n,r}$
Combination rule

VIEWPOINT

Deathday and Birthday

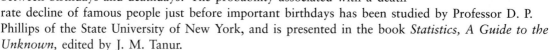

Can people really postpone death? If so, how much can the timing of death
be influenced by psychological, social, or other influential factors? One special event is a birthday. Will famous people try to postpone their deaths
until an important birthday? Both Thomas Jefferson and Samuel Adams
died on July 4, 1826, when the United States was celebrating its 50th birthday. Is this only a strange coincidence, or is there an unexpected connection
between birthdays and deathdays? The probability associated with a death
rate decline of famous people just before important birthdays has been studied by Professor D. P.
Phillips of the State University of New York, and is presented in the book *Statistics, A Guide to the
Unknown*, edited by J. M. Tanur.

CHAPTER REVIEW PROBLEMS

Medical: Bone Fractures

1. The American Academy of Orthopaedic Surgeons reported the age distribution of
   4.1 million patients with bone fractures treated in one year (*USA Today*). There
   were 1.6 million patients under 18 with bone fractures, 1.4 million patients between
   18 and 44 years of age inclusive, 0.6 million patients between 45 and 64 years old
   inclusive, and 0.5 million patients 65 and older.

(a) For each of the designated age groups, use relative frequencies to compute the probability that a randomly selected patient with a bone fracture is in that group.

(b) What is the probability that a randomly selected patient with a bone fracture is 44 years old or younger?

General: Dice and Coin

2. Suppose you throw a fair die and flip a fair coin. Let's represent the outcomes of 3 on the die and heads on the coin by *3H*.
   (a) One outcome is *3H*. What are the other outcomes? What is the sample space?
   (b) Are all outcomes in the sample space equally likely? Explain.
   (c) What is the probability of getting heads and a number less than 3?

General: Deck of 52 Cards

3. Two cards are drawn at random from a standard deck. (A standard deck has 52 cards: 13 hearts, 13 diamonds, 13 clubs, and 13 spades.)
   (a) Are the outcomes of the two cards independent? Why?
   (b) If the first card is replaced before the second is drawn, what is the probability that both cards will be hearts?
   (c) If the first card is not replaced before the second is drawn, what is the probability that both cards will be hearts?

General: Thumb Tacks

4. (a) Describe how you could use a relative frequency to estimate the probability that a thumbtack will land with its flat side down.
   (b) What is the sample space of outcomes for the thumbtack?
   (c) How would you make a probability assignment to this sample space if, when you drop 500 tacks, 340 land flat side down?

Management: Pay Raises

5. Does it pay to ask for a raise? A national survey of heads of households showed the percentage of those who asked for raises and the percentage who got them (*USA Today,* May 18, 1995). According to the survey, of the women interviewed, 24% had asked for raises, and of those women who had asked for raises, 45% received the raises. If a woman is selected at random from the survey population of women, find the following probabilities: *P*(woman asked for a raise); *P*(woman received raise, *given* she asked for one); *P*(woman asked for raise *and* received raise).

Management: Pay Raises

6. According to the same survey quoted in Problem 5, of the men interviewed, 20% had asked for raises and 59% of the men who had asked for raises *received* the raises. If a man is selected at random from the survey population of men, find the following probabilities: *P*(man asked for a raise); *P*(man received raise, *given* he asked for one); *P*(man asked for raise *and* received raise).

Business: Communications

7. A survey of employee communication managers at *Fortune 100* companies showed that by the year 2001, 42% anticipated that their companies would still use printed memos and postings for some workforce communications, 41% of the companies would use electronic mail, 21% would use face-to-face communications, and 18% would use video. Consider the percentage of companies listing each specified mode of communication as the probability that a company would use the specified mode. Look at the probabilities. Would you say that each company listed only one mode of workforce communication? Explain your answer.

Academic: Passing French

8. Class records at Rockwood College indicate that a student selected at random has a probability 0.77 of passing French 101. For the student who passes French 101,

the probability is 0.90 that he or she will pass French 102. What is the probability that a student selected at random will pass both French 101 and French 102?

General: Selection Process

9. There is money to send two of eight city council members to a conference in Honolulu. All want to go, so they decide to choose the members to go to the conference by a random process. How many different combinations of two council members can be selected from the eight who want to go to the conference?

General: Permutations and Combinations

10. Compute
   (a) $P_{7,2}$ (b) $C_{7,2}$ (c) $P_{3,3}$ (d) $C_{4,4}$

Marketing: Packaging

11. Freeze Dry Food, Inc. packages all its foods in clear plastic that is sealed. The quality control for the packaging process checks for three items: (1) that the weight shown is correct, (2) that the label is correct, and (3) that the package is properly sealed. These three processes can be done in any order. A computer-operated device directs the packages to the three inspection stations according to backlog in that area. If there is a larger backlog in one area, products are sent to one of the other two areas first. In how many different ways can a package be cycled through the three inspection stations?

Academic: Schedules

12. A student must satisfy the literature, social science, and philosophy requirements this semester. There are four literature courses to select from, three social science courses, and two philosophy courses. Make a tree diagram showing all the possible sequences of literature, social science, and philosophy courses.

Testing: Multiple Choice

13. There are five multiple-choice questions on an exam, each with four possible answers. Use the multiplication rule of counting to determine the number of possible answer sequences for the five questions. Only one of the sets can contain all five correct answers. If you are guessing, so that you are as likely to choose one sequence of answers as another, what is the probability of getting all five answers correct?

General: Flip a Coin

14. A coin is tossed six times. Use the multiplication rule of counting to determine the number of possible head–tail sequences that can occur.

DATA HIGHLIGHTS: GROUP PROJECTS

Break into small groups and discuss the following topics. Organize a brief outline in which you summarize the main points of your group discussion.

1. Look at the figure on the following page, Peeking at workers' e-mail. What group of people was surveyed? Estimate the probability that an executive selected at random from the survey population works at a company that uses e-mail. Estimate the probability that an executive peeks at employee e-mail, given that the company uses e-mail. Compute the probability that an executive from the survey population works at a company that uses e-mail *and* peeks at the employees' e-mail.

2. Consider the information given on the following page, Vulnerable knees. What is the probability that an orthopaedic case selected at random involves knee problems? Of

FIGURE 4-11

Source: Copyright 1996, USA TODAY. Reprinted by permission.

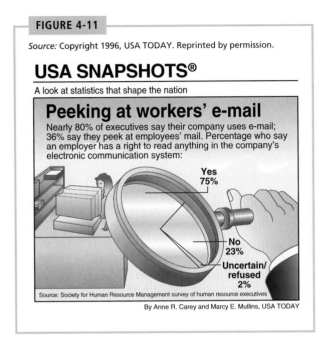

USA SNAPSHOTS®

A look at statistics that shape the nation

Peeking at workers' e-mail

Nearly 80% of executives say their company uses e-mail; 36% say they peek at employees' mail. Percentage who say an employer has a right to read anything in the company's electronic communication system:

Yes
75%

No
23%

Uncertain/
refused
2%

Source: Society for Human Resource Management survey of human resource executives

By Anne R. Carey and Marcy E. Mullins, USA TODAY

FIGURE 4-12

Source: Copyright 1997, USA TODAY. Reprinted by permission.

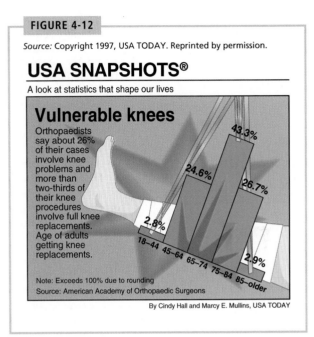

USA SNAPSHOTS®

A look at statistics that shape our lives

Vulnerable knees

Orthopaedists say about 26% of their cases involve knee problems and more than two-thirds of their knee procedures involve full knee replacements. Age of adults getting knee replacements.

43.3%

24.6%

26.7%

2.8%

2.9%

18–44 45–64 65–74 75–84 85–older

Note: Exceeds 100% due to rounding

Source: American Academy of Orthopaedic Surgeons

By Cindy Hall and Marcy E. Mullins, USA TODAY

those cases, estimate the probability that the case requires full knee replacement. Compute the probability that an orthopaedic case selected at random involves a knee problem *and* requires a full knee replacement. Next, look at the probability distribution for ages of patients requiring full knee replacement. Medicare insurance coverage begins when a person reaches age 65. What is the probability that the age of a person receiving a knee replacement is 65 or older?

⒧INKING CONCEPTS: WRITING PROJECTS

Discuss each of the following topics in class or review the topics on your own. Then write a brief but complete essay in which you summarize the main points. Please include formulas as appropriate.

1. Discuss the following concepts and give examples from everyday life where you encounter each concept. *Hint:* For instance, consider the "experiment" of arriving for class. Some possible outcomes are not arriving (that is, cutting class), arriving on time, and arriving late.
   (a) Statistical experiment.
   (b) Sample space.
   (c) Probability assignment to a sample space. In your discussion, be sure to include answers to the following questions.
      (i) Is there more than one valid way to assign probabilities to a sample space? Explain and give an example.

(ii) How can probabilities be estimated by relative frequencies? How can probabilities be computed if events are equally likely?

2. Discuss the concepts of mutually exclusive events and independent events. List several examples of each type of event from everyday life.

(a) If A and B are mutually exclusive events, does it follow that A and B *cannot* be independent events? Give an example to demonstrate your answer. (*Hint:* Discuss an election where only one person can win the election. Let A be the event that party A's candidate wins, and let B be the event that party B's candidate wins. Does the outcome of one event determine the outcome of the other event? Are A and B mutually exclusive events?)

(b) Discuss conditions under which $P(A \text{ and } B) = P(A) \cdot P(B)$ is true. Under what conditions is this not true?

(c) Discuss conditions under which $P(A \text{ or } B) = P(A) + P(B)$ is true. Under what conditions is this not true?

3. Although we learn a good deal about probability in this course, the main emphasis is on statistics. Write a few paragraphs in which you talk about the distinction between probability and statistics. In what types of problems would probability be the main tool? In what types of problems would statistics be the main tool? Give some examples of both types of problems. What kinds of outcomes or conclusions do we expect from each type of problem?

The problems in this section may be done using statistical computer software or calculators with statistical functions. Displays and suggestions are given for Minitab (Release 11), the TI-83 graphing calculator, and ComputerStat.

Demonstration of the Law of Large Numbers

Computers can be used to simulate experiments. In packages such as Minitab and Excel, programs using random-number generators can be designed (see the *Technology Guides*) to simulate activities such as tossing a die. In ComputerStat, such a program exists (menu selection: ➤**Descriptive Statistics** ➤**Simulate the Experiment of Tossing One Die**). The following printouts show the simulations for tossing a die 6, 12, 50, 500, 5000, 50,000, 500,000, and 1,000,000 times. Notice how the relative frequencies of the outcomes approach the theoretical probabilities of 1/6 or 0.16667 for each outcome. Do you expect the same results every time the simulation is done? Why or why not?

Results of tossing one die 6 times

Outcome	Number of Occurrences	Relative Frequency
•	0	.00000
⠢	1	.16667
⠪	2	.33333
⠼	0	.00000
⠫	1	.16667
⠿	2	.33333

Results of tossing one die 12 times

Outcome	Number of Occurrences	Relative Frequency
•	4	.33333
⠢	2	.16667
⠪	1	.08333
⠼	0	.00000
⠫	4	.33333
⠿	1	.08333

Results of tossing one die 50 times

Outcome	Number of Occurrences	Relative Frequency
•	8	.16000
⠢	7	.14000
⠪	9	.18000
⠼	8	.16000
⠫	11	.22000
⠿	7	.14000

Results of tossing one die 500 times

Outcome	Number of Occurrences	Relative Frequency
•	87	.17400
⠢	83	.16600
⠪	91	.18200
⠼	69	.13800
⠫	87	.17400
⠿	83	.16600

Results of tossing one die 5000 times

Outcome	Number of Occurrences	Relative Frequency
⚀	800	.16000
⚁	830	.16600
⚂	806	.16120
⚃	851	.17020
⚄	839	.16780
⚅	874	.17480

Results of tossing one die 50,000 times

Outcome	Number of Occurrences	Relative Frequency
⚀	8528	.17056
⚁	8354	.16708
⚂	8246	.16492
⚃	8414	.16828
⚄	8178	.16356
⚅	8280	.16560

Results of tossing one die 500,000 times

Outcome	Number of Occurrences	Relative Frequency
⚀	83644	.16729
⚁	83368	.16674
⚂	83398	.16680
⚃	83095	.16619
⚄	83268	.16654
⚅	83227	.16645

Results of tossing one die 1,000,000 times

Outcome	Number of Occurrences	Relative Frequency
⚀	166643	.16664
⚁	166168	.16617
⚂	167391	.16739
⚃	165790	.16579
⚄	167243	.16724
⚅	166765	.16677

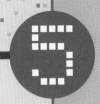

5

Introduction to Probability Distributions and the Binomial Distribution

Education is the key to unlock the golden door of freedom.

—George Washington Carver

George Washington Carver (1859–1943)

Carver was a winner of the Spingarn Medal for distinguished service in agricultural chemistry and the prestigious Roosevelt Medal for contributions to science. Carver was also a Fellow in the Royal Society of Arts in London, an honor given to very few Americans.

George Washington Carver won international fame for agricultural research. After graduating from Iowa State College, he was appointed a faculty member in the Iowa State Botany Department. Carver took charge of the greenhouse and started a fungus collection that later grew to more than 20,000 species. This collection brought him professional acclaim in the field of botany.

At the invitation of his friend Booker T. Washington, Carver joined the faculty of the Tuskegee Institute, where he spent the rest of his long and distinguished career. Carver's creative genius accounted for more than 300 inventions from peanuts, 118 inventions from sweet potatoes, and 75 inventions from pecans.

Gathering and analyzing data were important components of Carver's work. Methods you will learn in this course are widely used in research in every field, including agriculture.

5.1   # Introduction to Random Variables and Probability Distributions

For our purposes, we will say that a *statistical experiment* is any process by which an observation (or measurement) is obtained. Examples of statistical experiments are

1. Counting the number of eggs in a robin's nest

2. Measuring daily rainfall in inches

3. Counting the number of defective light bulbs in a case of bulbs

4. Measuring the weight of a polar bear cub in kilograms

Let x represent a quantitative variable that is measured in an experiment. We are interested in the numerical values that x can take on. So x = number of eggs in a robin's nest and x = weight of a polar bear cub in kilograms would be examples of such quantitative variables. Furthermore, we say that the quantitative variable x is a *random variable* because the value that x takes on in a given experiment is a chance or random outcome. We will study two types of random variables: *discrete random variables* and *continuous random variables.*

Discrete random variable

> **Definition**
>
> When the observations of a quantitative random variable can take on only a finite number of values or a countable number of values, we say that the variable is a *discrete random variable.*

We know what a *finite* number of values is, but what is a *countable* number of values? As an example, let the random variable x be the number of wells an oil prospector drills until the first productive well is found. Then x could be any of the values 1, 2, 3. . . . In theory, we have an infinite number of possibilities for the values of x, but the set of values of x corresponds to the set of counting numbers. Therefore, this type of infinity is called *countable,* and we say x has a countable number of values. This is an intuitive approach to the concept of a countable set. The reader interested in a more rigorous discussion is referred to the advanced text *Introduction to Mathematical Statistics* by Hogg and Craig (Macmillan Publishing).

In most of the cases we will consider, a *discrete random variable* will be the result of a count (the terms *countable* and *discrete,* however, have different mathematical meanings). For instance, the number of students in a certain section of a statistics course this term is a discrete random variable. The value must be a counting number such as 25, 57, or 135. The values 25.34 and 25½ are not possible. The cost of tuition to the nearest dollar or nearest cent is another example of a discrete random variable. In this case we are counting dollars or cents.

Continuous random
variable

> **Definition**
>
> When the observations of a quantitative random variable can take on any of the countless number of values in a line interval, we say that the variable is a *continuous random variable.*

For our purposes, we will see most *continuous random variables* occurring as the results of measurements. For example, the air pressure in an automobile tire represents a continuous random variable. The air pressure could in theory take on any value from 0 lb/in.^{2} (psi) to the bursting pressure of the tire. Values such as 20.126 psi, 20.12678 psi, and so forth are possible. Another example is the heights of students in your statistics class. The heights could in theory take on any value from a low of, say, 3 feet to a high of, say, 7.25 feet.

The distinction between discrete and continuous random variables is important because of the different mathematical techniques associated with the two kinds of random variables. Although we will not discuss these techniques at great length in this book, the distinction is very important in the study of advanced mathematical statistics.

In general, measurements of quantities such as length, weight, volume, temperature, and time yield continuous random variables. If the temperature changes from 12°C to 13°C, for example, it must take on all the temperature values between 12 and 13. Temperatures cannot just jump from one reading to the next. Discrete random variables often come from counts, such as the number of passing scores on an exam or the number of weeds in a garden.

GUIDED EXERCISE   1

Which of the following random variables are discrete and which are continuous?

(a) *Measure* the time it takes a student selected at random to register for the fall term.

⇨ Time can take on any value, so this is a continuous random variable.

(b) *Count* the number of bad checks drawn on Upright Bank on a day selected at random.

⇨ The number of bad checks can only be a whole number such as 0, 1, 2, 3, etc. This is a discrete variable.

(c) *Measure* the amount of gasoline needed to drive your car 200 miles.

⇨ We are measuring volume, which can assume any value, so this is a continuous random variable.

(d) Pick a random sample of 50 registered voters in a district and find the number who voted in the last county election.

⇨ This is a count, so the variable is discrete.

Probability distribution

A random variable has a probability distribution whether it is discrete or continuous. The *probability distribution* is simply an assignment of probabilities to the specific values of the random variable or to a range of values of the random variable.

1. The probability distribution of a *discrete* random variable has a probability assigned to *each* value of the random variable.

2. The sum of these probabilities must be 1.

Let's look at a discrete probability distribution and its graph.

EXAMPLE 1   Dr. Fidgit developed a test to measure boredom tolerance. He administered it to a group of 20,000 adults between the ages of 25 and 35. The possible scores were 0, 1, 2, 3, 4, 5, and 6, with 6 indicating the highest tolerance for boredom. The test results for this group are shown in Table 5-1.

(a) If a subject is chosen at random from this group, the probability that he or she will have a score of 3 is 6000/20,000, or 0.30. In a similar way, we can use the relative frequency to compute the probabilities for the other scores (Table 5-2). These probability assignments make up the probability distribution. Notice that the scores are mutually exclusive: no one subject has two scores. The sum of the probabilities of all the scores is 1.

Table 5-1 **Boredom Tolerance Test Scores for 20,000 Subjects**

Score	Number of Subjects
0	1400
1	2600
2	3600
3	6000
4	4400
5	1600
6	400

Table 5-2 **Probability Distribution of Scores on Boredom Tolerance Test**

Score x	Probability $P(x)$
0	0.07
1	0.13
2	0.18
3	0.30
4	0.22
5	0.08
6	0.02
	$\Sigma P(x) = 1$

(b) The graph of this distribution is simply a relative-frequency histogram (see Figure 5-1) in which the height of the bar over a score represents the probability of that score. Since each bar is one unit wide, the area of the bar over a score equals the height and thus represents the probability of that score. Since the sum of the probabilities is 1, the area under the graph is also 1.

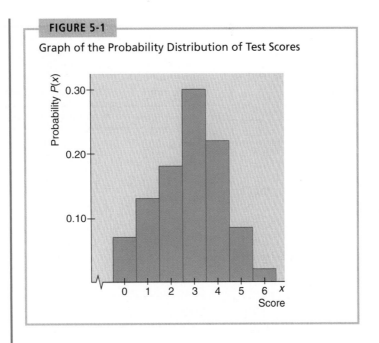

FIGURE 5-1

Graph of the Probability Distribution of Test Scores

(c) The Topnotch Clothing Company needs to hire someone with a score on the boredom tolerance test of 5 or 6 to operate the fabric press machine. Since the scores 5 and 6 are mutually exclusive, the probability that someone in the group who took the boredom tolerance test made either a 5 or a 6 is the sum

$$P(5 \text{ or } 6) = P(5) + P(6)$$
$$= 0.08 + 0.02$$
$$= 0.10$$

Notice that to find $P(5 \text{ or } 6)$ we could have simply added the *areas* of the bars over 5 and over 6. One out of 10 of the group who took the boredom tolerance test would qualify for the position at Topnotch Clothing.

GUIDED EXERCISE   2

One of the elementary tools of cryptanalysis (the science of code breaking) is to use relative frequencies of occurrence of different letters in the alphabet to break standard English alphabet codes. Large samples of plain text such as newspaper stories generally yield about the same relative frequencies for letters. A sample 1000 letters long yielded the information in Table 5-3.

Exercise continues

Exercise continued

(a) Use the relative frequencies to compute the omitted probabilities in Table 5-3.

Table 5-3   **Frequency of Letters in a 1000-Letter Sample**

Letter	Freq.	Prob.	Letter	Freq.	Prob.
A	73	_____	N	78	0.078
B	9	0.009	O	74	_____
C	30	0.030	P	27	0.027
D	44	0.044	Q	3	0.003
E	130	_____	R	77	0.077
F	28	0.028	S	63	0.063
G	16	0.016	T	93	0.093
H	35	0.035	U	27	_____
I	74	_____	V	13	0.013
J	2	0.002	W	16	0.016
K	3	0.003	X	5	0.005
L	35	0.035	Y	19	0.019
M	25	0.025	Z	1	0.001

Source: From *Elementary Cryptanalysis: A Mathematical Approach,* by Abraham Sinkov. Copyright © 1968 by Yale University. Reprinted by permission of Random House, Inc.

⟹ Table 5-4   **Completion of Table 5-3**

Letter	Relative Frequency	Probability
A	$\frac{73}{1000}$	0.073
E	$\frac{130}{1000}$	0.130
I	$\frac{74}{1000}$	0.074
O	$\frac{74}{1000}$	0.074
U	$\frac{27}{1000}$	0.027

(b) Do the probabilities of all the individual letters add up to 1?

⟹ Yes

(c) If a letter is selected at random from a newspaper story, what is the probability that the letter will be a vowel?

⟹ If a letter is selected at random,

$$P(a, e, i, o, \text{ or } u) = P(a) + P(e) + P(i) + P(o) + P(u)$$
$$= 0.073 + 0.130 + 0.074 + 0.074 + 0.027$$
$$= 0.378$$

Mean and standard deviation of a discrete probability distribution

A probability distribution can be thought of as a relative-frequency distribution based on a very large *n.* As such, it has a mean and standard deviation. If we are referring to the probability distribution of a *population,* then we use the Greek letters μ for the mean and σ for the standard deviation. When we see the Greek letters used, we know the information given is from the *entire population* rather than just a sample. If we have a sample probability distribution, we use $\bar{x}$ (*x* bar) and *s,* respectively, for the mean and standard deviation. For a given population, μ and σ are fixed numbers and are sometimes called the *parameters* of the population.

Definition

The *mean* and the *standard deviation of a discrete population* probability distribution are found by using these formulas:

Mean $\mu = \Sigma x P(x)$

Standard deviation $\sigma = \sqrt{\Sigma(x-\mu)^2 P(x)}$

where x is the value of a random variable,
$\qquad P(x)$ is the probability of that variable,
$\qquad$ the sum Σ is taken for all the values of the random variable.

Expected value

The mean of a probability distribution is often called the *expected value* of the distribution. This terminology reflects the idea that the mean represents a "central point" or "cluster point" for the entire distribution. Of course, the mean or expected value is an average value, and as such it *need not be a point of the sample space.*

EXAMPLE 2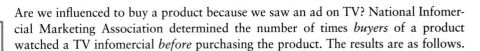

Are we influenced to buy a product because we saw an ad on TV? National Infomercial Marketing Association determined the number of times *buyers* of a product watched a TV infomercial *before* purchasing the product. The results are as follows.

Number of Times Buyers Saw Infomercial	1	2	3	4	5*
Percentage of Buyers	27%	31%	18%	9%	15%

*This category was 5 or more, but will be treated as 5 in this example.

We can treat the information shown as an estimate of the probability distribution because the events are mutually exclusive and the sum of the percentages is 100%. Compute the mean and standard deviation of the distribution.

SOLUTION: We put the data in the first two columns of a computation table and then fill in the other entries (see Table 5-5 on page 220). The average number of times a buyer views the infomercial before purchase is

$$\mu = \Sigma x P(x) = 2.54 \text{ (sum of column 3)}$$

To find the standard deviation, we take the square root of the sum of column 6:

$$\sigma = \sqrt{\Sigma(x - \mu)^2 P(x)} \approx \sqrt{1.869} \approx 1.37$$

Calculator Note Some calculators, including the TI-83, accept fractional frequencies. If yours does, you can get μ and σ directly by using techniques for grouped data and the calculator's STAT mode.

Table 5-5 Number of Times Buyers View Infomercial Before Making Purchase

x (number of viewings)	P(x)	xP(x)	x − μ	(x − μ)²	(x − μ)²P(x)
1	0.27	0.27	−1.54	2.372	0.640
2	0.31	0.62	−0.54	0.292	0.091
3	0.18	0.54	0.46	0.212	0.038
4	0.09	0.36	1.46	2.132	0.192
5	0.15	0.75	2.46	6.052	0.908

$$\mu = \Sigma xP(x) = 2.54 \qquad\qquad \Sigma(x - \mu)^2 P(x) = 1.869$$

GUIDED EXERCISE   3

At a carnival you pay \$2.00 to play a coin-flipping game with three fair coins. On each coin one side has the number 0 and the other side has the number 1. You flip the three coins at one time and you win \$1.00 for every 1 that appears on top. Are your expected earnings equal to the cost to play? We'll answer this question in several steps.

(a) In this game the random variable of interest counts the number of 1's that show. What is the sample space for the values of this random variable?

➡ The sample space is {0, 1, 2, 3}, since no 1's can come up, one 1 can appear, two 1's can appear, or three 1's can appear.

(b) There are eight equally likely outcomes for throwing three coins. They are 000, 001, 010, 011, 100, 101, _____, and _____.

➡ 110 and 111

(c) Complete Table 5-6.

Table 5-6

Number of 1's, x	Frequency	P(x)	xP(x)
0	1	0.125	0
1	3	0.375	_____
2	3	_____	_____
3	_____	_____	_____

➡ **Table 5-7 Completion of Table 5-6**

x	Frequency	P(x)	xP(x)
0	1	0.125	0
1	3	0.375	0.375
2	3	0.375	0.750
3	1	0.125	0.375

(d) The expected value is the sum

$$\mu = \Sigma xP(x)$$

Sum the appropriate column of Table 5-6 to find this value. Are your expected earnings less than, equal to, or more than the cost of the game?

➡ The expected value can be found by summing the last column of Table 5-7. The expected value is \$1.50. It cost \$2.00 to play the game; the expected value is less than the cost. The carnival is making money. In the long run the carnival can expect to make an average of about 50 cents per player.

We have seen probability distributions of discrete variables and the formulas for computing the means and standard deviations of discrete population probability distributions. Probability distributions of continuous random variables are similar except that the probability assignments are made to intervals of values rather than to specific values of the random variable. We will see an important example of a discrete probability distribution, the binomial distribution, in the next section and one of a continuous probability distribution in Chapter 6 when we study the normal distribution.

V I E W P O I N T

Net Vet/Electronic Zoo

Yes! You can get veterinary information about your pet, farm animal, or even exotic zoo animals right off the World Wide Web! You also can make a probability distribution for the random variable "number of hits" on a given Web page. For example, the horse page gets about 5% of the hits, whereas the fish page gets only 3%. If you want to complete the probability distribution, see Web site <http://vetnet.wustl.edu/about.html>, link to Statistics, and select Most Recent Web Trends.

SECTION 5.1 PROBLEMS

General: Random Variables

1. Which of the following are continuous variables, and which are discrete?
   (a) Number of traffic fatalities per year in the state of Florida
   (b) Distance a golf ball travels after being hit with a driver
   (c) Time required to drive from home to college on any given day
   (d) Number of ships in Pearl Harbor on any given day
   (e) Your weight before breakfast each morning

General: Random Variables

2. Which of the following are continuous variables, and which are discrete?
   (a) Speed of an airplane
   (b) Age of a college professor chosen at random
   (c) Number of books in the college bookstore
   (d) Weight of a football player chosen at random
   (e) Number of lightning strikes in Rocky Mountain National Park on a given day

General: Probability Distributions

3. Consider each distribution. Determine if it is a valid probability distribution or not, and explain your answer.

(a)

x	0	1	2
$P(x)$	0.25	0.60	0.15

(b)

x	0	1	2
$P(x)$	0.25	0.60	0.20

Marketing: Age Distribution

4. What is the age distribution of promotion-sensitive shoppers? A *supermarket super shopper* is defined as a shopper for whom at least 70% of the items purchased were on sale or purchased with coupons. The following table is based on information taken from *Trends in the United States* (Food Marketing Institute, Washington, D.C.).

Age range, years	18–28	29–39	40–50	51–61	62 and over
Midpoint x	23	34	45	56	67
Percentage of super shoppers	7%	44%	24%	14%	11%

For the 62 and over group, use the midpoint 67 years.
(a) Using the age midpoints x and the percentages of super shoppers, do we have a valid probability distribution? Explain.
(b) Use a histogram to graph the probability distribution of part a.
(c) Compute the expected age μ of a super shopper.
(d) Compute the standard deviation σ for ages of super shoppers.

Marketing: Income

5. What is the income distribution of super shoppers (see Problem 4)? In the following table, income units are in thousands of dollars, and each interval goes up to but does not include the given high value. The midpoints are given to the nearest thousand dollars.

Income range	5–15	15–25	25–35	35–45	45–55	55 or more
Midpoint x	10	20	30	40	50	60
Percentage of super shoppers	21%	14%	22%	15%	20%	8%

(a) Using the income midpoints x and the percentages of super shoppers, do we have a valid probability distribution? Explain.
(b) Use a histogram to graph the probability distribution of part a.
(c) Compute the expected income μ of a super shopper.
(d) Compute the standard deviation σ for the income of super shoppers.

Nursing: Age Distribution

6. What was the age distribution of nurses in Great Britain at the time of Florence Nightingale? Thanks to Florence Nightingale and the British census of 1851, we have the following information (based on data from the classic text *Notes on Nursing*, by Florence Nightingale). *Note:* In 1851 there were 25,466 nurses in Great Britain. Furthermore, Nightingale made a strict distinction between nurses and domestic servants.

Age range (yr)	20–29	30–39	40–49	50–59	60–69	70–79	80+
Midpoint x	24.5	34.5	44.5	54.5	64.5	74.5	84.5
Percentage of nurses	5.7%	9.7%	19.5%	29.2%	25.0%	9.1%	1.8%

(a) Using the age midpoints x and the percentages of nurses, do we have a valid probability distribution? Explain.

(b) Use a histogram to graph the probability distribution of part a.
(c) Find the probability that a British nurse selected at random in 1851 would be 60 years of age or older.
(d) Compute the expected age μ of a British nurse contemporary to Florence Nightingale.
(e) Compute the standard deviation σ for ages of nurses shown in the distribution.

Fishing:
Daily Creel Summary

7. The following data are based on information taken from *Daily Creel Summary*, published by the Paiute Indian Nation, Pyramid Lake, Nevada. Movie stars and U.S. presidents have fished Pyramid Lake. It is one of the best places in the lower 48 states to catch trophy cutthroat trout. In this table, x = number of fish caught in a 6-hour period. The percentage data are the percentages of fishermen who caught x fish in a 6-hour period while fishing from shore.

x	0	1	2	3	4 or more
%	44%	36%	15%	4%	1%

(a) Convert the percentages to probabilities and make a histogram of the probability distribution.
(b) Find the probability that a fisherman selected at random fishing from shore catches one or more fish in a 6-hour period.
(c) Find the probability that a fisherman selected at random fishing from shore catches two or more fish in a 6-hour period.
(d) Compute μ, the expected value of the number of fish caught per fisherman in a 6-hour period (round 4 or more to 4).
(e) Compute σ, the standard deviation of the number of fish caught per fisherman in a 6-hour period (round 4 or more to 4).

Vital Statistics: Insurance

8. Jim is a 60-year-old Anglo male in reasonably good health. He wants to take out a $50,000 term (that is, straight death benefit) life insurance policy until he is 65. The policy will expire on his 65th birthday. The probability of death in a given year is provided by the Vital Statistics Section of the *Statistical Abstract of the United States* (116th Edition).

x = age	60	61	62	63	64
P(death at this age)	0.01191	0.01292	0.01396	0.01503	0.01613

Jim is applying to Big Rock Insurance Company for his term insurance policy.

(a) What is the probability that Jim will die in his 60th year? Using this probability and the $50,000 death benefit, what is the expected loss to Big Rock Insurance?
(b) Repeat part a for years 61, 62, 63, and 64. What would be the total expected loss to Big Rock Insurance over the years 60 through 64?
(c) If Big Rock Insurance wants to make a profit of $700 above the expected total loss paid out for Jim's death, how much should it charge for the policy?
(d) If Big Rock Insurance Company charges $5000 for the policy, how much profit does the company expect to make?

Vital Statistics: Insurance

9. Sara is a 60-year-old Anglo female in reasonably good health. She wants to take out a $50,000 term (that is, straight death benefit) life insurance policy until she is 65.

The policy will expire on her 65th birthday. The probability of death in a given year is provided by the Vital Statistics Section of the *Statistical Abstract of the United States* (116th Edition).

x = age	60	61	62	63	64
P(death at this age)	0.00756	0.00825	0.00896	0.00965	0.01035

Sara is applying to Big Rock Insurance Company for her term insurance policy.

(a) What is the probability that Sara will die in her 60th year? Using this probability and the $50,000 death benefit, what is the expected loss to Big Rock Insurance?

(b) Repeat part a for years 61, 62, 63, and 64. What would be the total expected loss to Big Rock Insurance over the years 60 through 64?

(c) If Big Rock Insurance wants to make a profit of $700 above the expected total loss paid out for Sara's death, how much should it charge for the policy?

(d) If Big Rock Insurance Company charges $5000 for the policy, how much profit does the company expect to make?

Student Life: Raffle

10. The college student senate is sponsoring a spring break Caribbean cruise raffle. The proceeds are to be donated to the Samaritan Center for the Homeless. A local travel agency donated the cruise, valued at $2000. The students sold 2852 raffle tickets at $5 per ticket.

(a) Kevin bought six tickets. What is the probability that Kevin will win the spring break cruise to the Caribbean? What is the probability that Kevin will not win the cruise?

(b) Expected earnings can be found by multiplying the value of the cruise by the probability that Kevin will win. What are Kevin's expected earnings? Is this more or less than the amount Kevin paid for the six tickets? How much did Kevin effectively contribute to the Samaritan Center for the Homeless?

Section **5.2** **Binomial Probabilities**

Binomial Experiment

On a TV quiz show, each contestant has a try at the wheel of fortune. The wheel of fortune is a roulette wheel with 36 slots, one of which is gold. If the ball lands in the gold slot, the contestant wins $50,000. No other slot pays. What is the probability that the quiz show will have to pay the fortune to three contestants out of 100?

In this problem the contestant and the quiz show sponsors are concerned about only two outcomes from the wheel of fortune: the ball lands on the gold, or the ball does not land on the gold. This problem is typical of an entire class of problems that are characterized by the feature that there are exactly two possible outcomes (for each trial) of interest. These problems are called *binomial experiments,* or *Bernoulli experiments,* after the Swiss mathematician Jacob Bernoulli, who studied them extensively in the late 1600s.

Features of a binomial experiment

Features of a Binomial Experiment

1. There are a fixed number of trials. We denote this number by the letter n.

2. The n trials are independent and repeated under identical conditions.

3. Each trial has only two outcomes: success, denoted by S, and failure, denoted by F.

4. For each individual trial, the probability of success is the same. We denote the probability of success by p and that of failure by q. Since each trial results in either success or failure, $p + q = 1$ and $q = 1 - p$.

5. The central problem of a binomial experiment is to find the probability of r successes out of n trials.

EXAMPLE 3   ❯

Let's see how the wheel of fortune problem meets the criteria of a binomial experiment. We'll take the criteria one at a time.

1. Each of the 100 contestants has a trial at the wheel, so there are $n = 100$ trials in this problem.

2. Assuming that the wheel is fair, the trials are independent, since the result of one spin of the wheel has no effect on the results of other spins.

3. We are interested in only two outcomes on each spin of the wheel: either the ball lands on the gold, or it does not. Let's call landing on the gold *success* (S) and not landing on the gold *failure* (F). In general, the assignment of the terms *success* and *failure* to outcomes does not imply good or bad results. These terms are assigned simply for the user's convenience.

4. On each trial the probability p of success (landing on the gold) is 1/36, since there are 36 slots and only one of them is gold. Consequently, the probability of failure is

$$q = 1 - p = 1 - \frac{1}{36} = \frac{35}{36}$$

on each trial.

5. We want to know the probability of 3 successes out of 100 trials, so $r = 3$ in this example. It turns out that the probability the quiz show will have to pay the fortune to 3 contestants out of 100 is about 0.23. Later in this section we'll see how this probability was computed.

Anytime we make selections from a population *without replacement, we do not have independent trials.* However, replacement is often not practical. If the number of trials is quite small with respect to the population, we *almost* have independent trials, and we can say the situation is *closely approximated* by a binomial experiment. For instance, suppose we select 20 tuition bills at random from a collection of 10,000 bills issued at one college and observe if the bill is in error or not. If 600 of the 10,000 bills are in error, then the probability that the first one selected is in error is 600/10,000, or 0.0600. If the first is in error, then the probability that the second is in error is 599/9999, or 0.0599. Even if the first 19 bills selected are in error, the probability that the 20th is also in error is 581/9981, or 0.0582. All these probabilities round to 0.06, and we can say that the independence condition is approximately satisfied.

GUIDED EXERCISE   4

Let's analyze the following binomial experiment to determine p, q, n, and r.

According to the *Textbook of Medical Physiology*, 5th Edition, by Arthur Guyton, 9% of the population has blood type B. Suppose we choose 18 people at random from the population and test the blood type of each. What is the probability that 3 of these people have blood type B? (*Note:* Independence is approximated because 18 people is an extremely small sample with respect to the entire population.)

(a) In this experiment we are observing whether or not a person has type B blood. We will say we have a success if the person has type B blood. What is failure?

⇨ Failure occurs if a person does not have type B blood.

(b) The probability of success is 0.09, since 9% of the population has type B blood. What is the probability of failure, q?

⇨ The probability of failure is
$q = 1 - p$
$= 1 - 0.09 = 0.91$

(c) In this experiment there are $n = $ _____ trials.

⇨ In this experiment $n = 18$.

(d) We wish to compute the probability of 3 successes out of 18 trials. In this case $r = $ _____.

⇨ In this case $r = 3$.

Next we will see how to compute the probability of r successes out of n trials when we have a binomial experiment.

Computing Probabilities for a Binomial Experiment

The central problem of a binomial experiment is to find the probability of r successes out of n trials. Now we'll see how to find these probabilities.

A model with three trials

Suppose you are taking a timed final exam. You have three multiple-choice questions left to do. Each question has four suggested answers, and only one of the answers is correct. You have only 5 seconds left to do these three questions, so you decide to mark answers on the answer sheet without even reading the questions. Assuming that your answers are randomly selected, what are the probabilities that you get zero, one, two, and all three questions correct?

This is a binomial experiment. Each question can be thought of as a trial, so there are $n = 3$ trials. The possible outcomes on each trial are success S, indicating a correct response, and failure F, meaning a wrong answer. The trials are independent—the outcome of any one trial does not affect the outcome of the others.

What is the probability of success on any question? Since you are guessing and there are four answers from which to select, the probability of a correct answer is 0.25. The probability q of a wrong answer is then 0.75. In short, we have a binomial experiment with $n = 3$, $p = 0.25$, and $q = 0.75$.

Now what are the possible outcomes in terms of success or failure for these three trials? Let's use the notation SSF to mean success on the first question, success on the second, and failure on the third. There are eight possible combinations of S's and F's. They are

$$SSS \quad SSF \quad SFS \quad FSS \quad SFF \quad FSF \quad FFS \quad FFF$$

To compute the probability of each outcome, we can use the multiplication rule for independent events because the trials are independent. For instance, the probability of success on the first two questions and failure on the last is

$$P(SSF) = P(S) \cdot P(S) \cdot P(F) = p \cdot p \cdot q = p^2q = (0.25)^2(0.75) \approx 0.047$$

In a similar fashion, we can compute the probability of each of the eight outcomes. These are shown in Table 5-8, along with the number of successes r associated with each trial.

Now we can compute the probability of r successes out of three trials for $r = 0, 1, 2,$ or 3. Let's compute $P(1)$. The notation $P(1)$ stands for the probability of one success. For three trials, there are three different outcomes that show exactly one success. They are the outcomes SFF, FSF, and FFS. Since the outcomes are mutually exclusive, we can add the probabilities. So

Table 5-8 Outcomes for a Binomial Experiment with $n = 3$ Trials

Outcome	Probability of Outcome	r (number of successes)
SSS	$P(SSS) = P(S)P(S)P(S) = p^3 = (0.25)^3 \approx 0.016$	3
SSF	$P(SSF) = P(S)P(S)P(F) = p^2q = (0.25)^2(0.75) \approx 0.047$	2
SFS	$P(SFS) = P(S)P(F)P(S) = p^2q = (0.25)^2(0.75) \approx 0.047$	2
FSS	$P(FSS) = P(F)P(S)P(S) = p^2q = (0.25)^2(0.75) \approx 0.047$	2
SFF	$P(SFF) = P(S)P(F)P(F) = pq^2 = (0.25)(0.75)^2 \approx 0.141$	1
FSF	$P(FSF) = P(F)P(S)P(F) = pq^2 = (0.25)(0.75)^2 \approx 0.141$	1
FFS	$P(FFS) = P(F)P(F)P(S) = pq^2 = (0.25)(0.75)^2 \approx 0.141$	1
FFF	$P(FFF) = P(F)P(F)P(F) = q^3 = (0.75)^3 \approx 0.422$	0

$$P(1) = P(SFF \text{ or } FSF \text{ or } FFS) = P(SFF) + P(FSF) + P(FFS)$$
$$= pq^2 + pq^2 + pq^2$$
$$= 3pq^2$$
$$= 3(0.25)(0.75)^2$$
$$\approx 0.422$$

In the same way, we can find $P(0)$, $P(2)$, and $P(3)$. These values are shown in Table 5-9.

Table 5-9 $P(r)$ for $n = 3$ Trials, $p = 0.25$

r (number of successes)	$P(r)$ (probability of r successes in 3 trials)		$P(r)$ for $p = 0.25$
0	$P(0) = P(FFF)$	$= q^3$	0.422
1	$P(1) = P(SFF) + P(FSF) + P(FFS)$	$= 3pq^2$	0.422
2	$P(2) = P(SSF) + P(SFS) + P(FSS)$	$= 3p^2q$	0.141
3	$P(3) = P(SSS)$	$= p^3$	0.016

We have done quite a bit of work to determine your chances of $r = 0, 1, 2$, or 3 successes on three multiple-choice questions if you are just guessing. Now we see that there is only a small chance (about 0.016) that you will get them all correct.

The model we constructed in Table 5-9 to compute the probability of r successes out of three trials can be used for any binomial experiment with $n = 3$ trials. Simply change the values of p and q to fit the experiment. In Guided Exercise 5, we use this model again.

GUIDED EXERCISE   5

Maria is doing a study on the issue of the quarter system versus the semester system. To obtain faculty input, she mails out questionnaires to the faculty. The probability that a faculty member returns the completed questionnaire is 0.65. Three faculty members chosen at random from the foreign language department are sent questionnaires. Compute the probability that *exactly two* completed questionnaires are returned and the probability that *all three* are returned. We'll do these computations in steps.

(a) In this problem, what are the values of n, p, and q?   ➡ $n = 3, p = 0.65, q = 1 - 0.65 = 0.35$

(b) The probability that exactly two questionnaires will   ➡   We want $P(2)$, so $r = 2$:
be returned is $P(\underline{\hspace{1cm}})$. In this case $r = \underline{\hspace{1cm}}$. By
Table 5-9,
$$P(2) = 3p^2q \text{ for } n = 3 \text{ trials}$$
$$P(2) = 3(0.65)^2(0.35) = 0.444$$
Use this formula to compute $P(2)$.

Exercise continues

(c) Use the appropriate formula from Table 5-9 to compute the probability that all three question-naires will be returned.

⇨ $P(3) = p^3 = (0.65)^3 = 0.275$

General formula for binomial probability distribution

Table 5-9 can only be used as a model for computing the probability of r successes out of *three* trials. How can we compute the probability of 7 successes out of 10 trials? We can develop a table for $n = 10$, but this would be a tremendous task because there are 1024 possible combinations of successes and failures on 10 trials. Fortunately, mathematicians have given us a direct formula for computing the probability of r successes for any number of trials.

> **Formula for the Binomial Probability Distribution:**
>
> $$P(r) = C_{n,r} p^r q^{n-r}$$
>
> where $C_{n,r} = \dfrac{n!}{r!(n-r)!}$ is the *binomial coefficient*. Values of $C_{n,r}$ for various values of n and $r = 0, 1, 2, \ldots, n$ can be found in Table 2 in Appendix I.

Table for $C_{n,r}$

Those of you who studied the section on counting techniques (Section 4.3) will recognize the symbol $C_{n,r}$ as the symbol used for the number of combinations of n objects taken r at a time.

Table 2 in Appendix I gives the values of the binomial coefficient $C_{n,r}$ for selected values of n and r. However, you can compute $C_{n,r}$ directly from a formula or by using your calculator. (See Section 4.3, page 201, for the formula for $C_{n,r}$ and examples showing how to use that formula.)

In the meantime, let's look more carefully at the formula itself. There are two main parts. The expression $p^r q^{n-r}$ is the probability of getting one outcome with r successes and $n - r$ failures. The binomial coefficient $C_{n,r}$ counts the number of outcomes that have r successes and $n - r$ failures. For instance, in the case of $n = 3$ trials, we saw in Table 5-8 that the probability of getting an outcome with one success and two failures was pq^2. This is the value of $p^r q^{n-r}$ when $r = 1$ and $n = 3$. We also observed that there were three outcomes with one success and two failures, so $C_{3,1}$ is 3.

Now let's take a look at an application of the binomial distribution formula in Example 4.

EXAMPLE 4 ▷

Privacy is a concern for many users of the Internet. One survey showed that 59% of Internet users are somewhat concerned about the confidentiality of their e-mail. Based on this information, what is the probability that for a random sample of 10 Internet users, 6 are concerned about the privacy of their e-mail?

SOLUTION:

(a) This is a binomial experiment with 10 trials. If we assign success to an Internet user being concerned about the privacy of e-mail, the probability of success is 59%. We are interested in the probability of 6 successes. We have

$$n = 10 \qquad p = 0.59 \qquad q = 0.41 \qquad r = 6$$

By the formula,

$$
\begin{aligned}
P(6) &= C_{10,6}(0.59)^6(0.41)^{10-6} \\
&= 210(0.59)^6(0.41)^4 \qquad \text{Use Table 2 in Appendix I or a calculator.} \\
&\approx 210(0.0422)(0.0283) \quad \text{Use a calculator.} \\
&\approx 0.25
\end{aligned}
$$

There is a 25% chance that *exactly* 6 of the 10 Internet users are concerned about the privacy of e-mail.

(b) Many calculators have a built-in combinations function. On the TI-83 you can find the combinations function under the math menu. It is designated nCr. Figure 5-2 displays the process for computing $P(6)$ directly on the TI-83.

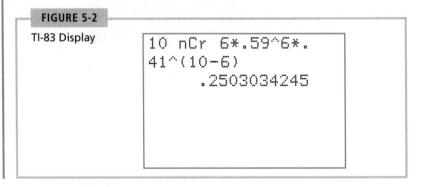

FIGURE 5-2

TI-83 Display

```
10 nCr 6*.59^6*.
41^(10-6)
        .2503034245
```

Table for *P(r)*

In many cases we will be interested in the probability of a range of successes rather than in the probability of an exact number of successes. For instance, we might wish to compute the probability that *at least* 6 of the 10 Internet users have concerns about the privacy of e-mail. In such a case, we need to use the addition rule for mutually exclusive events.

$$
\begin{aligned}
P(\textit{at least } 6 \text{ successes}) &= P(r \geq 6) \\
&= P(r = 6 \textit{ or } 7 \textit{ or } 8 \textit{ or } 9 \textit{ or } 10) \\
&= P(6) + P(7) + P(8) + P(9) + P(10)
\end{aligned}
$$

It would be quite a task to compute all the required probabilities by using the formula. Table 3 in Appendix I gives values of $P(r)$ for selected p and values of n through 20. To use the table, find the section labeled with your value of n. Then find the entry in the column headed by your value of p and the row labeled by the r value of interest.

Table 3 in Appendix I has only a limited selection of values for p. In fact, for Example 4 the probability of success, $p = 0.59$, is not available in the table. However, for other problems in this text you will find the specified values of p. Example 5 demonstrates the use of Table 3 (Appendix I) to find binomial probabilities.

EXAMPLE 5 ◆

Using the binomial distribution table to find $P(r)$

A biologist is studying a new hybrid tomato. It is known that the seeds of this hybrid tomato have probability 0.70 of germinating. The biologist plants 10 seeds.

(a) What is the probability that *exactly* 8 seeds will germinate?

SOLUTION: This is a binomial experiment with $n = 10$ trials. Each seed planted represents an independent trial. We'll say germination is success, so the probability of success on each trial is 0.70.

$$n = 10 \qquad p = 0.70 \qquad q = 0.30 \qquad r = 8$$

We wish to find $P(8)$, the probability of exactly eight successes.

In Table 3, Appendix I, find the section with $n = 10$. Then find the entry in the column headed by $p = 0.70$ and the row headed by the r value 8. This entry is 0.233.

$$P(8) = 0.233$$

(b) What is the probability that *at least* 8 seeds will germinate?

SOLUTION: In this case we are interested in the probability of 8 or more seeds germinating. This means we are to compute $P(r \geq 8)$. Since the events are mutually exclusive, we can use the addition rule

$$P(r \geq 8) = P(r = 8 \quad or \quad r = 9 \quad or \quad r = 10) = P(8) + P(9) + P(10)$$

We already know the value of $P(8)$. We need to find $P(9)$ and $P(10)$.

Use the same part of Table 3 in Appendix I, but find the entries in the rows headed by the r value 9 and the r value 10. Be sure to use the column headed by the p-value 0.70.

$$P(9) = 0.121 \quad and \quad P(10) = 0.028$$

Now we have all the parts necessary to compute $P(r \geq 8)$.

$$P(r \geq 8) = P(8) + P(9) + P(10)$$
$$= 0.233 + 0.121 + 0.028$$
$$= 0.382$$

In Guided Exercise 6 you'll practice using the formula for $P(r)$ in one part and then use Table 3 (Appendix I) for $P(r)$ values in the second part.

A rarely performed and somewhat risky eye operation is known to be successful in restoring the eyesight of 30% of the patients who undergo the operation. A team of surgeons has developed a new technique for this operation that has been successful for four of six operations. Does it seem likely that the new technique is much better than the old? We'll use the binomial probability distribution to answer this question. We'll compute the probability of at least four successes in six trials for the old technique.

(a) Each operation is a binomial trial. In this case, $n = $ _____, $p = $ _____, $q = $ _____, $r = $ _____.

➩ $n = 6, p = 0.30, q = 1 - 0.30 = 0.70, r = 4$

(b) Use your values of n, p, and q, as well as Table 2 in Appendix I (or your calculator), to compute $P(4)$ from the formula

$$P(r) = C_{n,r}p^r q^{n-r}$$

➩ $P(4) = C_{6,4}(0.30)^4(0.70)^2$
$$= 15(0.0081)(0.490)$$
$$\approx 0.060$$

(c) Compute the probability of *at least* four successes out of the six trials.

$$P(r \geq 4) = P(r = 4 \text{ or } r = 5 \text{ or } r = 6)$$
$$= P(4) + P(5) + P(6)$$

Use Table 3 in Appendix I to find values of $P(4)$, $P(5)$, and $P(6)$. Then use these values to compute $P(r \geq 4)$.

➩ To find $P(4)$, $P(5)$, and $P(6)$ in Table 3, we look in the section labeled $n = 6$. Then we find the column headed by $p = 0.30$. To find $P(4)$, we use the row labeled $r = 4$. For the values of $P(5)$ and $P(6)$, use the same column but change the row headers to $r = 5$ and $r = 6$, respectively.

$$P(r \geq 4) = P(4) + P(5) + P(6)$$
$$= 0.060 + 0.010 + 0.001$$
$$= 0.071$$

(d) Under the old operation technique, the probability that at least four patients out of six regain their eyesight is _____. Does it seem that the new technique is better than the old? Would you encourage the surgical team to do more work on the new technique?

➩ It seems the new technique is better than the old, since, by pure chance, the probability of four or more successes out of six trials is only 0.071 for the old technique. This means one of the following two things may be happening.

(i) The new method is no better than the old method, and our surgeons have encountered a relatively rare event (probability 0.071).

(ii) The new method is in fact better. We think it is worth encouraging the surgeons to do more work on the new technique.

Calculator Note The TI-83 calculator supports probability distributions. In the case of the binomial distribution, it has a **binompdf(n, p, r)** command that gives the binomial probability for a specified value of r and a cumulative distribution command **binomcdf(n, p, r)** that gives the cumulative total of the binomial probabilities for r or fewer successes. In both commands, you need to specify the number of trials n, the probability of success on a single trial p, and the number of successes r. To use the commands, press the **DISTR** key (that is, **2nd VARS**), scroll down to the binomial commands, and press **Enter**. For a binomial distribution with $n = 6$ trials, $p = 0.3$, and $r = 4$, the results are

```
binompdf(6,.3,4)

              .059535
binomcdf(6,.3,4)

              .989065
```

The probability that $r = 4$

The probability that $r \leq 4$

Common expressions and corresponding inequalities

Many times we are asked to compute the probability of a range of successes. For instance, in a binomial experiment with n trials, we may be asked to compute the probability of four or more successes. Table 5-10 shows how common English expressions such as "four or more successes" translate to inequalities involving r.

Table 5-10   Common English Expressions and Corresponding Inequalities (Consider a Binomial Experiment with n Trials and r Successes)

Expression	Inequality
Four or more successes	$r \geq 4$
At least four successes	That is, $r = 4, 5, 6, \ldots, n$
No fewer than four successes	
Not less than four successes	
Four or fewer successes	$r \leq 4$
At most four successes	That is, $r = 0, 1, 2, 3,$ or 4
No more than four successes	
The number of successes does not exceed four	
More than four successes	$r > 4$
The number of successes exceeds four	That is, $r = 5, 6, 7, \ldots, n$
Fewer than four successes	$r < 4$
The number of successes is not as large as four	That is, $r = 0, 1, 2, 3$

V I E W P O I N T

Lies! Lies!! Lies!!! The Psychology of Deceit

This is the title of an intriguing book by C. V. Ford, professor of psychiatry. The book recounts the true story of Floyd "Buzz" Fay, who was falsely convicted of murder on the basis of a failed polygraph examination. During his 2½ years of wrongful imprisonment, Buzz became a polygraph expert. He taught inmates, who freely confessed guilt, how to pass a polygraph examination. For more information on this topic, see Problem 7.

SECTION 5.2 PROBLEMS

In each of the following problems, the binomial distribution will be used. Please answer the following questions and then complete the problem.

(i) What makes up a trial? What is a success? What is a failure?

(ii) What are the values of *n, p,* and *q?*

General: Flip a Coin

1. A fair quarter is flipped three times. For each of the following probabilities, use the formula for the binomial distribution and a calculator to compute the requested probability. Next, look up the probability in Table 3 in Appendix I and compare the table result with the computed result.
   (a) Find the probability of getting exactly three heads.
   (b) Find the probability of getting exactly two heads.
   (c) Find the probability of getting two or more heads.
   (d) Find the probability of getting exactly three tails.

Academia: Multiple Choice Exam

2. Richard has just been given a 10-question multiple-choice quiz in his history class. Each question has 5 answers, of which only one is correct. Since Richard has not attended class recently, he doesn't know any of the answers. Assuming that Richard guesses on all 10 questions, find the indicated probabilities.
   (a) What is the probability that he will answer all questions correctly?
   (b) What is the probability that he will answer all questions incorrectly?
   (c) What is the probability that he will answer at least one of the questions correctly? Compute this probability in two ways. First, use the rule for mutually exclusive events and the probabilities shown in Table 3 in Appendix I. Then use the fact that $P(r \geq 1) = 1 - P(r = 0)$. Compare the two results. Should they be equal? Are they equal? If not, how do you account for the difference?
   (d) What is the probability that Richard will answer at least half the questions correctly?

Opinion Poll: Marriage

3. The percentage of American men who say they would marry the same woman if they had it to do all over again is 80%. The percentage of American women who say they would marry the same man again is 50% (*Harper's Index*).

(a) What is the probability that in a group of 10 married men, at least 7 will claim that they would marry the same woman again? What is the probability that less than half will say this?

(b) What is the probability that in a group of 10 married women, at least 7 will claim they would marry the same man again? What is the probability that less than half will say this?

Lifestyle: One-time Fling

4. The one-time fling. Have you ever purchased an article of clothing (dress, sports jacket, etc.), worn the item *once* to a party, and then returned the purchase? This is called a *one-time fling*. About 10% of all adults deliberately do a one-time fling and feel no guilt about it. (Source: *Are You Normal?* by Bernice Kanner, St. Martin's Press.) In a group of seven adult friends, what is the probability that

(a) no one has done a one-time fling?

(b) at least one person has done a one-time fling?

(c) no more than two people have done a one-time fling?

Sociology: Mother-in-Laws

5. The ★#@&#★ mother-in-law. Sociologists say that 90% of married women claim that their husbands' mothers are the biggest bones of contention in their marriages (sex and money are lower-rated areas of contention). (See the source in Problem 4.) Suppose that six married women are having coffee together one morning. What is the probability that

(a) all of them dislike their mothers-in-law?

(b) none of them dislikes her mother-in-law?

(c) at least four of them dislike their mothers-in-law?

(d) no more than three of them dislike their mothers-in-law?

Lifestyle: Neck Ties

6. A research team at Cornell University conducted a study showing that approximately 10% of all businessmen who wear ties wear them so tight that they actually reduce blood flow to the brain, diminishing cerebral functions (*Chances: Risk and Odds in Everyday Life,* by James Burke). At a board meeting of 20 businessmen, all of whom wear ties, what is the probability that

(a) at least one tie is too tight?

(b) more than two ties are too tight?

(c) no tie is too tight?

(d) at least 18 ties are *not* too tight?

Psychology:
Lie Detector Tests

7. Aldrich Ames is a convicted traitor who leaked American secrets to a foreign power. Yet Ames took routine lie detector tests and each time passed them. How can this be done? Recognizing control questions, employing unusual breathing patterns, biting one's tongue at the right time, pressing one's toes hard to the floor, and counting backward by sevens are countermeasures that are difficult to detect but can change the results of a polygraph examination (Source: *Lies! Lies!! Lies!!! The Psychology of Deceit,* by C. V. Ford, professor of psychiatry, University of Alabama.) In fact, it is reported in Professor Ford's book that after only 20 minutes of instruction by "Buzz" Fay (a prison inmate), 85% of those trained were able to pass the polygraph examination even when guilty of a crime. Suppose that a random sample of nine students (in a psychology laboratory) are told a "secret" and then given instructions on how to pass the polygraph examination without revealing their knowledge of the secret. What is the probability that

(a) all the students are able to pass the polygraph examination?
(b) more than half the students are able to pass the polygraph examination?
(c) no more than four of the students are able to pass the polygraph examination?
(d) all the students fail the polygraph examination?

Survey: Cellular Phones

8. According to an article appearing in *The Wall Street Journal*, about 35% of all U.S. households have cellular phones. Suppose that you are conducting a survey of customer satisfaction regarding cellular phones. If you called 11 households selected at random, what is the probability that
(a) every household has a cellular phone?
(b) more than 4 households have cellular phones?
(c) fewer than 5 do not have cellular phones?
(d) more than 7 do not have cellular phones?

Fishing: Northern Pike

9. Manitoba northern pike are hardy, tough fish. Using artificial lures with barbed treble hooks, it was found that the hooking mortality rate was only about 5%. This means that only 5% of pike that were caught and released died. (Source: *Proceedings of National Symposium on Catch and Release Fishing*, sponsored by Humboldt State University.) Suppose that a group of anglers caught and released 16 northern pike in Manitoba. What is the probability that
(a) none of the fish died?
(b) less than 3 of the fish died?
(c) all of the fish lived?
(d) more than 14 of the fish lived?

Business:
Accounting Records

10. Trevor is interested in purchasing the local hardware/sporting goods store in the small town of Dove Creek, Montana. After examining accounting records for the past several years, he found that the store has been grossing over $850 per day about 60% of the business days it is open. Estimate the probability the store will gross over $850
(a) at least 3 out of 5 business days.
(b) at least 6 out of 10 business days.
(c) less than 5 out of 10 business days.
(d) less than 6 out of the next 20 business days. If this actually happened, might it shake a person's confidence in the statement $p = 0.60$? Might it make a person suspect that p is less than 0.60? Explain.
(e) more than 17 out of the next 20 business days. If this actually happened, might a person suspect that p is greater than 0.60? Explain.

Advertising: Coffee

11. The Tasty Bean Coffee Company claims that its coffee is so good that you can distinguish it from any other coffee. Five different brands of coffee (one of them Tasty Bean) are set before tasters who are to pick the one that tastes the best. Suppose that there is really no difference in the way any of these coffees taste; however, each of four tasters picks one coffee anyway (not knowing which is which, because the coffee is in identical cups). What is the probability that
(a) all four tasters choose Tasty Bean?
(b) none of them chooses Tasty Bean?
(c) at least three choose Tasty Bean?

Psychology:
Marketing Personnel

12. Approximately 75% of all marketing personnel are extroverts, whereas about 60% of all computer programmers are introverts (*A Guide to the Development and Use of the Meyers-Briggs Type Indicator,* by Meyers and McCaulley).
    (a) At a meeting of 15 marketing personnel, what is the probability that 10 or more are extroverts? What is the probability that 5 or more are extroverts? What is the probability that all are extroverts?
    (b) In a group of 5 computer programmers, what is the probability that none are introverts? What is the probability that 3 or more are introverts? What is the probability that all are introverts?

Medical: Diabetes

13. People with diabetes may develop other health complications associated with the disease. The following information is based on a feature in *USA Today* entitled "A look at statistics that shape our lives." About 40% of all people with diabetes will also develop hypertension (blood pressure problems), and about 30% of people with diabetes will develop eye diseases. Suppose that you are the director of a health care center that has 10 people with diabetes and no other related health problems. Part of your duties is to monitor these patients for symptoms of new illnesses related to diabetes so that corrective measures can be started. What is the probability that
    (a) none of the diabetes patients will ever develop related hypertension?
    (b) less than 5 of the diabetes patients will ever develop related hypertension?
    (c) no more than 2 of the diabetes patients will ever develop a related eye disease?
    (d) at least 6 of the diabetes patients will never develop a related eye disease?

Lifestyle: Privacy

14. Are your finances, buying habits, medical records, and phone calls really private? A real concern for many adults is that computers and the Internet are reducing privacy. A survey conducted by Peter D. Hart Research Associates for the Shell Poll was reported in *USA Today*. According to the survey, 37% of the adults are concerned that employers are monitoring phone calls. Use the binomial distribution formula to calculate the probability that
    (a) out of 5 adults, none is concerned that employers are monitoring phone calls.
    (b) out of 5 adults, all are concerned that employers are monitoring phone calls.
    (c) out of 5 adults, exactly 3 are concerned that employers are monitoring phone calls.

Lifestyle: Privacy

15. According to the same poll quoted in Problem 14, 53% of the adults are concerned that Social Security numbers are used for general identification. For a group of 8 adults selected at random, we used Minitab to generate the binomial probability distribution and the cumulative binomial probability distribution (menu selections ►**Calc** ►**Probability Distributions** ►**Binomial**).

Number r	P(r)	P(<=r)
0	0.002381	0.00238
1	0.021481	0.02386
2	0.084781	0.10864
3	0.191208	0.29985
4	0.269521	0.56937
5	0.243143	0.81251
6	0.137091	0.94960
7	0.044169	0.99377
8	0.006226	1.00000

Find the probability that out of 8 adults selected at random

(a) at most 5 are concerned about Social Security numbers being used for identification. Do the problem by adding the probabilities P(r = 0) through P(r = 5). Is this the same as the cumulative probability P(r <= 5)?

(b) more than 5 are concerned about Social Security numbers being used for identification. First do the problem by adding the probabilities P(r = 6) through P(r = 8). Then do the problem by subtracting the cumulative probability P(r <= 5) from 1. Do you get the same results?

General: Control Charts

16. This problem will be referred to in the study of control charts (Section 6.1). In the binomial probability distribution, let the number of trials be n = 3, and let the probability of success be p = 0.0228. Use a calculator to compute
(a) the probability of two successes.
(b) the probability of three successes.
(c) the probability of two or three successes.

General: Symmetry in the Binomial Distribution Table

17. Study the binomial distribution table (Table 3, Appendix I). Notice that the probability of success on a single trial p ranges from 0.01 to 0.95. Some binomial distribution tables stop at 0.50 because of the symmetry in these tables. Let's look for that symmetry. Consider the section of the table with n = 5. Look at the numbers in the columns headed by p = 0.30 and p = 0.70. Do you detect any similarities? Consider the following probabilities for a binomial experiment with five trials.
(a) Compare P(3 successes) where p = 0.30 with P(2 successes) where p = 0.70.
(b) Compare P(3 or more successes) where p = 0.30 with P(2 or fewer successes) with p = 0.70.
(c) Find the value of P(4 successes) with p = 0.30. For what value of r is P(r successes) the same using p = 0.70?
(d) What column is symmetrical with the one headed by p = 0.20?

Section 5.3

The Mean and Standard Deviation of the Binomial Distribution

Graph of the binomial distribution

Any probability distribution may be represented in graphic form. How should we graph the binomial distribution? Remember, the binomial distribution tells us the probability of r successes out of n trials. Therefore, we'll place values of r along the horizontal axis and values of $P(r)$ on the vertical axis. The binomial distribution is a *discrete* probability distribution because r can assume only whole-number values such as 0, 1, 2, 3, and so on. Therefore, a histogram is an appropriate graph of a binomial distribution. Let's look at an example to see exactly how we'll make these histograms.

EXAMPLE 6 ➤

A waiter at the Green Spot Restaurant has learned from long experience that the probability that a lone diner will leave a tip is only 0.7. During one lunch hour he serves six people who are dining by themselves. Make a graph of the binomial probability distribution that shows the probabilities that 0, 1, 2, 3, 4, 5, and all 6 lone diners leave tips.

SOLUTION: This is a binomial experiment with $n = 6$ trials. *Success* is achieved when the lone diner leaves a tip, so the probability of success is 0.7 and that of failure is 0.3:

$$n = 6 \qquad p = 0.7 \qquad q = 0.3$$

We want to make a histogram showing the probabilities of r successes when r = 0, 1, 2, 3, 4, 5, and 6. It is easier to make the histogram if we first make a table of r values and the corresponding $P(r)$ values (Table 5-11). We'll use Table 3 in Appendix I to find the $P(r)$ values for $n = 6$ and $p = 0.70$.

To construct the histogram, we'll put r values on the horizontal axis and $P(r)$ values on the vertical axis. Our bars will be one unit wide and will be centered over the appropriate r value. The height of the bar over a particular r value gives the probability of that r value (see Figure 5-3).

The probability of a particular value of r is given not only by the height of the bar over that r value but also by the *area* of the bar. Each bar is only one unit wide, so the area (area = height times width) equals its height. Since the area of each bar represents the probability of the r value under it, the sum of the areas of the bars must be 1. In this example, the sum turns out to be 1.001. It is not exactly equal to 1 because of round-off error.

Table 5-11   Binomial Distribution for $n = 6$ and $p = 0.70$

r	$P(r)$
0	0.001
1	0.010
2	0.060
3	0.185
4	0.324
5	0.303
6	0.118

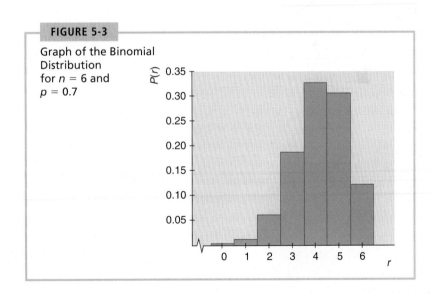

FIGURE 5-3

Graph of the Binomial Distribution for $n = 6$ and $p = 0.7$

Guided Exercise 7 illustrates another binomial distribution with $n = 6$ trials. The graph will be different from the one in Figure 5-3 because the probability of success p is different.

Jim enjoys playing basketball. He figures that he makes about 50% of the field goals he attempts during a game. Make a histogram showing the probabilities that Jim will make 0, 1, 2, 3, 4, 5, and 6 shots out of six attempted field goals.

(a) This is a binomial experiment with $n =$ _____ trials. In this situation we'll say success occurs when Jim makes an attempted field goal. What is the value of p?

⇨ In this example, $n = 6$ and $p = 0.5$.

(b) Use Table 3 in Appendix I to complete Table 5-12 of $P(r)$ values for $n = 6$ and $p = 0.5$.

Table 5-12

r	P(r)
0	0.016
1	0.094
2	0.234
3	_____
4	_____
5	_____
6	_____

⇨ **Table 5-13 Completion of Table 5-12**

r	P(r)
0	0.016
1	0.094
2	0.234
3	0.312
4	0.234
5	0.094
6	0.016

(c) Use the values of $P(r)$ given in Table 5-13 to complete the histogram in Figure 5-4.

FIGURE 5-4 Beginning of Graph of Binomial Distribution for $n = 6$ and $p = 0.5$

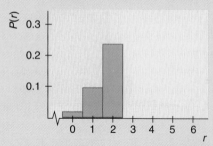

⇨ **FIGURE 5-5** Completion of Figure 5-4

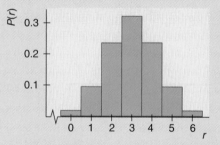

(d) The area of the bar over $r = 2$ is 0.234. What is the area of the bar over $r = 4$? How does the probability that Jim makes two field goals out of six compare with the probability that he makes four field goals out of six?

⇨ The area of the bar over $r = 4$ is also 0.234. Jim is as likely to make two out of six field goals attempted as he is to make four out of six.

In Example 6 and Guided Exercise 7 we see the graphs of two binomial distributions associated with $n = 6$ trials. The two graphs are different because the probabilities of success p are different in the two cases. In Example 6, $p = 0.7$ and the graph is skewed to the left—that is, the left tail is longer. In Guided Exercise 7, $p = 0.5$ and the graph is symmetrical—that is, if we fold it in half, the two halves coincide exactly. *Whenever p equals 0.5, the graph of the binomial distribution will be symmetrical no matter how many trials we have.* In Chapter 6 we will see that if the number of trials n is quite large, the binomial distribution is almost symmetrical even when p is not close to 0.5.

Mean and standard deviation of binomial probability distributions

Two other features that help describe the graph of any distribution are the balance point of the distribution and the spread of the distribution about that balance point. The *balance point* is the mean μ of the distribution, and the *measure of spread* that is most commonly used is the standard deviation σ. The mean μ is the *expected value* of the number of successes.

For the binomial distribution, there are two special formulas we can use to compute the mean μ and the standard deviation σ. These are easier to use than the general formulas in Section 5.1 for μ and σ of any discrete probability distribution.

For the Binomial Distribution:

$$\mu = np$$
$$\sigma = \sqrt{npq}$$

where n is the number of trials,
   p is the probability of success,
   q is the probability of failure ($q = 1 - p$).

EXAMPLE 7 ❯   Let's compute the mean and standard deviation of the distribution in Example 6, which describes the probabilities of lone diners leaving tips at the Green Spot Restaurant.

SOLUTION: In Example 6,
   $n = 6$ $p = 0.7$ $q = 0.3$
For the binomial distribution, the expected value is
   $\mu = np$
      $= 6(0.7) = 4.2$
The balance point of the distribution is at $\mu = 4.2$. The standard deviation is given by
   $\sigma = \sqrt{npq}$
      $= \sqrt{6(0.7)(0.3)}$
      $= \sqrt{1.26}$
      ≈ 1.12

The mean μ is not only the balance point of the distribution; it is also the *expected value* of r. Specifically, in Example 6, the waiter can expect 4.2 lone diners out of 6 to leave tips. (The waiter would probably round the expected value to 4 tippers out of 6.)

GUIDED EXERCISE   8

When Jim (from Guided Exercise 7) shoots field goals in basketball games, the probability that he makes a shot is only 0.5.

(a) The mean of the binomial distribution is the expected value of r successes out of n trials. Out of six throws, what is the expected number of goals Jim will make?

⟹ The expected value is the mean μ

$$\mu = np = 6(0.5) = 3$$

Jim can expect to make three goals out of six tries.

(b) For six trials, what is the standard deviation of the binomial distribution of the number of successful field goals Jim makes?

⟹ $\sigma = \sqrt{npq} = \sqrt{6(0.5)(0.5)} = \sqrt{1.5} \approx 1.22$

V I E W P O I N T

Kodiak Island, Alaska

Kodiak Island is famous for its giant brown bears. The sea surrounding the island is also famous for its king crab. The state of Alaska, Department of Fish and Game, has collected a huge amount of data regarding ocean latitude, longitude, and size of king crab. Of special interest to commercial fishing skippers is the size of crab. Those too small must be returned to the sea. Web site <http://lib.stat.cmu.edu/crab/> will give you locations and sizes of king crab catches near Kodiak Island. From this information it is possible to use the methods presented in this chapter and in Chapter 8 to estimate the proportion of legal crab in a sea skipper's catch.

SECTION 5.3 PROBLEMS

General: Histograms

1. Consider a binomial distribution with $n = 5$ trials. Use the probabilities given in Table 3 in Appendix I to make histograms showing the probabilities of $r = 0, 1, 2, 3, 4,$ and 5 successes for each of the p values given in parts a–c. Comment on the skewness of each distribution.

Binomial Probability
Distributions with $n = 6$
(Generated on the
TI-83 Calculator)

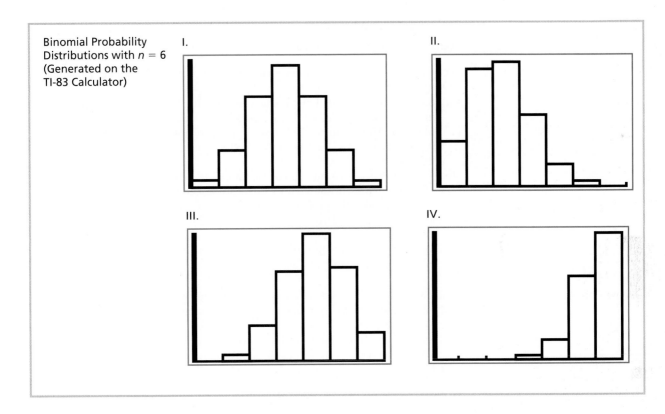

(a) The probability of success is $p = 0.50$.
(b) The probability of success is $p = 0.25$.
(c) The probability of success is $p = 0.75$.
(d) What is the relationship between the distributions in parts b and c?
(e) If the probability of success is $p = 0.73$, do you expect the distribution to be skewed to the right or to the left? Why?

General: Histograms

2. The figure above shows histograms for several binomial distributions with $n = 6$ trials. In parts a–d, match the given probability of success with the best graph.
   (a) $p = 0.30$ goes with graph _____.
   (b) $p = 0.50$ goes with graph _____.
   (c) $p = 0.65$ goes with graph _____.
   (d) $p = 0.90$ goes with graph _____.
   (e) In general, when the probability of success p is close to 0.5, would you say that the graph is more symmetrical or more skewed? In general, when the probability of success p is close to 1, would you say that the graph is skewed to the right or to the left? What about when p is close to 0?

Marketing: Photographs

3. Does the *kid factor* make a difference? If you are talking photography, the answer may be yes. The table on the top of page 244 is based on information from *American Demographics* (Vol. 19, No. 7).

Age of children in household, years	Under 2	None under 21
Percent of U.S. households that buy film	80%	50%

Let us say you are a market research person who interviews a random sample of 10 households.

(a) Suppose that the 10 households you interview are chosen to have children under the age of 2 years. Let r represent the number of such households that buy film. Make a histogram showing the probability distribution of r for $r = 0$ through $r = 10$. Find the mean and standard deviation of this probability distribution.

(b) Suppose that the 10 households are chosen to have no children under 21 years old. Let r represent the number of such households that buy film. Make a histogram showing the probability distribution of r for $r = 0$ through $r = 10$. Find the mean and standard deviation of this probability distribution.

(c) Compare the distributions shown in parts a and b. You are designing TV ads to sell film. Could you justify featuring ads that show parents taking pictures of toddlers? Explain your answer.

Quality Control: Inspection

4. The quality-control inspector of a production plant will reject a batch of syringes if two or more defectives are found in a random sample of eight syringes taken from the batch. Suppose the batch contains 1% defective syringes.

(a) Make a histogram showing the probabilities of $r = 0, 1, 2, 3, 4, 5, 6, 7$, and 8 defective syringes in a random sample of eight syringes.

(b) Find μ. What is the expected number of defective syringes the inspector will find?

(c) What is the probability that the batch will be accepted?

(d) Find σ.

Insurance: Liability Claims

5. The Mountain States Office of State Farm Insurance Company reports that approximately 85% of all automobile damage liability claims are made by people under 25 years of age. A random sample of five automobile insurance liability claims is under study.

(a) Make a histogram showing the probabilities that $r = 0$ to 5 claims are made by people under 25 years of age.

(b) Find the mean and standard deviation of this probability distribution. For samples of size 5, what is the expected number of claims made by people under 25 years of age?

Driving: Tailgating Cars

6. Do you tailgate the car in front of you? About 35% of all drivers will tailgate before passing, thinking they can make the car in front of them go faster (source: Bernice Kanner, *Are You Normal?* St. Martin's Press). Suppose that you are driving a considerable distance on a two-lane highway and are passed by 12 vehicles.

(a) Let r be the number of vehicles that tailgate before passing. Make a histogram showing the probability distribution of r for $r = 0$ through $r = 12$.

(b) Compute the expected number of vehicles out of 12 that will tailgate.

(c) Compute the standard deviation of this distribution.

7. Do you talk to machines? A recent survey for NCR Corporation showed that 47% of automated teller machine (ATM) users in the USA tell the ATM to hurry up.

(a) For a random sample of 11 people, the distribution of those telling the machine to hurry up is shown in the following Excel display (see the *Excel Technology Guide* that accompanies this text for details on how to create such a display). Is the probability of success (saying "hurry up" to the ATM) close to 50%? Is the histogram more skewed or more symmetrical? Compute the expected number out of 11 who tell the ATM to hurry up. Is the expected value μ close to the center of the graph?

r	P(r)
0	0.0009
1	0.0090
2	0.0401
3	0.1067
4	0.1892
5	0.2348
6	0.2083
7	0.1319
8	0.0585
9	0.0173
10	0.0031
11	0.0002

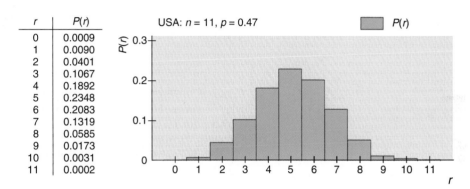

(b) In Great Britain, the ATM users are not so vocal or impatient. The probability is only 36% that a British ATM user tells the machine to hurry up. For a random sample of 11 British ATM users, the following display shows the probability distribution of telling the machine to hurry up. Is the probability of success less than 50%? Is the histogram skewed? In which direction? Compute the expected number out of 11 who tell the ATM to hurry up. Is the expected value on the left or right side of the graph?

r	P(r)
0	0.0074
1	0.0457
2	0.1284
3	0.2167
4	0.2438
5	0.1920
6	0.1080
7	0.0434
8	0.0122
9	0.0023
10	0.0003
11	0.0000

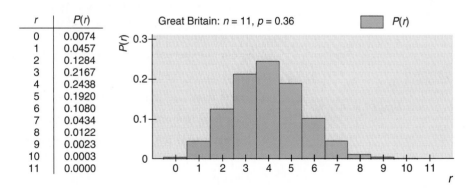

(c) What if we change success to "does not tell the ATM to hurry up." In Great Britain, what is the probability that an ATM user does not tell the machine to hurry up? The next display shows this distribution. Is this distribution skewed? In what direction? Compute the expected value. Does it lie on the left side or right side of the graph?

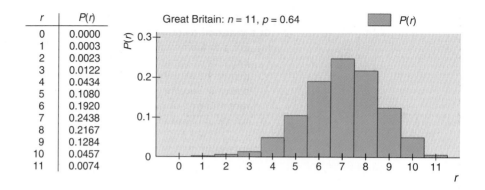

r	P(r)
0	0.0000
1	0.0003
2	0.0023
3	0.0122
4	0.0434
5	0.1080
6	0.1920
7	0.2438
8	0.2167
9	0.1284
10	0.0457
11	0.0074

(d) Compare the P(r) entries in the table in part b with those of the entries of the table of part c. How do the entries in part b for r = 0, 1, 2, 3 compare to those in part c for r = 11, 10, 9, and 8, respectively?

(e) Compute the standard deviations of each the distributions in parts (b) and (c). Are they the same? Why would you expect this?

SUMMARY

The concepts of discrete and continuous random variables were introduced in this chapter. We looked at the general probability distribution for a discrete random variable and saw how to compute the expected value μ and the standard deviation σ of a discrete probability distribution.

Then we turned our attention to a special discrete probability distribution, the binomial probability distribution. This distribution is used to determine the probability of outcomes from a binomial experiment.

A binomial experiment must meet the following criteria:

1. There are a fixed number of trials denoted by n.

2. The n trials are independent and repeated under identical conditions.

3. Each trial has only two outcomes: success S and failure F.

4. For each trial the probability p of success remains the same. The probability of failure is $q = 1 - p$.

5. The central problem is to find the probability of r successes out of n trials.

The formula for the binomial probability distribution is

$$P(r) = C_{n,r}p^r q^{n-r}$$

where $C_{n,r}$ is the binomial coefficient, as found in Table 2 in Appendix I. For certain values of p and n, $P(r)$ can be found directly in Table 3 in Appendix I.

The mean or expected value and standard deviation of the binomial distribution are given by the formulas

$$\mu = np$$
$$\sigma = \sqrt{npq} \qquad \text{where } q = 1 - p$$

V I E W P O I N T

What's Your Type?

Are students and professors *really* compatible? One way of answering this question is to look at Myers-Briggs Type Indicators for personality preferences. What is the probability that your professor is introverted and judgmental? What is the probability that you are extroverted and perceptive? Are most of the leaders in student government extroverted and judgmental? Is it true that members of Phi Beta Kappa have personality types more like the professors'? We will consider questions such as these in more detail in Chapter 8 (estimation) and Chapter 9 (hypothesis testing), where we will continue our work with binomial probabilities. In the meantime, you can find many answers regarding careers, probability, and personality types in *Applications of the Myers-Briggs Type Indicator in Higher Education*, edited by J. Provost and S. Anchors.

IMPORTANT WORDS & SYMBOLS

CHAPTER REVIEW PROBLEMS

Business: Automobile Leases

1. The Consumer Banker Association released a report showing the length of automobile leases for new automobiles. The results are

Lease Length in Months	Percent of Leases
13–24	12.7%
25–36	37.1%
37–48	28.5%
49–60	21.5%
More than 60	0.2%

(a) Using the midpoint of each class, and calling the midpoint of the last class 66.5 months, compute the expected lease term. Also find the standard deviation of the distribution.

(b) Sketch a graph of the probability for the duration of new auto leases.

Ecology: Wolves

2. Isle Royale, an island in Lake Superior, has provided an important study site for wolves and prey. In the National Park Service Scientific Monograph Series 11, *Wolf Ecology and Prey Relationships on Isle Royale,* Peterson gives results of many wolf–moose studies. Of special interest is the study of the number of moose killed by wolves. In the period from 1958 to 1974, there were 296 moose deaths identified as wolf kills. The age distribution of the kills is

Age of Moose in Years	Number Killed by Wolves
Calf (0.5 yr)	112
1–5	53
6–10	73
11–15	56
16–20	2

(a) For each age group, compute the probability that a moose in that age group is killed by a wolf.

(b) Consider all ages in a class equal to the class midpoint. Find the expected age of a moose killed by a wolf and the standard deviation of the ages.

Insurance: Property Damage

3. State Farm Insurance studies show that in Colorado, 55% of the auto insurance claims submitted for property damage are submitted by males under 25 years of age. Suppose 10 property damage claims involving automobiles are selected at random.

(a) Let r be the number of claims that are made by males under age 25. Make a histogram for the r-distribution probabilities.

(b) What is the probability that six or more claims are made by males under age 25?

(c) What is the expected number of claims made by males under age 25? What is the standard deviation of the r-probability distribution?

Driving: Speeding

4. Based on information from the *Denver Post*, it is estimated that 70% of cars on the Valley Highway (I-25) are going faster than the speed limit. A random sample of six cars is observed. What is the probability that
   (a) at least one is speeding?
   (b) none of the cars is speeding?

Prisons: Federal Inmates

5. According to *Harper's Index*, 50% of all federal inmates are serving time for drug dealing. A random sample of 16 federal inmates is selected.
   (a) What is the probability that 12 or more are serving time for drug dealing?
   (b) What is the probability that 7 or fewer are serving time for drug dealing?
   (c) What is the expected number of inmates serving time for drug dealing?

Airlines: On-time Flights

6. *Consumer Reports* rated airlines and found that 80% of the flights involved in the study arrived on time (that is, within 15 minutes of scheduled arrival time). Assuming that the on-time arrival rate is representative of the entire commercial airline industry, consider a random sample of 200 flights. What is the expected number that will arrive on time? What is the standard deviation of this distribution?

Agriculture: Grapefruit

7. It is estimated that 75% of a grapefruit crop is good; the other 25% have rotten centers that cannot be detected unless the grapefruit are cut open. The grapefruit are sold in sacks of 10. Let r be the number of good grapefruit in a sack.
   (a) Make a histogram of the probability distribution of r.
   (b) What is the probability of getting no more than one bad grapefruit in a sack? What is the probability of getting at least one good grapefruit in a sack?
   (c) What is the expected number of good grapefruit in a sack?
   (d) What is the standard deviation of the r probability distribution?

Health: Poison Ivy

8. Camp Wee-O-Wee has found that about 8% of young campers get poison ivy each season. If 273 children are registered for the summer season, about how many can be expected to get poison ivy?

Medical: MD Specialties

9. A survey has found that about 17% of all M.D.s in the United States have changed specialties at least once. In a city, with 600 M.D.s, what is the expected number who have changed specialties?

Restaurants: Reservations

10. The Orchard Café has found that about 5% of the parties who make reservations don't show up. If 82 party reservations have been made, how many can be expected to show up? Find the standard deviation of this distribution.

Marketing: Lawn Care

11. The We Care Lawn Service has found that about one out of five people will respond favorably to a certain telephone sales pitch. Suppose 15 people are called. Let r be the number who respond favorably.
    (a) Make a histogram for the r probability distribution.
    (b) What is the expected number of people who will respond favorably? Find the standard deviation of the r distribution.
    (c) What is the probability that at least three respond favorably? What is the probability that exactly three respond favorably?

Highway Department: Radar Traps

12. When David drives from Columbus to Cincinnati, the probability is 0.75 that at any given time he is going faster than the speed limit. If there are four radar traps (every car going over the speed limit is stopped and ticketed) between Columbus and Cincinnati, what is the probability David will be caught at least once (assume independence)?

Academic: True-False Tests

13. There are three true–false questions on a psychology test. Rita is out of time and just guesses, so the probability of a correct answer is 0.50.
    (a) Draw a histogram showing the probabilities of $r = 0, 1, 2$, and 3 correct answers for the three questions.
    (b) What is the expected number of correct answers?
    (c) Find the standard deviation of the r distribution.

DATA HIGHLIGHTS: GROUP PROJECTS

Break into small groups and discuss the following topics. Organize a brief outline in which you summarize the main points of your group discussion.

1. In the lottery game called Powerball, you could win a jackpot worth at least $10 million. Some jackpots have been more than $250 million! Powerball is a multistate lottery. To play Powerball, you purchase a $1 ticket. On the ticket you select five distinct numbers for white balls (numbered 1 through 49) and then one number for the red Powerball (numbered 1 through 42). The red Powerball number may be any of the numbers 1 through 42, including any of the numbers you selected for the white balls. Every Wednesday and Saturday there is a drawing. If your chosen numbers match those drawn, you win! The figure on page 251 shows all the prizes and the probability of winning each prize and specifies how many numbers on your ticket must match those drawn to win the prize. Many of the states that participate in Powerball maintain Web sites that display the results of each drawing, as well as a history of the results of previous drawings. One such site is the Connecticut Lottery site at <http://www.qgm.com/ctlott.html>.

    (a) Assume that the jackpot is $10 million and that there will be only one jackpot winner. The figure on page 251 lists the prizes and the probability of winning each prize. What is the probability of *not winning* any prize? Consider all the prizes and their respective probabilities and the prize of $0 (no win) and its probability. Use all these values to estimate your expected winnings μ if you buy one ticket. How much do you effectively contribute to the state where you purchased the ticket (ignoring the overhead cost of operating Powerball)?

    (b) Suppose that the jackpot rose to $25 million (and that there was to be only one winner). Compute your expected winnings if you buy one ticket. Does the probability of winning the jackpot change because the jackpot is higher?

    (c) Pretend that you are going to buy 10 Powerball tickets when the jackpot is $10 million. Use the random-number table (Table 1 of Appendix I) to select your numbers. Check the Connecticut Web site (or any other Powerball site) for the most recent drawing results to see if you would have won a prize.

Ways to Win Powerball

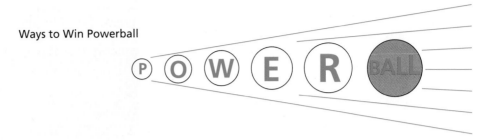

Match	Approximate Probability	Prize
5 white balls + **Powerball**	0.0000000125	Jackpot*
5 white balls	0.000000512	$100,000
4 white balls + **Powerball**	0.00000275	$5000
4 white balls	0.000113	$100
3 white balls + **Powerball**	0.000118	$100
3 white balls	0.0048	$7
2 white balls + **Powerball**	0.0017	$7
1 white ball + **Powerball**	0.0085	$4
0 white balls + **Powerball**	0.0135	$3
Overall chance of winning	0.0286 (one play)	

*The Jackpot will be divided equally (if necessary) among multiple winners
and is paid in 20 annual installments.

   (d) The probability of winning any prize is about 0.0286. Suppose that you decide
to buy five tickets. Use the binomial distribution to compute the probability of
winning (any prize) at least once. *Note:* You will need to use the binomial formula. Carry at least three digits after the decimal.

2. Would you like to travel in space, if given a chance? According to Opinion Research
for Space Day Partners, you are not alone. Forty-four percent of adults surveyed
agreed that they would travel in space if given a chance. Look at the figure on page
252, and use the information presented to answer the following questions.
   (a) According to this figure, the probability that an adult selected at random agrees
with the statement that humanity should explore planets is 64%. Round this
probability to 65%, and use this estimate with the binomial distribution table
(Table 3, Appendix I) to determine the probability that of 10 adults selected at
random, at least half agree that humanity should explore planets.
   (b) Does space exploration have an impact on daily life? Find the probability that
of 10 adults selected at random, at least 9 agree that space exploration does have
an impact on daily life. *Hint:* Use the formula for the binomial distribution.
   (c) In a room of 35 adults, what is the expected number who would travel in space
if given a chance? What is the standard deviation?

USA SNAPSHOTS®

A look at statistics that shape the nation

U.S. future on final frontier

Percentages of adults who agree with these statements about the USA's space program:

Youth should want to be astronauts	77%
Humanity should explore planets	64%
Space exploration impacts daily life	57%
Given a chance I'd travel in space	44%
Space will be colonized in my lifetime	18%

Source: Opinion Research for Space Day Partners

By Cindy Hall and Julie Stacey, USA TODAY

Source: Copyright 1997, USA TODAY. Reprinted by permission.

ⓛINKING CONCEPTS: WRITING PROJECTS

Discuss each of the following topics in class or review the topics on your own. Then write a brief but complete essay in which you summarize the main points. Please include formulas and graphs as appropriate.

1. Discuss what we mean by a binomial experiment. As you can see, a binomial process or binomial experiment involves a lot of assumptions. For example, all the trials are supposed to be independent and repeated under identical conditions. Is this always true? Can we always be completely certain the probability of success does not change from one trial to the next? In the real world there is almost nothing we can be absolutely sure about, so the *theoretical* assumptions of the binomial probability distribution often will not be completely satisfied. Does this mean we cannot use the binomial distribution to solve practical problems? Looking at this chapter, the answer that appears is that we can indeed use the binomial distribution even if all the assumptions are not *exactly* met. We find in practice that the conclusions are sufficiently accurate for our intended application. List three applications of the binomial distribution for which you think that, although some of the assumptions are not exactly met, there is still adequate reason to apply the binomial distribution.

2. Why do we need to learn the formula for the binomial probability distribution? Using the formula repeatedly can be very tedious. To cut down on tedious calculations, most people will use a binomial table such as the one in Appendix I.

   (a) However, there are many applications for which no table in *any* book is adequate. For instance, compute

   $$P(r = 3) \quad \text{where } n = 5 \text{ and } p = 0.735$$

Do you find the result in the table? Do the calculation by using the formula. List some other situations in which a table might not be adequate to solve a particular binomial distribution problem.

(b) The formula itself also has limitations. For instance, consider the difficulty of computing

$$P(r \geq 285) \quad \text{where } n = 500 \text{ and } p = 0.6$$

What are some of the difficulties you run into? Consider the calculation of $P(r = 285)$. You will be raising 0.6 and 0.4 to very high powers, which will give you very, very small numbers. Then you need to compute $C_{500,285}$, which is a very, very large number. When you combine extremely large and extremely small numbers in the same calculation, most accuracy is lost unless you carry a huge number of significant digits. If this isn't tedious enough, then consider the steps that you need in order to compute

$$P(r \geq 285) = P(r = 285) + P(r = 286) + \cdots + P(r = 500)$$

Does it seem clear that we need a better way to compute $P(r \geq 285)$? In Chapter 6 you will find a much better way to compute binomial probabilities when the number of trials is large.

3. In Chapter 2 we learned about means and standard deviations. In Section 5.1 we learned that a probability distribution also can have a mean and a standard deviation. Discuss what is meant by the expected value and standard deviation of a binomial distribution. How does this relate back to the material we learned in Chapter 2 and Section 5.1?

4. In Chapter 1 we looked at the shapes of distributions. Review the concepts of skewness and symmetry; then categorize the following distributions as to skewness or symmetry.
   (a) A binomial distribution with $n = 11$ trials and $p = 0.50$
   (b) A binomial distribution with $n = 11$ trials and $p = 0.10$
   (c) A binomial distribution with $n = 11$ trials and $p = 0.90$

In general, does it seem true that binomial probability distributions in which the probability of success is close to 0 are skewed right, whereas those with probability of success close to 1 are skewed left?

The problems in this section may be done using either statistical computer software or calculators that have statistical functions. Displays and suggestions are given for Minitab (Release 12), the TI-83 graphing calculator, ComputerStat, and Excel.

Binomial Distributions

Although tables of binomial probabilities can be found in most libraries, such tables are often inadequate. Either the value of p (the probability of success on a trial) you are looking for is not in the table, or the value of n (the number of trials) you are looking for is too large for the table. In Chapter 6 we will study the normal approximation to the binomial. This approximation is a great help in many practical applications. Even so, we sometimes use the formula for the binomial probability distribution on a computer or graphing calculator to compute the probability we want.

Minitab

For a user-specified probability of success, Minitab generates the binomial probability distribution of r successes out of n trials, as well as the cumulative binomial probability of at least r successes out of n trials. An example showing a Minitab display of binomial probabilities can be found in Problem 15 in Section 5.2.

To use the Minitab binomial distribution function, first enter the r values 0, 1, 2, . . . , n in a column. Then select ➤Calc ➤Probability Distribution ➤Binomial. In the dialogue box select Probability or Cumulative. Also enter the number of trials and the probability of success and the column containing the r values.

TI-83

The TI-83 calculator provides a number of probability distributions. Use the **DISTR** (**2nd VARS**) key to access the distributions. The **binompdf(n, p, r)** command gives the probability of r successes out of n trials when the probability of success on a single trial is p. The **binomcdf(n, p, r)** option gives the cumulative probability of *at least r* successes. See the screen print on page 233.

ComputerStat

Under the Probability Distributions menu, select Binomial Coefficients and Probability Distributions. This program prompts you to enter the number of trials and the probability of success on a single trial. Then it prints the probabilities and cumulative probabilities for $R = 0$ to N successes. A program option shows the graph of the distribution.

Excel

Excel also has a binomial distribution function that generates the binomial probability distribution of r successes out of n trials as well as the cumulative binomial probability of at least r successes out of n trials. Problem 7 in Section 5.3 shows an Excel display of binomial probabilities.

First enter the r values 0, 1, 2, . . . , n in a column on the worksheet. Menu choices: ➤paste function button f_x ➤Statistical ➤Binomdist. In the dialogue box, list the value of r or its cell address, the number of trials n, the probability of success p, and the logical operator false for $P(r)$ or true for $P(\text{at least } r \text{ successes})$.

APPLICATIONS

The following percentages were obtained over many years of observation by the U.S. Weather Bureau. All data listed are for the month of December.

Location	Long-Term Mean % of Clear Days in December
Juneau, Alaska	18%
Seattle, Washington	24%
Hilo, Hawaii	36%
Honolulu, Hawaii	60%
Las Vegas, Nevada	75%
Phoenix, Arizona	77%

Adapted from *Local Climatological Data*, U.S. Weather Bureau publication, "Normals, Means, and Extremes" Table.

In the locations listed, the month of December is a relatively stable month with respect to weather. Since weather patterns from one day to the next are more or less the same, it is reasonable to use a binomial probability model.

1. Let r be the number of clear days in December. Since December has 31 days, $0 \leq r \leq 31$. Using appropriate computer software or calculators available to you, find the probability $P(r)$ for each of the listed locations when $r = 0, 1, 2, \ldots, 31$.

2. For each location, what is the expected value of the probability distribution? What is the standard deviation?

You may find that using cumulative probabilities and appropriate subtraction of probabilities will make the solution of Applications 3 to 7 easier than adding probabilities.

3. Estimate the probability that Juneau will have at most 7 clear days in December.

4. Estimate the probability that Seattle will have from 5 to 10 (including 5 and 10) clear days in December.

5. Estimate the probability that Hilo will have at least 12 clear days in December.

6. Estimate the probability that Phoenix will have 20 or more clear days in December.

7. Estimate the probability that Las Vegas will have from 20 to 25 (including 20 and 25) clear days in December.

Computer Displays

Most statistical software packages will generate binomial probability distributions. The user may specify the number of successes and the probability of success on a single trial. In many packages there is an option for obtaining a cumulative distribution as well.

Minitab Display

See Section 5.2, Problem 15.

ComputerStat Display

The binomial distribution for number of trials = 6 and probability of success = 0.37 follows.

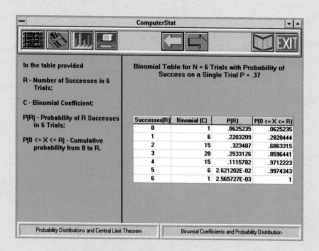

Successes(R)	Binomial (C)	P(R)	P(0 <= X <= R)
0	1	.0625235	.0625235
1	6	.2203209	.2828444
2	15	.323487	.6063315
3	20	.2533126	.8596441
4	15	.1115782	.9712223
5	6	2.621202E-02	.9974343
6	1	2.565727E-03	1

Binomial Table for N = 6 Trials with Probability of Success on a Single Trial P = .37

In the table provided

R - Number of Successes in 6 Trials;

C - Binomial Coefficient;

P(R) - Probability of R Successes in 6 Trials;

P(0 <= X <= R) - Cumulative probability from 0 to R.

Probability Distributions and Central Limit Theorem

Binomial Coefficients and Probability Distribution

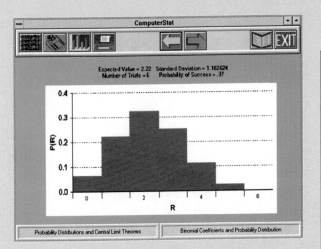

TI-83 Display

The TI-83 can be used to get the general shape of a binomial distribution. Enter the number of successes in list **L1** and the corresponding probabilities in list **L2**. Use the instruction **L2 = binompdf(n, p, L1)** to automatically enter the binomial probabilities in list **L2**. The following screens show the process used to obtain the histogram of a binomial distribution with $n = 6$ trials and probability of success $p = 0.37$.

L1 = number of successes, r
L2 = probability of r successes

L1	L2	L3	3
0	.06252	▬▬▬▬	
1	.22032		
2	.32349		
3	.25331		
4	.11158		
5	.02621		
6	.00257		
L3(1)=			

Select the histogram plot.

Set the window to include the highest frequency.

```
WINDOW
 Xmin=0
 Xmax=6
 Xscl=1
 Ymin=0
 Ymax=.4
 Yscl=1
 Xres=1
```

Plot the histogram.

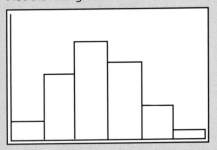

Excel Display

See Section 5.3, Problem 7.

6 Normal Distributions

One cannot escape the feeling that these mathematical formulas have an independent existence and an intelligence of their own, that they are wiser than we are, wiser even than their discoverers, that we get more out of them than was originally put into them.

—Heinrich Hertz

How can it be that mathematics, a product of human thought independent of experience, is so admirably adapted to the objects of reality?

—Albert Einstein

**Albert Einstein
(1879–1955)**

This brilliant German-born American physicist formulated the theory of relativity.

Heinrich Hertz was a pioneer in the study of radio waves. His work and the later work of Maxwell and Marconi led the way to modern radio, television, and radar. Albert Einstein is world renowned for his great discoveries in relativity and quantum mechanics. Everyone who has worked in both mathematics and real-world applications cannot help but marvel at how the "pure thought" of the mathematical sciences can predict and explain events in other realms. In this chapter we will study the single most important type of probability distribution in all of mathematical statistics: the normal distribution. Why is the normal distribution so important? Two of the reasons are that it applies to a wide variety of situations and other distributions tend to become normal under certain conditions.

Section 6.1 Graphs of Normal Probability Distributions

One of the most important examples of a continuous probability distribution is the *normal distribution*. This distribution was studied by the French mathematician Abraham de Moivre (1667–1754) and later by the German mathematician Carl Friedrich Gauss (1777–1855), whose work is so important that the normal distribution is sometimes called *Gaussian*. The work of these mathematicians provided the foundation on which much of the theory of statistical inference is based.

The applications of a normal probability distribution are so numerous that some mathematicians refer to it as "a veritable Boy Scout knife of statistics." However, before we can apply it, we must examine some of the properties of a normal distribution.

A rather complicated formula, presented in advanced statistics books, defines a normal distribution in terms of μ and σ, the mean and standard deviation of the population distribution. It is only through this formula that we can verify whether or not a distribution is normal. However, we can look at the graph of a normal distribution and get a good pictorial idea of some of the essential features of any normal distribution.

Normal curve

The graph of a normal distribution is called a *normal curve*. Its shape is very much like the cross section of a pile of dry sand. Because of its shape, blacksmiths would sometimes use piles of dry sand in the construction of molds for bells. Thus the normal curve is also called a *bell-shaped curve* (see Figure 6-1).

We see that a general normal curve is smooth and symmetrical about the vertical line over the mean μ. Notice that the highest point of the curve occurs over μ. If the distribution were graphed on a piece of sheet metal, cut out, and placed on a knife edge, the balance point would be at μ. We also see that the curve tends to level out and approach the horizontal axis (x axis) like a glider making a landing. However, in mathematical theory, such a glider would never quite finish its landing because a normal curve never touches the horizontal axis.

The parameter σ controls the spread of the curve. The curve is quite close to the horizontal axis at $\mu + 3\sigma$ and $\mu - 3\sigma$. Thus, if the standard deviation σ is large, the curve will be more spread out; if it is small, the curve will be more peaked. Figure 6-1 shows the normal curve cupped downward for an interval on either side of the mean μ. Then it begins to cup upward as we go to the lower part of the

FIGURE 6-1

A Normal Curve

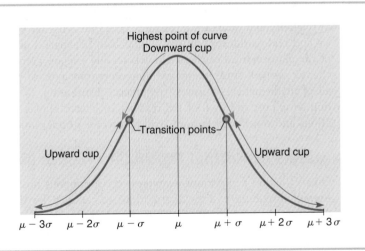

bell. The exact places where the transitions between the upward and downward cupping occur are above the points $\mu + \sigma$ and $\mu - \sigma$.

Let's summarize the important properties of a normal curve.

1. The curve is bell-shaped with the highest point over the mean μ.

2. It is symmetrical about a vertical line through μ.

3. The curve approaches the horizontal axis but never touches or crosses it.

4. The transition points between cupping upward and downward occur above $\mu + \sigma$ and $\mu - \sigma$.

GUIDED EXERCISE   1

Each of the curves in Figure 6-2 fails to be a normal curve. Give reasons why these curves are not normal curves.

FIGURE 6-2

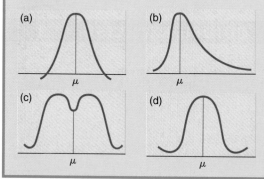

(a) A normal curve gets closer and closer to the horizontal axis, but it never touches it or crosses it.

(b) A normal curve must be symmetrical. This curve is not.

(c) A normal curve is bell-shaped with one peak. Because this curve has two peaks, it is not normal.

(d) The tails of a normal curve must get closer and closer to the x axis. In this curve the tails are going away from the x axis.

GUIDED EXERCISE   2

The points A, B, and C are indicated on the normal curve in Figure 6-3. One of these points is μ, one is $\mu + \sigma$, and one is $\mu - 2\sigma$.

FIGURE 6-3 A Normal Curve

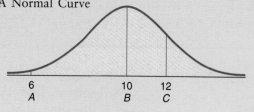

Exercise continues

Exercise continued

(a) Which point corresponds to the
mean? What is the value of μ?

⇨ The mean μ is under the peak of the normal curve.
The point B corresponds to the mean, so $\mu = 10$.

(b) Which point corresponds to $\mu + \sigma$?
Use the values of $\mu + \sigma$ and μ to
compute σ.

⇨ The point C where the curve changes from cupped
down to cupped up is one standard deviation σ
from the mean. The point C is $\mu + \sigma$. Since
$\mu + \sigma = 12$ and $\mu = 10$, $\sigma = 2$.

(c) Which point corresponds to
$\mu - 2\sigma$?

⇨ Since $\mu = 10$ and $\sigma = 2$, we see that

$$\mu - 2\sigma = 10 - 2(2) = 6$$

Point A corresponds to $\mu - 2\sigma$.

The parameters that control the shape of a normal curve are the mean μ and
the standard deviation σ. When both μ and σ are specified, a specific normal curve
is determined. Briefly, μ locates the balance point and σ determines the extent of
the spread.

GUIDED EXERCISE   3

Look at the normal curves
in Figure 6-4.

FIGURE 6-4

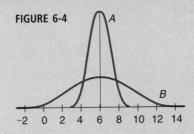

(a) Do these distributions have the same mean?
If so, what is it?

⇨ The means are the same, since both graphs
have the high point over 6. $\mu = 6$.

(b) One of the curves corresponds to a normal
distribution with $\sigma = 3$ and the other to
one with $\sigma = 1$. Which curve has which σ?

⇨ Curve A has $\sigma = 1$, and curve B has $\sigma = 3$.
(Since curve B is more spread out, it has the
larger σ value.)

Empirical rule

The total area under any normal curve studied in this book will *always* be 1.
The graph of the normal distribution is important because the portion of the *area*
under the curve above a given interval represents the *probability* that a measurement will lie in that interval.

In Section 2.2 we studied Chebyshev's theorem. This theorem gives us information about the *smallest* proportion of data that lie within 2, 3, or *k* standard deviations of the mean. This result applies to *any* distribution. However, for normal distributions, we can get a much more precise result, which is given by the *empirical rule.*

Empirical Rule

For a distribution that is symmetrical and bell-shaped (in particular, for a normal distribution):

Approximately 68.2% of the data values will lie within one standard deviation on each side of the mean.

Approximately 95.4% of the data values will lie within two standard deviations on each side of the mean.

Approximately 99.7% (or almost all) of the data values will be within three standard deviations on each side of the mean.

The preceding statement is called the *empirical rule* because, for symmetrical, bell-shaped distributions, the given percentages are observed in practice. Furthermore, for the normal distribution, the empirical rule is a direct consequence of the very nature of the distribution (see Figure 6-5). Notice that the empirical rule is a stronger statement than Chebyshev's theorem in that it gives *definite percentages,*

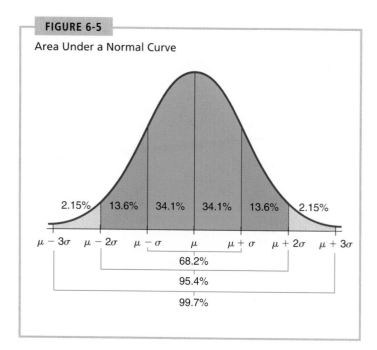

FIGURE 6-5

Area Under a Normal Curve

2.15%   13.6%   34.1%   34.1%   13.6%   2.15%

$\mu - 3\sigma$ $\mu - 2\sigma$ $\mu - \sigma$ μ $\mu + \sigma$ $\mu + 2\sigma$ $\mu + 3\sigma$

68.2%

95.4%

99.7%

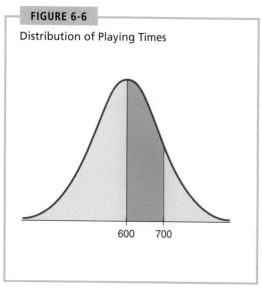

FIGURE 6-6

Distribution of Playing Times

600   700

not just lower limits. Of course, the empirical rule applies only to normal or symmetrical, bell-shaped distributions, whereas Chebyshev's theorem applies to all distributions.

EXAMPLE 1 ❯ The playing life of a Sunshine radio is normally distributed with mean $\mu = 600$ hours and standard deviation $\sigma = 100$ hours. What is the probability that a radio selected at random will last from 600 to 700 hours?

SOLUTION: The probability that the playing time will be between 600 and 700 hours is equal to the percentage of the total area under the curve that is shaded in Figure 6-6 on page 261. Since $\mu = 600$ and $\mu + \sigma = 600 + 100 = 700$, we see that the shaded area is simply the area between μ and $\mu + \sigma$. The area from μ to $\mu + \sigma$ is 34.1% of the total area. This tells us that the probability a Sunshine radio will last between 600 and 700 playing hours is 0.341.

GUIDED EXERCISE   4

The yearly wheat yield per acre on a particular farm is normally distributed with mean $\mu = 35$ bushels and standard deviation $\sigma = 8$ bushels.

(a) Shade the area under the curve in Figure 6-7 that represents the probability that an acre will yield between 19 and 35 bushels.

⇨ See Figure 6-8.

(b) Is the area the same as the area between $\mu - 2\sigma$ and μ?

⇨ Yes, since $\mu = 35$ and $\mu - 2\sigma = 35 - 2(8) = 19$.

FIGURE 6-8 Completion of Figure 6-7

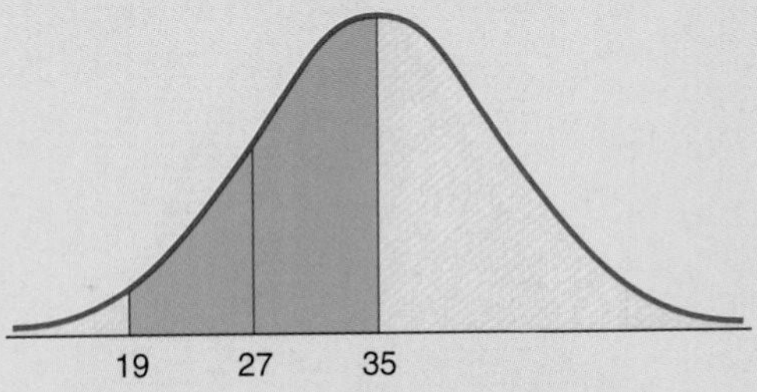

FIGURE 6-7

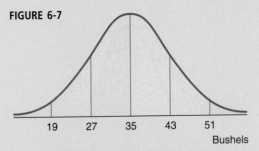

19   27   35   43   51
Bushels

(c) Use Figure 6-5 to find the percentage of area over the interval between 19 and 35.

⇨ The area between the values $\mu - 2\sigma$ and μ is 47.7% of the total area.

(d) What is the probability that the yield will be between 19 and 35 bushels per acre?

⇨ It is 47.7% of the total area, which is 1. Therefore, the probability is 0.477 that the yield will be between 19 and 35 bushels.

Control Charts

If we are examining data over a period of equally spaced time intervals or in some sequential order, then *control charts* are especially useful. Business managers and people in charge of production processes are aware that there exists an inherent amount of variability in any sequential set of data. For example, the sugar content of bottled drinks taken sequentially off a production line, the extent of clerical errors in a bank from day to day, advertising expenses from month to month, and even the number of new customers from year to year are examples of sequential data. There is a certain amount of variability in each.

A random variable x is said to be in *statistical control* if it can be described by the *same* probability distribution when it is observed at successive points in time. Control charts combine graphic and numerical descriptions of data with probability distributions.

Control charts were invented in the 1920s by Walter Shewhart at Bell Telephone Laboratories. Since a control chart is a *warning device,* it is not absolutely necessary that our assumptions and probability calculations be precisely correct. For example, the x distributions need not follow a normal distribution exactly. Any mound-shaped and more or less symmetrical distribution will be good enough.

How do we make a control chart? A control chart for a variable x is a plot of the observed x values (on the vertical scale) in time sequence order (the horizontal scale represents time). There is a center line at height μ and dashed control limits at $\mu \pm 2\sigma$ and at $\mu \pm 3\sigma$. How do we pick values for μ and σ? In most practical cases, values for μ (population mean) and σ (population standard deviation) are computed from past data for which the process we are studying was known to be *in control.* Methods for choosing sample sizes to fit given error tolerances can be found in Chapter 8.

Sometimes values for μ and σ are chosen as *target values.* That is, μ and σ values are chosen as set goals or targets that reflect the production level or service level at which a company hopes to perform. To be realistic, such target assignments for μ and σ should be reasonably close to actual data taken when the process was operating at a satisfactory production level. In Example 2 we will make a control chart; then we will discuss ways of analyzing it to see if a process or service is "in control."

EXAMPLE 2 ➤

Susan Tamara is director of personnel at the Antlers Lodge in Denali National Park, Alaska. Every summer Ms. Tamara hires many part-time employees from all over the United States. Most are college students seeking summer employment. Perhaps the biggest activity for Ms. Tamara's staff is that of "making up" the rooms each day. The lodge has 385 rooms, and from long experience Ms. Tamara has determined that by 3:30 P.M. each day the average number of rooms not made up is $\mu = 19.3$ with standard deviation $\sigma = 4.7$. Although the lodge has a policy that guest rooms be made up by 3:30 P.M., Ms. Tamara knows that there will always be a few rooms not made up by this time because there is a high personnel turnover and corresponding reassignment of staff to new jobs for which they are in a training period. Every 15 days Ms. Tamara has a general staff meeting. Before the meeting, she examines several control charts for the restaurant, the gift shop, and, of course, housekeeping. Table 6-1 shows the numbers of rooms that have not been ready by 3:30 P.M. over the past 15 days. Make a control chart for these data.

Before we make a control chart, we need to know a few things about the distribution of rooms not made up by 3:30 P.M. Ms. Tamara is aware from long experience that the distribution is symmetrical and bell-shaped. It is approximately normal, with mean $\mu = 19.3$ and standard deviation $\sigma = 4.7$. In addition, this distribution of x values is acceptable to the top administration of the Antlers Lodge.

Table 6-1 Numbers of Rooms Not Made Up by 3:30 P.M.

Day	1	2	3	4	5	6	7	8
x = number of rooms	11	20	25	23	16	19	8	25
Day	9	10	11	12	13	14	15	
x = number of rooms	17	20	23	29	18	14	10	

SOLUTION: A control chart for a variable x is a plot of the observed x values (vertical scale) in time sequence order (the horizontal scale represents time). Place horizontal lines at

The mean $\mu = 19.3$

The control limits $\mu \pm 2\sigma = 19.3 \pm 2(4.7)$ or 9.90 and 28.70

The control limits $\mu \pm 3\sigma = 19.3 \pm 3(4.7)$ or 5.20 and 33.40

Then plot the data from Table 6-1. (See Figure 6-9.)

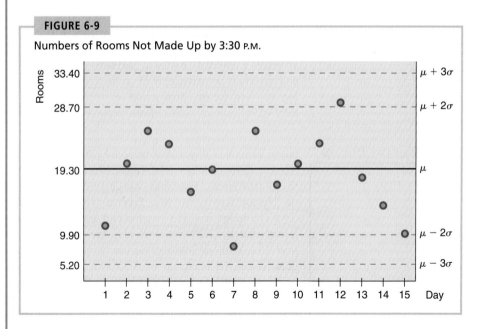

FIGURE 6-9

Numbers of Rooms Not Made Up by 3:30 P.M.

Once we have made a control chart, the main question is whether, as time goes on, the distribution of x values is staying the same or is changing. If the distribution is remaining more or less the same, we say it is in *statistical control*. If it is not, we say it is *out of control*.

Out-of-control warning signals

There are many popular methods used to set off a warning signal that a process is out of control. Remember, a random variable x is said to be *out of control* if successive time measurements of x indicate that it is no longer following the target probability distribution. We will assume that the target distribution is (approximately) normal and has (user-set) target values for μ and σ.

Three of the most popular warning signals are as follows.

1. **Out-of-Control Signal I:** *One point beyond the 3σ level.* What is the probability that signal I will be a false alarm? By the empirical rule, the probability that a point lies within 3σ of the mean is 0.997. The probability that signal I will give a false alarm is $1 - 0.997 = 0.003$. Remember, a false alarm means that the x distribution is really on the target distribution, and we simply have a very rare event (in this case, an event with a probability of 0.003). (See Figure 6-10a.)

2. **Out-of-Control Signal II:** *A run of nine consecutive points on one side of the center line (the line at target value μ).* What is the probability that signal II is a false alarm? If the x distribution and the target distribution are the same, then there is a 50% chance that an x value will lie above or below the center line at μ. Because the samples are (time) independent, the probability of a run of nine points on one side of the center line is $(0.5)^9 \approx 0.002$. If we consider both sides, this probability becomes 0.004. Therefore, the probability that signal II is a false alarm is approximately 0.004. (See Figure 6-10b.)

3. **Out-of-Control Signal III:** *At least two of three consecutive points beyond the 2σ level on the same side of the center line.* What is the probability that signal III will produce a false alarm? By the empirical rule, the probability that an x value will lie above the 2σ level is about 0.023. If we use the binomial probability distribution (with success being the point is above 2σ), then the probability of two or more successes out of three trials is

$$\frac{3!}{2!1!}(0.023)^2(0.977) + \frac{3!}{3!0!}(0.025)^3 \approx 0.002$$

If we take into account *both* above or below the center line, it follows that the probability that signal III is a false alarm is about 0.004. (See Figure 6-10c.)

SUMMARY

Type of Warning Signal	Probability of a False Alarm
Type I: Point beyond 3σ	0.003
Type II: Run of nine consecutive points all below center line μ or all above center line μ	0.004
Type III: At least two out of three consecutive points beyond 2σ	0.004

FIGURE 6-10

Out-of-Control Signals

(a) Out-of-Control Signal I:
One Point Beyond the
3σ Level

(b) Out-of-Control Signal II:
Run of Nine Consecutive
Points on One Side of the
Center Line

(c) Out-of-Control Signal III:
At Least Two of Three Consecutive
Points Beyond the 2σ Level on the
Same Side of the Center Line

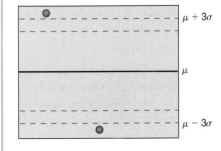
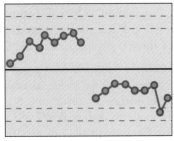
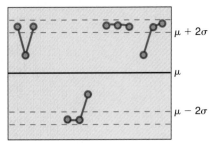

Remember, a control chart is only a warning device, and it is possible to get a false alarm. A false alarm happens when one (or more) of the out-of-control signals occurs, but the x distribution is really on the target or assigned distribution. In this case, we simply have a rare event (probability of 0.003 or 0.004). In practice, whenever a control chart indicates that a process is out of control, it is usually a good precaution to examine what is going on. If the process is out of control, corrective steps can be taken before things get a lot worse. The rare false alarm is a small price to pay if we can avert what might become real trouble.

From an intuitive point of view, signal I could be thought of as a blowup, something dramatically out of control. Signal II could be thought of as a slow drift out of control. Signal III is between a blowup and a slow drift.

EXAMPLE 3 ❯   Ms. Tamara of the Antlers Lodge examines the control chart for housekeeping. During the staff meeting, she makes recommendations about improving service or, if all is going well, gives her staff a well deserved "pat on the back." The most recent control chart for housekeeping is the one shown in Figure 6-11. Look at this control chart to determine if the housekeeping process is out of control or not.

SOLUTION: The x values are more or less evenly distributed about the mean $\mu = 19.3$. None of the points lies outside the $\mu \pm 3\sigma$ limit (i.e., above 33.40 or below 5.20 rooms). There is no run of nine consecutive points above or below μ. No two of three consecutive points are beyond the $\mu \pm 2\sigma$ limit (i.e., above 28.7 or below 9.90 rooms).

It appears that the x distribution is "in control." At the staff meeting, Ms. Tamara should tell her employees they are doing a reasonably good job and that they should keep up the fine work!

FIGURE 6-11

Numbers of Rooms Not
Made Up by 3:30 P.M.

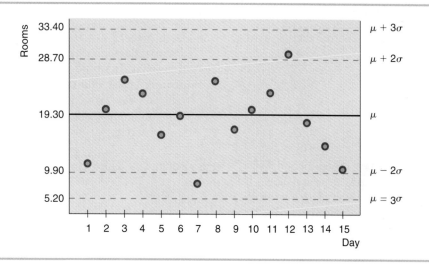

GUIDED EXERCISE   5

Over the next 15-day period, let's suppose that housekeeping again reports the numbers of rooms not made up by 3:30 P.M. to Ms. Tamara of the Antlers Lodge. The data in Table 6-2 show the results.

Table 6-2   Next 15-Day Report of Rooms Not Made Up by 3:30 P.M.

Day	1	2	3	4	5	6	7	8
x = number of rooms	25	8	23	15	26	24	31	21
Day	9	10	11	12	13	14	15	
x = number of rooms	27	20	25	21	27	11	16	

(a) We assume that we are still working with the symmetrical, bell-shaped distribution of x values, with mean $\mu = 19.3$ and $\sigma = 4.7$. Compute the "control limits" of $\mu \pm 2\sigma$ and $\mu \pm 3\sigma$. Draw a control chart showing the solid line at the mean and the control limits and plot the data for the 15-day period.

▷ **FIGURE 6-12** Next 15-Day Report of Rooms Not Made Up by 3:30 P.M.

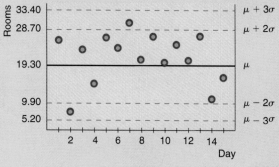

Exercise continues

(b) Interpret the control chart in part a.   ⟹ Days 5 to 13 are above $\mu = 19.3$. We have nine consecutive days on one side of the mean. This is a warning signal! It would appear that the mean μ is slowly drifting up beyond the target value of 19.3. The chart indicates that housekeeping is "out of control." Ms. Tamara should take corrective measures at her next staff meeting.

(c) Over another 15-day period Ms. Tamara obtains the data shown in Table 6-3 for housekeeping. Make a control chart using target values $\mu = 19.3$ and $\sigma = 4.7$.

⟹ **FIGURE 6-13** Third Housekeeping Data Report

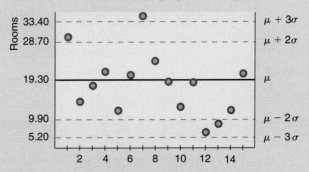

Table 6-3 Third Housekeeping Data Report

Day	1	2	3	4	5	6	7	8
Number of rooms	29	14	18	21	11	20	35	24

Day	9	10	11	12	13	14	15
Number of rooms	19	12	19	6	8	11	20

(d) Interpret the control chart in part c.   ⟹ On day 7 we have a data value beyond $\mu + 3\sigma$ (i.e., above 33.40). On days 11, 12, and 13 we have two of three data values beyond $\mu - 2\sigma$ (i.e., below 9.90). The occurrences on both these periods are out-of-control warning signals. Ms. Tamara might ask her staff about both these periods. There may be a lesson to be learned about day 7, when housekeeping apparently had a lot of trouble. Also, days 11, 12, and 13 were very good days. Perhaps a lesson could be learned about why things went so well.

VIEWPOINT

In Control? Out of Control?

If you care about quality, you also must care about control. Dr. Walter Shewhart invented control charts when he was working for Bell Laboratories. The great contribution of control charts is that they separate variation into two sources: (1) random or chance causes (in control) and (2) special or assignable causes (out of control). A process is said to be in *statistical control* when it is no longer afflicted with special or assignable causes. The performance of a process that is in statistical control is predictable. Predictability and quality control tend to be closely associated.

(Source: Adapted from the classic text *Statistical Methods from the Viewpoint of Quality Control*, by W. A. Shewhart, with Foreword by W. E. Deming, Dover Publications.)

SECTION 6.1 PROBLEMS

Normal Curves

1. Which, if any, of the curves in the figure below look(s) like a normal curve? If a curve is not a normal curve, tell why.

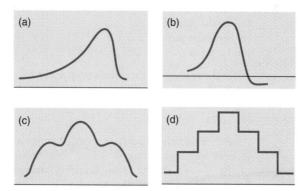

(a) (b) (c) (d)

Normal Curves

2. Look at the normal curve in the following figure, and find μ, $\mu + \sigma$, and σ.

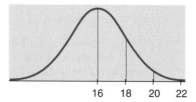

16 18 20 22

Normal Curves

3. Look at the two normal curves in the figures at the top of page 270. Which has the larger standard deviation? What is the mean of the curve in the figure on the left? What is the mean of the curve in the figure on the right?

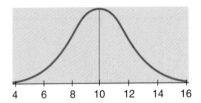

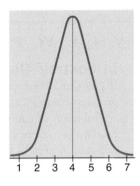

Normal Curves

4. Sketch a normal curve
   (a) with mean 15 and standard deviation 2.
   (b) with mean 15 and standard deviation 3.
   (c) with mean 12 and standard deviation 2.
   (d) with mean 12 and standard deviation 3.
   (e) Consider two normal curves. If the first one has a larger mean than the second one, must it have a larger standard deviation than the second one as well? Explain your answer.

Empirical Rule

5. What percentage of the area under the normal curve lies
   (a) to the left of μ?
   (b) between $\mu - \sigma$ and $\mu + \sigma$?
   (c) between $\mu - 3\sigma$ and $\mu + 3\sigma$?

Empirical Rule

6. What percentage of the area under a normal curve lies
   (a) to the right of μ?
   (b) between $\mu - 2\sigma$ and $\mu + 2\sigma$?
   (c) to the right of $\mu + 3\sigma$?

Vital Statistics: Heights of Coeds

7. Assuming that the heights of college women are normally distributed, with mean 65 in. and standard deviation 2.5 in. (based on information from *Statistical Abstract of the United States*, 118th Edition), answer the following questions. (*Hint:* Use Problems 5 and 6 and Figure 6-5.)
   (a) What percentage of women are taller than 65 in.?
   (b) What percentage of women are shorter than 65 in.?
   (c) What percentage of women are between 62.5 in. and 67.5 in.?
   (d) What percentage of women are between 60 in. and 70 in.?

Agriculture: Rhode Island Red Chicks

8. The incubation time for Rhode Island Red chicks is normally distributed with a mean of 21 days and a standard deviation of approximately 1 day (based on information from *World Book Encyclopedia*). Look at Figure 6-5 and answer the following questions. If 1000 eggs are being incubated, how many chicks do we expect will hatch
   (a) in 19 to 23 days?
   (b) in 20 to 22 days?
   (c) in 21 days or fewer?
   (d) in 18 to 24 days? (Assume all eggs eventually hatch.)

(*Note:* In this problem let us agree to think of a single day or a succession of days as a continuous interval of time.)

Archaeology:
Tree Ring Data

9. At Burnt Mesa Pueblo, archaeological studies have used the method of tree ring dating in an effort to determine when prehistoric people lived in the pueblo. Wood from several excavations gave a mean of (year) 1243 with a standard deviation of 36 years (*Bandelier Archaeological Excavation Project: Summer 1989 Excavations at Burnt Mesa Pueblo,* edited by Kohler, Washington State University Department of Anthropology). The distribution of dates was more or less mound-shaped and symmetrical about the mean. Use the empirical rule to
   (a) estimate a range of years centered about the mean in which about 68% of the data (tree ring dates) will be found.
   (b) estimate a range of years centered about the mean in which about 95% of the data (tree ring dates) will be found.
   (c) estimate a range of years centered about the mean in which almost all the data (tree ring dates) will be found.

Product Reliability:
Vending Machine

10. A vending machine automatically pours soft drinks into cups. The amount of soft drink dispensed into a cup is normally distributed with a mean of 7.6 oz and a standard deviation of 0.4 oz. Examine Figure 6-5 and answer the following questions.
   (a) Estimate the probability that the machine will overflow an 8-oz cup.
   (b) Estimate the probability that the machine will not overflow an 8-oz cup.
   (c) The machine has just been loaded with 850 cups. How many of these do you expect will overflow when served?

Physical Therapy:
Pain Threshold

11. "Effect of Helium-Neon Laser Auriculotherapy on Experimental Pain Threshold" is the title of an article in the journal *Physical Therapy* (Vol. 70, No. 1, pp. 24–30). In this article, laser therapy was discussed as a useful alternative to drugs in pain management of chronically ill patients. To measure pain threshold, a machine was used that delivered low-voltage direct current to different parts of the body (wrist, neck, and back). The machine measured current in milliamperes (mA). The pretreatment experimental group in the study had an average threshold of pain (pain was first detectable) at $\mu = 3.15$ mA with standard deviation $\sigma = 1.45$ mA. Assume that the distribution of threshold pain so measured in milliamperes is symmetrical and more or less mound-shaped. Use the empirical rule to
   (a) estimate a range of milliamperes centered about the mean in which about 68% of the experimental group will have a threshold of pain.
   (b) estimate a range of milliamperes centered about the mean in which about 95% of the experimental group will have a threshold of pain.

National Parks: Yellowstone

12. Yellowstone Park Medical Services (YPMS) provides emergency health care for park visitors. Such health care includes treatment for everything from indigestion and sunburn to more serious injuries. A recent issue of *Yellowstone Today* (National Park Service Publication) indicated that the average number of visitors treated each day by YPMS was 21.7. The estimated standard deviation was 4.2 (summer data). The distribution of numbers treated is approximately mound-shaped and symmetrical.

   (a) For a 10-day summer period, the following data show the number of visitors treated each day by YPMS.

Day	1	2	3	4	5	6	7	8	9	10
Number treated	25	19	17	15	20	24	30	19	16	23

Make a control chart for the daily number of visitors treated by YPMS, and plot the data on the control chart. Do the data indicate that the number of visitors treated by YPMS is "in control"? Explain your answer.

(b) For another 10-day summer period, the following data were obtained.

Day	1	2	3	4	5	6	7	8	9	10
Number treated	20	15	12	21	24	28	32	36	35	37

Make a control chart, and plot the data on the chart. Do the data indicate that the number of visitors treated by YPMS is "in control" or "out of control"? Explain your answer. Identify all out-of-control signals by type (I, II, or III). If you were the park superintendent, do you think YPMS might need some (temporary) extra help? Explain.

Travel Industry: Motels

13. The manager of Sun Motel has 316 rooms in Palo Alto, California. For observations over a period of time, she knows that on an average night 268 rooms will be rented. The long-term standard deviation is 12 rooms. This distribution is approximately mound-shaped and symmetrical. Minitab has a number of control chart options. Among them is the one called I-chart, which gives a control chart for individual measurements. The figures below show such control charts for two different 14-day periods (➤**Stat** ➤**Control Charts** ➤**I-MR**; in the options box, designate sigma positions at 2 and 3).

(a) Looking at the chart for the first two-week period, would you say the number of rooms rented during this period has been unusually low? Unusually high? About what was expected? Identify all the out-of-control signals by type (I, II, III). What might be some reasons for the "out-of-control" periods?

(b) Repeat part a for the second two-week period.

(a) First Two-Week Period

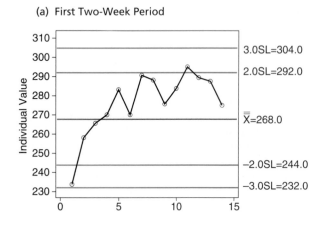

(b) Second Two-Week Period

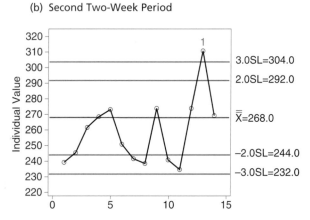

Environment: Air Pollution

14. The visibility standard index (VSI) is a measure of Denver air pollution that is reported each day in the *Rocky Mountain News*. The index ranges from 0 (excellent air) to 200 (very bad air). During winter months, when air pollution is higher, the index has a mean of about 90 (rated as fair) with a standard deviation of approximately 30. Suppose that for 15 days the following VSI was reported each day:

Day	1	2	3	4	5	6	7	8	9
VSI	80	115	100	90	15	10	53	75	80

Day	10	11	12	13	14	15
VSI	110	165	160	120	140	195

Make a control chart for the VSI, and plot the preceding data on the control chart. Identify all out-of-control signals (high or low) that you find in the control chart by type (I, II, or III).

Section **Standard Units and Areas Under the Standard Normal Distribution**

z Scores and Raw Scores

Normal distributions differ from one another in two ways: the mean μ may be located anywhere on the x axis, and the bell shape may be more or less spread according to the size of the standard deviation σ. The differences among the normal distributions cause difficulties when we try to compute the area under the curve in a specified interval and, hence, the probability that a measurement will fall in that interval.

It would be futile to try to set up a table of areas under the normal curve for each different μ and σ combination. We need a way to standardize the distributions so that we can use *one* table of areas for *all* normal distributions. We achieve this standardization by considering how many standard deviations a measurement lies from the mean. In this way we can compare a value in one normal distribution with a value in a different normal distribution. The next situation shows how this is done.

Suppose that Tina and Jack are in two different sections of the same course. Each section is quite large, and the scores on the midterm exams of each section follow a normal distribution. In Tina's section, the average (mean) was 64 and her score was 74. In Jack's section, the mean was 72 and his score was 82. Both Tina and Jack were pleased that their scores were each 10 points above the average of each respective section. However, the fact that each was 10 points above average does not really tell us how each did *with respect to the other students in the section*. In Figure 6-14 we see the normal distribution of grades for each section.

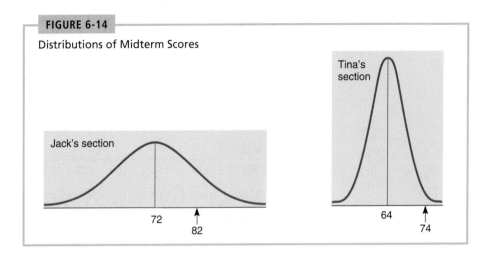

FIGURE 6-14

Distributions of Midterm Scores

Jack's section

Tina's section

72   82

64   74

Tina's 74 was higher than most of the other scores in her section, while Jack's 82 was only an upper-middle score in his section. Tina's score was far better with respect to her class than Jack's score with respect to his class.

Standard score

The preceding situation demonstrates that it is not sufficient to know the difference between a measurement (*x* value) and the mean of a distribution. We need also to consider the spread of the curve, or the standard deviation. What we really want to know is the number of standard deviations between a measurement and the mean. This "distance" takes both μ and σ into account.

There is a simple formula that we can use to compute the number *z* of standard deviations between a measurement *x* and the mean μ of a normal distribution with standard deviation σ:

$$\begin{pmatrix} \text{Number of standard deviations} \\ \text{between the measurement and} \\ \text{the mean} \end{pmatrix} = \begin{pmatrix} \dfrac{\text{difference between the}}{\text{measurement and the mean}} \\ \dfrac{}{\text{standard deviation}} \end{pmatrix}$$

Written in symbols, this formula is

$$z = \frac{x - \mu}{\sigma}$$

z Score

Definition The *z value* or *z score* tells us the number of standard deviations the original measurement is from the mean. The *z* value is in *standard units*.

The mean is a special value of a distribution. Let's see what happens when we convert $x = \mu$ to a *z* value:

$$z = \frac{x - \mu}{\sigma}$$

$$= \frac{\mu - \mu}{\sigma} \qquad for \; x = \mu$$

$$= 0$$

Table 6-4 *x* Values and Corresponding *z* Values

x Value in Original Distribution	Corresponding *z* Value or Standard Unit
$x = \mu$	$z = 0$
$x > \mu$	$z > 0$
$x < \mu$	$z < 0$

The mean of the original distribution is always zero, in standard units. This makes sense because the mean is zero standard deviations from itself.

An x value in the original distribution that is *above* the mean μ has a corresponding z value that is *positive*. Again, this makes sense because a measurement above the mean would be a positive number of standard deviations from the mean. Likewise, an x value *below* the mean has a *negative z* value. (See Table 6-4.)

> **Note**
>
> Unless otherwise stated, in the remainder of this book we will take the word *average* to be either the sample arithmetic mean $\bar{x}$ or the population mean μ.

EXAMPLE 4   A pizza parlor franchise specifies that the average (mean) amount of cheese on a large pizza should be 8 oz and the standard deviation only 0.5 oz. An inspector picks out a large pizza at random in one of the pizza parlors and finds that it is made with 6.9 oz of cheese. Assume that the amount of cheese on a pizza follows a normal distribution. If the amount of cheese is more than *three* standard deviations below the mean, the parlor will be in danger of losing its franchise. (Remember, in a normal distribution we are unlikely to find measurements more than three standard deviations from the mean, since 99.7% of all measurements fall within three standard deviations of the mean.)

How many standard deviations from the mean is 6.9? Is the pizza parlor in danger of losing its franchise?

SOLUTION: Since we want to know the number of standard deviations from the mean, we want to convert 6.9 to standard z units.

$$z = \frac{x - \mu}{\sigma}$$

$$= \frac{6.9 - 8}{0.5}$$

$$= -2.20$$

Therefore, the amount of cheese on the selected pizza is only -2.20 standard deviations from the mean. Note that the fact that z is negative indicates that the amount of cheese was 2.20 standard deviations *below* the mean. The parlor will not lose its franchise based on this sample.

GUIDED EXERCISE   6

A student has computed that it takes an average (mean) of 17 minutes with a standard deviation of 3 minutes to drive from home, park the car, and walk to an early morning class.

(a) One day it took the student 21 minutes to get to class. How many standard deviations from the average is that? Is the z value positive or negative? Explain why it should be either positive or negative.

⇨ The number of standard deviations from the mean is given by the z value:

$$z = \frac{x - \mu}{\sigma} = \frac{21 - 17}{3} \approx 1.33$$

The z value is positive. We would expect a positive z value, since 21 minutes is *more* than the mean of 17.

(b) Another day it took only 12 minutes for the student to get to class. What is this measurement in standard units? Is the z value positive or negative? Why should it be positive or negative?

⇨ The measurement in standard units is

$$z = \frac{x - \mu}{\sigma} = \frac{12 - 17}{3} \approx -1.67$$

Here the z value is negative, as we should expect, because 12 minutes is less than the mean of 17 minutes.

(c) Another day it took 17 minutes for the student to go from home to class. What is the z value? Why should you expect this answer?

⇨ In this case the value is

$$z = \frac{x - \mu}{\sigma} = \frac{17 - 17}{3} = 0.00$$

We expect this result because 17 minutes is the mean, and the z value of the mean is always zero.

Raw score

We have seen how to convert from x measurements to standard units z. We can easily reverse the process if we know μ and σ for the original distribution. For when we solve

$$z = \frac{x - \mu}{\sigma}$$

for x, we get

$$x = z\sigma + \mu$$

EXAMPLE 5 ▷

In Example 4 we talked about the amount of cheese required by a franchise for a large pizza. Again, the mean amount of cheese required is 8 oz with a standard deviation of 0.5 oz. The parlor can lose its franchise if the amount of cheese on its large pizza is less than three standard deviations below the mean. What is the minimum amount of cheese that can be placed on a large pizza according to the franchise?

SOLUTION: Here we need to convert $z = -3$ to information about x oz of cheese. We use the formula

$x = z\sigma + \mu = -3(0.5) + 8 = 6.5$ oz

The franchise will not approve a large pizza with less than 6.5 oz of cheese.

In many testing situations we hear the terms *raw score* and *z score*. The raw score is just the score in the original measuring units, and the *z* score is the score in standard units. Guided Exercise 7 illustrates these different units.

GUIDED EXERCISE   7

Marulla's *z* score on a college entrance exam is 1.3. If the raw scores have a mean of 480 and a standard deviation of 70 points, what is her raw score?

Here we are given *z*, σ, and μ. We need to find the raw score *x* corresponding to the *z* score 1.3.

$x = z\sigma + \mu$
$\quad = 1.3(70) + 480$
$\quad = 571$

Standard normal distribution

If the original distribution of *x values is normal*, then the corresponding *z values have a normal distribution as well*. The *z* distribution has a mean of 0 and a standard deviation of 1. The normal curve with these properties has a special name.

Definition The *standard normal distribution* is a normal distribution with mean $\mu = 0$ and standard deviation $\sigma = 1$ (Figure 6-15).

Any normal distribution of *x* values can be converted to the standard normal distribution by converting all *x* values to their corresponding *z* values. Let's look at the graphic interpretation of this transformation in Figure 6-16.

The resulting standard distribution will always have mean $\mu = 0$ and standard deviation $\sigma = 1$.

Areas Under the Standard Normal Curve

We have seen how to convert *any* normal distribution to the *standard* normal distribution. We can change any *x* value to a *z* value and back again. But what is the advantage of all this work? The advantage is that there are extensive tables that show the area under the standard normal curve for almost any interval along the *z* axis. The area is important because it is equal to the *probability* that the

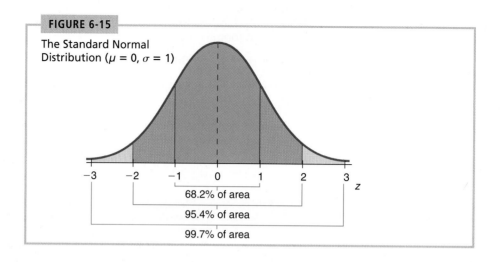

FIGURE 6-15

The Standard Normal
Distribution ($\mu = 0$, $\sigma = 1$)

68.2% of area

95.4% of area

99.7% of area

FIGURE 6-16

The Transformation of a
Normal Distribution to the
Standard Normal Distribution

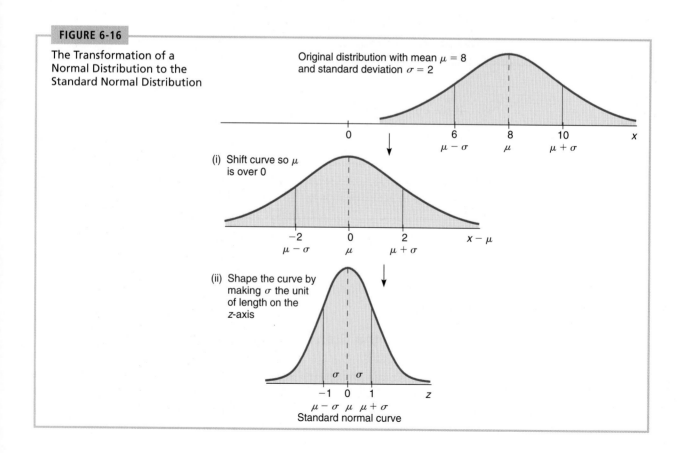

Original distribution with mean $\mu = 8$
and standard deviation $\sigma = 2$

(i) Shift curve so μ
is over 0

(ii) Shape the curve by
making σ the unit
of length on the
z-axis

Standard normal curve

measurement of an item selected at random falls in this interval. Thus the *standard* normal distribution can be a tremendously helpful tool.

For instance, Sunshine Stereo guarantees their cassette decks for a period of 2 years. The company statistician has computed that the cassette deck life is normally distributed with a mean of 2.3 years and a standard deviation 0.4 year. What is the probability that a cassette deck will stop working during the guarantee period?

To answer questions of this type, we convert the given normal distribution to the standard normal distribution. Then we use a table to find the area over the interval in question and, hence, the probability that an item selected at random will fall in that interval. Before we can carry out this plan, though, we must practice using Table 4 of Appendix I to find areas under the standard normal curve.

The empirical rule enables us to find certain areas under a standard normal curve. How do we find other areas under the standard normal curve? The most convenient way is to use a table. Because of the symmetry of the normal curve, it is possible to obtain all the areas we will need if we have only a table of areas from $z = 0$ to $z =$ some positive number. The following sequence of examples will show how this can be done using Table 4 of Appendix I.

EXAMPLE 6 ▷ Find the area under the standard normal curve between $z = 0$ and $z = 1$. This area is shown in Figure 6-17.

SOLUTION: We will use Table 4 of Appendix I. For convenience, we have included part of Table 4 in Table 6-5.

In the upper left corner of the table we see the letter z. The column under z gives us the units value and tenths for z. The other column headings indicate the hundredths value of z. The table entries give the areas under the normal curve from the mean $z = 0$ to a specified value of z. To find the area from $z = 0$ to $z = 1$, we observe that if $z = 1$, then the units value of z is 1 and the tenths value is 0. So we look in the column labeled z for 1.0. The area from $z = 0$ to $z = 1$ is given in the corresponding row of the column with heading 0.00 since $z = 1$ is the same as $z = 1.00$. The area we read from the table for $z = 1.00$ is 0.3413. It is the shaded value in Table 6-5.

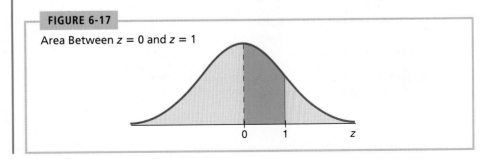

FIGURE 6-17

Area Between $z = 0$ and $z = 1$

Table 6-5 Portion of Table 4 (Appendix I) of Areas Under the Standard Normal Curve from z = 0 to the Indicated Value of z

z	0.00	0.01	0.02	0.03	0.04	0.05 . . .	0.09
.							
.							
.							
0.9	0.3159	0.3186	0.3212	0.3238	0.3264	0.3289 . . .	0.3389
1.0	0.3413	0.3438	0.3461	0.3485	0.3508	0.3531 . . .	0.3621
1.1	0.3643	0.3665	0.3686	0.3708	0.3729	0.3749 . . .	0.3830
.							
.							
.							
2.5	0.4938	0.4940	0.4941	0.4943	0.4945	0.4946 . . .	0.4952

GUIDED EXERCISE   8

In this exercise we will find the area under the standard normal curve from $z = 0$ to $z = 2.53$. We will use Table 6-5.

(a) Shade the area we are to find in Figure 6-18.   ⇨   See Figure 6-19.

FIGURE 6-18

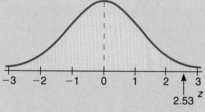

FIGURE 6-19 Completion of Figure 6-18

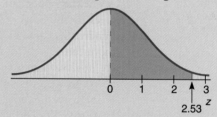

(b) For $z = 2.53$, the units value is _____ and the tenths value is _____, so we look in the column under z for the number _____.   ⇨   The units value is 2 and the tenths value is 0.5, so we look in the z column for the value 2.5.

(c) For $z = 2.53$, the hundredths value is _____, so we look in the column headings for the number 0.03.   ⇨   The hundredths value is 0.03.

(d) The area between $z = 0$ and $z = 2.53$ is given by the entry in the row beginning with 2.5 and in the column headed by 0.03. What is the area?   ⇨   The area is 0.4943.

Table 4 in Appendix I gives areas under the normal curve for regions *beginning* at $z = 0$ and extending to a specified positive z value. However, because the normal curve is symmetrical, we also can use the table directly to find areas beginning with a negative z value and extending to $z = 0$. Example 7 shows this process.

EXAMPLE 7 ➤ Find the area under the standard normal curve from $z = -2.34$ to $z = 0$.

SOLUTION: The area from $z = -2.34$ to $z = 0$ is the same as the area from $z = 0$ to $z = 2.34$. (See Figure 6-20.) From Table 4 (Appendix I), the area from 0 to 2.34 is 0.4904. Therefore, the area from $z = -2.34$ to $z = 0$ is also 0.4904.

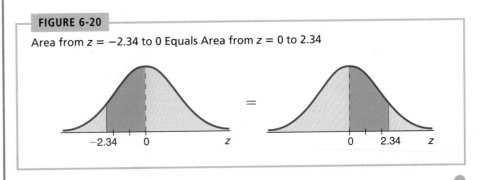

FIGURE 6-20

Area from $z = -2.34$ to 0 Equals Area from $z = 0$ to 2.34

To find areas other than those between a given z value and $z = 0$, we use Table 4 (Appendix I) together with addition or subtraction of areas we find in Table 4. Figure 6-21 shows how to combine areas. As you study the figure, notice that

1. For areas extending from one side of the mean $z = 0$ to the other side, we *add* areas found in Table 4.

2. For areas completely on one side of the mean $z = 0$ (but not bordering $z = 0$), we *subtract* areas found in Table 4.

3. The area extending from $z = 0$ and including the entire right half of the graph is 0.5000. Likewise, the area extending from $z = 0$ and including the entire left half of the graph is 0.5000.

EXAMPLE 8 ➤ Find the area under the standard normal curve in Figure 6-22 from $z = 1.00$ to $z = 2.70$.

Example 8 continues on pages 282 and 283.

FIGURE 6-21

Patterns for Finding Areas Under the Standard Normal Curve

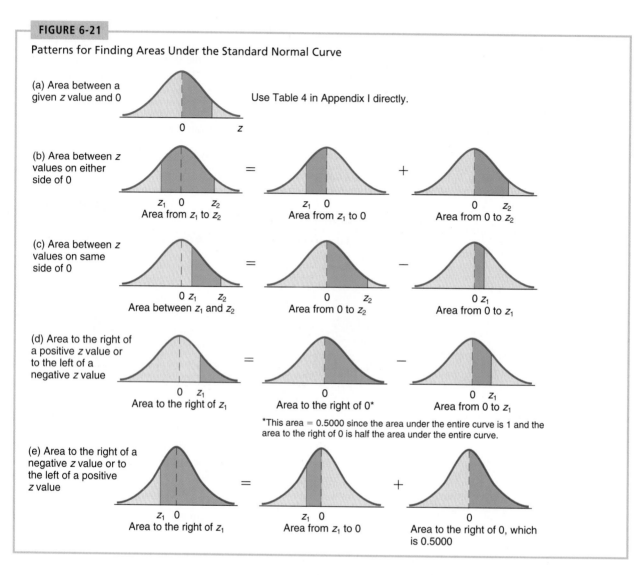

(a) Area between a given z value and 0

Use Table 4 in Appendix I directly.

(b) Area between z values on either side of 0

Area from z_1 to z_2 = Area from z_1 to 0 + Area from 0 to z_2

(c) Area between z values on same side of 0

Area between z_1 and z_2 = Area from 0 to z_2 − Area from 0 to z_1

(d) Area to the right of a positive z value or to the left of a negative z value

Area to the right of z_1 = Area to the right of 0* − Area from 0 to z_1

*This area = 0.5000 since the area under the entire curve is 1 and the area to the right of 0 is half the area under the entire curve.

(e) Area to the right of a negative z value or to the left of a positive z value

Area to the right of z_1 = Area from z_1 to 0 + Area to the right of 0, which is 0.5000

FIGURE 6-22

Area from $z = 1.00$ to $z = 2.70$

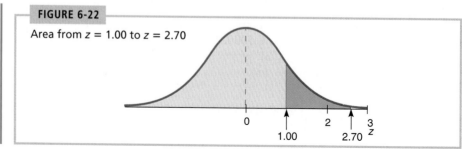

SOLUTION: The area we are trying to find lies entirely to the right of $z = 0$ and does not border $z = 0$. Therefore, we need to subtract component areas.

$$\left(\begin{array}{c}\text{Area from}\\1.00 \text{ to } 2.70\end{array}\right) = \left(\begin{array}{c}\text{area from}\\0 \text{ to } 2.70\end{array}\right) - \left(\begin{array}{c}\text{area from}\\0 \text{ to } 1.00\end{array}\right)$$

$$\downarrow \qquad\qquad \downarrow \qquad\qquad \downarrow$$

$$0.1552 \quad = \quad 0.4965 \quad - \quad 0.3413$$

The desired area is 0.1552.

GUIDED EXERCISE   9

Find the area from $z = -3.00$ to $z = 2.65$. First, we *draw the picture* (see Figure 6-23) and observe the location of the requested area. Next, we find component areas from Table 4 (Appendix I) and combine them appropriately.

FIGURE 6-23 Area Between $z = -3.00$ and $z = 2.65$

(a) Look at Figure 6-23. Should we add or subtract component areas?

⇨ Since the area extends from the left side of $z = 0$ to the right side, we add the component areas.

(b) Find the area under the standard normal curve between $z = 0$ and $z = 2.65$.

⇨ We look under the z column in Table 4 (Appendix I) until we find 2.6; then we stay in this row and move to the right until we are in the column headed by 0.05. The area from $z = 0$ to $z = 2.65$ is given by the entry 0.4960.

(c) Find the area under the standard normal curve between $z = -3.00$ and $z = 0$.

⇨ Since the area from $z = -3.00$ to $z = 0$ is the same as that from $z = 0$ to $z = 3.00$, we look down the z column until we find 3.0. Then we move to the right in this row until we are in the column headed by 0.00. This entry is 0.4987, which is the area from $z = -3.00$ to $z = 0$.

(d) Use parts b and c to find the area under the standard normal curve between $z = -3.00$ and $z = 2.65$ (Figure 6-23).

⇨ $$\left(\begin{array}{c}\text{Area from}\\-3.00 \text{ to } 2.65\end{array}\right) = \left(\begin{array}{c}\text{area from}\\-3.00 \text{ to } 0\end{array}\right) + \left(\begin{array}{c}\text{area from}\\0 \text{ to } 2.65\end{array}\right)$$

$$\downarrow \qquad\qquad \downarrow \qquad\qquad \downarrow$$

$$0.9947 \quad = \quad 0.4987 \quad + \quad 0.4960$$

The desired area is 0.9947.

EXAMPLE 9 ▷ Find the area under the standard normal curve to the left of $z = -0.94$.

SOLUTION: We sketch the area and notice that the area to the left of -0.94 is the same as the area to the right of 0.94. (See Figure 6-24.)

To find the area to the right of 0.94, we observe

$$\begin{pmatrix} \text{Area to the} \\ \text{right of 0.94} \end{pmatrix} = \begin{pmatrix} \text{area to the} \\ \text{right of 0} \end{pmatrix} - \begin{pmatrix} \text{area from} \\ \text{0 to 0.94} \end{pmatrix}$$

$$\qquad\qquad\qquad\downarrow\qquad\qquad\quad\downarrow$$

$$= \qquad 0.5000 \quad - \quad 0.3264$$

$$= \qquad 0.1736$$

FIGURE 6-24

Shaded Areas
Are Equal

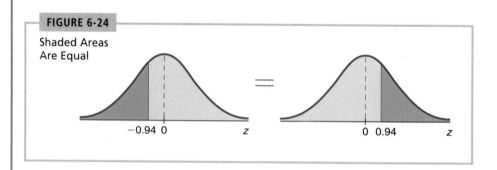

Notice that the area to the right of zero is one-half the area under the entire curve. Since the area under the entire curve (i.e., entire sample space) is 1, the area to the right of zero is $\frac{1}{2}$, or 0.5000. The area to the left of -0.94 equals that to the right of 0.94, so the desired area is 0.1736.

●

GUIDED EXERCISE   10

Let z be a random variable with a standard normal distribution. Find the probability $P(z \geq 1.15)$.

(a) $P(z \geq 1.15)$ refers to the probability that z values lie to the right of 1.15. Shade the corresponding area under the standard normal curve.

FIGURE 6-25 Area to Be Found

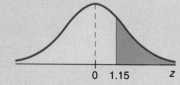

Exercise continues

Exercise continued

(b) The area to the right of $z = 0$ equals

_____.

⇨   The area to the right of $z = 0$ equals 0.5000.

(c) Find the area between $z = 0$ and $z = 1.15$.

⇨   From Table 4 of Appendix I, the area is 0.3749.

(d) Use the areas in parts b and c to find the area to the right of $z = 1.15$.

⇨   Area to the right of 1.15

$$= \begin{pmatrix} \text{area to the} \\ \text{right of 0} \end{pmatrix} - \begin{pmatrix} \text{area between} \\ \text{0 and 1.15} \end{pmatrix}$$

$$\qquad\quad \downarrow \qquad\qquad\qquad \downarrow$$

$$= \quad 0.5000 \quad - \quad 0.3749$$

$$= \quad 0.1251$$

We have practiced the skill of finding areas under the standard normal curve for various intervals along the z axis. This skill is important, since *the probability that z lies in an interval* is *given by the area* under the standard normal curve above that interval.

COMMENT If we look at Table 4 in Appendix I, we see the z values in the table stop at $z = 3.69$. What about z values larger than 3.69? If we have a z value larger than 3.69, we are so far out on the normal curve that for most practical applications we use $z = 3.69$ instead of the larger value. The corresponding area from 0 to z is between 0.4999 and 0.5000.

In the next section we will see how to use standard z scores and the standard normal distribution table to find areas under *any* normal curve.

V I E W P O I N T

Mighty Oaks from Little Acorns Grow

Just how big is that acorn? What if we compare it with other acorns? Is that oak tree taller than an average oak tree? How does it compare with other oak trees? What do you mean, this oak tree has a larger geographic range? Compared with what? Answers to questions such as these can be given only if we resort to *standardized statistical units.* Can you compare a single oak tree with an entire forest of oak trees? The answer is yes, if you use *standardized z-scores.* For more information about sizes of acorns, oak trees, and geographic locations, see Web site <http://lib.stat.cmu.edu/DASL>. Follow the links to Data Subjects, Biology, and Acorns.

SECTION 6.2 PROBLEMS

In these problems, assume that all the distributions are *normal*. In all problems in this chapter, *average* is always taken to be the arithmetic mean $\bar{x}$ or μ.

Physical Education:
First Aid Course

1. The college Physical Education Department offered an Advanced First Aid course last semester. The scores on the comprehensive final exam were normally distributed, and the z scores for some of the students were as follows.

    Robert, 1.10   Jan, 1.70   Susan, −2.00
    Joel, 0.00   John, −0.80 Linda, 1.60

    (a) Which of these students scored above the mean?
    (b) Which of these students scored on the mean?
    (c) Which of these students scored below the mean?
    (d) If the mean score was $\mu = 150$ with standard deviation $\sigma = 20$, what was the final exam score for each student?

Higher Education:
Professors

2. What do professors do with their time? They do research, teach classes, serve on academic committees, serve the student body (advise students, sponsor student clubs, attend student events), serve the community (consult, address civic groups), and a lot more. The specific answer depends on the individual professor and his or her special interests. Well, how much time does a professor spend on teaching activities? *The NEA Almanac of Higher Education,* published by the National Education Association, reports that the mean percentage of time professors spend on teaching activities is about $\mu = 51\%$ with standard deviation $\sigma = 25\%$. Find the standardized z values corresponding to the following professors' percentages of time allocated to teaching duties.

    (a) Dr. Taylor, 45%   (d) Ms. Simms, 65%
    (b) Mr. Patterson, 72%   (e) Dr. Adams, 33%
    (c) Dr. Smith, 75%   (f) Dr. Riley, 55%

Climate:
Temperatures in Honolulu

3. Data collected over a period of years show that the average daily temperature in Honolulu is $\mu = 73°F$ with standard deviation $\sigma = 5°F$ (U.S. Department of Commerce: *Environmental Data Service*). Convert each of the following intervals in °F to an interval of z values.

    (a) 53°F $< x <$ 93°F
    (b) $x <$ 65°F
    (c) 78°F $< x$

    Convert each of the following intervals of z values to an interval in °F.

    (d) 1.75 $< z$
    (e) $z <$ −1.90
    (f) −1.80 $< z <$ 1.65

Wild Life: Fawns

4. Fawns between 1 and 5 months old in Mesa Verde National Park have a body weight that is approximately normally distributed with mean $\mu = 27.2$ kilograms and standard deviation $\sigma = 4.3$ kilograms (based on information from *The Mule*

Deer of Mesa Verde National Park, by G. W. Mierau and J. L. Schmidt, Mesa Verde Museum Association). Let x be the weight of a fawn in kilograms. Convert each of the following x intervals to a z interval.

(a) $x < 30$

(b) $19 < x$

(c) $32 < x < 35$

Convert each of the following z intervals to an x interval.

(d) $-2.17 < z$

(e) $z < 1.28$

(f) $-1.99 < z < 1.44$

(g) If a fawn weighs 14 kilograms, would you say it is an unusually small animal? Explain using z values and Figure 6-15.

(h) If a fawn is unusually large, would you say that the z value for the weight of the fawn will be close to 0, -2, or 3? Explain.

Wild Life: Deer

5. The fall deer population in Mesa Verde National Park is approximately normally distributed with mean 4400 deer and standard deviation 620 deer (see reference in Problem 4). Let x be the random variable that represents the size of the deer population in Mesa Verde National Park in the fall of a given year. Convert each of the following x intervals to a z interval.

(a) $3300 < x$

(b) $x < 5400$

(c) $3500 < x < 5300$

Convert each of the following z intervals to an x interval.

(d) $-1.12 < z < 2.43$

(e) $z < 1.96$

(f) $2.58 < z$

(g) If the fall deer population were 2800 deer, would that be considered an unusually low number? If the fall population were 6300, would that be considered an unusually high population? Explain using z values and Figure 6-15.

Medical:
White Blood Cell Count

6. Let x = white blood cell (WBC) count per cubic millimeter of whole blood. Then x has a distribution that is approximately normal with mean $\mu = 7500$ and standard deviation $\sigma = 1750$ (based on information from *Diagnostic Tests with Nursing Implications,* edited by S. Loeb, Springhouse Press). Convert each of the following x intervals to a z interval.

(a) $9000 < x$

(b) $x < 6000$

(c) $3500 < x < 4500$

Convert each of the following z intervals to an x interval.

(d) $z < 1.15$

(e) $2.19 < z$

(f) $0.25 < z < 1.25$

(g) If someone had a WBC count of 2500, would that be considered unusually high or low? Explain using z values and Figure 6-15.

Medical:
Red Blood Cell Count

7. Let x = red blood cell (RBC) count in millions per cubic millimeter of whole blood. For healthy females, x has an approximately normal distribution with mean $\mu = 4.8$ and standard deviation $\sigma = 0.3$. (See reference in Problem 6.) Convert each of the following x intervals from laboratory tests to a z interval.
   (a) $4.5 < x$
   (b) $x < 4.2$
   (c) $4.0 < x < 5.5$

   Convert each of the following z intervals to an x interval.
   (d) $z < -1.44$
   (e) $1.28 < z$
   (f) $-2.25 < z < -1.00$
   (g) If a female had an RBC count of 5.9 or higher, would that be considered unusually high? Explain using z values and Figure 6-15.

Archaeology:
Tree Ring Data

8. Tree ring dates were used extensively in archaeological studies at Burnt Mesa Pueblo (*Bandelier Archaeological Excavation Project: Summer 1989 Excavations at Burnt Mesa Pueblo*, edited by Kohler, Washington State University Department of Anthropology). At one site on the mesa, tree ring dates (for many samples) gave a mean date μ_1 = year 1272 with standard deviation $\sigma_1 = 35$ years. At a second, removed site, the tree ring dates gave a mean of μ_2 = year 1122 with standard deviation $\sigma_2 = 40$ years. Assume that both sites had dates that were approximately normally distributed. In the first area an object was found and dated as x_1 = year 1250. In the second area another object was found and dated as x_2 = year 1234.
   (a) Convert both x_1 and x_2 to z values, and locate both these values under the standard normal curve in Figure 6-15.
   (b) Which of these two items is the more unusual as an archaeological find in its location?

In Problems 9–28, sketch the areas under the standard normal curve over the indicated intervals, and find the specified areas.

9. Between $z = 0$ and $z = 3.18$

10. Between $z = 0$ and $z = 2.92$

11. Between $z = 0$ and $z = -2.01$

12. Between $z = 0$ and $z = -1.93$

13. Between $z = -2.18$ and $z = 1.34$

14. Between $z = -1.40$ and $z = 2.03$

15. Between $z = 0.32$ and $z = 1.92$

16. Between $z = 1.42$ and $z = 2.17$

17. Between $z = -2.42$ and $z = -1.77$

18. Between $z = -1.98$ and $z = -0.03$

19. To the right of $z = 0$

20. To the left of $z = 0$

21. To the right of $z = 1.52$

22. To the right of $z = 0.15$

23. To the left of $z = -1.32$

24. To the left of $z = -0.47$

25. To the right of $z = -1.22$

26. To the right of $z = -2.17$

27. To the left of $z = 0.45$

28. To the left of $z = 0.72$

In Problems 29–48, let z be a random variable with a standard normal distribution. Find the indicated probability, and shade the corresponding area under the standard normal curve.

29. $P(0 \leq z \leq 1.62)$ 39. $P(z \leq 0)$

30. $P(0 \leq z \leq 0.54)$ 40. $P(z \geq 0)$

31. $P(-0.82 \leq z \leq 0)$ 41. $P(z \geq 1.35)$

32. $P(-2.37 \leq z \leq 0)$ 42. $P(z \geq 2.17)$

33. $P(-0.45 \leq z \leq 2.73)$ 43. $P(z \leq -0.13)$

34. $P(1.73 \leq z \leq 3.12)$ 44. $P(z \leq -2.15)$

35. $P(-2.18 \leq z \leq -0.42)$ 45. $P(z \geq -1.20)$

36. $P(-1.78 \leq z \leq -1.23)$ 46. $P(z \geq -1.50)$

37. $P(-1.20 \leq z \leq 2.64)$ 47. $P(z \leq 1.20)$

38. $P(-2.20 \leq z \leq 1.04)$ 48. $P(z \leq 3.20)$

Section **6.3**   **Areas Under Any Normal Curve**

Converting normal distributions to standard normal

In many applied situations, the original normal curve is not the standard normal curve. Generally, there will not be a table of areas available for the original normal curve. This does not mean that we cannot find the probability that a measurement x will fall in an interval from a to b. What we must do is *convert* original measurements x, a, and b to z values.

EXAMPLE 10 ❯

Let x have a normal distribution with $\mu = 10$ and $\sigma = 2$. Find the probability that an x value selected at random from this distribution is between 11 and 14. In symbols, find $P(11 \leq x \leq 14)$.

SOLUTION: Since probabilities correspond to areas under the distribution curve, we want to find the area under the x curve above the interval from $x = 11$ to $x = 14$. To do so, we will convert the x values to standard z values (see Figure 6-26) and then use Table 4 in Appendix I to find the corresponding area under the standard curve.

We use the formula

$$z = \frac{x - \mu}{\sigma}$$

FIGURE 6-26

The Interval $11 \leq x \leq 14$ Corresponds to the Interval $0.50 \leq z \leq 2.00$

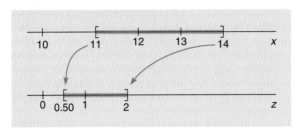

to convert the given x interval to a z interval.

$$z_1 = \frac{11 - 10}{2} = 0.50 \qquad (\text{Use } x = 11, \ \mu = 10, \ \sigma = 2.)$$

$$z_2 = \frac{14 - 10}{2} = 2.00 \qquad (\text{Use } x = 14, \ \mu = 10, \ \sigma = 2.)$$

The corresponding areas under the x and z curves are shown in Figure 6-27. From Figure 6-27 we see that

FIGURE 6-27

Corresponding Areas Under the x Curve and z Curve

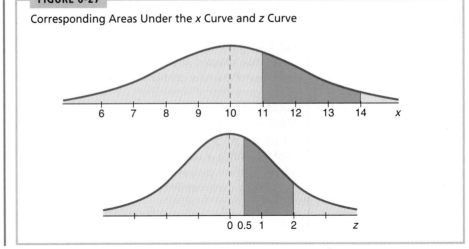

$$P(11 \leq x \leq 14) = P(0.50 \leq z \leq 2.00)$$

$$= P(0 \leq z \leq 2.00) - P(0 \leq z \leq 0.50)$$

$$= 0.4772 - 0.1915 \quad \text{(From Table 4, Appendix I)}$$

$$= 0.2857$$

The probability is 0.2857 that an x value selected at random from a normal distribution with mean 10 and standard deviation 2 lies between 11 and 14. ●

GUIDED EXERCISE   11

In Section 6.2 we talked about Sunshine Stereo cassette decks. The cassette deck life was normally distributed with a mean of 2.3 years and a standard deviation of 0.4 year. We wanted to know the probability that a cassette deck will break down during the guarantee period of 2 years.

(a) Let x represent the life of a cassette deck. The statement that the cassette deck breaks during the 2-year guarantee period means the life is less than 2 years, or $x \leq 2$. Convert this to a statement about z.

⇨ $z = \dfrac{x - \mu}{\sigma} = \dfrac{2 - 2.3}{0.4} = -0.75$

So $x \leq 2$ means $z \leq -0.75$.

(b) Indicate the area to be found in Figure 6-28. Does this area correspond to the probability that $z \leq -0.75$?

⇨ See Figure 6-29.
Yes, the shaded area does correspond to the probability that $z \leq -0.75$.

FIGURE 6-28

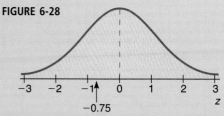

FIGURE 6-29 $z \leq -0.75$

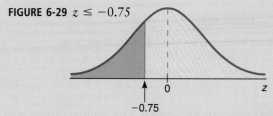

(c) The shaded portion of Figure 6-29 corresponds to the probability that $z \leq -0.75$. To compute this probability, we must do the subtraction of areas shown in Figure 6-30. Do that subtraction.

⇨ $P(z \leq -0.75) = P(z \leq 0) - P(-0.75 \leq z \leq 0)$

$\qquad\qquad\quad = \quad 0.5000 \quad - \quad 0.2734$

$\qquad\qquad\quad = \quad 0.2266$

Exercise continues

Exercise continued

FIGURE 6-30 $P(z \leq -0.75) = P(z \leq 0) - P(-0.75 \leq z \leq 0) = $ _____ $-$ _____ $=$ _____

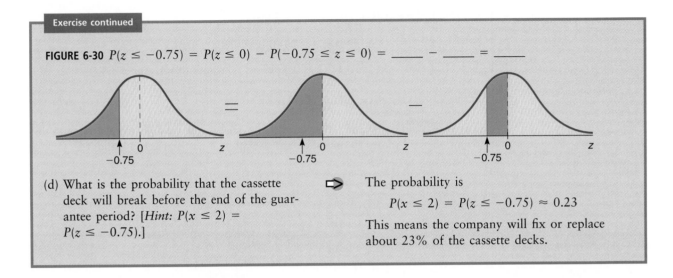

(d) What is the probability that the cassette deck will break before the end of the guarantee period? [*Hint:* $P(x \leq 2) = P(z \leq -0.75)$.]

⇨ The probability is

$$P(x \leq 2) = P(z \leq -0.75) \approx 0.23$$

This means the company will fix or replace about 23% of the cassette decks.

Calculator Note The TI-83 gives areas under any normal distribution and shades the specified area. Use the **DISTR (2nd VARS)** key and select **normalcdf(lower bound, upper bound, μ, σ)**. Figure 6-31(a) shows an example where we find the area between 11 and 14 under a normal distribution with $\mu = 10$ and $\sigma = 2$. To draw the normal curve, *first* set the window to accommodate the graph. Then use the **DISTR** key, select **DRAW**, and select **ShadeNorm(lower bound, upper bound, μ, σ)**. Figure 6-31(a) shows the command, and Figure 6-31(b) shows the graphic results. *Note:* If you do not specify μ or σ, it is assumed that the area is under the *standard* normal curve.

FIGURE 6-31

(a) Area between 11 and 14 under normal curve with $\mu = 10$, $\sigma = 2$

```
normalcdf(11,14,10,
2)
        .285874702
ShadeNorm(11,14,10,
2)
```

(b) Set window with **Xmin = 4, Xmax = 16, Ymin = −0.1**, and **Ymax = 0.3**

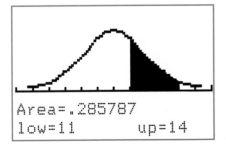

Finding *z* or *x*, given a
probability

Sometimes we need to find *z* or *x* values that correspond to a given area under the normal curve. This situation arises when we want to specify a guarantee period so that a given percentage of the total products produced by a company lasts at least as long as the duration of the guarantee period.

EXAMPLE 11 ➤

Magic Video Games, Inc. sells an expensive video computer games package. Because the package is so expensive, the company wants to advertise an impressive guarantee for the life expectancy of its computer control system. The guarantee policy will refund the full purchase price if a computer fails during the guarantee period. The research department has done tests which show that the mean life for the computer is 30 months, with standard deviation of 4 months. The computer life is normally distributed. How long can the guarantee period be if management does not want to refund the purchase price on more than 7% of the Magic Video packages?

SOLUTION: Let us look at the distribution of lifetimes for the computer control system, and shade the portion of the distribution in which the computer lasts fewer months than the guarantee period. (See Figure 6-32 on page 294.)

If a computer system lasts fewer months than the guarantee period, a full-price refund will have to be made. The lifetimes requiring a refund are in the shaded region in Figure 6-32. This region represents 7% of the total area under the curve.

We can use Table 4 in Appendix I to find the *z* value such that 7% of the total area under the *standard* normal curve lies to the left of the *z* value. Then we convert the *z* value to its corresponding *x* value to find the guarantee period.

The *z* value with 7% of the area to the left of it is the negative of the *z* value with 50% − 7% = 43% of the area between 0 and *z*. (See Figure 6-33 on page 294.)

We find the number *closest* to 0.4300 in the area region of Table 4 (Appendix I) and read the corresponding *z* value (see Table 6-6).

Table 6-6   Excerpt from Table 4 in Appendix I

z	...	0.04	0.05	0.06	0.07		0.08	0.09
.								
.								
.								
1.4	...	0.4251	0.4265	0.4279	0.4292	↑ 0.4300	0.4306	0.4319
.								
.								
.								

The value 0.4300 lies between area values 0.4292 and 0.4306, but it is closer to 0.4306. The corresponding *z* value is 1.48. Since we want the *z* value such that 7% of the total area is to the *left* of *z*, we use the symmetry of the curve and find the *z* value to be −1.48.

To translate this value back to an *x* value (in months) we use the formula

$$x = z\sigma + \mu$$
$$= -1.48(4) + 30 \qquad \text{(Use } \sigma = 4 \text{ months and } \mu = 30 \text{ months.)}$$
$$= 24.08 \text{ months}$$

The company can guarantee the Magic Video Games package for $x = 24$ months. For this guarantee period, it expects to refund the purchase price of no more than 7% of the video games packages. ●

COMMENT When we use Table 4 in Appendix I to find a z value corresponding to a given area, we usually use the nearest area value rather than interpolating between values. However, when the area value given is exactly halfway between two area values in the table, we use the z value halfway between the z values of the corresponding table areas. Table 6-7 shows an example in which the given area is halfway between two table areas. However, this interpolation convention is not always used, especially if the area is changing slowly, as it does at the tail ends of the distribution. *When the* z *value corresponding to an area is larger than, say, 2, the standard convention is to use the* z *value corresponding to the next larger area.* We see an example of this special case in Guided Exercise 1 in Chapter 8. ○

Table 6-7 Excerpt from Table 4 in Appendix I

z	...	0.04	$z = 1.645$ is halfway between 1.64 and 1.65 ↓	0.05	...
.					
.					
.					
1.6	...	0.4495	↑	0.4505	

0.4500 is exactly halfway between the area values 0.4495 and 0.4505

The area between $z = 0$ and $z = 1.645$ is 0.4500.

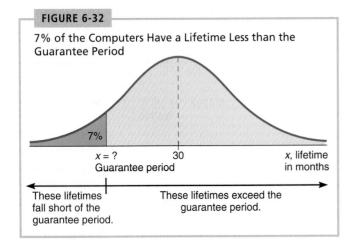

FIGURE 6-32

7% of the Computers Have a Lifetime Less than the Guarantee Period

7%

$x = ?$ 30 x, lifetime
Guarantee period   in months

These lifetimes fall short of the guarantee period.

These lifetimes exceed the guarantee period.

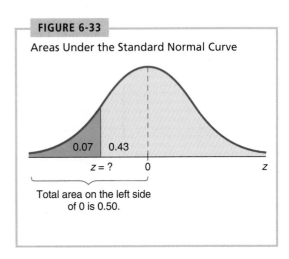

FIGURE 6-33

Areas Under the Standard Normal Curve

0.07 | 0.43

$z = ?$ 0 z

Total area on the left side of 0 is 0.50.

GUIDED EXERCISE   12

Find the z value such that 3% of the area under the standard normal curve lies to the left of z.

(a) Draw a standard normal curve; shade the region so that 3% of the area lies to the left of z.

▷ **FIGURE 6-34** 3% of the Total Area Lies to the Left of z

3% of area

$z = ?$ 0 z

(b) What portion of the area lies between this z value and 0?

▷ Since half the area lies to the left of zero, then $50\% - 3\% = 47\%$ lies between the z value and 0.

(c) Use Table 4 in Appendix I to find the z value such that 47% of the area under the standard normal curve is between 0 and z.

▷ The area value in Table 4 that is nearest to 0.4700 is 0.4699. The corresponding z value is 1.88.

(d) Use the symmetry of the standard normal curve to find the z value such that 3% of the area lies to the left of z. (That is, find the z value indicated in Figure 6-34.)

▷ 3% of the area lies to the left of $z = -1.88$.

VIEWPOINT

Want to Be an Archaeologist?

Each year about 4500 students work with professional archaeologists in scientific research at the Crow Canyon Archaeological Center, Cortez, Colorado. In fact, Crow Canyon was included in *The Princeton Review Guide to America's Top 100 Internships*. The nonprofit, multidisciplinary program at Crow Canyon enables students and laypeople with little or no background to get started in archaeological research. The only requirement is that you be interested in Native American culture and history. By the way, a knowledge of introductory statistics could come in handy in this internship. For more information about the program, see Web site <http://www.crowcanyon.org>.

SECTION 6.3 PROBLEMS

In Problems 1–10, assume that x has a normal distribution, with the specified mean and standard deviation. Find the indicated probabilities.

1. $P(3 \leq x \leq 6)$; $\mu = 4$; $\sigma = 2$

2. $P(10 \leq x \leq 26)$; $\mu = 15$; $\sigma = 4$

3. $P(50 \leq x \leq 70)$; $\mu = 40$; $\sigma = 15$

4. $P(7 \leq x \leq 9)$; $\mu = 5$; $\sigma = 1.2$

5. $P(8 \leq x \leq 12)$; $\mu = 15$; $\sigma = 3.2$

6. $P(40 \leq x \leq 47)$; $\mu = 50$; $\sigma = 15$

7. $P(x \geq 30)$; $\mu = 20$; $\sigma = 3.4$

8. $P(x \geq 120)$; $\mu = 100$; $\sigma = 15$

9. $P(x \geq 90)$; $\mu = 100$; $\sigma = 15$

10. $P(x \geq 2)$; $\mu = 3$; $\sigma = 0.25$

In Problems 11–20, find the z value described and sketch the area described.

11. Find $z \geq 0$ such that 45% of the standard normal curve lies between 0 and z.

12. Find $z \geq 0$ such that 47.5% of the standard normal curve lies between 0 and z.

13. Find $z \leq 0$ such that 42% of the standard normal curve lies between z and 0.

14. Find $z \leq 0$ such that 33% of the standard normal curve lies between z and 0.

15. Find z such that 6% of the standard normal curve lies to the left of z.

16. Find z such that 5.2% of the standard normal curve lies to the left of z.

17. Find z such that 8% of the standard normal curve lies to the right of z.

18. Find z such that 5% of the standard normal curve lies to the right of z.

19. Find the z value such that 98% of the standard normal curve lies between $-z$ and z.

20. Find the z value such that 95% of the standard normal curve lies between $-z$ and z.

Medical: Blood Glucose

21. The level of blood glucose and diabetes are closely related. Let x be a random variable measured in milligrams of glucose per deciliter ($\frac{1}{10}$ liter) of blood. After a 12-hour fast, the random variable x will have a distribution that is approximately normal with mean $\mu = 85$ and standard deviation $\sigma = 25$ (*Diagnostic Tests with Nursing Implications*, edited by S. Loeb, Springhouse Press). *Note:* After 50 years of age, both the mean and standard deviation tend to increase. What is the probability that for an adult (under 50 years old) after a 12-hour fast
(a) x is more than 60?
(b) x is less than 110?
(c) x is between 60 and 110?
(d) x is greater than 140 (borderline diabetes starts at 140)?

Medical: Blood Protoplasm

22. Porphyrin is a pigment in blood protoplasm and other body fluids that is significant in body energy and storage. Let x be a random variable that represents the number

of milligrams of porphyrin per deciliter of blood. In healthy adults, x is approximately normally distributed with mean $\mu = 38$ and standard deviation $\sigma = 12$ (see reference in Problem 21). What is the probability that

(a) x is less than 60?

(b) x is greater than 16?

(c) x is between 16 and 60?

(d) x is more than 60 (which may indicate an infection, anemia, or another type of illness)?

Education: SAT/ACT

23. For a given population of high school seniors, the Scholastic Aptitude Test (SAT) in mathematics has a mean score of 500 with a standard deviation of 100. Another widely used test is the American College Testing (ACT) exam. The mathematics portion of the ACT has a mean of 18 and a standard deviation of 6. (For more information, see Web site <http://www.collegeboard.org>.) Both SAT and ACT scores are normally distributed. What is the probability that a randomly selected high school senior's score on the mathematics part of the SAT will be

(a) more than 675?

(b) less than 450?

(c) between 450 and 675?

What is the probability that a randomly selected high school senior's score on the mathematics part of the ACT will be

(d) more than 28?

(e) more than 12?

(f) between 12 and 28?

Education: SAT/ACT

24. Please refer to the SAT and ACT information in Problem 23.

(a) Suppose that an engineering school honors program will accept only high school seniors with mathematics SAT or ACT scores in the top 10%. What is the minimum SAT score in mathematics for this program? What is the minimum ACT score in mathematics for this program?

(b) Suppose that an engineering school will accept only high school seniors with mathematics SAT or ACT scores in the top 20%. What is the minimum SAT score in mathematics for this program? What is the minimum ACT score in mathematics for this program?

(c) Suppose that an engineering school will accept only high school seniors with mathematics SAT or ACT scores in the top 60%. What is the minimum SAT score in mathematics for this program? What is the minimum ACT score in mathematics for this program?

Anthropology: Hopi Pottery

25. Thickness measurements of ancient prehistoric Native American pot shards discovered in a Hopi village were approximately normally distributed with a mean of 5.1 mm and a standard deviation of 0.9 mm. (Source: *Homol'oviII: Archaeology of an Ancestral Hopi Village, Arizona,* edited by E. C. Adams and K. A. Hays, University of Arizona Press.) For a randomly found shard, what is the probability that the thickness is

(a) less than 3.0 mm?

(b) more than 7.0 mm?

(c) between 3.0 mm and 7.0 mm?

Quality Control:
Guarantee Period

26. Accrotime is a company that manufactures quartz crystal watches. Accrotime researchers have shown that the watches have an average life of 28 months before certain electronic components deteriorate, causing the watch to become unreliable. The standard deviation of watch lifetimes is 5 months, and the distribution of lifetimes is normal.
   (a) If Accrotime guarantees a full refund on any defective watch for 2 years after purchase, what percentage of total production will the company expect to replace?
   (b) If Accrotime does not want to make refunds on more than 12% of the watches it makes, how long should the guarantee period be (to the nearest month)?

Quality Control:
Guarantee Period

27. Quick Start Company makes 12-volt car batteries. After many years of product testing, the company knows that the average life of a Quick Start battery is normally distributed, with a mean of 45 months and a standard deviation of 8 months.
   (a) If Quick Start guarantees a full refund on any battery that fails within the 36-month period after purchase, what percentage of its batteries will the company expect to replace?
   (b) If Quick Start does not want to replace more than 10% of its batteries under the full-refund guarantee policy, for how long should the company guarantee the batteries (to the nearest month)?

Sales: Product Replacement

28. *Consumer Reports* gave information about the ages at which various household products are replaced. For example, color TVs are replaced at an average age of $\mu = 8$ years after purchase, and the (95% of data) range was from 5 to 11 years. Thus the range was $11 - 5 = 6$ years. Let x be the age (years) at which a color TV is replaced. Assume that x has a distribution that is approximately normal.
   (a) Read the empirical rule (Section 6.1); then explain why approximately four standard deviations will cover a 95% range of data values centered at the mean. Since the range is 6 years, explain why 1.5 years would be a good approximation for the standard deviation of x values.
   (b) What is the probability that someone will keep a color TV more than 5 years before replacement?
   (c) What is the probability that someone will keep a color TV less than 10 years before replacement?
   (d) Assume that the average life of a color TV is 8 years with a standard deviation of 1.5 years before it breaks. Suppose that a company guarantees color TVs and will replace a TV that breaks while under guarantee with a new one. However, the company does not want to replace more than 10% of the TVs under guarantee. For how long should the guarantee be made (round to the nearest tenth of a year)?

Veterinary Science: Horses

29. The resting heart rate for an adult horse should average about $\mu = 46$ beats per minute with (95% of data) range from 22 to 70 beats per minute, based on information from *The Merck Veterinary Manual* (a classic reference used in most veterinary colleges). Let x be a random variable that represents the resting heart rate for an adult horse. Assume that x has a distribution that is approximately normal.
   (a) Estimate the standard deviation of the x distribution. (*Hint:* See Problem 28.)
   (b) What is the probability that the heart rate is less than 25 beats per minute?
   (c) What is the probability that the heart rate is greater than 60 beats per minute?

(d) What is the probability that the heart rate is between 25 and 60 beats per minute?

(e) A horse whose resting heart rate is in the upper 10% of the probability distribution of heart rates may have a secondary infection or illness that needs to be treated. What is the heart rate corresponding to the upper 10% cutoff point of the probability distribution?

Insurance: Satellites

30. A relay microchip in a telecommunications satellite has a life expectancy that follows a normal distribution with a mean of 90 months and a standard deviation of 3.7 months. When this computer-relay microchip malfunctions, the entire satellite is useless. A large London insurance company is going to insure the satellite for 50 million dollars. Assume that the only part of the satellite in question is the microchip. All other components will work indefinitely.

(a) For how many months should the satellite be insured to be 99% confident that it will last beyond the insurance date?

(b) If the satellite is insured for 84 months, what is the probability that it will malfunction before the insurance coverage ends?

(c) If the satellite is insured for 84 months, what is the expected loss to the insurance company?

(d) If the insurance company charges $3 million for 84 months of insurance, how much profit does the company expect to make?

General: Normal Curves

31. The figure below shows the graphs (Minitab-generated) for three normal distributions. The mean for each distribution is $\mu = 0$. The standard deviations are 1.5, 1, and 0.5, respectively.

(a) Using the color, identify which graph has $\sigma = 1.5$, which has $\sigma = 1$, and which has $\sigma = 0.5$

(b) Consider the normal curves between $x = -2$ and $x = 2$. Under which graph is the area the largest? The smallest?

(c) Consider the normal curves to the right of $x = 2$. Under which graph is this area the largest? The smallest?

Normal Distributions with mu = 0 and sigma = 1.5, 1, 0.5

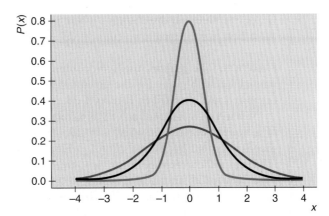

Section **6.4**   # Normal Approximation to the Binomial Distribution

The probability that a new vaccine will protect adults from cholera is known to be 0.85. It is administered to 300 adults who must enter an area where the disease is prevalent. What is the probability that more than 280 of these adults will be protected from cholera by the vaccine?

This question falls in the category of a binomial experiment with number of trials n equal to 300, the probability of success p equal to 0.85, and the number of successes r greater than 280. It is possible to use the formula for the binomial distribution to compute the probability that r is greater than 280. However, this approach would involve a number of tedious and long calculations. There is an easier way to do this problem, for under the conditions stated below the normal distribution can be used to approximate the binomial distribution.

Criteria $np > 5$ and $nq > 5$ Again, let p be the probability of success and let q be the probability of failure in a single binomial trial. Let n be the number of trials in the binomial experiment. If n, p, and q are such that *both* $np > 5$ and $nq > 5$, then the normal probability distribution with $\mu = np$ and $\sigma = \sqrt{npq}$ will be a good approximation to the binomial distribution. As n gets larger, the approximation becomes better.

SUMMARY Consider the binomial distribution with

n = number of trials

r = number of successes

p = probability of success

q = probability of failure = $1 - p$

If $np > 5$ and $nq > 5$
then r has a binomial distribution that is approximated by a *normal* distribution with

$$\mu = np \quad \text{and} \quad \sigma = \sqrt{npq}$$ ○

EXAMPLE 12 ❯   Graph the binomial distributions where $p = 0.25$, $q = 0.75$, and the number of trials is first $n = 3$, then $n = 10$, $n = 25$, and finally $n = 50$.

SOLUTION: The authors used the computer program ComputerStat to obtain the binomial distributions for the given values of p, q, and n. The results have been organized and graphed in Figures 6-35 to 6-38.

When $n = 3$, the outline of the histogram does not even begin to take the shape of a normal curve. But when $n = 10$, 25, or 50, it does begin to take a normal shape, indicated by the red curves in Figures 6-36 to 6-38 on page 301.

From a theoretical point of view, the histograms in Figures 6-36 to 6-38 would have bars for all values of r from $r = 0$ to $r = n$. However, in the construction of these histograms, the bars of height less than 0.001 unit have been omitted—that is, in this example probabilities less than 0.001 have been rounded to 0.   ●

FIGURE 6-35

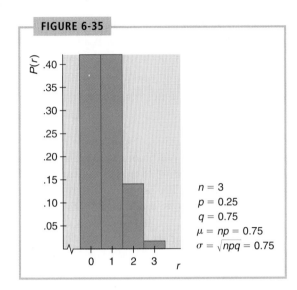

$n = 3$
$p = 0.25$
$q = 0.75$
$\mu = np = 0.75$
$\sigma = \sqrt{npq} = 0.75$

FIGURE 6-36

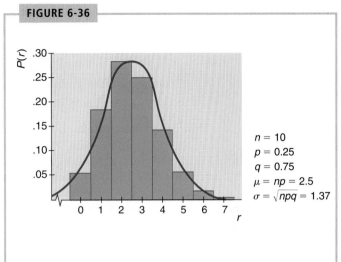

$n = 10$
$p = 0.25$
$q = 0.75$
$\mu = np = 2.5$
$\sigma = \sqrt{npq} = 1.37$

FIGURE 6-37

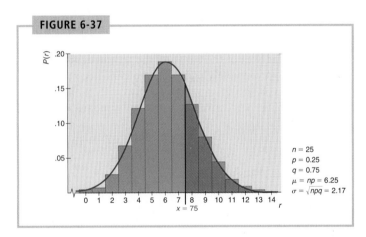

$x = 75$

$n = 25$
$p = 0.25$
$q = 0.75$
$\mu = np = 6.25$
$\sigma = \sqrt{npq} = 2.17$

FIGURE 6-38

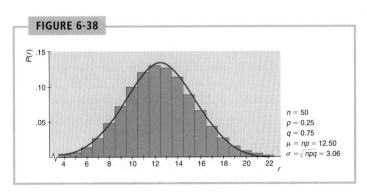

$n = 50$
$p = 0.25$
$q = 0.75$
$\mu = np = 12.50$
$\sigma = \sqrt{npq} = 3.06$

EXAMPLE 13 ❯

The owner of a new apartment building must install 25 water heaters. From past experience in other apartment buildings, she knows that Quick Hot is a good brand. It is guaranteed for 5 years only, but from her past experience she knows that the probability it will last 10 years is 0.25.

(a) What is the probability that 8 or more of the 25 water heaters will last at least 10 years?

SOLUTION: In this example, $n = 25$ and $p = 0.25$, so Figure 6-37 represents the probability distribution we will use. Let r be the binomial random variable corresponding to the number of successes out of $n = 25$ trials. We want to find $P(r \geq 8)$ by using the normal approximation. This probability is represented graphically (Figure 6-37) by the area of the bar over 8 and all bars to the right of the bar over 8.

Let x be a normal random variable corresponding to a normal distribution with $\mu = np = 25(0.25) = 6.25$ and $\sigma = \sqrt{npq} = \sqrt{25(0.25)(0.75)} \approx 2.17$. This normal curve is represented by the red line in Figure 6-37. The area under the normal curve from $x = 7.5$ to the right is approximately the same as the area of the bars from the bar over $r = 8$ to the right. It is important to notice that we start with $x = 7.5$ because the bar over $r = 8$ really starts at $x = 7.5$.

The area of the bars and the area under the corresponding red (normal) curve are approximately equal, so we conclude that $P(r \geq 8)$ is approximately equal to $P(x \geq 7.5)$.

When we convert $x = 7.5$ to standard units, we get

$$z = \frac{x - \mu}{\sigma} = \frac{7.5 - 6.25}{2.17} \approx 0.58$$

The probability we want is

$$P(x \geq 7.5) = P(z \geq 0.58) = 0.5000 - P(0 \leq z \leq 0.58)$$
$$= 0.5000 - 0.2190$$
$$= 0.2810$$

(b) How does the result in part a compare with the result we can obtain by using the formula for the binomial probability distribution with $n = 25$ and $p = 0.25$?

SOLUTION: Using the binomial distribution program of ComputerStat, the authors computed that $P(r \geq 8) = 0.2735$. This means that the probability is approximately 0.27 that 8 or more water heaters will last at least 10 years.

(c) How do the results of parts a and b compare?

SOLUTION: The error of approximation is the difference between the approximate value (0.2810) and the true value (0.2735). The error is only $0.2810 - 0.2735 = 0.0075$, which is negligible for most practical purposes.

We knew in advance that the normal approximation to the binomial probability would be good, since $np = 25(0.25) = 6.25$ and $nq = 25(0.75) = 18.75$ are both greater than 5. These are the conditions that assure us that the normal approximation will be sufficiently close to the binomial probability for most practical purposes.

Continuity correction: converting r values to x values

Remember that when using the normal distribution to approximate the binomial, we are computing the areas under bars. The bar over r goes from $r - 0.5$ to $r + 0.5$. If r is a *left* endpoint of an interval, we *subtract* 0.5 to get the corresponding normal variable x. If r is a *right* endpoint of an interval, we *add* 0.5 to get the corresponding variable x. For instance, $P(6 \leq r \leq 10)$, where r is a binomial variable, is approximated by $P(5.5 \leq x \leq 10.5)$, where x is the corresponding normal variable (see Figure 6-39). Adding or subtracting 0.5 from r to obtain a corresponding range of normal values x is called the *continuity correction*. It is needed because we are approximating a discrete probability distribution by a continuous probability distribution.

FIGURE 6-39

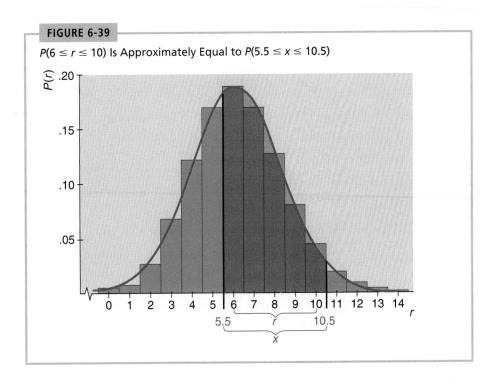

$P(6 \leq r \leq 10)$ Is Approximately Equal to $P(5.5 \leq x \leq 10.5)$

GUIDED EXERCISE 13

From many years of observation, a biologist knows the probability is only 0.65 that any given Arctic tern will survive the migration from its summer nesting area to its winter feeding grounds. A random sample of 500 Arctic terns were banded at their summer nesting area. Use the normal approximation to the binomial and the following steps to find the probability that between 310 and 340 of the banded Arctic terns will survive the migration. Let r be the number of surviving terns.

(a) To approximate $P(310 \leq r \leq 340)$, we use the normal curve with $\mu = \underline{\hspace{1cm}}$ and $\sigma = \underline{\hspace{1cm}}$.

⇨ We use the normal curve with
$$\mu = np = 500(0.65) = 325$$
$$\sigma = \sqrt{npq} = \sqrt{500(0.65)(0.35)} \approx 10.67$$

(b) $P(310 \leq r \leq 340)$ is approximately equal to $P(\underline{\hspace{1cm}} \leq x \leq \underline{\hspace{1cm}})$, where x is a variable from the normal distribution described in part a.

⇨ Since 310 is the left endpoint, we subtract 0.5, and since 340 is the right endpoint, we add 0.5. Consequently,
$$P(310 \leq r \leq 340) \approx P(309.5 \leq x \leq 340.5)$$

(c) Convert the condition $309.5 \leq x \leq 340.5$ to a condition in standard units.

⇨ Since $\mu = 325$ and $\sigma = 10.67$, the condition $309.5 \leq x \leq 340.5$ becomes
$$\frac{309.5 - 325}{10.67} \leq z \leq \frac{340.5 - 325}{10.67}$$
or $-1.45 \leq z \leq 1.45$

(d) $P(310 \leq r \leq 340) = P(309.5 \leq x \leq 340.5)$
$= P(-1.45 \leq z \leq 1.45)$
$= \underline{\hspace{1cm}}$

⇨ $P(-1.45 \leq z \leq 1.45) = 2P(0 \leq z \leq 1.45)$
$= 2(0.4265)$
$= 0.8530$

which is approximately the probability we were seeking.

(e) Will the normal distribution make a good approximation to the binomial for this problem? Explain your answer.

⇨ Since
$$np = 500(0.65) = 325 \quad \text{and}$$
$$nq = 500(0.35) = 175$$
are both greater than 5, the normal distribution will be a good approximation to the binomial.

VIEWPOINT

Sunspots, Tree Rings, and Statistics

Ancient Chinese astronomers recorded extreme sunspot activity with a peak around 1200 A.D. Mesa Verde tree rings in the period between 1276 and 1299 were unusually narrow, indicating a drought and/or a severe cold spell in the region at that time. A cooling trend could have narrowed the window of frost-free days below the approximately 80 days needed for cultivation of aboriginal corn and beans. Is this the reason the ancient Anasazi dwellings in Mesa Verde were abandoned? Is there a connection with the extreme sunspot activity? Much research and statistical work continue to be done on this topic.

Reference: *Prehistoric Astronomy in the Southwest,* by J. McKim Malville and C. Putnam, Department of Astronomy, University of Colorado.

SECTION 6.4 PROBLEMS

Note: When we say *between a* and *b,* we mean every value from *a* to *b, including a* and *b.*

Health Science: Lead

1. A decade ago, high levels of lead in the blood put 88% of children at risk. A concerted effort was made to remove lead from the environment. Now, according to the *Third National Health and Nutrition Examination Survey* (*NHANES III*) conducted by the Centers for Disease Control, only 9% of children in the United States are at risk of high blood-lead levels.
   (a) In a random sample of 200 children taken a decade ago, what is the probability that 50 or more had high blood-lead levels?
   (b) In a random sample of 200 children taken now, what is the probability that 50 or more have high blood-lead levels?

Insurance: Deductible

2. Do you try to *pad* an insurance claim to cover your deductible? About 40% of all U.S. adults will try to pad their insurance claims. (Source: *Are You Normal?,* by Bernice Kanner, St. Martin's Press.) Suppose that you are the director of an insurance adjustment office. Your office has just received 128 insurance claims to be processed in the next few days. What is the probability that
   (a) half or more of the claims have been padded?
   (b) less than 45 of the claims have been padded?
   (c) from 40 to 64 (including 40 and 64) of the claims have been padded?
   (d) more than 80 of the claims have *not* been padded?

Insurance: Accident Claims

3. What about those other drivers on the road? *Consumer Reports* (Vol. 64, No. 9) indicates that 11% of its readers who filed auto accident claims did so because they

were hit by uninsured or underinsured drivers. For a random sample of 300 auto accident claims, what is the probability that

(a) 50 or more are filed because of being hit by uninsured or underinsured motorists?

(b) 30 or fewer are filed for this reason?

(c) between 30 and 50 are filed for this reason?

(d) In the solution to this problem, what were n, p, and q? Does it appear that both np and nq are larger than 5? Why is this an important consideration?

Health: Eyesight

4. Afraid of the dark? Will a night light help? Maybe, but it may cause other problems. A recent study published in *Nature* indicates that 34% of the infants (under age 2) who sleep in rooms with night lights develop myopia or nearsightedness. For a random sample of 200 infants who sleep in rooms with night lights, what is the probability that

(a) at least 80 will develop myopia?

(b) fewer than 30 will develop myopia?

(c) between 30 and 80 will develop myopia?

(d) In the solution to this problem, what were n, p, and q? Does it appear that both np and nq are larger than 5? Why is this an important consideration?

Vital Statistics: 90th Birthday

5. It is estimated that 3.5% of the general population will live past their 90th birthdays (*Statistical Abstract of the United States*, 118th Edition). In a graduating class of 753 high school seniors, what is the probability that

(a) 15 or more will live beyond their 90th birthdays?

(b) 30 or more will live beyond their 90th birthdays?

(c) between 25 and 35 will live beyond their 90th birthdays?

(d) more than 40 will live beyond their 90th birthdays?

Fishing: Billfish

6. Ocean fishing for billfish is very popular in the Cozumel Region of Mexico. In *World Record Game Fishes* (published by the International Game Fish Association), it was stated that in the Cozumel Region about 44% of strikes (while trolling) resulted in catches. Suppose that on a given day a fleet of fishing boats got a total of 24 strikes. What is the probability that the number of fish caught was

(a) 12 or less?

(b) 5 or more?

(c) between 5 and 12 (including 5 and 12)?

(d) In the solution to this problem, what were n, p, and q? Does it appear that both np and nq are larger than 5? Why is this an important consideration?

Marketing: Grocery Stores

7. The *Denver Post* stated that 80% of all new products introduced in grocery stores fail (are taken off the market) within 2 years. If a grocery store chain introduces 66 new products, what is the probability that within 2 years

(a) 47 or more will fail?

(b) 58 or fewer will fail?

(c) 15 or more will succeed?

(d) fewer than 10 will succeed?

Crime: Murder

8. What are the chances that a person who was murdered actually knew the murderer? The answer to this question explains why a lot of police detective work begins with

relatives and friends of the victim. About 64% of the people who are murdered actually knew the persons who did the murders (*Chances: Risk and Odds in Everyday Life,* by James Burke). Suppose that a detective file in New Orleans has 63 current unsolved murders. What is the probability that
(a) at least 35 of the victims knew their murderers?
(b) at most 48 of the victims knew their murderers?
(c) fewer than 30 victims did *not* know their murderers?
(d) more than 20 victims did *not* know their murderers?

Lifestyle: Acquaintances

9. Old Friends Information Service is a California company that finds addresses for people who have lost track of each other. Old Friends claims to be 70% successful in reuniting people (*Wall Street Journal*). In December, Old Friends had 430 requests for addresses of lost acquaintances. What is the probability that the number of addresses found was
(a) more than 280?
(b) at least 320?
(c) between 280 and 320?
(d) In the solution to this problem, what were *n, p,* and *q*? Does it appear that both *np* and *nq* are larger than 5? Why is this an important consideration?

Anthropology: Pottery

10. Santa Fe black on white is a style of pottery that occurs in about 61% of the pot shards found in the Bandelier National Monument area (*Bandelier Archaeological Excavation Project: Summer 1990 Excavations at Burnt Mesa Pueblo,* edited by Kohler, Washington State University Department of Anthropology). At one excavation site, 8641 pot shards have been found that have not yet been cleaned and identified. What is the probability that
(a) less than 5200 are Santa Fe black on white?
(b) more than 5400 are Santa Fe black on white?
(c) between 5200 and 5400 are Santa Fe black on white?
(d) In the solution to this problem, what were *n, p,* and *q*? Does it appear that both *np* and *nq* are larger than 5? Why is this an important consideration?

Law: Bar Exam

11. Over the years, it has been observed that of all the lawyers who take the state bar exam, only 57% pass (information from the National Conference on Bar Examiners, referenced in *The Book of Odds,* by Shook and Shook, Signet). Suppose that this year 850 lawyers are going to take the Ohio bar exam. What is the probability that
(a) 540 or more will pass?
(b) 500 or fewer will pass?
(c) between 485 and 525 will pass?

Marketing: Coupons

12. More than 200 billion grocery coupons are distributed each year for discounts exceeding $84 billion. However, according to a report in *USA Today,* only 3.2% of the coupons are redeemed. If a company distributes 5000 coupons, what is the probability that
(a) more than 100 are redeemed?
(b) fewer than 200 are redeemed?
(c) between 100 and 200 are redeemed?

SUMMARY

In this chapter we have examined graphs of normal distributions, control charts, standard units, z scores, and areas under the standard normal curve.

It is important to remember that a normal distribution with mean μ and standard deviation σ can be transformed into the standard normal distribution with mean 0 and standard deviation 1 by the formula

$$z = \frac{x - \mu}{\sigma}$$

Values of z and Table 4 in Appendix I can be used to obtain the area under the standard normal curve from 0 to z (for positive z values). Other areas can be obtained from Table 4 in Appendix I and the symmetry of the normal curve. Areas under the normal curve represent probabilities that a z value will fall in the interval over which the area lies.

Given a probability associated with an interval on the standard normal curve, we can use Table 4 in Appendix I to obtain the associated z values. When x follows a normal distribution, we can find the x values associated with a given probability by (1) finding the corresponding z values and (2) converting to x values by using the formula

$$x = z\sigma + \mu$$

In the last section we saw that we can use the normal distribution to approximate the binomial distribution if $np > 5$ and $nq > 5$, where n is the number of trials, p is the probability of success on a single trial, and $q = 1 - p$.

IMPORTANT WORDS & SYMBOLS

Section 6.1
Normal distributions
Normal curves
Upward cup and downward cup on normal curves
Symmetry of normal curves
Empirical rule
Control chart
Out-of-control signals

Section 6.2
Standard units
z value or z score

Standard normal distribution ($\mu = 0$ and $\sigma = 1$)
Raw score, x
Areas under the standard normal curve

Section 6.3
Areas under any normal curve

Section 6.4
Normal approximation to the binomial distribution
Continuity correction

VIEWPOINT

Nenana Ice Classic

The Nenana Ice Classic is a betting pool offering a large cash prize to the lucky winner who can guess the time, to the nearest minute, of the ice breakup on the Tanana River at the town of Nenana, Alaska. Official breakup time is defined when the surging river dislodges a tripod on the ice. This breaks an attached line and stops a clock set to Yukon Standard Time. The event is so popular that the first state legislature of Alaska (1959) made the Nenana Ice Classic an official statewide lottery. Since 1918, the earliest breakup was April 20, 1940, at 3:27 P.M., and the latest recorded breakup was May 20, 1964, at 11:41 A.M. Want to make a statistical guess predicting when the ice will break up? Breakup times from 1918 to 1996 are recorded in *The Alaska Almanac,* published by Alaska Northwest Books, Anchorage.

CHAPTER REVIEW PROBLEMS

1. Given that z is the standard normal variable (with mean 0 and standard deviation 1), find
   (a) $P(0 \leq z \leq 1.75)$
   (b) $P(-1.29 \leq z \leq 0)$
   (c) $P(1.03 \leq z \leq 1.21)$
   (d) $P(z \geq 2.31)$
   (e) $P(z \leq -1.96)$
   (f) $P(z \leq 1.00)$

2. Given that z is the standard normal variable (with mean 0 and standard deviation 1), find
   (a) $P(0 \leq z \leq 0.75)$
   (b) $P(-1.50 \leq z \leq 0)$
   (c) $P(-2.67 \leq z \leq -1.74)$
   (d) $P(z \geq 1.56)$
   (e) $P(z \leq -0.97)$
   (f) $P(z \leq 2.01)$

3. Given that x is a normal variable with mean $\mu = 47$ and standard deviation $\sigma = 6.2$, find
   (a) $P(x \leq 60)$ (b) $P(x \geq 50)$ (c) $P(50 \leq x \leq 60)$

4. Given that x is a normal variable with mean $\mu = 110$ and standard deviation $\sigma = 12$, find
   (a) $P(x \leq 120)$ (b) $P(x \geq 80)$ (c) $P(108 \leq x \leq 117)$

5. Find z such that 5% of the area under the standard normal curve lies to the right of z.

6. Find z such that 1% of the area under the standard normal curve lies to the left of z.

7. Find z such that 95% of the area under the standard normal curve lies between $-z$ and z.

8. Find z such that 99% of the area under the standard normal curve lies between $-z$ and z.

Nursing: Licensing Exam

9. On a practical nursing licensing exam, the mean score is 79 and the standard deviation is 9 points.
   (a) What is the standardized score of a student with a raw score of 87?
   (b) What is the standardized score of a student with a raw score of 79?
   (c) Assuming the scores follow a normal distribution, what is the probability that a score selected at random is above 85?

Aptitude Test: Mechanics

10. On an auto mechanic aptitude test, the mean score is 270 points and the standard deviation is 35 points.

   (a) If a student has a standardized score of 1.9, how many points is that?
   (b) If a student has a standardized score of −0.25, how many points is that?
   (c) Assuming the scores follow a normal distribution, what is the probability that a student will get between 200 and 340 points?

Environment: Aluminum Cans

11. One environmental group did a study of recycling habits in a California community. They found that 70% of the aluminum cans sold in the area were recycled.
   (a) If 400 cans are sold today, what is the probability that 300 or more will be recycled?
   (b) Of the 400 cans sold, what is the probability that between 260 and 300 will be recycled?

Quality Control: CD Players

12. Future Electronics makes compact disc players. Its research department has found that the lives of the laser beam devices are normally distributed with mean 5000 hours and standard deviation 450 hours.
   (a) Find the probability that a laser beam device will wear out in 5000 hours or less.
   (b) Future Electronics wants to place a guarantee on the players so that no more than 5% fail during the guarantee period. Because the laser pickup is the part most likely to wear out first, the guarantee period will be based on the life of the laser beam device. How many playing hours should the guarantee cover? (Round to the next playing hour.)

Business: Delivery Service

13. Express Courier Service has found that the delivery times for packages are normally distributed with mean 14 hours and standard deviation 2 hours.
   (a) For a package selected at random, what is the probability that it will be delivered in 18 hours or less?
   (b) What should be the guaranteed delivery time on all packages in order to be 95% sure that a given package will be delivered within this time? (*Hint:* Note that 5% of the packages will *not* be delivered within the guarantee time period.)

Aircraft: Landing Gear

14. Hydraulic pressure in the main cylinder of the landing gear of a commercial jet is very important for a safe landing. If the pressure is not high enough, the landing gear may not lower properly. If it is too high, the connectors in the hydraulic line may spring a leak.

   In-flight landing tests show that the actual pressure in the main cylinders is a variable with mean 819 pounds per square inch and standard deviation 23 pounds per square inch. Assume that these values for the mean and standard deviation are considered safe values by engineers.

   (a) For nine consecutive test landings, the pressure in the main cylinder is recorded as follows:

Landing number	1	2	3	4	5	6	7	8	9
Pressure	870	855	830	815	847	836	825	810	792

Make a control chart for the pressure in the main cylinder of the hydraulic landing gear, and plot the data on the control chart. Looking at the control chart, would you say the pressure is "in control" or "out of control"? Explain your answer. Identify any out-of-control signals by type (I, II, or III).

(b) For 10 consecutive test landings, the pressure was recorded on another plane as follows:

Landing number	1	2	3	4	5	6	7	8	9	10
Pressure	865	850	841	820	815	789	801	765	730	725

Make a control chart and plot the data on the chart. Would you say that the pressure is "in control" or not? Explain your answer. Identify any out-of-control signals by type (I, II, or III).

Sales: Electronic Scanners

15. Instead of hearing the jingle of prices being rung up manually on cash registers, we now hear the beep of prices being scanned by electronic scanners. How accurate is price scanning? There are errors, and according to a *Denver Post* article, when the error occurs in the store's favor, it is larger than when it occurs in the customer's favor. An investigation of large discount stores by the Colorado state inspectors showed that the average error in the store's favor was $2.66. Assume that the distribution of scanner errors is more or less mound-shaped. If the standard deviation of scanner errors (in the store's favor) is $0.85, use the empirical rule to
    (a) estimate a range of scanner errors centered about the mean in which 68% of the errors will lie.
    (b) estimate a range of scanner errors centered about the mean in which 95% of the errors will lie.
    (c) estimate a range of scanner errors centered about the mean in which almost all the errors will lie.

Customer Service: Complaints

16. The Customer Service Center in a large New York department store has determined that the amount of time spent with a customer with a complaint is normally distributed with a mean of 9.3 minutes and a standard deviation of 2.5 minutes. What is the probability that for a randomly chosen customer with a complaint the amount of time spent resolving the complaint will be
    (a) less than 10 minutes?
    (b) more than 5 minutes?
    (c) between 8 and 15 minutes?

Medical: Blood Type

17. Blood type AB is found in only 3% of the population (*Textbook of Medical Physiology*, by A. Guyton, M.D.). If 250 people are chosen at random, what is the probability that
    (a) 5 or more will have this blood type?
    (b) between 5 and 10 will have this blood type?

Communication: Unlisted Numbers

18. How easy is it to get in contact with a person in Sacramento, California? If you don't know the telephone number, it could be difficult. The data from Survey Sam-

pling of Fairfield, Connecticut reported in *American Demographics* show that 68% of the telephone-owning households in Sacramento have unlisted numbers. For a random sample of 150 Sacramento households with telephones, what is the probability that

(a) 100 or more have unlisted numbers?

(b) fewer than 100 have unlisted numbers?

(c) between 50 and 65 (including 50 and 65) have *listed* numbers?

DATA HIGHLIGHTS: GROUP PROJECTS

Break into small groups and discuss the following topics. Organize a brief outline in which you summarize the main points of your group discussion.

1. Examine the following figure, "Median Age at First Marriage." Government documents and the Census Bureau show that the age at (first) marriage for U.S. citizens is approximately normally distributed for both men and women. For men, the average age is about 27 years, and for women, the average is about 24.5 years. For both sexes the standard deviation is about 2.5 years.

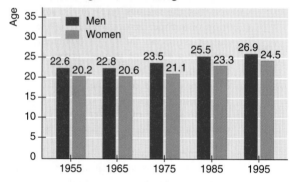

Median Age at First Marriage

(a) If the distribution is symmetrical and mound-shaped (such as the normal distribution), why would you expect the median, mean, and mode to be equal?

(b) Consider the age at first marriage in 1995. What is the probability that a man selected at random was over age 30 at the time of his (first) marriage? What is the probability that he was under age 20? What is the probability that he was between 20 and 30?

(c) Consider the age at first marriage in 1995. What is the probability that a woman selected at random was over age 28 at the time of her (first) marriage? What is the probability that she was under age 18? What is the probability that she was between 18 and 28?

(d) At what age were only 10% of eligible men (who had never been married before) left? At what age were only 5% left?

(e) At what age were only 10% of eligible women (who had never been married before) left? At what age were only 5% left?

(f) The Census Bureau tracks data regarding marriages and age at first marriage. The *Statistical Abstract of the United States* is published each year and contains tables giving this information. The information also may be available on the Census Bureau Web site at www.census.gov/. Look for marriage under the subject index, and follow the links to see if the table is available. Using either the *Abstract* or the Census Bureau Web site, find the most recent information on median age at first marriage. Although the median age changes from year to year, the standard deviation usually does not change much. Under the assumption that the age at first marriage follows a distribution that is approximately normal, the mean age at first marriage equals the median age. Using a standard deviation of about 2.5 years, repeat parts b through e for the most recent year for which median age at first marriage is available.

2. Examine the following figure, "Shopping by the clock" (*USA Today*).

Source: Copyright 1992, USA TODAY. Reprinted with permission.

(a) Notice that 52% of the people who go to shopping centers spend less than 1 hour. However, the figure also states that people spend an average of 69 minutes on each shopping center visit. How could *both* these claims be correct? Write a brief, complete essay in which you discuss mean, median, and mode in the context of symmetrical distribution. Also discuss mean, median, and mode for distributions that are skewed left and skewed right, and for general distributions. Then answer the question: How could 52% of the people spend less than 1 hour in the shopping center while the average (we don't know which average was used) time spent is 69 minutes?

(b) Ala Moana Shopping Center in Honolulu is sometimes advertised as the largest shopping center for 2500 miles and the best shopping center in the middle of the Pacific Ocean. There are many interesting Hawaiian, Asian, and other ethnic shops in Ala Moana, so this center is a favorite of tourists. Suppose that a tour group of 75 people has just arrived at the shopping center. The International Council of Shopping Centers indicates that 86% (52% plus 34%) of the people will spend 2 hours or less in a shopping center. Let us assume this applies to our tour group.

   (i)   For the tour group of 75 people, what is the expected number who finish shopping in 2 hours or less? What is the standard deviation?

   (ii) For the tour group of 75 people, what is the probability that 55 or more will finish shopping in 2 hours or less?

   (iii) For the tour group, what is the probability that 70 or more will finish shopping in 2 hours or less?

   (iv) What is the probability that between 50 and 70 (including 50 and 70) people in the tour group will finish shopping in 2 hours or less?

   (v)   If you are a tour director, how could you use this information to plan an appropriate length of time for the stop at Ala Moana shopping center? You do not want to frustrate your group by allowing too little time for shopping, but you also do not want to spend too much time at the center.

ⓛINKING CONCEPTS: WRITING PROJECTS

Discuss each of the following topics in class or review the topics on your own. Then write a brief but complete essay in which you summarize the main points. Please include formulas and graphs as appropriate.

1. If you look up the word *normal* in a dictionary, you will find it is synonymous with the words *standard* and *usual*. Consider the very wide and general applications of the normal probability distribution. Comment on why good synonyms for *normal probability distribution* might be *standard probability distribution* or *usual probability distribution*. List at least three random variables from everyday life for which you think the normal probability distribution could be applicable.

2. Why are standard *z* values so important? Is it true that *z* values have no units of measurement? Why would this be desirable for comparing data sets with *different* units of measurement? How can we assess differences in quality or performance by simply comparing *z* values under a standard normal curve? Examine the formula for computing standard *z* values. Notice that it involves *both* the mean and standard deviation. Recall that in Chapter 2 we commented that the mean of a data collection is not entirely adequate to describe the data; you need the standard deviation as well. Discuss this topic again in light of what you now know about normal distributions and standard *z* values.

3. If you look up the word *empirical* in a dictionary, you will find it means relying on experiment and observation rather than on theory. Discuss the empirical rule in this context. The empirical rule certainly applies to the normal distribution, but does it also apply to a wide variety of other distributions that are not *exactly* (theoretically) normal? Discuss the terms *mound-shaped* and *symmetrical*. Draw several sketches of distributions that are mound-shaped *and* symmetrical. Draw sketches of distributions that are not mound-shaped *or* not symmetrical. To which distributions will the empirical rule apply?

4. Most companies that manufacture products have divisions of quality control or quality assurance. The purpose of the quality control division is to make reasonably certain that the products manufactured are up to company standards. Write a brief essay in which you describe how the statistics you have learned so far could be applied to an industrial application (such as control charts and the Antlers Lodge example).

5. From Chapter 5 you no doubt remember we promised you a much better way to compute binomial probabilities when the number of trials is large.
   (a) Using the methods of Section 6.4, compute $P(r \geq 285)$ when $n = 500$ and $p = 0.6$.
   (b) Refer back to Problem 2(b) under Linking Concepts in Chapter 5. The preceding method of computing $P(r \geq 285)$ is much easier, isn't it?
   (c) The normal approximation to the binomial distribution is another example of the importance of the normal distribution in statistics. Including Chapter 6, list general applications of the normal distribution we have made so far in this text.

USING TECHNOLOGY

The problems in this section may be done using statistical computer software or calculators with statistical functions. Displays and suggestions are given for Minitab (Release 12), Excel, the TI-83 graphing calculator, and ComputerStat.

APPLICATIONS

1. The average earnings of city government employees in October 1995 are given for a sample of 18 cities in the United States (*Statistical Abstract of the United States,* 118th Edition). Use the data shown, or find similar data in a more recent edition of the *Abstract* to complete the following.

City	Average Earnings for October 1995
Baltimore	3010
Chicago	3757
Cleveland	2805
Dallas	2876
Detroit	2908
Houston	2459
Indianapolis	2490
Los Angeles	4133
Memphis	2679
Milwaukee	2793
New York	3454
Philadelphia	3222
Phoenix	3281
San Antonio	2539
San Diego	3472
San Francisco	3710
San Jose	4700
Washington, D.C.	3269

Source: United States Department of Commerce, Bureau of the Census. *Statistical Abstract of the United States,* 118th Ed. Washington: GPO, 1998.

(a) To compare the earnings from one city to another, we will look at z values for each city. Use the sample mean and sample standard deviation to compute the z values. You may do this "by hand" using a calculator, or you may use a computer software package.

In Minitab, enter the salaries in column C1. To generate z values and store the results in column C2, use the following menu choices: ➤**Calc** ➤**Standardize**. In the dialogue box, select C1 for input column and C2 for the output column. Then choose the option "subtract mean and divide by standard deviation." Minitab will compute the sample mean $\bar{x}$ and standard deviation s for the salaries and then compute z using the formula $z = (x - \bar{x})/s$.

In Excel, you need to compute the sample mean $\bar{x}$ and standard deviation s ahead of time. To do so, enter the salary data in column A. To compute $\bar{x}$, use the menu choices ➤**Paste function f$_x$** ➤**Statistical** ➤**AVERAGE**. To compute s, use the menu choices ➤**Paste function f$_x$** ➤**Statistical** ➤**STDEV**. Finally, to generate the z values, use the menu choices ➤**Paste function f$_x$** ➤**Statistical** ➤**STANDARDIZE**. Remember to highlight the cell in which you want each value to appear before you use the commands.

On the TI-83, enter the salary data in list L$_1$. Then use ➤**STAT** ➤**CALC** ➤**1-Var Stats** to compute the standard mean $\bar{x}$ and the standard deviation s. Then go back to ➤STAT ➤EDIT. Arrow up to the header label L_2. At the $L_2 =$ prompt, type in $(x - \bar{x})/s$ and press Enter. You will find the $\bar{x}$ and s symbols under the ➤VARS ➤Statistics . . . keys. The z values will then appear in list L_2.

(b) Look at the z values for each salary. Which are above average? Which are below average?

(c) Which salaries are within one standard deviation of the mean?

2. The standard normal probability distribution is very important in all of statistics. In Table 4 in Appendix I we have listed some standard normal probabilities. What if you wanted a more accurate table (that is, one with more significant digits displayed), with more entries? How could you use a computer to find probabilities in a standard normal table? The complete answer to this question is quite technical and requires mathematics beyond the scope of this text. However, the basic formulas are very accurate and can be used by anyone. These formulas can be found in the following reference:

> Abramowitz and Stegun, *Handbook of Mathematical Functions*. Washington: National Bureau of Standards, 1968.

We suggest that interested readers consult this reference.

(a) Determine whether your software or calculator produces areas under the standard normal curve, and then generate a brief table for z values $-3.5, -3, -2.5, -2, -1.5, -1, -0.5, 0, 0.5, 1, 1.5, 2, 2.5, 3,$ and 3.5.

In the case of software such as Minitab, Excel, and the TI-83 calculator, the technology generates areas under the normal curve. However, the area given is to the left of the specified z value, as shown in the figure.

Area to the Left of z

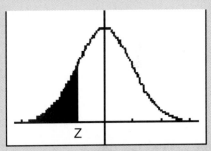

To use Minitab to generate the table of areas, enter the values -3.5 to 3.5 in increments of 0.5 in column C1. Then use the menu choices ➤Calc ➤Probability Distributions ➤Normal. In the resulting dialogue box, choose Cumulative Probability. For the standard normal distribution, use the default value of 0 for the mean and 1 for the standard deviation. Then select column C1 as the input column and list C2 for the optional storage. The cumulative area to the left of the specified z values will appear in column C2.

To use Excel to generate the table of areas, enter the values -3.5 to 3.5 in increments of 0.5 in column A. Select the first row of column B. Then use the menu choices ➤Paste function f_x ➤Statistical ➤NORMDIST. In the dialogue box, use A1 for x, 0 for the mean, 1 for the standard deviation, and type "true" in the box for Cumulative. Again, be sure to highlight the cell in which you want the result to appear before using the command.

On the TI-83, use the key choice ➤DISTR ➤Normalcdf (lower bound, upper bound, μ, σ). Use -10 for the lower bound, the z value for the upper bound, 0 for μ, and 1 for σ. Figure 6-31 in Section 6.3 shows a display of the basic command.

(b) Look at the entries in the technology-generated table of areas to the left of z under the standard normal distribution. Use Table 4 in Appendix I and the techniques shown in this chapter to find the areas to the left of $z = 1.50$, $z = -0.50$, $z = -2.50$, and $z = -3.00$. Compare these results with the results generated by your computer or calculator.

Computer and Calculator Displays

The following displays show you some examples of features relating to the normal distribution that are available through computer software packages and graphing calculators.

TI-83 Display

The TI-83 has several built-in probability density functions. Among them is one for normal distributions. We use this function to graph normal distributions. To begin, press the **Y=** key. Then press the **DISTR** key and select **normalpdf(X, μ, σ)**. Enter the

```
Plot1 Plot2 Plot3
\Y1 ∎normalpdf(X,0,
 .5)
\Y2 ∎normalpdf(X,0,1)
\Y3 ∎normalpdf(X,0,
 2)∎
\Y4 =
```

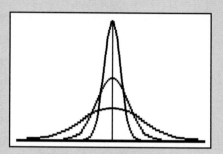

appropriate values of μ and σ. The display shows the graphs of three normal distributions, all with $\mu = 0$, the tallest with $\sigma = 0.5$, the middle with $\sigma = 1$, and the shortest with $\sigma = 2$. Set the window with Xmin $= -6$ and Xmax $= 6$, and then use **ZoomFit** from the **Zoom** key.

ComputerStat Display

A Binomial Distribution Suitable for Normal Approximation

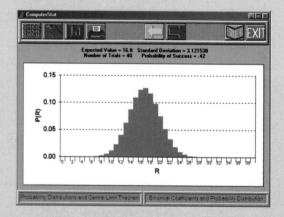

Use the menu option Binomial Coefficients and Probability Distributions. Enter 40 for the number of trials and 0.42 for the probability of success.

318

7 Introduction to Sampling Distributions

No one wants to learn by mistakes, but we cannot learn enough from success to go beyond the state of the art. . . . Such is the nature not only of science and engineering, but of all human endeavors.

—Henry Petroski

Experience isn't what happens to you. It's what you make out of what happens to you.

—Aldous Huxley

Aldous Huxley (1894–1963)

This British novelist and critic wrote about the theme of the human being confronted by the modern world.

Henry Petroski is a professor of engineering at Duke University, and Aldous Huxley is a well-known modern writer. Both seem to imply that experience, mistakes, information, and life itself are closely related. Life and uncertainty appear to be inseparable. Only those who are no longer living can escape chance happenings. Mistakes are bound to occur. However, not all mistakes are bad. The discovery of penicillin was a "mistake" when mold (penicillin) was accidentally introduced into a bacteria culture by Alexander Fleming in 1928.

Most of the really important decisions in life involve incomplete information. In one lifetime we simply cannot experience *everything*. Nor should we even want to. This is one reason why *experience by way of sampling* is so important. Statistics can help you have the experiences and yet maintain some control over mistakes. Remember, it is what you make out of experience (sample data) that is of real value.

In this chapter we will study how information from samples relates to information about populations. We cannot be certain that the information from a sample reflects corresponding information about the entire population, but we can describe likely differences. Study this and later chapters carefully. We believe that your effort will be rewarded by helping you appreciate the joy and wonder of living in an uncertain universe.

Section **7.1**

Sampling Distributions

Let us begin with some common statistical terms. Most of these terms have been discussed before, but this is a good time to review them.

From a statistical point of view, a *population* can be thought of as a set of measurements (or counts), either existing or conceptual. We discussed populations at some length in Chapter 1. A *sample* is a subset of measurements from the population. For our purposes, the most important samples are *random samples,* which were discussed in Section 1.2.

Parameter

A *parameter* is a numerical descriptive measure of a *population*. Examples of population parameters are the population mean μ, the population variance σ^2, the population standard deviation σ, and the population proportion of successes p in a binomial distribution.

Statistic

A *statistic* is a numerical descriptive measure of a *sample* (usually a random sample). Examples of statistics are the sample mean $\bar{x}$, the sample variance s^2, the sample standard deviation s, and the sample estimate $\hat{p} = r/n$ (read $\hat{p}$ as "p hat") for the proportion of successes in a binomial distribution. Notice that for a given population a specified parameter is a *fixed* quantity, while the statistic might vary depending on which sample has been selected.

Often we do not have access to all the measurements of an entire population because of constraints on time, money, effort, etc. So we must use measurements from a sample instead. In such cases, we will use a statistic (such as $\bar{x}$, s, or $\hat{p}$) to make *inferences* about corresponding population parameters (e.g., μ, σ, or p). The principal types of inferences we will make are

1. to *estimate* the value of a population parameter.

2. to formulate a *decision* about the value of a population parameter.

Sampling distribution

In order to evaluate the reliability of our inferences, we will need to know the probability distribution for the statistic we are using. Such a probability distribution is called a *sampling distribution*. Perhaps an example will help clarify this discussion.

EXAMPLE 1 ➤

Pinedale, Wisconsin is a rural community with a children's fishing pond. Posted rules say that all fish under 6 inches must be returned to the pond, only children under 12 years old may fish, and a limit of five fish may be kept per day. Susan is a college student who was hired by the community last summer to make sure the rules were obeyed and to see that the children were safe from accidents. The pond contains only rainbow trout and has been well stocked for many years. Each child has no difficulty catching his or her limit of five trout.

As a project for her biometrics class, Susan kept a record of the lengths (to the nearest inch) of all trout caught last summer. Hundreds of children visited the pond and caught their limit of five trout, so Susan has a lot of data. To make Table 7-1 (page 322), Susan selected 100 children at random and listed the length of each of the five trout caught by each child in the sample. Then, for each child, she listed the mean length of the five trout that child caught.

Now let us turn our attention to the following question: What is the average (mean) length of a trout taken from the Pinedale children's pond last summer?

SOLUTION: We can get an idea of the average length by looking at the far right column of Table 7-1. But just looking at 100 of the $\bar{x}$ values doesn't tell us much. Let's organize our $\bar{x}$ values into a frequency table. We used a class width of 0.38 to make Table 7-2 (page 323).

COMMENT Techniques of Section 2.3 dictate a class width of 0.4. However, this choice results in the tenth class being beyond the data. Consequently, we shortened the class width slightly and also started the first class with a value slightly lower than the smallest data value.

The far right column in Table 7-2 contains relative frequencies $f/100$. Recall that the relative frequencies may be thought of as probabilities, so we effectively have a probability distribution. Since $\bar{x}$ represents the mean length of trout (based on samples of five trout caught by each child), we estimate the probability of $\bar{x}$ falling into each class by using the relative frequencies. Figure 7-1 is a relative-frequency or probability distribution of the $\bar{x}$ values.

The bars in Figure 7-1 represent our estimated probabilities of $\bar{x}$ values based on the data in Table 7-1. The bell-shaped curve represents the theoretical probability distribution that would be obtained if the number of children (i.e., number of $\bar{x}$ values) were much larger.

Figure 7-1 represents a *probability sampling distribution* for the sample mean $\bar{x}$ of trout lengths based on random samples of size 5. We see that the distribution is mound-shaped and even somewhat bell-shaped. Irregularities are due to the small number of samples used (only 100 sample means) and the rather small sample size (5 trout per child). These irregularities would become less obvious and even disappear if the sample of children became much larger, if we used a larger number of classes in Figure 7-1, and if the number of trout used in each sample became larger. In fact, the curve would eventually become a perfect bell-shaped curve. We will discuss this property at some length in the next section, which introduces the *central limit theorem*.

Table 7-1 Length Measurements of Trout Caught by a Random Sample of 100 Children at the Pinedale Children's Pond

Sample	Length (to nearest inch)					$\bar{x}$ = Sample Mean	Sample	Length (to nearest inch)					$\bar{x}$ = Sample Mean
1	11	10	10	12	11	10.8	51	9	10	12	10	9	10.0
2	11	11	9	9	9	9.8	52	7	11	10	11	10	9.8
3	12	9	10	11	10	10.4	53	9	11	9	11	12	10.4
4	11	10	13	11	8	10.6	54	12	9	8	10	11	10.0
5	10	10	13	11	12	11.2	55	8	11	10	9	10	9.6
6	12	7	10	9	11	9.8	56	10	10	9	9	13	10.2
7	7	10	13	10	10	10.0	57	9	8	10	10	12	9.8
8	10	9	9	9	10	9.4	58	10	11	9	8	9	9.4
9	10	10	11	12	8	10.2	59	10	8	9	10	12	9.8
10	10	11	10	7	9	9.4	60	11	9	9	11	11	10.2
11	12	11	11	11	13	11.6	61	11	10	11	10	11	10.6
12	10	11	10	12	13	11.2	62	12	10	10	9	11	10.4
13	11	10	10	9	11	10.2	63	10	10	9	11	7	9.4
14	10	10	13	8	11	10.4	64	11	11	12	10	11	11.0
15	9	11	9	10	10	9.8	65	10	10	11	10	9	10.0
16	13	9	11	12	10	11.0	66	8	9	10	11	11	9.8
17	8	9	7	10	11	9.0	67	9	11	11	9	8	9.6
18	12	12	8	12	12	11.2	68	10	9	10	9	11	9.8
19	10	8	9	10	10	9.4	69	9	9	11	11	11	10.2
20	10	11	10	10	10	10.2	70	13	11	11	9	11	11.0
21	11	10	11	9	12	10.6	71	12	10	8	8	9	9.4
22	9	12	9	10	9	9.8	72	13	7	12	9	10	10.2
23	8	11	10	11	10	10.0	73	9	10	9	8	9	9.0
24	9	12	10	9	11	10.2	74	11	11	10	9	10	10.2
25	9	9	8	9	10	9.0	75	9	11	14	9	11	10.8
26	11	11	12	11	11	11.2	76	14	10	11	12	12	11.8
27	10	10	10	11	13	10.8	77	8	12	10	10	9	9.8
28	8	7	9	10	8	8.4	78	8	10	13	9	8	9.6
29	11	11	8	10	11	10.2	79	11	11	11	13	10	11.2
30	8	11	11	9	12	10.2	80	12	10	11	12	9	10.8
31	11	9	12	10	10	10.4	81	10	9	10	10	13	10.4
32	10	11	10	11	12	10.8	82	11	10	9	9	12	10.2
33	12	11	8	8	11	10.0	83	11	11	10	10	10	10.4
34	8	10	10	9	10	9.4	84	11	10	11	9	9	10.0
35	10	10	10	10	11	10.2	85	10	11	10	9	7	9.4
36	10	8	10	11	13	10.4	86	7	11	10	9	11	9.6
37	11	10	11	11	10	10.6	87	10	11	10	10	10	10.2
38	7	13	9	12	11	10.4	88	9	8	11	10	12	10.0
39	11	11	8	11	11	10.4	89	14	9	12	10	9	10.8
40	11	10	11	12	9	10.6	90	9	12	9	10	10	10.0
41	11	10	9	11	12	10.6	91	10	10	8	6	11	9.0
42	11	13	10	12	9	11.0	92	8	9	11	9	10	9.4
43	10	9	11	10	11	10.2	93	8	10	9	9	11	9.4
44	10	9	11	10	9	9.8	94	12	11	12	13	10	11.6
45	12	11	9	11	12	11.0	95	11	11	9	9	9	9.8
46	13	9	11	8	8	9.8	96	8	12	8	11	10	9.8
47	10	11	11	11	10	10.6	97	13	11	11	12	8	11.0
48	9	9	10	11	11	10.0	98	10	11	8	10	11	10.0
49	10	9	9	10	10	9.6	99	13	10	7	11	9	10.0
50	10	10	6	9	10	9.0	100	9	9	10	12	12	10.4

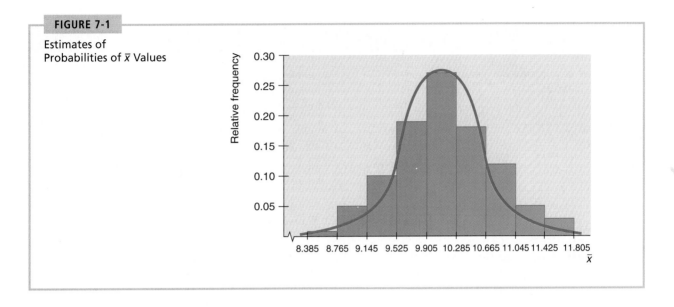

FIGURE 7-1

Estimates of
Probabilities of $\bar{x}$ Values

Table 7-2 Frequency Table for 100 Values of $\bar{x}$

Class	Class Limits Lower	Class Limits Upper	f = Frequency	$f/100$ = Relative Frequency
1	8.39	8.76	1	0.01
2	8.77	9.14	5	0.05
3	9.15	9.52	10	0.10
4	9.53	9.90	19	0.19
5	9.91	10.28	27	0.27
6	10.29	10.66	18	0.18
7	10.67	11.04	12	0.12
8	11.05	11.42	5	0.05
9	11.43	11.80	3	0.03

There are other sampling distributions besides the $\bar{x}$ distribution. In the chapters ahead we will see that other statistics have different sampling distributions. However, the $\bar{x}$ sampling distribution is very important. It will serve us well in our inferential work in Chapters 8 and 9 on estimation and testing.

Let us summarize the information about sampling distributions in the following exercise.

(a) What is a population parameter? Give an example.

⇨ A population parameter is a numerical descriptive measure of a population. Examples are μ, σ, and p. (There are many others.)

(b) What is a sample statistic? Give an example.

⇨ A sample statistic or a statistic is a numerical descriptive measure of a sample. Examples are $\bar{x}$, s, and $\hat{p}$.

(c) What is a sampling distribution?

⇨ A sampling distribution is a probability distribution for the sample statistic we are using.

(d) In Table 7-1, what makes up the members of the sample? What is the sample statistic corresponding to each sample? What is the sampling distribution? To which population parameter does this sampling distribution correspond?

⇨ There are 100 samples, each of which has five trout lengths. The first sample of five trout has lengths 11, 10, 10, 12, and 11. The sample statistic is the sample mean $\bar{x} = 10.8$. The sampling distribution is shown in Figure 7-1. This sampling distribution relates to the population mean μ of all lengths of trout taken from the Pinedale children's pond (i.e., trout over 6 inches long).

(e) Where will sampling distributions be used in our study of statistics?

⇨ Sampling distributions will be used for statistical inference. (Chapter 8 will concentrate on the method of inference called *estimation*. Chapter 9 will concentrate on a method of inference called *testing*.)

VIEWPOINT

"Chance Favors the Prepared Mind"
—Louis Pasteur

It also has been said that a discovery is nothing more than an accident that meets a prepared mind. Sampling can be one of the best forms of preparation. In fact, sampling may be the primary way we humans venture into the unknown. Probability sampling distributions can provide new information for the sociologist, scientist, or economist. In addition, ordinary human sampling of life can help writers and artists develop preferences, style, and insight. Ansel Adams became famous for photographing lyrical, unforgettable landscapes such as "Moonrise, Hernandez, New Mexico." Adams claimed that he was a strong believer in the quote by Pasteur. In fact, he claims that the Hernandez photograph was just such a favored chance happening that his prepared mind readily grasped. During his lifetime, Adams made over $25 million from sales and royalties on the Hernandez photograph.

This is a good time to review several important concepts, some of which we have studied earlier. Please write out a careful but brief answer to each of the following questions.

1. What is a population? Give three examples.

2. What is a random sample from a population? (*Hint:* See Section 1.2.)

3. What is a population parameter? Give three examples.

4. What is a sample statistic? Give three examples.

5. What is the meaning of the term *statistical inference*? What types of inferences will we make about population parameters?

6. What is a sampling distribution?

7. How do frequency tables, relative frequencies, and histograms using relative frequencies help us understand sampling distributions?

8. How can relative frequencies be used to help us estimate probabilities occurring in sampling distributions?

9. Give an example of a specific sampling distribution we studied in this section. Outline other possible examples of sampling distributions from areas such as business administration, economics, finance, psychology, political science, sociology, biology, medical science, sports, engineering, chemistry, linguistics, and so on.

Section **7.2**

The Central Limit Theorem

In Section 7.1 we began a study of the distribution of $\bar{x}$ values, where $\bar{x}$ was the (sample) mean length of five trout caught by children at the Pinedale children's fishing pond. Let's consider this example again in the light of a very important theorem of mathematical statistics.

Properties of $\bar{x}$ distribution, assuming x has normal distribution

THEOREM 7.1 Let x be a random variable with a *normal distribution* whose mean is μ and standard deviation is σ. Let $\bar{x}$ be the sample mean corresponding to random samples of size n taken from the x distribution. Then the following are true:

(a) The $\bar{x}$ distribution is a *normal distribution*.

(b) The mean of the $\bar{x}$ distribution is μ.

(c) The standard deviation of the $\bar{x}$ distribution is $\sigma/\sqrt{n}$.

We conclude from Theorem 7.1 that when x has a normal distribution, the $\bar{x}$ distribution will be normal *for any sample size n*. Furthermore, we can convert the $\bar{x}$ distribution to the standard normal z distribution using the following formulas.

$$\mu_{\bar{x}} = \mu$$

$$\sigma_{\bar{x}} = \frac{\sigma}{\sqrt{n}}$$

$$z = \frac{\bar{x} - \mu_{\bar{x}}}{\sigma_{\bar{x}}} = \frac{\bar{x} - \mu}{\sigma/\sqrt{n}} = \frac{\sqrt{n}(\bar{x} - \mu)}{\sigma}$$

where n is the sample size,
μ is the mean of the x distribution, and
σ is the standard deviation of the x distribution.

Theorem 7.1 is a wonderful theorem. It says that the $\bar{x}$ distribution will be normal provided the x distribution is normal. The sample size n could be 2, 3, 4, or any (fixed) sample size we wish. Furthermore, the mean of the $\bar{x}$ distribution is μ (same as for the x distribution), but the standard deviation is $\sigma/\sqrt{n}$ (which is, of course, smaller than σ). The next example illustrates Theorem 7.1.

EXAMPLE 2 ▷ Suppose that a team of biologists has been studying the Pinedale children's fishing pond. Let x represent the length of a single trout taken at random from the pond. This group of biologists has determined that x has a normal distribution with mean $\mu = 10.2$ inches and standard deviation $\sigma = 1.4$ inches.

(a) What is the probability that a *single trout* taken at random from the pond is between 8 and 12 inches long?

SOLUTION: We use the methods of Chapter 6 with $\mu = 10.2$ and $\sigma = 1.4$ to get

$$z = \frac{x - \mu}{\sigma} = \frac{x - 10.2}{1.4}$$

Therefore,

$$P(8 < x < 12) = P\left(\frac{8 - 10.2}{1.4} < z < \frac{12 - 10.2}{1.4}\right)$$

$$\approx P(-1.57 < z < 1.29)$$

$$= 0.4418 + 0.4015 = 0.8433$$

Therefore, the probability is about 0.8433 that a *single* trout taken at random is between 8 and 12 inches long.

(b) What is the probability that the *mean length* $\bar{x}$ of five trout taken at random is between 8 and 12 inches?

SOLUTION: If we let $\mu_{\bar{x}}$ represent the mean of the $\bar{x}$ distribution based on samples of size 5, then Theorem 7.1 part b tells us that

$$\mu_{\bar{x}} = \mu = 10.2$$

If $\sigma_{\bar{x}}$ represents the standard deviation of the $\bar{x}$ distribution based on samples of size 5, then Theorem 7.1 part c tells us that

$$\sigma_{\bar{x}} = \sigma/\sqrt{n} = 1.4/\sqrt{5} \approx 0.63$$

To create a standard z variable from $\bar{x}$, we subtract $\mu_{\bar{x}}$ and divide by $\sigma_{\bar{x}}$:

$$z = \frac{\bar{x} - \mu_{\bar{x}}}{\sigma_{\bar{x}}} = \frac{\bar{x} - \mu}{\sigma/\sqrt{n}} = \frac{\bar{x} - 10.2}{0.63}$$

To standardize the interval $8 < \bar{x} < 12$, we use 8 and then 12 in place of $\bar{x}$ in the preceding formula for z.

$$8 < \bar{x} < 12$$

$$\frac{8 - 10.2}{0.63} < z < \frac{12 - 10.2}{0.63}$$

$$-3.49 < z < 2.86$$

Theorem 7.1 part a tells us that $\bar{x}$ has a normal distribution. Therefore,

$$P(8 < \bar{x} < 12) \approx P(-3.49 < z < 2.86)$$

$$= 0.4998 + 0.4979 = 0.9977$$

The probability is about 0.9977 that the mean length based on a sample size of 5 is between 8 and 12 inches.

(c) Looking at the results of parts a and b, we see that the probabilities (0.8433 and 0.9977) are quite different. Why is this the case?

SOLUTION: According to Theorem 7.1, both x and $\bar{x}$ have normal distributions, and both have the same mean of 10.2 inches. The difference is in the standard deviation for x and $\bar{x}$. The standard deviation of the x distribution is $\sigma = 1.4$. The standard deviation of the $\bar{x}$ distribution is

$$s_{\bar{x}} = s/\sqrt{n} = 1.4/\sqrt{5} \approx 0.63$$

The standard deviation of $\bar{x}$ is less than half the standard deviation of x. Figure 7-2 (on page 328) shows the distributions of x and $\bar{x}$.

Looking at Figure 7-2(a) and (b), we see that both curves use the same scale on the horizontal axis. The means are the same, and the shaded area is above the interval from 8 to 12 on each graph. It becomes clear that the smaller standard deviation of the $\bar{x}$ distribution has the effect of gathering together much more of the total probability into the region over its mean. Therefore, the region from 8 to 12 has a much higher probability for the $\bar{x}$ distribution. ●

Standard error of the mean

Theorem 7.1 describes the distribution of a particular statistic: namely, the distribution of sample means $\bar{x}$. The standard deviation of a statistic is referred to as the *standard error* of that statistic. For the $\bar{x}$ sampling distribution, the standard error of $\bar{x}$ is $\sigma_{\bar{x}}$ or $\sigma/\sqrt{n}$. In other words, the standard error of the mean is $\sigma/\sqrt{n}$. In Minitab, the output

SE MEAN

FIGURE 7-2

General Shapes of the
x and $\bar{x}$ Distributions

(a) The x distribution with $\mu = 10.2$ and $\sigma = 1.4$

(b) The $\bar{x}$ distribution with $\mu_{\bar{x}} = 10.2$ and $\sigma_{\bar{x}} = 0.63$ for samples of size $n = 5$

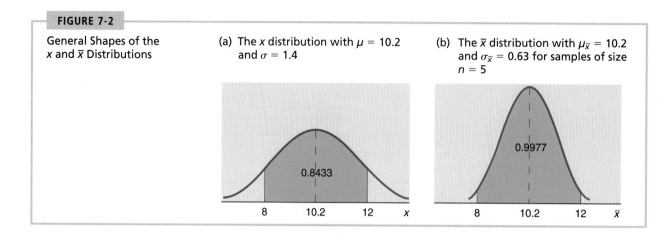

refers to the standard error of the $\bar{x}$ distribution. The expression *standard error* appears commonly on computer printouts and refers to the standard deviation of the sampling distribution being used.

Theorem 7.1 gives complete information about the $\bar{x}$ distribution, provided the original x distribution is known to be normal. What happens if we don't have information about the shape of the original x distribution? The *central limit theorem* tells us what to expect.

Central limit theorem

THEOREM 7.2 THE CENTRAL LIMIT THEOREM If x has *any* distribution with mean μ and standard deviation σ, then the sample mean $\bar{x}$ based on a random sample of size n will have a distribution that approaches the distribution of a normal random variable with mean μ and standard deviation $\sigma/\sqrt{n}$ as n increases without limit. ○

The central limit theorem is indeed surprising. It says that x can have *any* distribution whatever, but as the sample size gets larger and larger, the distribution of $\bar{x}$ will approach a *normal* distribution. From this relation, we begin to appreciate the scope and significance of the normal distribution.

In the central limit theorem, the degree to which the distribution of $\bar{x}$ values fits a normal distribution depends on both the selected value of n and the original distribution of x values. A natural question is: How large should the sample size be if we want to apply the central limit theorem? After a great deal of theoretical as well as empirical study, statisticians agree that if n is 30 or larger, the $\bar{x}$ distribution will appear to be normal and the central limit theorem will apply. However, this rule should not be applied blindly. If the x distribution is definitely not symmetrical about its mean, then the $\bar{x}$ distribution also will display a lack of symmetry. In such a case, a sample size larger than 30 may be required to get a reasonable approximation to the normal.

In practice, it is a good idea, when possible, to make a histogram of sample x values. If the histogram is approximately mound-shaped, and if it is more or less symmetrical, then we may be assured that, for all practical purposes, the $\bar{x}$ distribution will be well approximated by a normal distribution and the central limit theorem will apply when the sample size is 30 or larger. The main thing to remember is that in almost all practical applications, a sample size of 30 or more is adequate for the central limit theorem to hold. However, in a few rare applications you may need a sample size larger than 30 to get good results.

Let's summarize this information for convenient reference:

For almost all x distributions, if we use a random sample of size 30 or larger, the $\bar{x}$ distribution will be approximately normal, and the larger the sample size becomes, the closer the $\bar{x}$ distribution gets to the normal. Furthermore, we may convert the $\bar{x}$ distribution to a standard normal distribution using the formulas shown below.

$$\mu_{\bar{x}} = \mu$$

$$\sigma_{\bar{x}} = \frac{\sigma}{\sqrt{n}}$$

$$z = \frac{\bar{x} - \mu_{\bar{x}}}{\sigma_{\bar{x}}} = \frac{\bar{x} - \mu}{\sigma/\sqrt{n}} = \frac{\sqrt{n}(\bar{x} - \mu)}{\sigma}$$

where n is the sample size ($n \geq 30$),
μ is the mean of the x distribution, and
σ is the standard deviation of the x distribution.

GUIDED EXERCISE 2

(a) Suppose x has a *normal* distribution with mean $\mu = 18$ and standard deviation $\sigma = 3$. If we draw random samples of size 5 from the x distribution and $\bar{x}$ represents the sample mean, what can you say about the $\bar{x}$ distribution? How could you standardize the $\bar{x}$ distribution?

Since the x distribution is given to be *normal*, the $\bar{x}$ distribution also will be normal even though the sample size is much less than 30. The mean is $\mu_{\bar{x}} = \mu = 18$. The standard deviation is

$$\sigma_{\bar{x}} = \sigma/\sqrt{n} = 3/\sqrt{5} \approx 1.3$$

We could standardize $\bar{x}$ as follows:

$$z = \frac{\bar{x} - \mu}{\sigma/\sqrt{n}}$$

$$z = \frac{\bar{x} - 18}{1.3}$$

Exercise continues

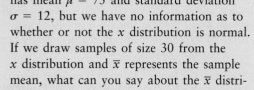

Exercise continued

(b) Suppose we know that the x distribution has mean $\mu = 75$ and standard deviation $\sigma = 12$, but we have no information as to whether or not the x distribution is normal. If we draw samples of size 30 from the x distribution and $\bar{x}$ represents the sample mean, what can you say about the $\bar{x}$ distribution? How could you standardize the $\bar{x}$ distribution?

⇨ Since the sample size is large enough, the $\bar{x}$ distribution will be approximately a normal distribution. The mean of the $\bar{x}$ distribution is

$$\mu_{\bar{x}} = \mu = 75$$

The standard deviation of the distribution is

$$\sigma_{\bar{x}} = \sigma/\sqrt{n} = 12/\sqrt{30} \approx 2.2$$

We could standardize $\bar{x}$ as follows:

$$z = \frac{\bar{x} - \mu}{\sigma/\sqrt{n}} \approx \frac{\bar{x} - 75}{2.2}$$

(c) Suppose you did not know that x had a normal distribution. Would you be justified in saying that the $\bar{x}$ distribution is approximately normal if the sample size was $n = 8$?

⇨ No, the sample size should be 30 or larger if we don't know that x has a normal distribution.

Now let's look at an example that demonstrates the use of the central limit theorem in a decision-making process.

EXAMPLE 3 ❯ A certain strain of bacteria occurs in all raw milk. Let x be the bacteria count per milliliter of milk. The health department has found that if the milk is not contaminated, then x has a distribution that is more or less mound-shaped and symmetrical. The mean of the x distribution is $\mu = 2500$, and the standard deviation is $\sigma = 300$. In a large commercial dairy, the health inspector takes 42 random samples of the milk produced each day. At the end of the day, the bacteria counts in the 42 samples are averaged to obtain the sample mean bacteria count $\bar{x}$.

(a) Assuming that the milk is not contaminated, what is the distribution of $\bar{x}$?

SOLUTION: The sample size is $n = 42$. Since this value exceeds 30, the central limit theorem applies, and we know that $\bar{x}$ will be approximately normal with mean

$$\mu_{\bar{x}} = \mu = 2500$$

and standard deviation

$$\sigma_{\bar{x}} = \sigma/\sqrt{n} = 300/\sqrt{42} \approx 46.3$$

(b) Assuming that the milk is not contaminated, what is the probability that the average bacteria count $\bar{x}$ for one day is between 2350 and 2650 bacteria per milliliter?

SOLUTION: We convert the interval

$$2350 \le \bar{x} \le 2650$$

to a corresponding interval on the standard z *axis*.

$$z = \frac{\bar{x} - \mu}{\sigma/\sqrt{n}} \approx \frac{\bar{x} - 2500}{46.3}$$

$\bar{x} = 2350$ converts to $z = \dfrac{2350 - 2500}{46.3} \approx -3.24$

$\bar{x} = 2650$ converts to $z = \dfrac{2650 - 2500}{46.3} \approx 3.24$

Therefore,

$$
\begin{aligned}
P(2350 \le \bar{x} \le 2650) &\approx P(-3.24 \le z \le 3.24) \\
&= 2P(0 \le z \le 3.24) \\
&= 2(0.4994) \\
&= 0.9988
\end{aligned}
$$

The probability is 0.9988 that $\bar{x}$ is between 2350 and 2650.

(c) At the end of each day, the inspector must decide to accept or reject the accumulated milk that has been held in cold storage awaiting shipment. Suppose that the 42 samples taken by the inspector have a mean bacteria count $\bar{x}$ that is *not* between 2350 and 2650. If you were the inspector, what would be your comment on this situation?

SOLUTION: The probability that $\bar{x}$ is between 2350 and 2650 is very high. If the inspector finds that the average bacteria count for the 42 samples is not between 2350 and 2650, then it is reasonable to conclude that there is something wrong with the milk. If $\bar{x}$ is less than 2350, you might suspect someone added chemicals to the milk to artificially reduce the bacteria count. If $\bar{x}$ is above 2650, you might suspect some other kind of biologic contamination.

GUIDED EXERCISE   3

In mountain country, major highways sometimes use tunnels instead of long, winding roads over high passes. However, too many vehicles in a tunnel at the same time can cause a hazardous situation. Traffic engineers are studying a long tunnel in Colorado. If x represents the time for a vehicle to go through the tunnel, it is known that the x distribution has mean $\mu = 12.1$ minutes and standard deviation $\sigma = 3.8$ minutes under ordinary traffic conditions. From a histogram of x values, it was found that the x distribution is mound-shaped with some symmetry about the mean.

Exercise continues

> **Exercise continued**
>
> Engineers have calculated that, *on average,* vehicles should spend from 11 to 13 minutes in the tunnel. If the time is less than 11 minutes, traffic is moving too fast for safe travel in the tunnel. If the time is more than 13 minutes, there is a problem of bad air (too much carbon monoxide and other pollutants).
>
> Under ordinary conditions, there are about 50 vehicles in the tunnel at any one time. What is the probability that the mean time for 50 vehicles in the tunnel will be from 11 to 13 minutes?
>
> We will answer this question in steps.
>
> (a) Let $\bar{x}$ represent the sample mean based on samples of size 50. Describe the $\bar{x}$ distribution.   ⇨   From the central limit theorem we expect the $\bar{x}$ distribution to be approximately normal with mean
>
> $$\mu_{\bar{x}} = \mu = 12.1$$
>
> and standard deviation
>
> $$\sigma_{\bar{x}} = \frac{\sigma}{\sqrt{n}} = \frac{3.8}{\sqrt{50}} \approx 0.54$$
>
> (b) Find $P(11 < \bar{x} < 13)$.   ⇨   We convert the interval
>
> $$11 < \bar{x} < 13$$
>
> to a standard z interval and use the standard normal probability table to find our answer. Since
>
> $$z = \frac{\bar{x} - \mu}{\sigma/\sqrt{n}} \approx \frac{\bar{x} - 12.1}{0.54}$$
>
> then $\bar{x} = 11$ converts to
>
> $$z = \frac{11 - 12.1}{0.54} \approx -2.04$$
>
> and $\bar{x} = 13$ converts to
>
> $$z = \frac{13 - 12.1}{0.54} \approx 1.67$$
>
> Therefore,
>
> $$P(11 < \bar{x} < 13) \approx P(-2.04 < z < 1.67)$$
> $$= 0.4793 + 0.4525$$
> $$= 0.9318$$
>
> (c) Comment on your answer for part b.   ⇨   It would seem that about 93% of the time there should be no safety hazard for average traffic flow.

V I E W P O I N T

Chaos!

Is there a different side to random sampling? Can sampling be used as a weapon? According to the *Wall Street Journal*, the answer could be yes. The acronym for **C**reate **H**avoc **A**round **O**ur **S**ystem is *CHAOS*. The Association of Flight Attendants (AFA) is a union that successfully used **CHAOS** against Alaska Airlines in 1994 as a negotiation tool. **CHAOS** involves a small sample of random strikes—a few flights at a time—instead of a mass walkout. The president of the AFA claims that by striking randomly, "we take control of the schedule." The entire schedule becomes unreliable, and that is something management cannot tolerate. In 1986, TWA flight attendants struck in a mass walkout, and all were permanently replaced. Using **CHAOS**, only a few jobs are put at risk, and these are usually not lost. It appears that random sampling can be used as a weapon.

SECTION 7.2 PROBLEMS

In these problems, the word *average* refers to the arithmetic mean $\bar{x}$ or μ, as appropriate.

General: Exploring Sampling Distributions

1. Suppose that x has a distribution with $\mu = 15$ and $\sigma = 14$.
   (a) If a random sample of size $n = 49$ is drawn, find $\mu_{\bar{x}}, \sigma_{\bar{x}}$, and $P(15 \le \bar{x} \le 17)$.
   (b) If a random sample of size $n = 64$ is drawn, find $\mu_{\bar{x}}, \sigma_{\bar{x}}$, and $P(15 \le \bar{x} \le 17)$.
   (c) Why should you expect the probability of part b to be higher than that of part a? (*Hint:* Consider the standard deviations in parts a and b.)

General: Exploring Sampling Distributions

2. Suppose that x has a distribution with $\mu = 100$ and $\sigma = 48$.
   (a) If a random sample of size $n = 81$ is drawn, find $\mu_{\bar{x}}, \sigma_{\bar{x}}$, and $P(92 \le \bar{x} \le 100)$.
   (b) If a random sample of size $n = 121$ is drawn, find $\mu_{\bar{x}}, \sigma_{\bar{x}}$, and $P(92 \le \bar{x} \le 100)$.
   (c) Again, comment on the differences in the probabilities in parts a and b. Why do you expect the differences?

General: Exploring Sampling Distributions

3. Suppose that x has a distribution with $\mu = 25$ and $\sigma = 3.5$.
   (a) If random samples of size $n = 9$ are selected, can we say anything about the $\bar{x}$ distribution of sample means?
   (b) If the original x distribution is normal, can we say anything about the $\bar{x}$ distribution from samples of size $n = 9$? Find $P(23 \le \bar{x} \le 26)$.

General: Graphical Interpretation

4. (a) The figure at the top of the following page (Minitab-generated) shows two sampling distributions for $\bar{x}$ taken from the same normal population of x values. One distribution came from samples of size 49 and the other from samples of size 100. Which curve goes with sample size 49? Which goes with sample size 100?

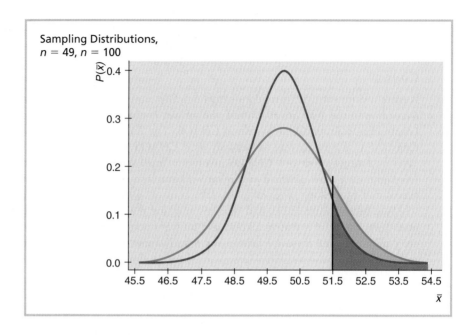

Sampling Distributions,
$n = 49$, $n = 100$

(b) Consider the area under each graph to the right of 51.5. For which sample size is the area to the right of 51.5 greater?

(c) Consider the statement "The smaller the sample size, the more likely it is that the sample mean will be greater than a specified value to the right of the mean μ of the population from which the sample is taken." Do you think this statement is true based on your answer to part b? Explain.

Transportation: Coal

5. Coal is carried from a mine in West Virginia to a power plant in New York in hopper cars on a long train. The automatic hopper car loader is set to put 75 tons of coal in each car. The actual weights of coal loaded into each car are *normally distributed* with mean $\mu = 75$ tons and standard deviation $\sigma = 0.8$ ton.

   (a) What is the probability that one car chosen at random will have less than 74.5 tons of coal?

   (b) What is the probability that 20 cars chosen at random will have a mean load weight $\bar{x}$ of less than 74.5 tons of coal?

   (c) Suppose that the weight of coal in one car was less than 74.5 tons. Would that fact make you suspect that the loader had slipped out of adjustment? Suppose the weight of coal in 20 cars selected at random had an average $\bar{x}$ less than 74.5 tons. Would that fact make you suspect that the loader had slipped out of adjustment? Why?

**Vital Statistics:
Heights of Men**

6. The heights of 18-year-old men are approximately *normally distributed*, with mean 68 inches and standard deviation 3 inches (based on information from *Statistical Abstract of the United States*, 118th Edition).

   (a) What is the probability that an 18-year-old man selected at random is between 67 and 69 inches tall?

   (b) If a random sample of nine 18-year-old men is selected, what is the probability that the mean height $\bar{x}$ is between 67 and 69 inches?

(c) Compare your answers for parts a and b. Was the probability in part b much higher? Why would you expect this?

Medical: Blood Glucose

7. Let x be a random variable that represents level of glucose in the blood (milligrams per deciliter of blood) after a 12-hour fast. Assume for people under 50 years old that x has a distribution that is approximately normal with mean $\mu = 85$ and an estimated standard deviation $\sigma = 25$ (based on information from *Diagnostic Tests with Nursing Applications*, edited by S. Loeb, Springhouse). A test result $x < 40$ is an indication of severe excess insulin, and medication is usually prescribed.
   (a) What is the probability that on a single test $x < 40$?
   (b) Suppose that a doctor uses the average $\bar{x}$ for two tests taken about a week apart. What can we say about the probability distribution of $\bar{x}$? (*Hint:* See Theorem 7.1.) What is the probability that $\bar{x} < 40$?
   (c) Repeat part b for $n = 3$ tests taken a week apart.
   (d) Repeat part b for $n = 5$ tests taken a week apart.
   (e) Compare your answers for parts a, b, c, and d. Did the probabilities decrease as n increased? Explain what this might mean if you were a doctor or a nurse. If a patient had a test result of $\bar{x} < 40$ based on five tests, explain why you could assume that either this was an extremely rare event or (more likely) the person had a case of excess insulin.

Medical: White Blood Cells

8. Let x be a random variable that represents white blood cell count per cubic millimeter of whole blood. Assume that x has a distribution that is approximately normal with mean $\mu = 7500$ and estimated standard deviation $\sigma = 1750$ (see reference in Problem 7). A test result of $x < 3500$ is an indication of leukopenia. This indicates bone marrow depression that may be the result of a viral infection.
   (a) What is the probability that on a single test $x < 3500$?
   (b) Suppose that a doctor uses the average $\bar{x}$ for two tests taken about a week apart. What can we say about the probability distribution of $\bar{x}$? What is the probability that $\bar{x} < 3500$?
   (c) Repeat part b for $n = 3$ tests taken a week apart.
   (d) Compare your answers for parts a, b, and c. How did the probabilities change as n increased? If a person had $\bar{x} < 3500$ based on three tests, what conclusion would you draw as a doctor or a nurse?

Environmental Studies: Deer

9. Let x be a random variable that represents weights in kilograms (kg) of healthy adult female deer (does) in December in Mesa Verde National Park. Then x has a distribution that is approximately normal with mean $\mu = 63.0$ kg and standard deviation $\sigma = 7.1$ kg. (Source: *The Mule Deer of Mesa Verde National Park*, by G. W. Mierau and J. L. Schmidt, Mesa Verde Museum Association.) Suppose that a doe that weighs less than 54 kg is considered undernourished.
   (a) What is the probability that a single doe captured at random in December (weighed and released) is undernourished?
   (b) If the park has about 2200 does, what number do you expect to be undernourished in December?
   (c) To estimate the health of the December doe population, park rangers use the rule that the average weight of $n = 50$ does should be more than 60 kg. If the average weight is less than 60 kg, it is thought that the entire population of does might be undernourished. What is the probability the average weight $\bar{x}$ for a random sample of 50 does is less than 60 kg (assume a healthy population)?

(d) Compute the probability that $\bar{x} < 64.2$ kg for 50 does (assume a healthy population). Suppose park rangers captured, weighed, and released 50 does in December, and the average weight was $\bar{x} = 64.2$ kg. Do you think the doe population is undernourished or not? Explain.

Finance: Mutual Funds

10. Templeton World is a mutual fund that invests in both U.S. and foreign markets. Let x be a random variable that represents the monthly percentage return for the Templeton World fund. Based on information from the *Morningstar Guide to Mutual Funds* (available in most libraries), x has mean $\mu = 1.6\%$ and standard deviation $\sigma = 0.9\%$.

(a) Templeton World fund has over 250 stocks that combine together to give the overall monthly percentage return x. We can consider the monthly return of the stocks in the fund to be a sample from the population of monthly returns of all world stocks. Then we see that the overall monthly return x for Templeton World fund is itself an average return computed using all 250 stocks in the fund. Why would this indicate that x has an approximately normal distribution? Explain. (*Hint:* See discussion following Theorem 7.2 on page 328.)

(b) After 6 months, what is the probability that the *average* monthly percentage return $\bar{x}$ will be between 1 and 2%? (*Hint:* See Theorem 7.1, and assume that x has a normal distribution as based on part a.)

(c) After 2 years, what is the probability that $\bar{x}$ will be between 1 and 2%?

(d) Compare your answers for parts b and c. Did the probability increase as n (number of months) increased? Why would you expect this to happen?

(e) If after 2 years the average monthly percentage return $\bar{x}$ was less than 1%, would that tend to shake your confidence in the statement that $\mu = 1.6\%$? Might you suspect that μ had slipped below 1.6%? Explain.

General:
Sampling Distributions

11. (a) If we have a distribution of x values that is more or less mound-shaped and somewhat symmetrical, what is the size of sample needed in order to claim that the distribution of sample means $\bar{x}$ from random samples of that size is approximately normal?

(b) If the original distribution of x values is known to be normal, do we need to make any restriction about sample size in order to claim that the distribution of sample means $\bar{x}$ taken from random samples of a given size is normal?

Section 7.3   **Class Project Illustrating the Central Limit Theorem**

As we have seen in the preceding two sections, the value of a sample statistic such as $\bar{x}$ can vary from sample to sample. The central limit theorem describes the distribution of the sample statistic $\bar{x}$ when samples are sufficiently large (usually of size 30 or more).

In this section we explore a sampling distribution of $\bar{x}$ values computed for samples drawn from the uniform distribution of digits 0 through 9. The in-class project follows.

STEP 1: Generate Random Samples of Specified Size n from Digits 0 Through 9

Have each student use the random-number table (Table 1 in Appendix I) or a random-number generator on a calculator or computer to select a random sample of $n = 5$ digits from the list 0, 1, 2, 3, 4, 5, 6, 7, 8, 9. In our samples we allow repetition of digits. Therefore, 3, 6, 1, 6, and 8 would be a random sample of size $n = 5$ digits. Be sure each student uses the same sample size. It is best to have at least 50 samples, so ask each student to create an appropriate number of samples so that there are at least 50 samples in all. It is important that students using a random-number table select the starting place in the table at random.

STEP 2: Compute the Sample Mean $\bar{x}$ of the Digits in Each Sample

Have each student compute $\bar{x}$ for each of his or her samples of size n.

STEP 3: Compute the Sample Mean of the Means (i.e., $\bar{x}_{\bar{x}}$)

Make a list of all the values of the sample means computed in step 2. Compute the mean of this list. Notice that this number represents the mean of the means. Also compute the standard deviation s of the list of sample means. Notice the variation in the sample means. How likely is it that any single sample provides a perfect estimate for the population mean of digits from 0 through 9? (Note that the population mean μ of the digits from 0 through 9 is 4.5.) In Chapter 8 we will see how to use the information from *one* sample as well as information about the distribution of sample means $\bar{x}$ to estimate the population parameter μ.

STEP 4: Compare the Sample Distribution of $\bar{x}$ Values with a Normal Distribution Having the Mean and Standard Deviation Computed in Step 3.

(a) Use the values of $\bar{x}$ and s computed in step 3 to create the intervals shown in column 1 of Table 7-3.

(b) Again have each student read his or her mean (from step 2). Tally the sample means computed by the students in step 2 to determine how many fall in each interval of column 2. Then compute the percent of data in each interval and record the results in column 3.

Table 7-3 Frequency Table of Sample Means

1. Interval	2. Frequency	3. Percent	4. Hypothetical Normal Distribution
$\bar{x} - 3s$ to $\bar{x} - 2s$	Tally the sample	Compute	2 or 3%
$\bar{x} - 2s$ to $\bar{x} - s$	means computed	percents	13 or 14%
$\bar{x} - s$ to $\bar{x}$	by the students	from	About 34%
$\bar{x}$ to $\bar{x} + s$	in step 2 and	column 2	About 34%
$\bar{x} + s$ to $\bar{x} + 2s$	place here.	and place	13 or 14%
$\bar{x} + 2s$ to $\bar{x} + 3s$		here.	2 or 3%

(c) The percentages listed in column 4 are those from a normal distribution (see Figure 6-5 on page 261 showing the empirical rule). Compare the percentages in column 3 with those in column 4. How do the sample percentages compare with the hypothetical normal distribution?

STEP 5: Create a Histogram

Using the frequency table in step 4, construct a histogram of the sample means that has six bars. Some bars may be of zero height depending on how your table turned out. Looking at your histogram, would you say it is approximately mound-shaped and symmetrical? Does it seem to give the general outline of a normal curve?

STEP 6: Determine How This Project Relates to the Central Limit Theorem

As an outside class project, have students repeat steps 1 and 2 for sample sizes $n = 10, 20, 30,$ and 40. For each sample size, ask one student to gather the resulting $\bar{x}$ data and compile the mean and standard deviation of the sample means (i.e., complete step 3). For each sample size, ask another student to create a frequency table as in step 4. For each sample size, ask another student to create the histogram as in step 5. Have students present the final results in class. Finally, have a class discussion about the $\bar{x}$ distributions for different sample sizes. Do the results tend to illustrate the central limit theorem? What are some of the limitations of this experiment?

Computer Demonstration   For samples of different sizes, the basic process outlined in steps 1 to 6 can be done much more quickly on a computer. Using ComputerStat, select the menu option Probability Distributions and Central Limit Theorem. Then choose Central Limit Theorem Demonstration. Simply follow screen instructions and the computer will do the rest. Note that the computer samples from all real numbers between 0 and 9, not just integers. Variations on this project for Minitab and Excel are shown in the Using Technology section of this chapter.

In Sections 7.2 and 7.3 we have looked at the sampling distributions for a mean $\bar{x}$. In Chapter 8 we will examine the sampling distribution of a proportion $\hat{p}$.

ⓈECTION 7.3 PROBLEMS

1. Consider the experiment of rolling a die. The possible outcomes are 1, 2, 3, 4, 5, and 6. Create random samples that simulate the experiment of rolling the die $n = 10$ times. Perform steps 1–5 of the class project for this experiment.

2. Based on the class project and your knowledge of the central limit theorem, write an essay on the statement: "The larger your sample size, the more likely it is that the sample mean $\bar{x}$ is close to the population mean μ of the distribution."

SUMMARY

Sampling distributions give us the basis for inferential statistics. By studying the distribution of sample statistics, we can learn about a population parameter.

The central limit theorem describes the sampling distribution of sample means taken from samples of size n. It tells us that for increasing sample size n, the distribution of sample means $\bar{x}$ approaches a normal distribution with mean $\mu_{\bar{x}} = \mu$ and standard deviation $\sigma_{\bar{x}} = \sigma/\sqrt{n}$. The values of μ and σ are the population mean and standard deviation, respectively, of the original x distribution.

IMPORTANT WORDS & SYMBOLS

Section 7.1
Population parameter
Statistic
Sampling distribution

Section 7.2
Central limit theorem
Standard error of the mean
$\mu_{\bar{x}}$
$\sigma_{\bar{x}}$

VIEWPOINT

Why Wait? Apply Now for a College Loan!

The cost of education is high. The cost of not having an education is higher. What can you expect? What about tuition and student fees? What about room and board? What is the total cost for 1 year at college? Perhaps some averages based on random samples of colleges could be useful. For more information, see Web site <http://www.usnews.com/>. Use the Index and follow the links to Best Values—Stiker Prices for the geographic regions of the colleges of interest.

CHAPTER REVIEW PROBLEMS

Lifestyle: Sleep

1. Let x be a random variable representing the amount of sleep each adult in New York City got last night. Consider a sampling distribution of sample means $\bar{x}$.
   (a) As the sample size becomes increasingly large, what distribution does the $\bar{x}$ distribution approach?
   (b) As the sample size becomes increasingly large, what value will the mean $\mu_{\bar{x}}$ of the $\bar{x}$ distribution approach?
   (c) What value will the standard deviation $\sigma_{\bar{x}}$ of the sampling distribution approach?
   (d) How do the two $\bar{x}$ distributions for sample sizes $n = 50$ and $n = 100$ compare?

General:
Sampling Distributions

2. If x has a normal distribution with mean $\mu = 15$ and standard deviation $\sigma = 3$, describe the distribution of $\bar{x}$ values for sample size n, where $n = 4$, $n = 16$, and $n = 100$. How do the $\bar{x}$ distributions compare for the various sample sizes?

Management: Personnel

3. The personnel office at a large electronics firm regularly schedules job interviews and maintains records of the interviews. From the past records, they have found that the length of a first interview is normally distributed with mean $\mu = 35$ minutes and standard deviation $\sigma = 7$ minutes.
(a) What is the probability that a first interview will last 40 minutes or longer?
(b) Nine first interviews are usually scheduled per day. What is the probability that the average length of time for the nine interviews will be 40 minutes or longer?

Medical: Muscle Relaxant

4. A new muscle relaxant is available. Researchers at the firm developing the relaxant have done studies indicating that the time lapse between administration of the drug and beginning effects of the drug is normally distributed with mean $\mu = 38$ minutes and standard deviation $\sigma = 5$ minutes.
(a) The drug is administered to one patient selected at random. What is the probability that the time it takes to go into effect is 35 minutes or less?
(b) The drug is administered to a random sample of 10 patients. What is the probability that the average time before it is effective for all 10 patients is 35 minutes or less?
(c) Comment on the differences of the results in parts a and b.

Psychology: IQ

5. Assume that IQ scores are normally distributed with a standard deviation of 15 points and a mean of 100 points. If 100 people are chosen at random, what is the probability that the sample mean of IQ scores will not differ from the population mean by more than 2 points?

Climate: Temperatures

6. Let x be a random variable that represents daily high temperatures (in degrees Fahrenheit) in January. The following information is based on a report from the U.S. Department of Commerce Environmental Data Services. For Miami, Florida, the mean of the x distribution is $\mu = 76$ and the standard deviation is approximately $\sigma = 1.9$. For Fairbanks, Alaska, the mean of the x distribution is $\mu = 0$ with approximate standard deviation $\sigma = 5.3$. Assume that x has a normal distribution.
(a) For one day chosen at random in January, what is the probability that the high temperature in Miami will be less than 77°F? What is the probability the high temperature in Fairbanks will be less than 3°F?
(b) If we choose $n = 7$ days in January, what can we say about the probability distribution of $\bar{x}$, the average high temperature? What is the probability that $\bar{x}$ is less than 77°F for Miami? Less than 3°F for Fairbanks?
(c) Suppose that we cannot assume that x has a normal distribution, but we can say that the distribution is approximately symmetrical and mound-shaped. In this case, what can we say about the $\bar{x}$ probability distribution? If we use all 31 days in January, what is the probability that $\bar{x} < 77$°F in Miami? That $\bar{x} < 3$°F in Fairbanks?

DATA HIGHLIGHTS: GROUP PROJECTS

Break into small groups and discuss the following topics. Organize a brief outline in which you summarize the main points of your group discussion.

Iris setosa is a beautiful wildflower that is found in such diverse places as Alaska, the Gulf of St. Lawrence, much of North America, and even English meadows and parks. R. A. Fisher, with his colleague Dr. Edgar Anderson, studied these flowers extensively. Dr. Anderson described how he collected information on irises:

> I have studied such irises as I could get to see, in as great detail as possible, measuring iris standard after iris standard and iris fall after iris fall, sitting squat-legged with record book and ruler in mountain meadows, in cypress swamps, on lake beaches, and in English parks. [Anderson, E., "The Irises of the Gaspé Peninsula," *Bulletin, American Iris Society, 59*:2–5, 1935]

The data in the table below were collected by Dr. Anderson and were published by his friend and colleague R. A. Fisher in a paper entitled "The Use of Multiple Measurements in Taxonomic Problems" (*Annals of Eugenics,* part II, 179–188, 1936). These data are also available at Web site <http://lib.stat.cmu.edu/DASL/> under famous data sets.

Let x be a random variable representing petal length. Using the TI-83 calculator, it was found that the sample mean is $\bar{x} = 1.46$ cm and the sample standard deviation is $s = 0.17$ cm. The figure below shows a histogram for the given data generated on a TI-83 calculator.

(a) Examine the histogram for petal lengths. Would you say that the distribution is approximately mound-shaped and symmetrical? Our sample has only 50 irises; if many thousands of irises had been used, do you think that the distribution would have looked even more like a normal curve? Let x be the petal length of *Iris setosa*. Research has shown that x has an approximately normal distribution with mean $\mu = 1.5$ cm and standard deviation $\sigma = 0.2$ cm.

Petal Length in Centimeters for *Iris setosa*

1.4	1.4	1.3	1.5	1.4
1.7	1.4	1.5	1.4	1.5
1.5	1.6	1.4	1.1	1.2
1.5	1.3	1.4	1.7	1.5
1.7	1.5	1	1.7	1.9
1.6	1.6	1.5	1.4	1.6
1.6	1.5	1.5	1.4	1.5
1.2	1.3	1.4	1.3	1.5
1.3	1.3	1.3	1.6	1.9
1.4	1.6	1.4	1.5	1.4

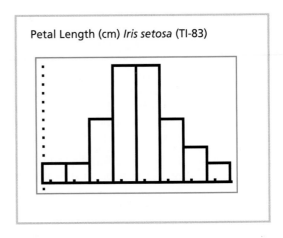

Petal Length (cm) *Iris setosa* (TI-83)

(b) Use the empirical rule with $\mu = 1.5$ and $\sigma = 0.2$ to get an interval in which approximately 68% of the petal lengths will fall. Repeat this for 95% and 99.7%. Examine the raw data and compute the percentage of the raw data that actually falls in each of these intervals (the 68% interval, the 95% interval, and the 99.7% interval). Compare your computed percentages with those given by the empirical rule.

(c) Compute the probability that a petal length is between 1.3 and 1.6 cm. Compute the probability that the petal length is greater than 1.6 cm.

(d) Suppose that a random sample of 30 irises is obtained. Compute the probability that the average petal length for this sample is between 1.3 and 1.6 cm. Compute the probability that the average petal length is greater than 1.6 cm.

(e) Compare your answers for parts c and d. Do you notice any differences? Why would you expect these differences?

LINKING CONCEPTS: WRITING PROJECTS

Discuss each of the following topics in class or review the topics on your own. Then write a brief but complete essay in which you summarize the main points. Please include formulas and graphs as appropriate.

1. Most people would agree that increased information should give better predictions. Discuss how sampling distributions actually accomplish better predictions by using more information. Examine Theorem 7.1 again. Suppose that x is a random variable with a *normal* distribution. Then $\bar{x}$, the sample mean based on random samples of size n, also will have a normal distribution for *any* value of $n = 1, 2, 3, \ldots$.

   What happens to the standard deviation of the $\bar{x}$ distribution as n (the sample size) increases? Consider the following table for different values of n.

n	1	2	3	4	10	50	100
$\sigma/\sqrt{n}$	1σ	0.71σ	0.58σ	0.50σ	0.32σ	0.14σ	0.10σ

   In this case, "increased information" means a larger sample size n. Give a brief explanation of why a large *standard deviation* usually results in poor statistical predictions, whereas a small standard deviation usually results in much better predictions. Since the standard deviation of the sampling distribution $\bar{x}$ is $\sigma/\sqrt{n}$, we can decrease the standard deviation by increasing n. In fact, if we look at the preceding table, we see that if we use a sample of size $n = 4$, we cut the standard deviation of $\bar{x}$ to 50% of the standard deviation σ of x. If we were to use a sample of size $n = 100$, we would cut the standard deviation of $\bar{x}$ to 10% of the standard deviation σ of x.

Give the preceding discussion some thought and explain why you expect to get much better predictions for μ by using $\bar{x}$ from a sample of size n rather than by just using x. Write a brief essay in which you explain why sampling distributions are an important tool in statistics.

2. In a way, the central limit theorem can be thought of as a kind of "grand central station." It is a connecting hub or center for a great deal of statistical work. We will use it extensively in Chapters 8, 9, and 10. Put in a very elementary way, it says that as the sample size n increases, the mean $\bar{x}$ will always approach a normal distribution no matter where the original x variable came from. For most people, it is the complete generality of the central limit theorem that is so awe-inspiring: it applies to practically everything. List and discuss at least three variables from everyday life for which you expect that the variable x itself does *not* follow a normal or bell-shaped distribution. Then discuss what would happen to the sampling distribution $\bar{x}$ as the sample size increased. Sketch diagrams of the $\bar{x}$ distributions as the sample size n increases.

Computer Displays

ComputerStat

The program Central Limit Theorem Demonstration in ComputerStat parallels the activities of the class project in Section 7.3. The program draws 100 random samples of size n from the uniform distribution of real numbers in the interval from 0 to 9. The user may select the sample size n to be any integer between 2 and 50. The program then computes the mean of each of the 100 samples and shows the distribution of sample means. The percentages of sample means in the intervals marked by 1, 2, and 3 standard deviations from the mean of the means are compared with the percentages predicted by a normal distribution with mean $\mu_{\bar{x}}$ and $\sigma_{\bar{x}}$. Finally, a histogram of the sample means is presented. The figure below shows the graph that is generated when samples of size $n = 10$ are used.

You can use the program Central Limit Theorem Demonstration to do the class project in Section 7.3. The only difference is that the ComputerStat program samples from all real numbers between 0 and 9 rather than sampling only from integers between 0 and 9.

Minitab

One way to demonstrate the central limit theorem through Minitab is to follow a four-step procedure.

FIGURE 7-3
ComputerStat Sampling Distribution Display

(a) $n = 10$

(b) $n = 35$

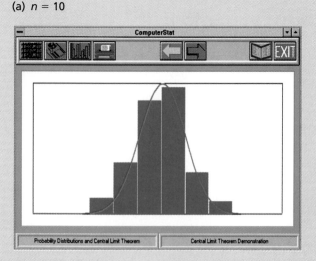

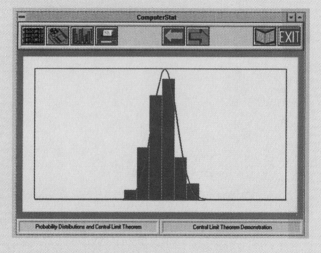

344

Step 1: Generate 500 random samples of size 5 from a normal distribution with mean $\mu = 10$ and standard deviation $\sigma = 2$. Menu selection: **➤Calc ➤Random Data ➤Normal.** In the dialogue box, specify 500 as the number of rows to be generated. Store them in columns C1–C5. Use 10 as the mean and 2 as the standard deviation. Note that you may vary the number of random samples, the size of the samples, and the distribution from which you are sampling.

Step 2: Compute the mean for each of the 500 samples. Use the menu selection **➤Calc ➤Row Statistics.** In the dialogue box, select the mean, use columns C1–C5 for the input, and store the results in column C6.

Step 3: Create a histogram of the means in column C6. Use the menu selection **➤Graph ➤Histogram.** In the dialogue box, use C6 for the X values of graph 1, keep the option of bars, and keep the option of midpoints.

Step 4: Find the sample mean and sample standard deviation of the distribution of sample means that are stored in column C6. Use the menu selection **➤Stat ➤Basic Statistics ➤Display Descriptive Statistics.** In the dialogue box, select C6 for the variable.

Once you have completed the four steps, compare the sample results with the predicted theoretical results. By the theory, the distribution of $\bar{x}$ values from samples of size 5 taken from a normal distribution with mean $\mu = 10$ and standard deviation $\sigma = 2$ should have a mean of 10 and a standard deviation of $2/\sqrt{5}$, or about 0.894.

FIGURE 7-4

Minitab Display

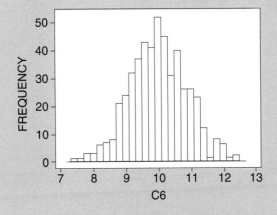

Descriptive Statistics

VARIABLE	N	MEAN	MEDIAN	TRMEAN	STDEV	SEMEAN
C6	500	9.9928	9.9841	9.9949	0.8653	0.0387

VARIABLE	Minimum	Maximum	Q1	Q3
C6	7.4100	12.4520	9.3991	10.5870

345

Excel Display

In Excel we can also demonstrate the central limit theorem by means of a four-step procedure.

Step 1: Generate 500 random samples of size 5 from a normal distribution with mean $\mu = 10$ and standard deviation $\sigma = 2$. Use menu selection ➤**Tools** ➤**Data Analysis** ➤**Random Number Generator.** In the dialogue box, declare 5 variables and 500 random numbers, and select normal distribution with mean 10 and standard deviation 2. List the output range as A2:E501.

Step 2: Compute the mean of each row of 5 numbers. Highlight the F2 position. Use the menu selection ➤**Paste function f_x** ➤**Statistical** ➤**Average.** In the dialogue box, use numbers A2:E2. Then use the cursor to fill the entire F column.

Step 3: Create a histogram. Use the menu choices ➤**Tools** ➤**Data Analysis** ➤**Histogram.** The input columns are F1:F501. Put the output in a new workbook and move the graph back to your data sheet.

Step 4: Find the sample mean and sample standard deviation of the distribution of sample means that are stored in column F. Label "mean =" in G1 and "sta dev =" in G2. Highlight H1 and use menu choices ➤**Paste function f_x** ➤**Statistical** ➤**Average** for the data in F2:F501. Highlight H2 and use menu choices ➤**Paste function f_x** ➤**Statistical** ➤**Stadev** for the data in F2:F501.

FIGURE 7-5
Excel Display

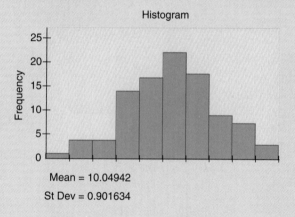

Mean = 10.04942

St Dev = 0.901634

Estimation

We dance round in a ring and suppose,
But the Secret sits in the middle and knows.

—Robert Frost,
"The Secret
Sits"*

**Robert Lee Frost
(1874–1963)**

This celebrated American poet
drew poetic symbols largely from
common experiences observed in
his rural New England

In Chapter 1 we said that statistics is the study of how to collect, organize, analyze, and interpret numerical data. That part concerned with analysis, interpretation, and forming conclusions about the source of the data is called *statistical inference*. Problems of statistical inference require us to draw a *sample* of observations from a larger *population*. A sample usually contains incomplete information, so in a sense we must "dance round in a ring and suppose." Nevertheless, conclusions about the population can be obtained from sample data by use of statistical estimates. This chapter introduces you to several widely used methods of estimation.

Section 8.1

Estimating μ with Large Samples

An estimate of a population parameter given by a single number is called a *point estimate* of that parameter. It should be no great surprise that we use $\bar{x}$ (the sam-

*Source: From *The Poetry of Robert Frost*, edited by Edward Connery Lathem. Copyright 1942 by Robert Frost. Copyright © 1969 by Henry Holt and Company, Inc. Copyright © 1970 by Lesley Frost Ballantine. Reprinted by permission of Henry Holt and Company, Inc.

Basic terminology

ple mean) as a point estimate for μ (the population mean) and s (the sample standard deviation) as a point estimate for σ (the population standard deviation). In this section we will discuss estimates of μ and σ from large samples ($n \geq 30$).

Statistical theory and empirical results show that if a distribution is approximately mound-shaped and symmetrical, then when the sample size is 30 or larger, we are safe, for most practical purposes, if we estimate σ by s. The error resulting from taking the population standard deviation σ to be equal to the sample standard deviation s is negligible. However, when the sample size is less than 30, we will use special small sample methods, which we will study in Section 8.2.

> For large samples of size $n \geq 30$,
>
> $$\sigma \approx s$$
>
> is a good estimate, for most practical purposes.

Using $\overline{x}$ to estimate μ is not quite so simple, even when we have a large sample size. The *error of estimate* is the magnitude of the difference between the point estimate and the true parameter value. Using $\overline{x}$ as a point estimate for μ, the error of estimate is the magnitude of $\overline{x} - \mu$. If we use absolute value notation, we can indicate the error of estimate for μ by the notation $|\overline{x} - \mu|$.

We cannot say exactly how close $\overline{x}$ is to μ when μ is unknown. Therefore, the exact error of estimate is unknown when the population parameter is unknown. Of course, μ is usually not known or there would be no need to estimate it. In this section we will use the language of probability to give us an idea of the size of the error of estimate when we use $\overline{x}$ as a point estimate of μ.

First, we need to learn about *confidence levels*. The reliability of an estimate will be measured by the confidence level.

Finding the critical value

Suppose we want a confidence level of c (see Figure 8-1). Theoretically, you can choose c to be any value between 0 and 1, but usually c is equal to a number such

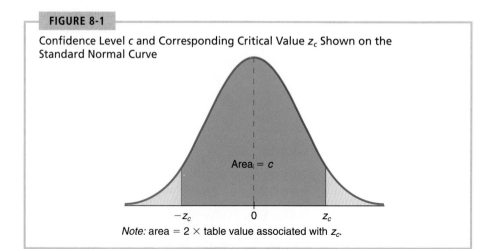

FIGURE 8-1

Confidence Level c and Corresponding Critical Value z_c Shown on the Standard Normal Curve

Area = c

$-z_c$ 0 z_c

Note: area = 2 × table value associated with z_c.

as 0.90, 0.95, or 0.99. In each case, the value z_c is the number such that the area under the standard normal curve falling between $-z_c$ and z_c is equal to c. The value z_c is called the *critical value* for a confidence level of c.

The area under the normal curve from $-z_c$ to z_c is the probability that the standardized normal variable z lies in that interval. This means that

$$P(-z_c < z < z_c) = c$$

EXAMPLE 1 ▶

Let us use Table 4 in Appendix I to find a number $z_{0.95}$ such that 95% of the area under the standard normal curve lies between $-z_{0.95}$ and $z_{0.95}$. That is, we will find $z_{0.95}$ such that

$$P(-z_{0.95} < z < z_{0.95}) = 0.95$$

SOLUTION: Table 4 in Appendix I gives the area under the normal curve from the mean 0 to any point z. The condition $P(-z_{0.95} < z < z_{0.95}) = 0.95$ is the same as the condition $2P(0 < z < z_{0.95}) = 0.95$, since the standard normal curve is symmetrical about the mean 0. When we divide both sides of the last equation by 2, we get

$$P(0 < z < z_{0.95}) = \frac{0.95}{2} = 0.4750$$

Note that 0.4750 is an entry in Table 4 in Appendix I. Table 8-1 is an excerpt from that table.

We will use Table 8-1 to find $z_{0.95}$. From Table 8-1 we see that $z_{0.95} = 1.96$, so the probability is 0.95 that the standardized statistic z lies between -1.96 and 1.96. In symbols, we have

$$P(-1.96 < z < 1.96) = 0.95$$

●

Table 8-1   Areas Under the Standard Normal Curve from 0 to z (Excerpt from Table 4, Appendix I)

z	0.00	0.01	0.02	0.03	0.04	0.05	0.06	0.07 . . .
0.0								
.								
.								
.								
1.8	0.4641	0.4649	0.4656	0.4664	0.4671	0.4678	0.4686	0.4693
1.9	0.4713	0.4719	0.4726	0.4732	0.4738	0.4744	0.4750	0.4756

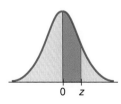

Area is 0.4750 for
$z = 1.96$

GUIDED EXERCISE 1

(a) Is it true that the condition

$$P(-z_{0.99} < z < z_{0.99}) = 0.99$$

is equivalent to the condition

$$2P(0 < z < z_{0.99}) = 0.99?$$

Why?

⟹ It is true that the conditions are equivalent because the standard normal curve is symmetrical about its mean 0.

(b) Use the information in part a and Table 4 in Appendix I to find the value of $z_{0.99}$.

⟹ To complete the computation, we divide both sides of the equation

$$2P(0 < z < z_{0.99}) = 0.99$$

by 2 and get the equivalent equation

$$P(0 < z < z_{0.99}) = \frac{0.99}{2} = 0.4950$$

We look up the area 0.4950 in Table 4 in Appendix I and then find the z value that produces that area. The value 0.4950 is not in the table; however, the values 0.4949 and 0.4951 are in the table. Even though 0.4950 is exactly halfway between these two values, the two values are so close together we use the higher value 0.4951. This gives us

$$z_{0.99} = 2.58$$

The results of Example 1 and Guided Exercise 1 will be used a great deal in our later work. For convenience, Table 8-2 gives some levels of confidence and corresponding critical values z_c. The same information is provided in Table 4(b) of Appendix I.

Table 8-2 Some Levels of Confidence and Their Corresponding Critical Values

Level of Confidence c	Critical Value z_c
0.75, or 75%	1.15
0.80, or 80%	1.28
0.85, or 85%	1.44
0.90, or 90%	1.645
0.95, or 95%	1.96
0.98, or 98%	2.33
0.99, or 99%	2.58

Error of estimate

An estimate is not very valuable unless we have some kind of measure of how "good" it is. Now that we have studied confidence levels and critical values, the language of probability can give us an idea of the size of the error of estimate caused by using the sample mean $\bar{x}$ as an estimate for the population mean.

Remember that $\bar{x}$ is a random variable. Each time we draw a sample of size n from a population, we can get a different value for $\bar{x}$. According to the central limit theorem, if the sample size is large, then $\bar{x}$ has a distribution that is approximately normal with mean $\mu_{\bar{x}} = \mu$, the population mean we are trying to estimate. The standard deviation is $\sigma_{\bar{x}} = \sigma/\sqrt{n}$.

This information, together with our work on confidence levels, leads us to the probability statement

$$P\left(-z_c\frac{\sigma}{\sqrt{n}} < \bar{x} - \mu < z_c\frac{\sigma}{\sqrt{n}}\right) = c \tag{1}*$$

Equation (1) uses the language of probability to give us an idea of the size of the error of estimate for the corresponding confidence level c. In words, Equation (1) says that the probability is c that our point estimate $\bar{x}$ is within a distance $\pm z_c(\sigma/\sqrt{n})$ of the population mean μ. This relationship is shown in Figure 8-2.

The *error of estimate* using $\bar{x}$ as a point estimate for μ is $|\bar{x} - \mu|$. In most practical problems, μ is unknown, so the error of estimate is also unknown. However, Equation (1) allows us to compute an *error tolerance E*, which serves as a bound on the error of estimate. Using a $c\%$ level of confidence, we can say the point estimate $\bar{x}$ differs from the population mean μ by a maximal error tolerance of

$$E = z_c\frac{\sigma}{\sqrt{n}}$$

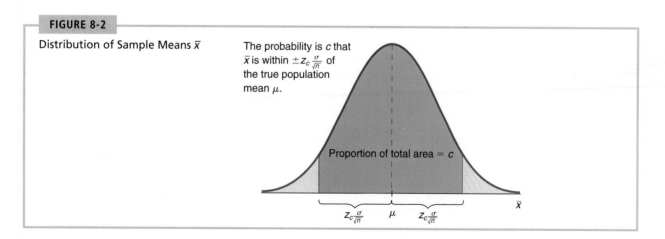

FIGURE 8-2

Distribution of Sample Means $\bar{x}$

The probability is c that $\bar{x}$ is within $\pm z_c\frac{\sigma}{\sqrt{n}}$ of the true population mean μ.

Proportion of total area = c

$z_c\frac{\sigma}{\sqrt{n}}$ μ $z_c\frac{\sigma}{\sqrt{n}}$

$\bar{x}$

*To derive this equation, we start with the probability statement $P(-z_c < z < z_c) = c$. Since $n \geq 30$, we can use the central limit theorem and replace z by $(\bar{x} - \mu)/(\sigma/\sqrt{n})$. Finally, multiply all parts of the inequality by $(\sigma/\sqrt{n})$ to obtain Equation (1).

Using the symbol E in place of $z_c(\sigma/\sqrt{n})$, we can rewrite Equation (1) as

$$P(-E < \bar{x} - \mu < E) = c \qquad (2)$$

Applying a little algebra to Equation (2) produces

$$P(\bar{x} - E < \mu < \bar{x} + E) = c \qquad (3)$$

Confidence intervals
(large sample)

Equation (3) says that there is a chance of c that the population mean μ lies in the interval from $\bar{x} - E$ to $\bar{x} + E$. We say the interval from $\bar{x} - E$ to $\bar{x} + E$ is a *c confidence interval*.

---

c Confidence Interval for μ (Large Samples, n ≥ 30)

$$\bar{x} - E < \mu < \bar{x} + E \qquad (4)$$

where $\bar{x}$ = sample mean

$E = z_c \dfrac{\sigma}{\sqrt{n}}$ if the population standard deviation σ is known

$E \approx z_c \dfrac{s}{\sqrt{n}}$ if σ is not known and we use the sample standard deviation s as an approximation for σ

c = confidence level $(0 < c < 1)$

z_c = critical value (see Table 8-2 or Table 4(b) of Appendix I for commonly used values)

n = sample size $(n \geq 30)$

EXAMPLE 2   In this example we will create a confidence interval for the average time it takes Julia to jog 2 miles. She has been jogging over a period of several years, during which time her physical condition has remained constantly good. Usually she jogs 2 miles per day. During the past year Julia has sometimes recorded her times for running 2 miles. She has a sample of 90 of these times. For these 90 times the mean was $\bar{x} = 15.60$ minutes and the standard deviation was $s = 1.80$ minutes. Let μ be the mean jogging time for the entire distribution of Julia's 2-mile running times (taken over the past year). Find a 0.95 confidence interval for μ.

SOLUTION: The interval from $\bar{x} - E$ to $\bar{x} + E$ will be a 95% confidence interval for μ. In this case, $c = 0.95$, so $z_c = 1.96$ (see Table 8-2). The sample size $n = 90$ is large enough that we may approximate σ as $s = 1.80$ minutes. Therefore,

$$E \approx z_c \frac{s}{\sqrt{n}}$$

$$E = 1.96\left(\frac{1.80}{\sqrt{90}}\right)$$

$$E \approx 0.37$$

Using Equation (4), the given value of $\bar{x}$, and our computed value for E, we get the 95% confidence interval for μ.

$$\bar{x} - E < \mu < \bar{x} + E$$
$$15.60 - 0.37 < \mu < 15.60 + 0.37$$
$$15.23 < \mu < 15.97$$

We conclude that there is a 95% chance that the population mean μ of jogging times for Julia is between 15.23 and 15.97 minutes.

A few comments are in order about the general meaning of the term *confidence interval*. It is important to realize that the endpoints $\bar{x} \pm E$ are really statistical *variables*. Equation (3) says that we have a chance c of obtaining a sample such that the interval, once it is computed, will contain the parameter μ. Of course, after the confidence interval is numerically fixed, it either does or does not contain μ. So the probability is 1 or 0 that the interval, when it is fixed, will contain μ. A nontrivial probability statement can be made only about variables, not about constants. Therefore, Equation (3) really says that if we repeat the experiment many times and get lots of confidence intervals (for the same sample size), then the proportion of all intervals that will turn out to contain the mean μ is c.

In Figure 8-3 the horizontal lines represent 0.90 confidence intervals for various samples of the same size from a distribution. Some of these intervals contain μ, and others do not. Since the intervals are 0.90 confidence intervals, about 90% of all such intervals should contain μ. For each sample the interval goes from $\bar{x} - E$ to $\bar{x} + E$.

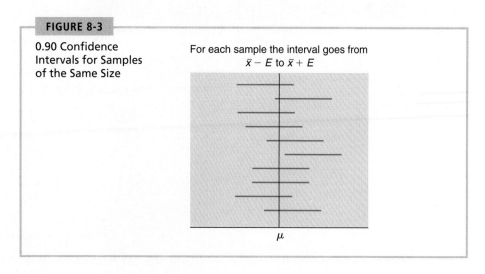

FIGURE 8-3

0.90 Confidence Intervals for Samples of the Same Size

For each sample the interval goes from $\bar{x} - E$ to $\bar{x} + E$

μ

COMMENT Please see "Using Technology" at the end of this chapter for a computer demonstration of this discussion about confidence intervals.

GUIDED EXERCISE· 2

Walter usually meets Julia at the track. He prefers to jog 3 miles. While Julia kept her record, he also kept one for his time for jogging 3 miles. For his 90 times, the mean was $\bar{x} = 22.50$ minutes and the standard deviation was $s = 2.40$ minutes. Let μ be the mean jogging time for the entire distribution of Walter's 3-mile running times over the past several years. How can we find a 0.99 confidence interval for μ?

(a) What is the value of $z_{0.99}$? (See Table 8-2.) ⟹ $z_{0.99} = 2.58$

(b) Since the sample size is large, what can we use for σ? ⟹ $\sigma \approx s = 2.40$

(c) What is the value of E? ⟹ $E = z_c \dfrac{\sigma}{\sqrt{n}} \approx 2.58 \left(\dfrac{2.40}{\sqrt{90}} \right) = 0.65$

(d) What are the endpoints of a 0.99 confidence interval for μ? ⟹ The endpoints are given by

$$\bar{x} - E \approx 22.50 - 0.65$$
$$= 21.85$$
$$\bar{x} + E \approx 22.50 + 0.65$$
$$= 23.15$$

GUIDED EXERCISE 3

A large loan company specializes in making automobile loans for used cars. The board of directors wants to estimate the average amount lent for cars during the past year. The company takes a random sample of 225 customer files for this period. The mean amount lent for this sample of 225 loans is $\bar{x} = \$8200$ and the standard deviation is $s = \$750$. Let μ be the mean of all car loans made over the past year.

Find a 0.95 confidence interval for μ. ⟹ Since $n = 225$ is a large sample, we take $\sigma \approx s = 750$. From Table 8-2, we see that $z_{0.95} = 1.96$. Then

$$E \approx z_c \dfrac{s}{\sqrt{n}} = 1.96 \left(\dfrac{750}{\sqrt{225}} \right) = 98$$
$$\bar{x} - E \approx 8200 - 98 = \$8102$$
$$\bar{x} + E \approx 8200 + 98 = \$8298$$

The interval from \$8102 to \$8298 is a 0.95 confidence interval for μ.

We have said that a sample of size 30 or larger is a large sample. In this section we
indicated two important reasons why our methods require large samples.

What are these reasons?   *Reason 1:* Our methods require $\bar{x}$ to have approximately a normal
distribution. We know from the central limit theorem that this
will be the case for large samples.

Reason 2: Unless we somehow know σ, our methods require us to
approximate σ with the sample standard deviation s. This approximation will be good only if the sample size is large.

Calculator Note The TI-83 fully supports confidence intervals. To access confidence interval choices, press the **STAT** key and select **TESTS.** Choose option
7:ZInterval, to compute a confidence interval for a mean based on a large
sample. There are two options permitted for inputting data. Choose the **Inpt: Stats**
if you know $\bar{x}$ and have an estimate for σ. For example, the following screens show
the results for Example 2, where Julia sampled 90 jogging times and found that
$\bar{x} = 15.60$ and $s = 1.80$. Using $s = 1.80$ as our estimate for σ, we get the following results for a 95% confidence interval.

Select **Inpt: Stats;** enter 1.80 for σ, 15.60   Select **Calculate.** The 95% confidence
for $\bar{x}$, 90 for **n**, 95 or 0.95 for   interval is shown.
C-Level.

```
ZInterval
 Inpt:Data Stats
 σ:1.8
 x̄:15.6
 n:90
 C-Level:95
 Calculate
```

```
ZInterval
 (15.228,15.972)
 x̄=15.6
 n=90

 ■
```

The mathematical notation (15.228, 15.972) means the interval $15.228 < \mu <$
15.972. Notice that the lower bound is listed first and the upper bound is listed last.

If you select **Inpt: Data,** you still need to input the value or estimate σ. In addition, you specify the list that contains the data and the frequency of each data value
(default 1) or the list that contains the frequency. Again, you specify the confidence
level and press **Calculate.** The calculator will compute $\bar{x}$ and n for the data in your
specified list and the designated confidence interval.

When we use samples to estimate the mean of a population, we generate a small error. However, samples are useful even when it is possible to survey the entire population because the use of a sample may yield savings of time or effort in collecting data.

V I E W P O I N T

Couch Potato

Oh well, no one's perfect. Some of the ways we risk our health are obvious: eating fat-filled snacks, binge drinking, smoking, driving with seat belts unbuckled, avoiding regular checkups, skipping exercise, and so on. On average, who is best and who is worst? One way to answer this question is to look at population averages and confidence intervals in geographic locations. *American Demographics* (Vol. 19, No. 8) reports that the healthiest segments of the U.S. population are those who live in Alaska and Minnesota. At the other extreme, the least healthy segments of our population are those who reside in West Virginia and Mississippi. For more information, see Web site <http://www.demographic.com>.

SECTION 8.1 PROBLEMS

Medical: Vitamin E

1. In an article exploring blood serum levels of vitamins and lung cancer risks (*The New England Journal of Medicine*, November 13, 1986), the mean serum level of vitamin E in the control group was 11.9 mg/liter with a standard deviation of 4.30 mg/liter. There were 196 patients in the control group. (These patients were free of all cancer, except possibly skin cancer, in the subsequent 8 years.) Using this information, find a 95% confidence interval for the mean serum level of vitamin E in all persons similar to the control group.

Medical: Vitamin E

2. In the article cited in Problem 1, 99 patients who were cancer-free at the time the blood was drawn later developed lung cancer. For these patients, the mean blood serum level of vitamin E was 10.5 mg/liter with a standard deviation of 3.2 mg/liter. Using this information, find a 95% confidence interval for the mean serum level of vitamin E in the population of all persons with similar lung cancer risks.

Marketing: Motel

3. Irv and Nancy are thinking about buying the Rockwood Motel located on Interstate 70. Before they make up their minds, they want to estimate the average number of vehicles that go by the motel each day in the summer. Fortunately, the highway department has been counting vehicles on I-70 near the motel. A random sample of 36 summer days shows a sample average of 16,000 cars per day with a standard deviation of 2400 cars. Find a 0.90 confidence interval for the population mean number of cars per summer day going past the Rockwood Motel.

Leisure: Hot Air Balloon

4. How hot is the air in the top (crown) of a hot air balloon? Information from *Ballooning: The Complete Guide to Riding the Winds,* by Wirth and Young claims that the air in the crown should be an average of 100°C for a balloon to be in a state of equilibrium. However, the temperature does not need to be exactly 100°C. What is a reasonable and safe range of temperatures? This may vary with the size, (decorative) shape, and outside temperature of the balloon. All balloons have a temperature gauge in the crown. Suppose that 56 readings (for a balloon in equilibrium) gave a mean temperature $\bar{x} = 97°C$ with sample standard deviation $s = 17°C$.
   (a) Compute a 95% confidence interval for the average temperature for which this balloon will be in a steady-state equilibrium.
   (b) If the average temperature in the crown of the balloon goes above the high end of your confidence interval, do you expect the balloon to go up or down? Explain.

Animal Behavior: Coyotes

5. A special feature of camping in the western United States is listening to coyotes howl at night. Just how long is one coyote howl? Since coyotes tend to hunt (mice, rabbits, etc.) in groups, it can seem that howls go on for a long time. This is the result of a collection of coyotes howling together. Individual coyotes tend to howl for a shorter period of time. The howls may help individual members of the pack locate each other in the dark. The following data are based on information from *Coyotes: Biology, Behavior, and Management,* edited by M. Bekoff, University of Colorado (Academic Press).
   (a) Based on a random sample of 102 coyotes, it was found that the mean duration of howls was $\bar{x} = 1.2$ seconds with sample standard deviation $s = 0.4$ second. Compute a 99% confidence interval for the population mean duration μ of all coyote howls.
   (b) For the 102 coyotes in the sample, the mean frequency of howls was $\bar{x} = 609$ Hz (hertz) with sample standard deviation $s = 248$ Hz. Compute a 95% confidence interval for the population mean frequency μ of all coyote howls.

Botany: Iris

6. Dr. Edgar Anderson was a botanist who collected vast amounts of data for several species of wild iris (see Data Highlights in Chapter 7). *Iris virginica,* a lovely wildflower spread over most of the American continent and much of Europe, is one of the species Dr. Anderson studied. His friend R. A. Fisher published the data in a paper entitled "The Use of Multiple Measurements in Taxonomic Problems," *Annals of Eugenics* 7, part II:179–188, 1936. For a sample of 50 *Iris virginica:*
   (a) The sample mean petal length was $\bar{x} = 5.55$ cm with sample standard deviation $s = 0.57$ cm. Compute an 85% confidence interval for the population mean petal length.
   (b) The sample mean petal width was $\bar{x} = 2.03$ cm with sample standard deviation $s = 0.27$ cm. Compute a 90% confidence interval for the population mean petal width.

Business: Profit per Employee

7. Jobs and productivity! How do banks rate? One way to answer this question is to examine annual profits per employee. *Forbes Top Companies,* edited by J. T. Davis (John Wiley & Sons), gave the following data about annual profits per employee (in units of one thousand dollars per employee) for representative companies in

financial services. Companies such as Wells Fargo, First Bank System, Key Banks, Norwest Banks, and so on were included.

42.9	43.8	48.2	60.6	54.9	55.1	52.9	54.9	42.5	33.0	33.6
36.9	27.0	47.1	33.8	28.1	28.5	29.1	36.5	36.1	26.9	27.8
28.8	29.3	31.5	31.7	31.1	38.0	32.0	31.7	32.9	23.1	54.9
43.8	36.9	31.9	25.5	23.2	29.8	22.3	26.5	26.7		

(a) Use a calculator with mean and sample standard deviation keys to verify that for the preceding data $\bar{x} \approx 36.0$ and $s \approx 10.2$.

(b) Let us say that the preceding data are representative of the entire sector of (successful) financial services corporations. Find a 75% confidence interval for μ, the average annual profit per employee for all successful banks.

(c) Let us say that you are the manager of a local bank with a large number of employees. Suppose the annual profits per employee are less than 30 thousand dollars per employee. Do you think that this might be somewhat low compared with other successful financial institutions? Explain by referring to the confidence interval you computed in part b.

(d) Suppose the annual profits are more than 40 thousand dollars per employee. As manager of the bank, would you feel somewhat better? Explain by referring to the confidence interval you computed in part b.

(e) Repeat parts b, c, and d for a 90% confidence level.

Business: Profit per Employee   8. Jobs and productivity! How do retail stores rate? One way to answer this question is to examine annual profits per employee. The following data give annual profits per employee (in units of one thousand dollars per employee) for companies in retail sales. (See reference in Problem 7.) Companies such as Gap, Nordstrom, Circuit City, Dillards, JCPenney, Sears, Wal-Mart, Office Depot, and Toys 'Я' Us are included.

4.4	6.5	4.2	8.9	8.7	8.1	6.1	6.0	2.6	2.9	8.1	−1.9
11.9	8.2	6.4	4.7	5.5	4.8	3.0	4.3	−6.0	1.5	2.9	4.8
−1.7	9.4	5.5	5.8	4.7	6.2	15.0	4.1	3.7	5.1	4.2	

(a) Use a calculator with mean and sample standard deviation keys to verify that for the preceding data $\bar{x} \approx 5.1$ and $s \approx 3.8$.

(b) Let us say that the preceding data are representative of the entire sector of retail sales companies. Find an 80% confidence interval for μ, the average annual profit per employee for retail sales.

(c) Let us say that you are the manager of a retail store with a large number of employees. Suppose the annual profits per employee are less than 3 thousand dollars per employee. Do you think that this might be low compared with other retail stores? Explain by referring to the confidence interval you computed in part b.

(d) Suppose the annual profits are more than 6.5 thousand dollars per employee. As store manager, would you feel somewhat better? Explain by referring to the confidence interval you computed in part b.

(e) Repeat parts b, c, and d for a 95% confidence interval.

Calories: Chocolate Chip
Cookies

9. *Consumer Reports* gave the following data on calories in 30-gram servings of choco-late chip cookies. Both fresh-baked (such as Duncan Hines and Pillsbury) and pack-aged cookies (such as Pepperidge Farm and Nabisco) were included.

153	152	146	138	130	146	149	138	168
147	140	156	155	163	153	155	160	145
138	150	135	155	156	150	146	129	127
171	148	155	132	155	127	150	110	

(a) Use a calculator with mean and standard deviation keys to verify that the sam-ple mean number of calories is $\bar{x} = 146.5$ with sample standard deviation $s = 12.7$ calories.

(b) We take the point of view that the preceding data are representative of the pop-ulation of all chocolate chip cookies. Find an 80% confidence interval for the mean calories μ in 30-gram servings of all chocolate chip cookies. Find the length of this interval.

(c) Repeat part b using a 90% confidence interval.

(d) Repeat part b using a 99% confidence interval.

(e) Compare the lengths from parts b, c, and d. Comment on how these lengths change as c, the confidence level, increases.

Pro-football Players: Age

10. How old are professional football players? Consider the following positions: defen-sive end, defensive guard, defensive tackle, offensive guard, offensive end, offensive tackle. A random sample of professional players in the given positions was taken from the San Diego Chargers, Kansas City Chiefs, Denver Broncos, Los Angeles Raiders, and Seattle Seahawks. The following information gives age in years and was obtained from *The Sports Encyclopedia, Pro Football* (11th Edition).

26	24	25	36	26	32	31	34
32	27	32	23	24	29	30	30
29	33	25	32	24	25	28	22
22	24	23	33	32	31	26	28
32	25	22	28	25	29	25	27

(a) Use a calculator with mean and sample standard deviation keys to verify that for the preceding data $\bar{x} \approx 27.8$ years and $s \approx 3.8$ years.

(b) Let us say that the preceding data are representative of the entire population of professional football players holding the designated positions. Compute an 80% confidence interval for μ, the population mean age of all professional football players in the designated positions.

(c) Let us say that you are the head coach of a professional football team. Suppose that a rookie end, tackle, or guard wants to join your team. This person is 33 years old. Do you think that this age might be somewhat old for a rookie in a starting position? Explain by referring to the confidence interval you computed in part b.

(d) Repeat parts b and c for a 99% confidence interval.

Medical: Sleep

11. How long does it take to fall asleep at night? This depends on what happened the night before. Alexander Borbely is a professor at the University of Zurich Medical School and director of the Sleep Laboratory. The following is adapted from the book, *Secrets of Sleep*, by Professor Borbely.

    (a) Suppose that a random sample of 38 college students was kept awake all night and the next day (24 hours total). The mean time for this group to go to sleep the next night was $\bar{x}$ = 2.5 minutes with sample standard deviation s = 0.7 minute. Compute a 90% confidence interval for the mean time of all such (sleep-deprived) students to fall asleep. What is the length of this interval?

    (b) Suppose that a random sample of 38 college students had a normal (8-hour) sleep and a normal (16-hour) day. The mean time for this group to go to sleep the next night was $\bar{x}$ = 15.2 minutes with sample standard deviation s = 4.8 minutes. Compute a 90% confidence interval for the mean time to go to sleep for all people in this (normal) group. What is the length of this interval?

    (c) Suppose that a random sample of 38 college students stayed in bed at least 12 hours and then after another 12 hours went back to bed. The mean time for this group to fall asleep was $\bar{x}$ = 25.7 minutes with sample standard deviation s = 8.3 minutes. Compute a 90% confidence interval for all people in this group to fall asleep. What is the length of this interval?

    (d) Compare the lengths of the intervals in parts a, b, and c. As the sample standard deviations got larger, did the intervals get longer? Why would you expect this from the method of calculation? Explain.

Climate: Temperatures

12. The U.S. Department of Commerce Environmental Data Service keeps track of average monthly temperatures in various U.S. cities. Data for average temperature (°F) in Phoenix, Arizona for the month of January over the past 40 years yield the following Minitab-generated (►Stat ►Basic Statistics ►1—Sample z) printouts for confidence intervals.

Z Confidence Intervals
The assumed sigma = 3.04

VARIABLE	N	MEAN	STDEV	SEMEAN	90.0 % CI
TEMP	40	51.158	3.040	0.481	(50.367 , 51.948)

VARIABLE	N	MEAN	STDEV	SEMEAN	99.0 % CI
TEMP	40	51.158	3.040	0.481	(49.919 , 52.396)

    (a) Notice that the confidence interval is given in interval notation below the confidence level CI. Looking at the computer displays, which confidence interval is longer, the 90% confidence interval or the 99% confidence interval?

    (b) If someone told you that the earth was heating up and the average January temperature in Phoenix was now 53°F, what might you think about such a claim? Is it possible that a few more years of observation might be needed before such a claim could be made? Explain.

Section **8.2**   # Estimating μ with Small Samples

For samples of size 30 or larger we can approximate the population standard deviation σ by s, the sample standard deviation. Then we can use the central limit theorem to find bounds on the error of estimate and confidence intervals for μ.

There are many practical and important situations, however, where large samples are simply not available. Suppose an archaeologist discovers only seven fossil skeletons from a previously unknown species of miniature horse. Reconstructions of the skeletons of these seven miniature horses show their mean shoulder height to be $\overline{x} = 46.1$ cm. Let μ be the mean shoulder height for this entire species of miniature horse. How can we find the maximal error of estimate $|\overline{x} - \mu|$? How can we find a confidence interval for μ? We will return to this problem later in this section.

Student's t distribution

To avoid the error involved in replacing σ by s, i.e., approximating σ by s, when the sample size is small (less than 30), we introduce a new variable called *Student's t variable*. The t variable and its corresponding distribution, called *Student's t distribution*, were discovered in 1908 by W. S. Gosset. He was employed as a statistician by Guinness brewing company, a company that frowned on the publication of research by its employees, so Gosset published his research under the pseudonym Student. Gosset was the first to recognize the importance of developing statistical methods for obtaining reliable information from small samples. It might be more fitting to call this *Gosset's t distribution*; however, in the literature of mathematical statistics it is known as *Student's t distribution*.

The t variable is defined by the following formula:

$$t = \frac{\overline{x} - \mu}{\dfrac{s}{\sqrt{n}}} \tag{5}$$

where $\overline{x}$ is the mean of a random sample of n measurements, μ is the population mean of the x distribution, and s is the sample standard deviation.

COMMENT You should note that our t variable is just like

$$z = \frac{\overline{x} - \mu}{\dfrac{\sigma}{\sqrt{n}}}$$

except that we replace σ with s. Unlike our methods for large samples, σ cannot be approximated by s when the sample size is less than 30 and we cannot use the normal distribution. However, we will be using the same methods as in Section 8.1 to find the maximal error of estimate and to find confidence intervals, but we use the Student's t distribution.   O

If many random samples of size n are drawn, then we get many t values from Equation (5). These t values can be organized into a frequency table, and a histogram can be drawn, thereby giving us an idea of the shape of the t distribution (for a given n).

Fortunately, all this work is unnecessary because mathematical theorems can be used to obtain a formula for the t distribution. However, it is important to observe that these theorems say that the shape of the t distribution depends only on n, provided the basic variable x has a normal distribution. So *when we use the t distribution, we will assume that the x distribution is normal.*

Degrees of freedom

Table 5 in Appendix I gives values of the variable t corresponding to what we call the number of *degrees of freedom,* abbreviated *d.f.* For the methods used in this section, the number of degrees of freedom is given by the formula

$$d.f. = n - 1 \tag{6}$$

where *d.f.* stands for the degrees of freedom and n is the sample size being used.

Each choice for *d.f.* gives a different t distribution. However, for *d.f.* larger than about 30, the t distribution and the standard normal z distribution are almost the same.

The graph of a t distribution is always symmetrical about its mean, which (as for the z distribution) is 0. The main observable difference between a t distribution and the standard normal z distribution is that a t distribution has somewhat thicker tails.

Figure 8-4 shows a standard normal z distribution and Student's t distribution with *d.f.* $= 3$ and *d.f.* $= 5$.

Using Table 5 in Appendix I to find critical values for confidence intervals

Table 5 in Appendix I gives various t values for different degrees of freedom *d.f.* We will use this table to find critical values t_c for a c confidence level. In other words, we want to find t_c such that an area equal to c under the t distribution for a given number of degrees of freedom falls between $-t_c$ and t_c. In the language of probability, we want to find t_c such that

$$P(-t_c < t < t_c) = c$$

This probability corresponds to the area shaded in Figure 8-5.

FIGURE 8-4

A Standard Normal Distribution and Student's t Distribution with *d.f.* $= 3$ and *d.f.* $= 5$

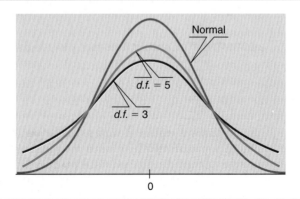

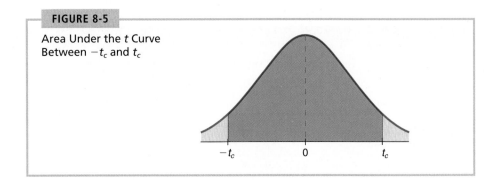

FIGURE 8-5

Area Under the t Curve
Between $-t_c$ and t_c

Table 5 in Appendix I has been arranged so that c is one of the column headings and the degrees of freedom $d.f.$ are the row headings. To find t_c for any specific c, we find the column headed by that c value and read down until we reach the row headed by the appropriate number of degrees of freedom $d.f.$ (You will notice two other column headings: α' and α''. We will use these later, but for the time being, ignore them.)

EXAMPLE 3 ◗   Use Table 8-3 (an excerpt from Table 5 in Appendix I) to find the critical value t_c for a 0.99 confidence level for a t distribution with sample size $n = 5$.

**Table 8-3   Student's t Distribution
(Excerpt from Table 5, Appendix I)**

c	. . . 0.90	0.95	0.98	0.99
α'	—	—	—	—
α''	—	—	—	—
$d.f.$				
⋮				
3	. . . 2.353	3.182	4.541	5.841
4	. . . 2.132	2.776	3.747	4.604
⋮				
7	. . . 1.895	2.365	2.998	3.499
8	. . . 1.860	2.306	2.896	3.355

Source: Table 5, Appendix I, was generated by Minitab.

SOLUTION:

(a) First, we find the column with c heading 0.99. This is the last column.

(b) Next, we compute the number of degrees of freedom: $d.f. = n - 1 = 5 - 1 = 4$.

(c) We read down the column under the heading $c = 0.99$ until we reach the row headed by 4 (under $d.f.$). The entry is 4.604. Therefore, $t_{0.99} = 4.604$. ●

GUIDED EXERCISE   5

Use Table 5 of Appendix I to find t_c for a 0.90 confidence level of a t distribution with sample size $n = 9$.

(a) We find the column headed by $c =$ _____. This is the _____ (first, second, third, fourth, fifth, sixth) column.

     ⇨ $c = 0.90$. This is the fourth column.

(b) The degrees of freedom are given by

$d.f. = n - 1 =$ _____.

     ⇨ $d.f. = n - 1 = 9 - 1 = 8$

(c) Read down the column found in part a until you reach the entry in the row headed by $d.f. = 8$. The value of $t_{0.90}$ is _____ for a sample size of 9.

     ⇨ $t_{0.90} = 1.860$ for a sample size $n = 9$.

(d) Find t_c for a 0.95 confidence level of a t distribution with sample size $n = 9$.

     ⇨ $t_{0.95} = 2.306$ for a sample of size $n = 9$.

Maximal error of estimate

       In Section 8.1 we found bounds $\pm E$ on the error of estimate for a c confidence level. Using the same basic approach, we arrive at the conclusion that

$$E = t_c \frac{s}{\sqrt{n}}$$

is the maximal error of estimate for a c confidence level with small samples (i.e., $|\bar{x} - \mu| < E$ with probability c). The analogue of Equation (1) in Section 8.1 is

$$P\left(-t_c \frac{s}{\sqrt{n}} < \bar{x} - \mu < t_c \frac{s}{\sqrt{n}}\right) = c \tag{7}$$

COMMENT Comparing Equation (7) with Equation (1) in Section 8.1, it becomes evident that we are using the same basic method on the t distribution that we used on the z distribution.

       Likewise, for small samples from normal populations, Equation (3) in Section 8.1 becomes

$$P(\bar{x} - E < \mu < \bar{x} + E) = c \tag{8}$$

where $E = t_c(s/\sqrt{n})$. Let us organize what we have been doing in a convenient summary.

Confidence interval for μ (small sample)

SUMMARY For small samples ($n < 30$) taken from a normal population where σ is unknown, a c confidence interval for the population mean μ is as follows:

---

c Confidence Interval for μ (Small Sample)

$$\bar{x} - E < \mu < \bar{x} + E \qquad\qquad (9)$$

where $\bar{x}$ = sample mean

$$E = t_c \frac{s}{\sqrt{n}}$$

c = confidence level $(0 < c < 1)$

t_c = critical value for confidence level c,
    and degrees of freedom $d.f. = n - 1$
    taken from t distribution

n = sample size (small samples, $n < 30$)

s = sample standard deviation

COMMENT In our applications of the Student's t distribution we have made the basic assumption that x has a normal distribution. However, the same methods apply even if x is only approximately normal. In fact, the main requirement for using the Student's t distribution is that the distribution of x values be reasonably symmetrical and mound-shaped. If this is the case, then the methods we employ with the t distribution can be considered valid for most practical applications. ○

EXAMPLE 4 ❯ Let's return to our archaeologist and the newly discovered (but extinct) species of miniature horse discussed at the beginning of this section. There are only seven known existing skeletons with shoulder heights (in centimeters) 45.3, 47.1, 44.2, 46.8, 46.5, 45.5, and 47.6. For this sample data the mean is $\bar{x} = 46.14$ and the sample standard deviation is $s = 1.19$. Let μ be the mean shoulder height (in centimeters) for this entire species of miniature horse, and assume that the population of shoulder heights is approximately normal.

Find a 99% confidence interval for μ, the mean shoulder height of the entire population of such horses.

SOLUTION: In this case, $n = 7$, so $d.f. = n - 1 = 7 - 1 = 6$. For $c = 0.99$, Table 5 in Appendix I gives $t_{0.99} = 3.707$ (for $d.f. = 6$). The sample standard deviation is $s = 1.19$.

$$E = t_c \frac{s}{\sqrt{n}} = (3.707)\frac{1.19}{\sqrt{7}} = 1.67$$

The 99% confidence interval is

$$\bar{x} - E < \mu < \bar{x} + E$$
$$46.14 - 1.67 < \mu < 46.14 + 1.67$$
$$44.5 < \mu < 47.8$$

GUIDED EXERCISE   6

A company has a new process for manufacturing large artificial sapphires. The production of each gem is expensive, so the number available for examination is limited. In a trial run, 12 sapphires are produced. The mean weight for these 12 gems is $\bar{x} = 6.75$ carats, and the sample standard deviation is $s = 0.33$ carats. Let μ be the mean weight for the distribution of all sapphires produced by the new process.

(a) What is $d.f.$ for this setting?

⇨ $d.f. = n - 1$ where n is the sample size. Since $n = 12$, $d.f. = 12 - 1 = 11$.

(b) Use Table 5 in Appendix I to find $t_{0.95}$.

⇨ Using Table 5 with $d.f. = 11$ and $c = 0.95$, we find $t_{0.95} = 2.201$.

(c) Find E.

⇨ $E = t_{0.95} \dfrac{s}{\sqrt{n}} = (2.201)\dfrac{0.33}{\sqrt{12}} \approx 0.21$

(d) Find a 95% confidence interval for μ.

⇨ $\bar{x} - E < \mu < \bar{x} + E$

$6.75 - 0.21 < \mu < 6.75 + 0.21$

$6.54 < \mu < 6.96$

(e) What assumption about the distribution of all sapphires had to be made to obtain these answers?

⇨ The population of artificial sapphire weights is approximately normal.

Calculator Note The TI-83 computes confidence intervals for the mean based on the Student's t distribution. Press the **STAT** key, select **TESTS**, and use option **8:TInterval**. As with **ZInterval** discussed in Section 8.1, there are two ways to input data. If you select **Inpt: Stats**, then enter the values of $\bar{x}$, the sample standard deviation s_x, the sample size n, and the confidence level. If you select **Inpt: Data**, specify the list containing the raw data, the frequency (default 1) or list containing the frequencies of the raw data values, and the confidence level. The calculator computes $\bar{x}$, s, and n from your list of data and uses the t distribution to produce the confidence interval. Let's use the **Inpt: Data** option to find a 99% confidence interval for the mean shoulder height of extinct miniature horses (data presented in Example 4). Put the raw data in list L_1 and then select **TInterval**.

Enter **L₁** for **List,** 1 for **Freq,** and 99 or 0.99 for **C-Level.**

```
TInterval
 Inpt:Data Stats
 List:L1
 Freq:1
 C-Level:99
 Calculate
```

Select **Calculate.** The 99% confidence interval is shown.

```
TInterval
 (44.475,47.81)
 x̄=46.14285714
 Sx=1.190038015
 n=7
```

The 99% confidence interval is $44.475 < \mu < 47.81$.

We have several formulas for confidence intervals for the population mean μ. How do we choose an appropriate one? We need to look at the sample size, the distribution of the original population, and whether or not the population standard deviation σ is known. There are essentially four cases for which we have the tools to find $c\%$ confidence intervals for the mean μ.

Summary: Confidence Intervals for the Mean

Large Sample Case

$n \geq 30$

1. If σ is not known, then a c confidence interval for μ is

$$\bar{x} - z_c \frac{s}{\sqrt{n}} < \mu < \bar{x} + z_c \frac{s}{\sqrt{n}}$$

2. If σ is known, then a $c\%$ confidence interval for μ is

$$\bar{x} - z_c \frac{\sigma}{\sqrt{n}} < \mu < \bar{x} + z_c \frac{\sigma}{\sqrt{n}}$$

Small Sample Case

$n < 30$

If the population is approximately normal and σ is not known, then a $c\%$ confidence interval for μ is

$$\bar{x} - t_c \frac{s}{\sqrt{n}} < \mu < \bar{x} + t_c \frac{s}{\sqrt{n}} \qquad \text{Use } d.f. = n - 1.$$

For Any Sample Size

If the population *is normal* and σ is known, then for any sample size (large or small) a $c\%$ confidence interval for μ is

$$\bar{x} - z_c \frac{\sigma}{\sqrt{n}} < \mu < \bar{x} + z_c \frac{\sigma}{\sqrt{n}}$$

V I E W P O I N T

Earthquakes

California, Washington, Nevada, and even Yellowstone National Park all have
earthquakes. Some earthquakes are severe. All earthquakes bring fear and anxiety to people living near the quake. Is San Francisco due for a really big quake
like the 1906 major earthquake? How big are the sizes of recent earthquakes
compared with really big earthquakes? What is the duration of an earthquake?
How long is the time span between major earthquakes? One way to answer
questions such as these is to use existing data to estimate confidence intervals on the average size,
duration, and time interval between quakes. Recent data sets for computing such confidence intervals
can be found at the National Earthquake Information Service of the U.S. Geological Survey at Web
site <http://www.seismo.usbr.gov/>.

S E C T I O N 8 . 2 P R O B L E M S

In all the following problems, assume that the population of *x* values has an approximately normal distribution.

1. Use Table 5 in Appendix I to find t_c for a 0.95 confidence level when the sample
   size is 18.

2. Use Table 5 in Appendix I to find t_c for a 0.99 confidence level when the sample
   size is 4.

3. Use Table 5 in Appendix I to find t_c for a 0.90 confidence level when the sample
   size is 22.

4. Use Table 5 in Appendix I to find t_c for a 0.95 confidence level when the sample
   size is 12.

Leisure: Camping

5. Just how expensive is camping equipment? How about a large-sized tent? A random sample of large tents listed in *Consumer Reports: Special Outdoor Issue* (Vol.
   62, No. 6) gave the following prices ($). The tents listed include brand names such
   as L.L. Bean, Sears, Coleman, Wal-Mart, and so forth.

115	140	80	150	250	230
110	135	110	210	120	130

   (a) Use a calculator with mean and sample standard deviation keys to verify that
       $\overline{x} \approx \$148.33$ and $s \approx \$53.02$.

(b) Using the given data as representative of the population of all prices of large-sized tents, find a 90% confidence interval for the mean price μ of all such tents.

Franchise: Candy Store

6. Do you want to own your own candy store? With some interest in running your own business and a decent credit rating, you can probably get a bank loan on startup costs for franchises such as Candy Express, The Fudge Company, Karmel Corn, Rocky Mountain Chocolate Factory, and so on. Startup costs (units in $1000) for a random sample of candy stores are given below. (Source: *Entrepreneur Magazine*, Vol. 23, No. 10.)

95	173	129	95	75	94	116	100	85

(a) Use a calculator with mean and sample standard deviation keys to verify that $\bar{x} \approx 106.9$ thousand dollars and $s \approx 29.4$ thousand dollars.
(b) Find a 90% confidence interval for the population average startup costs μ for candy store franchises.

Wild Life: Mountain Lions

7. How much do wild mountain lions weigh? *The 77th Annual Report of the New Mexico Department of Game and Fish*, edited by Bill Montoya, gave the following information. Adult wild mountain lions (18 months or older) captured and released for the first time in the San Andres Mountains gave the following weights (lb):

68	104	128	122	60	64

(a) Use a calculator with mean and sample standard deviation keys to verify that $\bar{x} \approx 91.0$ lb and $s \approx 30.7$ lb.
(b) Find a 75% confidence interval for the population average weight μ of all adult mountain lions in the specified region.

Calories: French Fries

8. How many calories are there in 3 ounces of french fries? It depends on where you get them. *Good Cholesterol Bad Cholesterol*, by Roth and Streicher, gives the data from eight popular fast-food restaurants. The data (in calories) are

222	255	254	230	249	222	237	287

Use these data to find a 99% confidence interval for the mean calorie count in 3 ounces of french fries obtained from fast-food restaurants.

IRS: Tax on Tips

9. André is head waiter at a famous gourmet restaurant in San Francisco. The Internal Revenue Service is doing an audit of his tax return this year. In particular, the IRS wants to know the average amount André gets for a tip. In an effort to satisfy the IRS, André took a random sample of eight credit card receipts, each of which indicated his tip. The results were

$10.00	$12.75	$11.93	$11.15	$15.70	$14.50	$9.10	$13.65

(a) Use a calculator to verify that the sample mean is $12.35 and the sample standard deviation is $2.25.
(b) Find a 90% confidence interval for the population mean of tips received by André.

Archaeology: Excavation

10. Suppose that you are an anthropologist/historian doing research at an excavation site. How deep do you need to dig to locate artifacts of historical significance? The answer may depend on how long ago the events you hope to study occurred. If the events are relatively recent, you may not need to dig very deep. In 1881 the Lincoln County Cattle Baron Wars were in progress in the New Mexico Territory, and one special outlaw, Billy the Kid, killed two deputies in his escape from the Lincoln County Jail. In 1985 the anthropologist/historian Yvonne Oakes was commissioned by the Museum of New Mexico to study the now famous Lincoln County Courthouse and Jail area. (Source: *Archaeological Testing at Three Historic Sites at Lincoln State Monument,* by Y. R. Oakes, Museum of New Mexico.)

   (a) The depths (in inches) below grade at which significant artifacts were found in a random sample of trenches are

10	10.5	14	14	4	4.5	5	12	12	16
9	6	8	9	10	9	11	11	12	

   Use a calculator with mean and sample standard deviation keys to verify that $\bar{x} \approx 9.8$ inches and $s \approx 3.3$ inches. Compute a 90% confidence interval for the population mean depth μ at which significant artifacts will be found.

   (b) The depth (in inches) to sterile ground (no more artifacts found) for seven excavation trenches are

17	19	21.5	16	14	16	16

   Use a calculator with mean and sample standard deviation keys to verify that $\bar{x} \approx 17.1$ inches and $s \approx 2.5$ inches. Compute an 80% confidence interval for the population mean depth μ to sterile ground.

Vital Statistics: Heights of Men

11. The distribution of the heights of 18-year-old men in the United States is approximately normal with mean 68 inches and standard deviation 3 inches (U.S. Census Bureau). In Minitab we can simulate the drawing of random samples of size 20 from this population (➤Calc ➤Random Data ➤Normal with 20 rows from a distribution with mean 68 and standard deviation 3). Then we can have Minitab compute a 95% confidence interval and draw a boxplot of the data ➤Stat ➤Basic Statistics ➤1—Sample t, with boxplot selected in the graphs). The boxplots and confidence intervals for four different samples are shown in the figures on the following page. The four confidence intervals are

VARIABLE	N	MEAN	STDEV	SEMEAN	95.0 % CI
Sample 1	20	68.050	2.901	0.649	(66.692 , 69.407)
Sample 2	20	67.958	3.137	0.702	(66.490 , 69.426)
Sample 3	20	67.976	2.639	0.590	(66.741 , 69.211)
Sample 4	20	66.908	2.440	0.546	(65.766 , 68.050)

   (a) Examine the figure (parts a to d). How do the boxplots for the four samples differ? Why would you expect the boxplots to differ?

   (b) Examine the 95% confidence intervals for the four samples shown in the above printout. Do the intervals differ in length? Do the intervals all contain the expected population mean of 68 inches? If we draw more samples, do you expect all of the resulting 95% confidence intervals to contain $\mu = 68$? Why or why not?

95% Confidence Intervals
for Mean Height of
18-Year-Old Men
(Sample Size 20)

(a) Boxplot of Sample 1
(with 95% *t*-confidence interval for the mean)

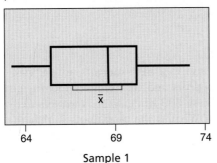

Sample 1

(b) Boxplot of Sample 2
(with 95% *t*-confidence interval for the mean)

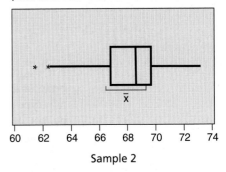

Sample 2

(c) Boxplot of Sample 3
(with 95% *t*-confidence interval for the mean)

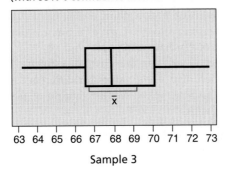

Sample 3

(d) Boxplot of Sample 4
(with 95% *t*-confidence interval for the mean)

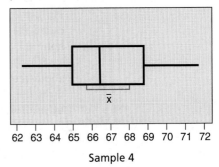

Sample 4

Section **8.3**

Estimating *p* in the Binomial Distribution

The binomial distribution is completely determined by the number of trials *n* and the probability *p* of success in a single trial. For most experiments, the number of trials is chosen in advance. Then the distribution is completely determined by *p*. In this section we will consider the problem of estimating *p* under the assumption that *n* has already been selected.

Basic criteria

Again, we are employing what are called *large sample methods*. We will assume that the normal curve is a good approximation to the binomial distribution and, when necessary, we will use sample estimates for the standard deviation. Empirical studies have shown that these methods are quite good provided that *both*

$$np > 5 \quad \text{and} \quad nq > 5 \quad \text{where } q = 1 - p$$

Let *r* be the number of successes out of *n* trials in a binomial experiment. We will take the sample proportion of successes $\hat{p}$ (read "*p* hat") = *r*/*n* as our *point estimate* for *p*, the population proportion of successes.

Point estimate for p	Point estimate for q
$\hat{p} = \dfrac{r}{n}$	$\hat{q} = 1 - \hat{p}$

For example, suppose that 800 students are selected at random from a student body of 20,000 students and that each of them is given a shot to prevent a certain type of flu. These 800 students are then exposed to the flu, and 600 of them do not get the flu. What is the probability p that the shot will be successful for any single student selected at random from the entire population of 20,000 students? We estimate p for the entire student body by computing r/n from the sample of 800 students. The value $\hat{p} = r/n$ is 600/800, or 0.75. The value $\hat{p} = 0.75$ is then the point estimate for p.

Error of estimate   The difference between the actual value of p and the estimate $\hat{p}$ is the size of our error caused by using $\hat{p}$ as a point estimate for p. The magnitude of $\hat{p} - p$ is called the *error of estimate* for $\hat{p} = r/n$ as a point estimate for p. In absolute value notation, the error of estimate is $|\hat{p} - p|$.

To compute the bounds for the error of estimate, we need some information about the distribution of $\hat{p} = r/n$ values for different samples of the same size n. It turns out that, for large samples, the distribution of $\hat{p}$ values is well approximated by a *normal curve* with

$$\text{mean } \mu = p \quad \text{and} \quad \text{standard error } \sigma = \sqrt{pq/n}$$

Since the distribution of $\hat{p} = r/n$ is approximately normal, we use features of the standard normal distribution to find the bounds for the difference $\hat{p} - p$. Recall that z_c is the number such that an area equal to c under the standard normal curve falls between $-z_c$ and z_c. Then, in terms of the language of probability,

$$P\left(-z_c\sqrt{\frac{pq}{n}} < \hat{p} - p < z_c\sqrt{\frac{pq}{n}}\right) = c \tag{10}*$$

Equation (10) says that the chance is c that the numerical difference between $\hat{p}$ and p is between $-z_c\sqrt{pq/n}$ and $z_c\sqrt{pq/n}$. With the c confidence level, our estimate $\hat{p}$ differs from p by no more than

$$E = z_c\sqrt{pq/n}$$

---

*Recall from Section 6.4 that when n is large, the binomial distribution of the number of successes r is approximately normal with mean $\mu = np$ and standard deviation $\sigma = \sqrt{npq}$. Therefore, $z = (r - np)/\sqrt{npq}$. Dividing the numerator and denominator by n shows that $\hat{p} = r/n$ has a normal distribution with mean $\mu = p$ and standard deviation $\sigma = \sqrt{pq/n}$. Beginning with the equation $P(-z_c < z < z_c) = c$, replacing z by $(\hat{p} - p)/\sqrt{pq/n}$, and multiplying all parts of the inequality by $\sqrt{pq/n}$, we obtain Equation 10.

Confidence interval for p

As in Section 8.1, we call E the *maximal error tolerance* of the error of estimate $|\hat{p} - p|$ for a confidence level c.

To find a c confidence interval for p, we will use E in place of the expression $z_c\sqrt{pq/n}$ in Equation (10). Then we get

$$P(-E < \hat{p} - p < E) = c \qquad (11)$$

Some algebraic manipulation produces the mathematically equivalent statement

$$P(\hat{p} - E < p < \hat{p} + E) = c \qquad (12)$$

Equation (12) says that the probability is c that p lies in the interval from $\hat{p} - E$ to $\hat{p} + E$. Therefore, the interval from $\hat{p} - E$ to $\hat{p} + E$ is the c confidence interval for p that we wanted to find.

There is one technical difficulty in computing the c confidence interval for p. The expression $E = z_c\sqrt{pq/n}$ requires that we know the values of p and q. In most situations, we do not know the actual values of p or q, so we use our point estimates

$$p \approx \hat{p} \qquad \text{and} \qquad q = 1 - p \approx 1 - \hat{p}$$

to estimate E. These estimates are safe for most practical purposes, since we are dealing with large sample theory ($np > 5$ and $nq > 5$).

For convenient reference, we'll summarize the information about c confidence intervals for p, the probability of success in a binomial distribution.

SUMMARY Consider a binomial distribution where n = number of trials, r = number of successes out of the n trials, p = probability of success on each trial, and q = probability of failure on each trial.

If n, p, and q are such that

$$np > 5 \qquad \text{and} \qquad nq > 5$$

then a c confidence interval for p is

$$\hat{p} - E < p < \hat{p} + E$$

where $\hat{p} = \dfrac{r}{n}$

$$E \approx z_c\sqrt{\frac{\hat{p}(1 - \hat{p})}{n}}$$

z_c = critical value for confidence level c taken from a normal distribution (see Table 8-2 or Table 4(b) of Appendix I).

EXAMPLE 5 ➤ Let's return to our flu shot experiment described at the beginning of this section. Suppose that 800 students were selected at random from a student body of 20,000 and given shots to prevent a certain type of flu. All 800 students were exposed to the flu, and 600 of them did not get the flu. Let p represent the probability that the shot will be successful for any single student selected at random from the entire population of 20,000. Let q be the probability that the shot is not successful.

(a) What is the number of trials n? What is the value of r?

SOLUTION: Since each of the 800 students receiving the shot may be thought of as a trial, then $n = 800$, and $r = 600$ is the number of successful trials.

(b) What are the point estimates for p and q?

SOLUTION: We estimate p by the sample point estimate

$$\hat{p} = \frac{r}{n} = \frac{600}{800} = 0.75$$

We estimate q by

$$\hat{q} = 1 - \hat{p} = 1 - 0.75 = 0.25$$

(c) Would it seem that the number of trials is large enough to justify a normal approximation to the binomial?

SOLUTION: Since $n = 800$, $p \approx 0.75$, and $q \approx 0.25$, then

$$np \approx (800)(0.75) = 600 > 5 \quad \text{and} \quad nq \approx (800)(0.25) = 200 > 5$$

A normal approximation is certainly justified.

(d) Find a 99% confidence interval for p.

SOLUTION:

$z_{0.99} = 2.58$ (see Table 8-2)

$$E \approx z_{0.99}\sqrt{\frac{\hat{p}(1-\hat{p})}{n}}$$

$$\approx 2.58\sqrt{\frac{(0.75)(0.25)}{800}}$$

$$\approx 0.0395$$

The 99% confidence interval is then

$$\hat{p} - E < p < \hat{p} + E$$
$$0.75 - 0.0395 < p < 0.75 + 0.0395$$
$$0.71 < p < 0.79$$

GUIDED EXERCISE   7

A random sample of 188 books purchased at a local bookstore showed that 66 of the books were murder mysteries. Let p represent the proportion of books sold by this store that are murder mysteries.

(a) What is a point estimate for p?

➡ $\hat{p} = \dfrac{r}{n} = \dfrac{66}{188} = 0.35$

(b) Find a 90% confidence interval for p.

➡ $E = z_c \sqrt{\dfrac{\hat{p}(1 - \hat{p})}{n}}$

$= 1.645 \sqrt{\dfrac{(0.35)(1 - 0.35)}{188}}$

$= 0.0572$

The confidence interval is
$$\hat{p} - E < p < \hat{p} + E$$
$$0.35 - 0.0572 < p < 0.35 + 0.0572$$
$$0.29 < p < 0.41$$

(c) What is the meaning of the confidence interval you just computed?

➡ If we had computed the interval for many different sets of 188 books, we would find that about 90% of the intervals actually contained p, the population proportion of mysteries. Consequently, we can be 90% confident that our interval is one of the intervals that contains the unknown value p.

(d) To compute the confidence interval, we used a normal approximation. Does this seem justified?

➡ $n = 188$
$p \approx 0.35$ and $q \approx 0.65$
Since $np \approx 65.8 > 5$ and $nq \approx 122.2 > 5$, the approximation is justified.

It is interesting to note that our sample point estimate $\hat{p} = r/n$ and the confidence interval for the population proportion p do not depend on the *size of the population*. In our bookstore example, it made no difference how many books the store sold. On the other hand, the *size of the sample* does affect the accuracy of a statistical estimate. In the next section we will study the effect of sample size on the reliability of our estimate.

Calculator Note The TI-83 calculator also produces confidence intervals for proportions. First, press the **STAT** key, select **TESTS**, and choose option **A:1-PropZInt**. You need to enter the number of successes, designated as x on the TI-83, the number of trials n, and the confidence level. Then enter **Calculate**. The following screens show the results for Guided Exercise 7, where we found a

90% confidence interval for the proportion of mystery books sold at a local bookstore based on information that from a random sample of 188 books sold, 66 were mysteries.

Enter 66 for **x**; 188 for **n**; 90 or 0.90 for the confidence level.

Enter **Calculate**. The results are shown.

```
1-PropZInt
 x:66
 n:188
 C-Level:90
 Calculate
```

```
1-PropZInt
 (.29381,.40832)
 p=.3510638298
 n=188
```

The 90% confidence interval is $0.29381 < p < 0.40832$. Notice that the point estimate $\hat{p} \approx 0.35$ is also given.

Margin of error

Newspapers frequently report the results of opinion polls. In articles that are more complete, a statement about the margin of error accompanies the poll results. The *margin of error* is the maximal error of estimate E for a confidence interval. Usually a 95% confidence interval is assumed.

General Interpretation of Poll Results

1. When a poll states the results of a survey, the proportion reported to respond in the designated manner is $\hat{p}$, the sample estimate of the population proportion.

2. The *margin of error* is the maximal error E of a 95% confidence interval for p.

3. A 95% confidence interval for the population proportion p is

   poll report $\hat{p}$ − margin of error $E < p <$ poll report $\hat{p}$ + margin of error E

Some articles clarify the meaning of the margin of error further by saying that it is an error due to sampling. For instance, the following comments accompany results of a political poll reported in the *Wall Street Journal*.

How Poll Was Conducted

The *Wall Street Journal*/NBC News poll was based on nationwide telephone interviews of 1508 adults conducted last Friday through Tuesday by the polling organizations of Peter Hart and Robert Teeter.

The sample was drawn from 315 randomly selected geographic points in the continental U.S. Each region was represented in proportion to its population. Households were selected by a method that gave all telephone numbers . . . an equal chance of being included.

One adult, 18 years or older, was selected from each household by a procedure to provide the correct number of male and female respondents.

Chances are 19 of 20 that if all adults with telephones in the U.S. had been surveyed, the findings would differ from these poll results by no more than 2.6 percentage points in either direction.

GUIDED EXERCISE   8

Read the last paragraph of the article, "How Poll Was Conducted."

(a) What confidence level corresponds to the phrase "chances are 19 of 20 that if . . ."?

⇨ $\dfrac{19}{20} = 0.95$

A 95% confidence interval is being discussed.

(b) The article indicates that everyone in the sample was asked the question, "Which party, the Democratic Party or the Republican Party, do you think would do a better job handling . . . education?" Possible responses were Democrats, neither, both, or Republicans. The poll reported that 32% of the respondents said "Democrats." Does 32% represent the sample statistic $\hat{p}$ or the population parameter p for the proportion of adults responding "Democrats"?

⇨ 32% represents the sample statistic $\hat{p}$ because 32% represents the percentage of the adults in the *sample* who responded "Democrats."

(c) Continue reading the last paragraph of the article. It goes on to state, ". . . if all adults with telephones in the U.S. had been surveyed, the findings would differ from these poll results by no more than 2.6 percentage points in either direction." Use this information together with parts a and b to find a 95% confidence interval for the proportion p of the specified population who would respond "Democrats" to the question.

⇨ The value "2.6 percentage points" represents the margin of error. Since the margin of error is equivalent to E, the maximal error of estimate for a 95% confidence interval, the confidence interval is

$$32\% - 2.6\% < p < 32\% + 2.6\%$$
$$29.4\% < p < 34.6\%$$

The poll indicates that at the time of the poll, between 29.4% and 34.6% of the specified population think Democrats would do a better job handling education.

VIEWPOINT

"Band-Aid Surgery"

Faster recovery time and less pain: sounds great. An alternative surgical technique called *laparoscopic* ("Band-Aid") *surgery* involves small incisions in which tiny video cameras and long surgical instruments are maneuvered. Instead of a 10-inch incision, surgeons might use four little stabs about $\frac{1}{2}$-inch in length. However, not every such surgery is successful. An article in the Health Section of the *Wall Street Journal* recommends using a surgeon who has done at least 50 such surgeries. Then the prospective patient should ask about the *rate of conversion*—that is, the proportion p of times the surgeon has been forced by complications to switch in midoperation to conventional surgery. A confidence interval for the proportion p would be useful patient information.

SECTION 8.3 PROBLEMS

For all these problems, carry at least four digits after the decimal in your calculations.

Myers-Briggs: Actors

1. Isabel Myers was a pioneer in the study of personality types. The following information is taken from *A Guide to the Development and Use of the Myers-Briggs Type Indicator,* by Myers and McCaulley (Consulting Psychologists Press). In a random sample of 62 professional actors, it was found that 39 were extroverts.
   (a) Let p represent the proportion of all actors who are extroverts. Find a point estimate for p.
   (b) Find a 95% confidence interval for p. Give a brief interpretation of the meaning of the confidence interval you have found.
   (c) Do you think the conditions $np > 5$ and $nq > 5$ are satisfied in this problem? Explain why this would be an important consideration.

Myers-Briggs: Judges

2. In a random sample of 519 judges, it was found that 285 were introverts (see reference in Problem 1).
   (a) Let p represent the proportion of all judges who are introverts. Find a point estimate for p.
   (b) Find a 99% confidence interval for p. Give a brief interpretation of the meaning of the confidence interval you have found.
   (c) Do you think the conditions $np > 5$ and $nq > 5$ are satisfied in this problem? Explain why this would be an important consideration.

Navajo Reservation: Hogans

3. A random sample of 5222 permanent dwellings on the entire Navajo Indian Reservation showed that 1619 were traditional Navajo hogans (*Navajo Architecture: Forms, History, Distributions,* by Jett and Spencer, University of Arizona Press).
   (a) Let p be the proportion of all permanent dwellings on the entire Navajo Reservation that are traditional hogans. Find a point estimate for p.
   (b) Find a 99% confidence interval for p. Give a brief interpretation of the confidence interval.

(c) Do you think that $np > 5$ and $nq > 5$ are satisfied in this problem? Explain why this would be an important consideration.

Archaeology: Pottery

4. Santa Fe black on white is a type of pottery commonly found at archaeological excavations in Bandelier National Monument. At one excavation site a sample of 592 potsherds was found, of which 360 were identified as Santa Fe black on white (*Bandelier Archaeological Excavation Project: Summer 1990 Excavations at Burnt Mesa Pueblo and Casa del Rito*, edited by Kohler and Root, Washington State University).

(a) Let p represent the population proportion of Santa Fe black on white potsherds at the excavation site. Find a point estimate for p.

(b) Find a 95% confidence interval for p. Give a brief statement of the meaning of the confidence interval.

(c) Do you think that $np > 5$ and $nq > 5$ are satisfied in this problem? Why do you think this is important?

Physicians: Charity Care

5. A random sample of 5792 physicians in Colorado showed that 3139 provided at least some charity care (i.e., treated poor people at no cost). These data are based on information from *State Health Care Data: Utilization, Spending, and Characteristics* (American Medical Association).

(a) Let p represent the proportion of all Colorado physicians who provide some charity care. Find a point estimate for p.

(b) Find a 99% confidence interval for p. Give a brief explanation of the meaning of your answer in the context of this problem.

(c) Is use of the normal approximation to the binomial justified in this problem? Explain.

Criminal Justice: Escaped Convicts

6. Case studies showed that out of 10,351 convicts who escaped from U.S. prisons, only 7867 were recaptured (*The Book of Odds*, by Shook and Shook, Signet).

(a) Let p represent the proportion of all escaped convicts who will eventually be recaptured. Find a point estimate for p.

(b) Find a 99% confidence interval for p. Give a brief statement of the meaning of the confidence interval.

(c) Is use of the normal approximation to the binomial justified in this problem? Explain.

Ecology: Fish Survival

7. In a combined study of northern pike, cutthroat trout, rainbow trout, and lake trout, it was found that 26 out of 855 fish died when caught and released using barbless hooks on flies or lures. All hooks were removed from the fish. (Source: *A National Symposium on Catch and Release Fishing*, Humbolt State University Press.)

(a) Let p represent the proportion of all pike and trout that die (i.e., p is the mortality rate) when caught and released using barbless hooks. Find a point estimate for p.

(b) Find a 99% confidence interval for p, and give a brief explanation of the meaning of the interval.

(c) Is use of the normal approximation to the binomial justified in this problem? Explain.

Pharmacy: Errors

8. Does a doctor's illegible handwriting cause problems for pharmacists? According to a report in *USA Today*, the percentage of pharmacists who say a doctor's illegible handwriting causes problems in errors or safety is 85%. Let's assume that this report is based on sample data.

   (a) Compute a 95% confidence interval for the population proportion of pharmacists who say a doctor's illegible handwriting causes errors or safety problems under the assumption that the sample estimate $\hat{p}$ is based on a random sample of 100 pharmacists.

   (b) Repeat part a under the assumption that the sample consists of 1000 pharmacists.

   (c) Compare the confidence intervals computed in parts a and b. What effect does increasing the sample size have on the length of the confidence interval?

Surgeons: Knees

9. According to the American Academy of Orthopaedic Surfeons, 26.6% of all orthopaedic treatments are to the knee. Let's assume that this report is based on sample data.

   (a) Compute a 95% confidence interval for the actual proportion p of orthopaedic treatments that are to the knee under the assumption that the sample estimate $\hat{p}$ is based on a random sample of 1000 general orthopaedic case treatments.

   (b) Repeat part a under the assumption that the sample consists of 10,000 general orthopaedic case treatments.

   (c) Compare the confidence intervals computed in parts a and b. What effect does increasing the sample size have on the length of the confidence interval?

Newspaper Reporting: Margin of Error

10. If you were writing a newspaper report on the results of the poll of pharmacists regarding the percentage that feel that a doctor's illegible handwriting can cause errors or safety problems (see Problem 8, part a), what would you state as the margin of error for a sample size of 100? Follow the convention that the margin of error is based on a 95% confidence interval.

Supermarkets: Customer Loyalty

11. In a marketing survey, a random sample of 730 shoppers revealed that 628 remained loyal to their favorite supermarkets during the past year (i.e., did not switch stores). (Source: *Trends in the United States: Consumer Attitudes and the Supermarket*, The Research Department, Food Marketing Institute.)

    (a) Let p represent the proportion of all shoppers who remain loyal to their favorite supermarkets. Find a point estimate for p.

    (b) Find a 95% confidence interval for p. Give a brief explanation of the meaning of the interval.

    (c) As a news writer, how would you report the survey results regarding the percentage of supermarket shoppers who remained loyal to their favorite supermarkets during the past year? What is the margin of error based on a 95% confidence interval?

12. In a marketing survey, a random sample of 1001 supermarket shoppers revealed that 273 always stock up on an item when they find that item at a real bargain price. (See reference in Problem 11.)
    (a) Let p represent the proportion of all supermarket shoppers who always stock up on an item when they find a real bargain. Find a point estimate for p.
    (b) Find a 95% confidence interval for p. Give a brief explanation of the meaning of the interval.
    (c) As a news writer, how would you report the survey results on the percentage of supermarket shoppers who stock up on an item when they find the item at a real bargain? What is the margin of error based on a 95% confidence interval?

13. How many people would give up some income to gain more leisure time? According to a *USA Today*/CNN/Gallup nationwide telephone poll, only 14% of adults would give up income for time. The margin of sampling errors was plus or minus 3 percentage points. Following the convention that the margin of error is based on a 95% confidence interval, find a 95% confidence interval for the percentage of the population that would give up income to gain more leisure time.

14. A *New York Times*/CBS poll asked the question, "What do you think is the most important problem facing this country today?" Nineteen percent of the respondents answered "crime and violence." The margin of sampling error was plus or minus 3 percentage points. Following the convention that the margin of error is based on a 95% confidence interval, find a 95% confidence interval for the percentage of the population that would respond "crime and violence" to the question asked by the pollsters.

Section 8.4

Choosing the Sample Size

In the design stages of statistical research projects, it is a good idea to decide in advance on the confidence level you wish to use and to select the *maximum* error of estimate E you want for your project. How you choose to make these decisions depends on the requirements of the project and the practical nature of the problem. Whatever specifications you make, the next step is to determine the sample size. In this section we will assume that the distribution of sample means $\bar{x}$ is approximately normal and, when necessary, we will approximate σ by the sample standard deviation s. These methods are technically justifiable since the sizes of our samples will be at least 30.

Sample size for estimating μ

Let's say that at a confidence level of c, we want our point estimate $\bar{x}$ for μ to be in error either way by less than some quantity E. In other words, E is the maximum error of estimate we can tolerate. Using the language of probability, we want the following to be true:

$$P(-E < \bar{x} - \mu < E) = c \tag{13}$$

This is essentially the same as Equation (1) in Section 8.1. Let's compare them.

$$P(-E < \bar{x} - \mu < E) = c \tag{13}$$

$$P\left(-z_c \frac{\sigma}{\sqrt{n}} < \bar{x} - \mu < z_c \frac{\sigma}{\sqrt{n}}\right) = c \tag{1}$$

From this comparison, we see that we want E to be

$$E = z_c \frac{\sigma}{\sqrt{n}}$$

Solving this equation for n, we get

$$n = \left(\frac{z_c \sigma}{E}\right)^2 \tag{14}$$

To compute n from Equation (14), we must know the value of σ. If the value of σ is not previously known, we do a preliminary sampling to approximate it. For most practical purposes, a preliminary sample of size 30 or larger will give a sample standard deviation s, which we may use to approximate σ.

EXAMPLE 6 ❯ A wildlife study is designed to find the mean weight of salmon caught by an Alaskan fishing company. As a preliminary study, a random sample of 50 freshly caught salmon is weighed. The sample standard deviation of the weights of these 50 fish is $s = 2.15$ lb. How large a sample should be taken to be 99% confident that the sample mean $\bar{x}$ is within 0.20 lb of the true mean weight μ?

SOLUTION: In this problem, $z_{0.99} = 2.58$ (see Table 8-2) and $E = 0.20$. The preliminary study of 50 fish is large enough to permit a good approximation of σ by $s = 2.15$. Therefore, Equation (14) becomes

$$n = \left(\frac{z_c \sigma}{E}\right)^2 \approx \left(\frac{(2.58)(2.15)}{0.20}\right)^2 = 769.2$$

In determining sample size, any fractional value of n is always rounded to the *next higher whole number.* We conclude that a sample size of 770 will be enough to satisfy the specifications. Of course, a sample size larger than 770 would also work. ●

EXAMPLE 7 ❯ A certain company makes light fixtures on an assembly line. An efficiency expert wants to determine the mean time it takes an employee to assemble the switch on one of these fixtures. A preliminary study used a random sample of 45 observations and found that the sample standard deviation was $s = 78$ seconds. How many more observations are necessary for the efficiency expert to be 95% sure that the point estimate $\bar{x}$ will be "off" from the true mean μ by at most 15 seconds?

SOLUTION: In this example we approximate σ by $s = 78$. We use $z_{0.95} = 1.96$ (see Table 8-2 or Table 4(b) of Appendix I). The maximum error of estimate is specified to be $E = 15$ seconds. Equation (14) gives us

$$n = \left(\frac{z_c \sigma}{E}\right)^2 = \left(\frac{(1.96)(78)}{15}\right)^2 = 103.9$$

The efficiency expert should use a sample of minimum size 104. Since the preliminary study has 45 observations, an additional $104 - 45 = 59$ observations are necessary.   ●

GUIDED EXERCISE   9

A large state university has over 1800 faculty members. The dean of faculty wants to estimate the average teaching experience (in years) of the faculty members. A preliminary random sample of 60 faculty members yields a sample standard deviation of $s = 3.4$ years. The dean wants to be 99% confident that the sample mean $\bar{x}$ does not differ from the population mean by more than half a year. How large a sample should be used? Let's answer this question in parts.

(a) What value can we use to approximate σ? Why can we do this?

⟹ $s = 3.4$ years is a good approximation because a preliminary sample of 60 is fairly large.

(b) What is $z_{0.99}$? (*Hint:* See Table 8-2.)

⟹ $z_{0.99} = 2.58$

(c) What is E for this problem?

⟹ $E = 0.5$ year

(d) Which is the correct formula for n:

$$\left(\frac{z_c \sigma}{n}\right)^2, \quad \left(\frac{z_c \sigma}{E}\right)^2, \quad \text{or} \quad \left(\frac{z_c E}{\sigma}\right)^2$$

⟹ $n = \left(\dfrac{z_c \sigma}{E}\right)^2$

(e) Use the formula for n to find the minimum sample size. Should your answer be rounded up or down to a whole number?

⟹ $n = \left(\dfrac{(2.58)(3.4)}{0.5}\right)^2 = (17.544)^2 = 307.8$

Always round n up to the next whole number. Our final answer $n = 308$ is the minimum size.

Sample size for estimating $\hat{p} = r/n$

(If you omitted the binomial distribution, omit the rest of this section.)

Next, we will determine the minimum sample size when we use the sample proportion $\hat{p} = r/n$ as a point estimate for p in a binomial distribution. We will use the methods of normal approximation (large samples) discussed in Section 8.3. Suppose for a confidence level c we want the estimate $\hat{p} = r/n$ for p to be in error either way by less than some quantity E. Using the language of probability, we want the following to be true.

$$P(-E < \hat{p} - p < E) = c \tag{15}$$

Let's compare this with Equation (10) in Section 8.3. For convenience, they both are written together:

$$P(-E < \hat{p} - p < E) = c \tag{15}$$

$$P\left(-z_c\sqrt{\frac{pq}{n}} < \hat{p} - p < z_c\sqrt{\frac{pq}{n}}\right) = c \tag{10}$$

The comparison of the two equations gives a formula for E:

$$E = z_c\sqrt{\frac{pq}{n}}$$

Solving this equation for n, we get

$$n = pq\left(\frac{z_c}{E}\right)^2$$

Since $q = 1 - p$, our equation for n can be written

$$n = p(1 - p)\left(\frac{z_c}{E}\right)^2 \tag{16}$$

Equation (16) cannot be used unless we already have a preliminary estimate for p. To get around this difficulty, we use a result from algebra showing that the maximum value of $p(1 - p)$ is $\frac{1}{4}$. Therefore, *when we have no preliminary estimate for p*, we use the formula

$$n = \frac{1}{4}\left(\frac{z_c}{E}\right)^2 \tag{17}$$

Since Equation (17) may make the sample size unnecessarily large, we can say the probability is *at least* (and possibly more than) c that the point estimate $\hat{p} = r/n$ for p will be in error either way by less than the quantity E.

EXAMPLE 8 ➤   A company is in the business of selling wholesale popcorn to grocery stores. The company buys directly from farmers. A buyer for the company is examining a large amount of corn from a certain farmer. Before the purchase is made, the buyer wants to estimate p, the probability that a kernel will pop.

Suppose that a random sample of n kernels is taken and r of these kernels pop. The buyer wants to be 95% sure that the point estimate $\hat{p} = r/n$ for p will be in error either way by less than 0.01.

(a) If no preliminary study is made to estimate p, how large a sample should the buyer use?

SOLUTION: In this case we use Equation (17) with $z_{0.95} = 1.96$ (see Table 8-2 or Table 4(b) of Appendix I) and $E = 0.01$.

$$n = \frac{1}{4}\left(\frac{z_c}{E}\right)^2 = \frac{1}{4}\left(\frac{1.96}{0.01}\right)^2 = 0.25(38{,}416) = 9604$$

We would need a sample of $n = 9604$ kernels.

(b) A preliminary study showed that p was approximately 0.86. If the buyer uses the results of the preliminary study, how large a sample should be used?

SOLUTION: In this case we use Equation (16) with $p \approx 0.86$. Again, from Table 8-2, $z_{0.95} = 1.96$, and from the problem, $E = 0.01$.

$$n = p(1 - p)\left(\frac{z_c}{E}\right)^2 = (0.86)(0.14)\left(\frac{1.96}{0.01}\right)^2 = 4625.29$$

The sample size should be at least $n = 4626$ kernels. This sample is less than half the sample size necessary without the preliminary study.

GUIDED EXERCISE   10

In Indianapolis, the Department of Public Health wants to estimate the proportion of children (grades 1–8) who require corrective lenses for their vision. A random sample of n children is taken, and r of these children are found to require corrective lenses. Let p be the true proportion of children requiring corrective lenses. The health department wants to be 99% sure that the point estimate $\hat{p} = r/n$ for p will be in error either way by less than 0.03.

(a) If no preliminary study is made to estimate p, how large a sample should the health department use? Let's answer this question in parts.

(i) Which formula shall we use: (16) ⇨   (i) We use Equation (17) because we do not have an
     or (17)?   estimate for p.

(ii) What is the value of E, and   ⇨   (ii) $E = 0.03$ and $z_{0.99} = 2.58$ (see Table 8-2).
     what is the value of z_c in this
     problem?

(iii) What is the value of n?   ⇨   (iii) $n = \frac{1}{4}\left(\frac{z_c}{E}\right)^2 = \frac{1}{4}\left(\frac{2.58}{0.03}\right)^2 = 1849$

                                            So without a preliminary study to find p, we will
                                            need a sample size of at least $n = 1849$ children.

(b) A preliminary random sample of 100 children indicates that 23 require corrective lenses. Using the results of this preliminary study, how large a sample should the health department use? Again, let's answer this question in parts.

Exercise continues

Exercise continued

(i) Which formula should we use: (16) or (17)?

 (i) We use Equation (16) because we have an estimate of p from a preliminary study.

(ii) What are the values of E and z_c for this problem?

 (ii) $E = 0.03$ and $z_{0.99} = 2.58$

(iii) What approximate value shall we use for p?

 (iii) $p \approx 0.23$

(iv) What is the value of n?

 (iv) $n = p(1 - p)\left(\dfrac{z_c}{E}\right)^2$

$$= (0.23)(0.77)\left(\dfrac{2.58}{0.03}\right)^2$$

$$= 1309.83$$

Therefore, the sample size should be at least 1310 children.

VIEWPOINT

Profiles in Crime

What proportion of the U.S. population will eventually spend time in a prison? (Answer: About 5.1%.) What proportion of federal prison inmates have at least some college education? (Answer: About 28%.) If a person is released from prison, what is the probability that he or she will commit a felony and be returned to prison within 3 years? (Answer: About 41%.) What is the gender and age distribution of prison inmates? How accurate are these statistics? Sample size plays an important role in statistical accuracy. Methods of this section and data from the Bureau of Justice Statistics can help you determine the implied level of accuracy. See Web site <http://www.ojp.usdoj.gov/> and follow the links to justice statistics.

SECTION 8.4 PROBLEMS

For each of the following problems,

(i) identify whether we are going to estimate a population mean μ or a population proportion p.

(ii) write down the appropriate sample size formula(s) for the problem, and then solve the problem.

Environmental Studies: Yellowstone Park

1. Test plots in Yellowstone National Park were used to study tree reproduction of lodgepole pine (*Yellowstone Vegetation*, by D. G. Despain, Roberts Rinehart). In plots of 50 square meters, the number x of new lodgepole saplings was counted. In

a given region, based on long observation, the standard deviation of x values is estimated at $\sigma = 44$. How many 50-square-meter plots should be used in such a region to be 95% sure the sample mean $\bar{x}$ of lodgepole saplings is within 10 saplings of the population mean μ of lodgepole saplings in all 50-square-meter plots in this area?

Environmental Studies: Yellowstone Park

2. Root depth for grasses and shrubs in a type of soil known as glacial outwash was studied in Yellowstone National Park by D. G. Despain (see reference in Problem 1). Let x be a random variable representing root depth in this type of soil. It was found that the standard deviation of x values is approximately $\sigma = 8.94$ in. In a proposed study region of glacial outwash, how many plants should be carefully dug up and studied to be 90% sure that the sample mean root depth $\bar{x}$ is within 0.5 in. of the population mean root depth?

Pro-basketball Players: Weights

3. The NBA "All-Time Player Directory" of *The Official NBA Basketball Encyclopedia* (Villard Books) gave the weight of each basketball player. A random sample of 56 basketball players from the directory gave a sample standard deviation $s = 26.58$ pounds for the weights of players. How many more basketball players from the "All-Time Player Directory" should be included in the sample to be 90% sure that the sample mean player weight $\bar{x}$ is within 4 pounds of the population mean μ?

Pro-basketball Players: Heights

4. A random sample of 41 basketball players from the "All-Time Player Directory" (see reference in Problem 3) gave a sample standard deviation for the heights of players $s = 3.32$ inches. How many more basketball players from the "All-Time Player Directory" should be included in the sample to be 95% sure that the sample mean $\bar{x}$ is within 0.75 inch of the population mean μ of all players listed in the NBA encyclopedia?

Anthropology: Pottery

5. A random sample of 83 reconstructed clay vessels from the Turquoise Ridge archaeological site indicated the standard deviation of diameters to be approximately 5.5 cm (based on information from *Anthropological Paper Number 118,* University of Utah Press). How many more such clay vessels must be found and reconstructed to be 95% sure that the sample mean $\bar{x}$ of diameters is within 1 cm of the population mean μ of all such reconstructed clay vessels at this archaeological site?

Environmental Studies: Bighorn Sheep

6. A random sample of 37 adult male desert bighorn sheep indicated the standard deviation of the sheep weights to be 15.8 lb. (Source: *The Bighorn of Death Valley,* Fauna of the National Parks of the United States Monograph Number 6, U.S. Government Printing Office.) How many more such adult male desert bighorn sheep should be included in the sample to be 90% sure that the sample mean weight $\bar{x}$ is within 2.5 lb of the population mean weight μ of all such bighorn sheep in this region?

Airlines: Phone Reservations

7. When customers phone airlines to make reservations, they usually find it irritating if they are kept on hold for a long time. In an effort to determine how long phone customers are kept on hold, one airline took a random sample of 167 phone calls and determined the length of time (in minutes) each caller was kept on hold. The sample standard deviation was 3.8 minutes. How many more phone customers should be included in the sample to be 99% sure that the sample mean $\bar{x}$ of hold times is within 30 seconds of the population mean μ of hold times?

8. Some drivers engage in unsafe behaviors such as running traffic lights that are turning red. Suppose you want to estimate the proportion p of drivers in your area who will admit to running lights that are turning red.

(a) If you have no preliminary estimate for p, how many people should you include in a random sample to be 90% sure that the point estimate $\hat{p}$ will be within a distance of 0.05 from p?

(b) Answer part a if you use the preliminary estimate from the National Highway Traffic Safety Administration survey that nationally 52% of drivers admit to running lights that are turning red.

9. About what proportion p of drivers admit that they exceed the posted speed limits on interstate highways by 20 mph or more?

(a) If you have no preliminary estimate for p, how many people should you include in a random sample to be 95% sure that the point estimate $\hat{p}$ will be within a distance of 0.05 from p?

(b) Answer part a if you use the preliminary estimate from the National Highway Traffic Safety Administration survey that nationally 14% of drivers admit to speeding by 20 mph or more on interstate highways.

10. How hard is it to reach a businessperson by phone? Let p be the proportion of calls to businesspeople for which the caller reaches the person being called on the *first* try.

(a) If you have no preliminary estimate for p, how many business phone calls should you include in a random sample to be 80% sure that the point estimate $\hat{p}$ will be within a distance of 0.03 from p?

(b) *The Book of Odds,* by Shook and Shook (Signet), reports that businesspeople can be reached by a single phone call approximately 17% of the time. Using this (national) estimate for p, answer part a.

11. What percentage of the campus student body is female? Let p be the proportion of women students on your campus.

(a) If no preliminary study is made to estimate p, how large a sample is needed to be 99% sure that a point estimate $\hat{p}$ will be within a distance of 0.05 from p?

(b) The *Statistical Abstract of the United States* (118th Edition) indicates that approximately 55% of college students are females. Answer part a using this estimate for p.

12. The National Council of Small Businesses is interested in the proportion of small businesses that declared Chapter 11 bankruptcy last year. Since there are so many small businesses, the National Council intends to estimate the proportion from a random sample. Let p be the proportion of small businesses that declared Chapter 11 bankruptcy last year.

(a) If no preliminary sample is taken to estimate p, how large a sample is necessary to be 95% sure that a point estimate $\hat{p}$ will be within a distance of 0.10 from p?

(b) In a preliminary random sample of 38 small businesses, it was found that 6 had declared Chapter 11 bankruptcy. How many *more* small businesses should be included in the sample to be 95% sure that a point estimate $\hat{p}$ will be within a distance of 0.10 from p?

Sociology: Information
Overload

13. Do you think society suffers from information overload? What proportion p of fellow students think that society suffers from information overload?
    (a) If you have no preliminary estimate for p, how many people should you include in a random sample to be 95% sure that the point estimate $\hat{p}$ will be within a distance of 0.03 from p?
    (b) A recent survey in *USA Today* reported that 61% of adults nationwide believe that society suffers from information overload. Answer part a on the basis of this preliminary estimate for p.

Margin of Error: Registered
Voters

14. Suppose you are working for a local polling firm. Your project involves interviewing a random sample of registered voters to determine the percentage favoring the use of state lottery proceeds for park improvements. The firm wants to ensure that the margin of error is no more than 3 percentage points either way. Assuming that there is no preliminary estimate for the percentage of registered voters favoring this use of lottery funds, what is the *minimum* number of respondents required? (*Hint:* Convert the margin of error to decimal form and use the *New York Times*/CBS convention that the margin of error is based on a 95% confidence level.)

SUMMARY

We have studied point estimates and interval estimates. For point estimates we found E, the maximal error of estimate, and for interval estimates we found the interval endpoints. In each case, E or endpoints were determined by four factors: the confidence level c, the sample estimate for μ or p, the sample standard deviation s, and the sample size n.

For large samples ($n \geq 30$), we used the normal distribution, the central limit theorem, and sometimes the normal approximation to the binomial distribution. For small samples ($n < 30$), we made use of Student's t distribution. To choose the sample size in applications, we found formulas for n such that, with probability c, the sample estimate for μ or p is in error by less than a preassigned number E.

IMPORTANT WORDS & SYMBOLS

Section 8.1
Error of estimate $|\bar{x} - \mu|$
Confidence level c
Critical value z_c
Point estimate for μ
E, the maximal error of estimate
Interval estimate for μ
Large samples, $n \geq 30$
c confidence interval

Section 8.2
Small samples, $n < 30$
Student's t variable

Critical value t_c
Degrees of freedom $(d.f.)$

Section 8.3
Point estimate for p, $\hat{p}$
Error of estimate $|\hat{p} - p|$
Interval estimate for p
Margin of error

Section 8.4
Sample size n

VIEWPOINT

All Systems Go?

On January 28, 1986, the Space Shuttle *Challenger* caught fire and blew up only seconds after launch. A great deal of good engineering went into the design of the *Challenger*. However, when a system has several confidence levels operating at once, it can happen, in rare cases, that risks will increase rather than cancel out. Diane Vaughn is a professor of sociology at Boston College and author of the book *The Challenger Launch Decision* (University of Chicago Press). Her book contains an excellent discussion of risks, the normalization of deviants, and cost/safety trade-offs. Vaughn's book is described as "a remarkable and important analysis of how social structures can induce consequential errors in a decision process" (Robert K. Merton, Columbia University).

CHAPTER REVIEW PROBLEMS

Categorize each problem according to (a) parameter being estimated, proportion *p*, or mean μ and (b) large sample or small sample. Then solve the problem.

General: Terms

1. In your own words, carefully explain the meanings of the following terms: point estimate, critical value, maximal error of estimate, confidence level, confidence interval, large samples, and small samples.

Auto Insurance: Claims

2. Anystate Auto Insurance Company took a random sample of 370 insurance claims paid out during a 1-year period. The average claim paid was $750 with a standard deviation of $150. Find 0.90 and 0.99 confidence intervals for the mean claim payment.

Psychology: Need for Closure

3. Three experiments investigating the relation between need for cognitive closure and persuasion were reported in "Motivated Resistance and Openness to Persuasion in the Presence or Absence of Prior Information," by A. W. Kruglanski (*Journal of Personality and Social Psychology*, Vol. 65, No. 5, pp. 861–874). Part of the study involved administering a "need for closure scale" to a group of students enrolled in an introductory psychology course. The "need for closure scale" has scores ranging from 101 to 201. For the 73 students in the highest quartile of the distribution, the mean score was $\bar{x} = 178.70$ with sample standard deviation $s = 7.81$. These students were all classified as high on their need for closure. Assume that the 73 students represent a random sample of all students who are classified as high on their need for closure. Find a 95% confidence interval for the population mean score μ on the "need for closure scale" for all students with a high need for closure.

Psychology: Need for Closure

4. How large a sample is needed in Problem 3 if we wish to be 99% confident that the sample mean score is within 2 points of the population mean score for students who are high on the need for closure?

Archaeology: Excavations

5. The Wind Mountain archaeological site is located in southwestern New Mexico. Wind Mountain was home to an ancient culture of prehistoric Native Americans called Anasazi. A random sample of excavations at Wind Mountain gave the following depths (in centimeters) from the present-day surface grade to the location of

significant archaeological artifacts. (Source: *Mimbres Mogollon Archaeology,* by A. Woosley and A. McIntyre, University of New Mexico Press.)

85	45	120	80	75	55	65	60
65	95	90	70	75	65	68	

(a) Use a calculator with mean and sample standard deviation keys to verify that $\bar{x} \approx 74.2$ cm and $s \approx 18.3$ cm.

(b) Compute a 95% confidence interval for the mean depth μ at which archaeological artifacts from the Wind Mountain excavation site can be found.

Archaeology: Pottery

6. Shards of clay vessels were put together to reconstruct rim diameters of the original ceramic vessels at the Wind Mountain archaeological site (see source in Problem 5). A random sample of ceramic vessels gave the following rim diameters (in centimeters).

15.9	13.4	22.1	12.7	13.1	19.6	11.7	13.5	17.7	18.1

(a) Use a calculator with mean and sample standard deviation keys to verify that $\bar{x} \approx 15.8$ cm and $s \approx 3.5$ cm.

(b) Compute an 80% confidence interval for the population mean μ of rim diameters for such ceramic vessels found at the Wind Mountain archaeological site.

Business: Work Force Issues

7. The National Study of the Changing Work Force conducted an extensive survey of 2958 hourly wage and salaried workers on issues ranging from relationships with their bosses to household chores. The data were gathered through hour-long telephone interviews with a nationally representative sample (*Wall Street Journal*). In response to the question, "What does success mean to you?" 1538 responded, "Personal satisfaction from doing a good job." Let p be the population proportion of all hourly wage and salaried workers who would respond the same way to the stated question. Find a 90% confidence interval for p.

Business: Work Force Issues

8. How large a sample is needed in Problem 7 if we wish to be 95% confident that the sample percentage of those equating success with personal satisfaction is within 1% of the population percentage? (*Hint:* Use $p \approx 0.52$ as a preliminary estimate.)

Archaeology: Pottery

9. Three circle, red on white is one distinctive pattern painted on ceramic vessels of the Anasazi period found at the Wind Mountain archaeological site (see source for Problem 5). At one excavation a sample of 167 potsherds indicated that 68 were of the three circle, red on white pattern.

(a) Find a point estimate $\hat{p}$ for the proportion of all ceramic potsherds at this site that are of the three circle, red on white pattern.

(b) Compute a 95% confidence interval for the population proportion p of all ceramic potsherds with this distinctive pattern found at the site.

Archaeology: Pottery

10. Consider the three circle, red on white pattern discussed in Problem 9. How many ceramic potsherds must be found and identified if we are to be 95% confident that the sample proportion $\hat{p}$ of such potsherds is within 6% of the population proportion of vessels with the three circle, red on white pattern found at this excavation site? (*Hint:* Use the results of Problem 9 as a preliminary estimate.)

TV: Commercials

11. During a television miniseries, what is the average length of time between commercial breaks? A random sample of 20 such periods was selected from miniseries that

were aired on commercial television stations last year. The times between commercial breaks were (to the nearest minute)

5	7	8	14	13
10	9	8	11	12
14	11	9	10	6
8	12	5	11	8

(a) Use a calculator to verify that the sample mean of the times between commercial breaks is $\bar{x} = 9.55$ min with standard deviation $s = 2.72$ min.
(b) Find a 95% confidence interval for the mean length of time between commercial breaks.

Animal Behavior: Coyotes

12. The book *Coyotes, Biology, Behavior and Management*, edited by Marc Bekoff, Academic Press, contains many interesting facts about coyote populations and behavior. One study involved the spring coyote density (that is, coyotes per square kilometer) in the National Elk Refuge near Grand Teton National Park in Wyoming. A sample of 5 years showed the following coyote density (coyotes per square kilometer) in the refuge.

0.51   0.58   0.46   0.54   0.54

Assuming the spring coyote density is approximately normal, find a 95% confidence interval for the coyote density on the National Elk Refuge.

DATA HIGHLIGHTS: GROUP PROJECTS

Break into small groups and discuss the following topics. Organize a brief outline in which you summarize the main points of your group discussion.

1. Garrison Bay is a small bay in Washington state. A popular recreational activity in the bay is clam digging. For several years, this harvest has been monitored and the size distribution of clams recorded. Data for lengths and widths of little neck clams (*Protothaca staminea*) were recorded by a method of systematic sampling in a study done by S. Scherba and V. F. Gallucci ("The Application of Systematic Sampling to a Study of Infaunal Variation in a Soft Substrate Intertidal Environment," *Fishery Bulletin* 74:937–948). The data in the tables below give lengths and widths for 35 little neck clams.

Length of Little Neck Clams (mm)

530	517	505	512	487	481	485	479	452	468
459	449	472	471	455	394	475	335	508	486
474	465	420	402	410	393	389	330	305	169
91	537	519	509	511					

Width of Little Neck Clams (mm)

494	477	471	413	407	427	408	430	395	417
394	397	402	401	385	338	422	288	464	436
414	402	383	340	349	333	356	268	264	141
77	498	456	433	447					

(a) Use a calculator to compute the sample means and sample standard deviations for lengths and for widths. Compute the coefficient of variation for each.

(b) Compute a 95% confidence interval for the population mean length of all Garrison Bay little neck clams.

(c) How many more little neck clams would be needed in a sample if you wanted to be 95% sure the sample mean length was within a maximal error of 10 mm of the population mean length?

(d) Compute a 95% confidence interval for the population mean width of all Garrison Bay little neck clams.

(e) How many more little neck clams would be needed in a sample if you wanted to be 95% sure the sample mean width was within a maximal error of 10 mm of the population mean width?

2. Examine the following figure, "Clocks roll back Sunday" (*USA Today*).

USA SNAPSHOTS®

A look at statistics that shape the nation

Clocks roll back Sunday

This Sunday clocks in most of the USA will be moved back one hour to standard time — not the first choice of most people. Those who prefer:

Daylight saving time 66%

Standard time 28%

No preference 6%

Source: Hilton Time Surveys of 1,024 adults

By Sam Ward and Marcy E. Mullins, USA TODAY

Source: Copyright 1993, USA TODAY. Reprinted with permission.

(a) Of the 1024 adults surveyed, 66% were reported to favor daylight saving time. How many people in the sample preferred daylight saving time? Using the statistic $\hat{p} = 0.66$ and sample size $n = 1024$, find a 95% confidence interval for the proportion of people p who favor daylight saving time. How could you report this information in terms of a margin of error?

(b) Look at the figure to find the sample statistic $\hat{p}$ for the proportion of people preferring standard time. Find a 95% confidence interval for the population proportion p of people who favor standard time. Report the same information in terms of a margin of error.

3. Examine the following figure, "Coupons: Limited clipping" (*USA Today*).

USA SNAPSHOTS®

A look at statistics that shape our lives

Coupons: Limited clipping

■ Merchandise coupons distributed
■ Merchandise coupons redeemed

Number

310 billion

7.7 billion

Value

$177.9 billion

$4.5 billion

Source: NCH Promotional Services
By Elys A. McLean, USA TODAY

Source: Copyright 1993, USA TODAY. Reprinted with permission.

(a) Use the figure to estimate the percentage of merchandise coupons that were redeemed. Also estimate the percentage of the dollar value of the coupons that were redeemed. Are these numbers approximately equal?

(b) Suppose that you are a marketing executive working for a national chain of toy stores. You wish to estimate the percentage of coupons that will be redeemed for the toy stores. How many coupons should you check to be 95% sure that the percentage of coupons redeemed is within 1% of the population proportion of all coupons redeemed for the toy store?

(c) Use the results of part a as a preliminary estimate for p, the percentage of coupons that are redeemed, and redo part b.

(d) Suppose that you sent out 937 coupons and found that 27 were redeemed. Explain why you could be 95% confident that the proportion of such coupons redeemed in the future would be between 1.9% and 3.9%.

(e) Suppose that the dollar value of a collection of coupons was $10,000. Use the data in the figure to find the expected value and standard deviation of the dollar value of the redeemed coupons. What is the probability that between $225 and $275 (out of the $10,000) is redeemed?

ⓛINKING CONCEPTS: WRITING PROJECTS

Discuss each of the following topics in class or review the topics on your own. Then write a brief but complete essay in which you summarize the main points. Please include formulas and graphs as appropriate.

1. In this chapter we have studied confidence intervals. Carefully read the following statements about confidence intervals.

(a) Once the endpoints of the confidence interval are numerically fixed, the parameter in question (either μ or p) does or does not fall inside the "fixed" interval.

(b) A given fixed interval either does or does not contain the parameter μ or p; therefore, the probability is 1 or 0 that the parameter is in the interval.

Next, read the following statements. Then discuss all four statements in the context of what we actually mean by a confidence interval.

(c) Nontrivial probability statements can be made only about variables, not about constants.

(d) The confidence level c represents the proportion of all (fixed) intervals that would contain the parameter if we repeated the process many, many times.

2. Throughout Chapter 8 we have used the normal distribution, the central limit theorem, or Student's t distribution.

(a) Give a brief outline describing how confidence intervals for means use the central limit theorem or Student's t distribution in their basic construction.

(b) Give a brief outline describing how the normal approximation to the binomial distribution is used in the construction of confidence intervals for a proportion p.

(c) Give a brief outline describing how the sample size for a predetermined error tolerance and level of confidence is determined from the normal distribution or the central limit theorem.

3. When results of a survey or a poll are published, the sample size is usually given as well as the margin of error. For example, suppose the *Honolulu Star Bulletin* reported that they surveyed 385 Honolulu residents and 78% said they favored mandatory jail sentences for people convicted of driving under the influence of drugs or alcohol (with a margin of error of 3 percentage points in either direction). Usually the confidence level of the interval is not given, but it is standard practice to use the margin of error for a 95% confidence interval when no other confidence level is given.

(a) The paper reported a point estimate of 78% with a margin of error of $\pm 3\%$. Write this information in the form of a confidence interval for p, the population proportion of residents favoring mandatory jail sentences for people convicted of driving under the influence. What is the assumed confidence level?

(b) The margin of error is simply the error due to using a sample instead of the entire population. It does not take into account the bias that might be introduced by the wording of the question, by the truthfulness of the respondents, or by other factors. Suppose the question was asked in this fashion: "Considering the devastating injuries suffered by innocent victims in auto accidents caused by drunken or drugged drivers, do you favor a mandatory jail sentence for those convicted of driving under the influence of drugs or alcohol?" Do you think the wording of the question would influence the respondents? Do you think the population proportion of those favoring mandatory jail sentences is accurately represented by a confidence interval based on responses to such a question? Explain your answer.

Suppose the question had been: "Considering the existing overcrowding of our prisons, do you favor a mandatory jail sentence for people convicted of driving under the influence of drugs or alcohol?" Do you think the population proportion of those favoring mandatory jail sentences is accurately represented by a confidence interval based on responses to such a question? Explain your answer.

The problems in this section may be done using statistical computer software or calculators with statistical functions. Displays and suggestions are given for Minitab (Release 12), the TI-83 graphing calculator, ComputerStat, and Excel.

Finding a Confidence Interval for a Population Mean μ (Small Sample)

Most statistical software packages for computers and some graphing calculators have commands for generating confidence intervals for data you have entered.

Minitab

To generate a 90% confidence interval for the mean based on the sample data in column C1, first use the data window to enter the data in a column. Then follow the menu choices ➤**Stat** ➤**Basic Statistics** ➤**1-Sample t.** In the dialogue box, select the column containing your data, select the confidence interval, and enter the percent confidence level.

TI-83

Confidence intervals on the TI-83 are accessed by using the **STAT** key. Then select **TESTS.** Option **8:TInterval** gives confidence intervals for μ based on the t distribution. See the Calculator Note in Section 8.2 for specific details on this option.

ComputerStat

Under the menu **Confidence Intervals,** select the program **Confidence Intervals for a Population Mean *MU*.** Then follow the instructions on the screen to enter your data and your confidence level.

Excel

Excel computes confidence intervals for the mean. However, regardless of sample size, it uses the normal distribution. The output is always the value of the maximal error of estimate E. Recall that for a normal distribution $E = z_c \sigma / \sqrt{n}$. Then, to construct the confidence interval the user must compute (or have Excel compute) the mean of the data. Finally, the user finds the lower and upper bounds of the confidence interval by adding and subtracting E from the mean. The menu choices ➤f_x (paste function) ➤**Statistical** ➤**Confidence produce** E. When asked for the value of alpha, use the value 100% − confidence level %. Enter the standard deviation (either s or σ) and the sample size.

Another option for finding the maximal error of estimate E is to use the menu options ➤**Tools** ➤**Data Analysis** ➤**Descriptive Statistics.** Select the cells containing the data and check the summary statistics and confidence level boxes. Then specify the confidence level. The maximal error of estimate E will be computed by using the sample standard deviation of your data and the normal distribution. The output will show the mean and the value of E.

APPLICATION 1

Cryptanalysis, the science of breaking codes, makes extensive use of language patterns. The frequency of various letter combinations is an important part of the study. A letter combination consisting of a single letter is a monograph, while combinations consisting of two letters are called digraphs, and those with three letters are called trigraphs. In the English language the most frequent digraph is the letter combination TH.

The characteristic rate of a letter combination is a measurement of its rate of occurrence. To compute the characteristic rate, count the number of occurrences of a given letter combination and divide by the number of letters in the text. For instance, to estimate the characteristic rate of the digraph TH, you could select a newspaper text and pick a random starting place. From that place mark off 2000 letters and count the number of times that TH occurs. Then divide the number of occurrences by 2000.

The characteristic rate of a digraph can vary slightly depending on the style of the author; so to estimate an overall characteristic frequency, you want to consider several samples of newspaper text by different authors. Suppose you did this with a random sample of 15 articles and found the characteristic rate of the digraph TH in the articles. The results follow.

0.0275	0.0230	0.0300	0.0255
0.0280	0.0295	0.0265	0.0265
0.0240	0.0315	0.0250	0.0265
0.0290	0.0295	0.0275	

(a) Find a 95% confidence interval for the mean characteristic rate of the digraph TH.

(b) Repeat part a for a 90% confidence interval. *Note:* you do not need to enter the data again; simply use the option to rerun the program with data you have already entered.

(c) Repeat part a for an 80% confidence interval.

(d) Repeat part a for a 70% confidence interval.

(e) Repeat part a for a 60% confidence interval.

(f) For each confidence interval in parts a–e, compute the length of the given interval. Do you notice a relation between the confidence level and the length of the interval?

A good reference for cryptanalysis is a book by Sinkov:

Sinkov, Abraham, *Elementary Cryptanalysis,* New York: Random House, 1968

This book gives other common digraphs and trigraphs.

Finding a Confidence Interval for a Proportion *p*

Minitab

For versions of Minitab earlier than Release 12, no direct commands exist for finding a confidence interval for *p*. In Release 12, use the menu selection ➤ **Stat** ➤ **Basic Statistics** ➤ **1 Proportion.** Select summarized Data. Enter the number of trials, the number of successes, and the confidence level.

TI-83

Press the **STAT** key and select **TESTS.** Option **A1-PropZInt** generates confidence intervals for a proportion *p*. See the Calculator Note in Section 8.3 for details.

ComputerStat

Choose **Confidence Intervals** from the main menu. Then select **Confidence Intervals for P the Probability of Success in a Binomial Distribution** and follow the instructions on the screen.

Excel

Excel has no direct commands for computing confidence intervals for proportions.

APPLICATION 2

There must be nurses on duty in hospitals around the clock. Therefore, many nurses work various shifts. Tasto, Colligan, et al. did a study of the health consequences of shift work. Their results are published in the following government document:

United States Department of Health, Education, and Welfare, NIOSH Technical Report, Tasto, Colligan, et al., *Health Consequences of Shift Work,* Washington: GPO, 1978, p. 25.

They used a large random sample of nurses on various shifts in 12 hospitals. Part of the report concerns the number of sick days taken by nurses on various shifts. A random sample of 315 day-shift nurses showed that 62 took no sick days during a six-month period. During that same period a random sample of 309 nurses on rotating duty showed that 51 took no sick days.

(a) We wish to estimate the proportion p of rotating-shift nurses who take no sick days in a six-month period. Find a c% confidence interval for p when $c = 98, 90, 85, 75$, and 60. Notice that you enter the data only once and rerun the program with the same data for the different c values.

(b) For each confidence interval in part a, compute the length of the interval. Do you notice a relation between the confidence level and the length of the interval?

(c) We wish to estimate the proportion p of day-shift nurses who take no sick days in a six-month period. Find a c% confidence interval for p when $c = 99, 95, 80, 70$, and 60. Is there a relation between the confidence level and the length of the interval?

Computer Displays

TI-83 Display

Student's *t* Distribution with 5 Degrees of Freedom

Press the **DISTR** (**2nd VARS**) key and then select option **4:tpdf(x,d.f.)** to get the height of the graph of the t distribution with the designated degrees of freedom at the point $t = x$. To get the cumulative area between two values of t, use option **5:tcdf(lower t value, upper t value, d.f.).** Use the **DRAW** option of **DISTR** and

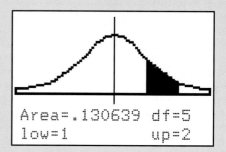

select **Shade_t(lower t value, upper t value, d.f.)** to draw the t distribution. **Shade_t(2,3,5)** shades the area between $t = 2$ and $t = 3$ under a t distribution with 5 degrees of freedom.

ComputerStat Display of Confidence Interval Demonstration

Select the menu item **Confidence Intervals,** and then choose **Confidence Interval Demonstration.** This program draws random samples of a designated size from a uniform distribution of numbers between 0 and 1. Then it creates 90% confidence intervals for the population mean μ. A graph of the intervals shows which ones actually contain μ.

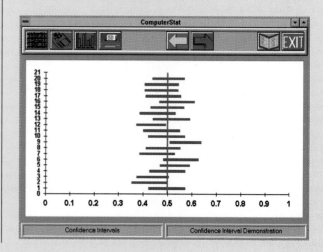

398

9 Hypothesis Testing Involving One Population

"Would you tell me, please, which way I ought to go from here?"
"That depends a good deal on where you want to get to,"
said the Cat.
"I don't much care where—" said Alice.
"Then it doesn't matter which way you go," said the Cat.

—Lewis Carroll
Alice's Adventures in Wonderland

Charles Lutwidge Dodgson (1832–1898)

Using the pseudonym Lewis Carroll, this English mathematician and author wrote *Alice's Adventures in Wonderland*.

Charles Dodgson was an English mathematician who loved to write children's stories in his free time. The dialogue above between Alice and the Cheshire Cat occurs in the masterpiece *Alice's Adventures in Wonderland*, written by Dodgson under the pen name Lewis Carroll. These lines relate to our study of hypothesis testing. Statistical tests cannot answer all of life's questions. They cannot always tell us "where to go," but after this decision is made on other grounds, they can help us find the best way to get there.

Section 9.1 Introduction to Hypothesis Testing

In Chapter 1 we emphasized the fact that a statistician's most important job is to draw inferences about populations based on samples taken from those populations. Most statistical inference centers around the parameters of a population (often the mean or probability of success in a binomial trial). Methods for drawing inferences about parameters are of two types: either we make decisions concerning the value of the parameter, or we actually estimate the value of the parameter. When we

399

estimate the value (or location) of a parameter, we are using methods of estimation as studied in Chapter 8. Decisions concerning the value of a parameter are obtained by hypothesis testing, the topic we shall study in this chapter.

Students often ask which method should be used on a particular problem—that is, should the parameter be estimated, or should we test a hypothesis involving the parameter? The answer lies in the practical nature of the problem and the questions posed about it. Some people prefer to test theories concerning the parameters. Others prefer to express their inferences as estimates. Both estimation and hypothesis testing are found extensively in the literature of statistical applications.

Null hypothesis

Our first step is to establish a working hypothesis about the population parameter in question. This hypothesis is called the *null hypothesis*, denoted by the symbol H_0. The value specified in the null hypothesis is often a historical value, a claim, or a production specification. For instance, if the average height of a professional male basketball player was 6.5 feet 10 years ago, we might use a null hypothesis H_0: $\mu = 6.5$ feet for a study involving the average height of this year's professional male basketball players. If television networks claim that the average length of time devoted to commercials in a 60-minute program is 12 minutes, we would use H_0: $\mu = 12$ minutes as our null hypothesis in a study regarding the length of time devoted to commercials. Finally, if a repair shop claims that it should take an average of 25 minutes to install a new muffler on a passenger automobile, we would use H_0: $\mu = 25$ minutes as the null hypothesis for a study of how well the repair shop is conforming to specified times for a muffler installation.

Alternate hypothesis

Any hypothesis that differs from the null hypothesis is called an *alternate hypothesis*. An alternate hypothesis is constructed in such a way that it is the one to be accepted when the null hypothesis must be rejected. The alternate hypothesis is denoted by the symbol H_1. For instance, if we believe the average height of professional male basketball players is greater than it was 10 years ago, we would use an alternate hypothesis H_1: $\mu > 6.5$ feet with the null hypothesis H_0: $\mu = 6.5$ feet.

EXAMPLE 1 ▷

A car manufacturer advertises that its new subcompact models get 47 miles per gallon (mpg). Let μ be the mean of the mileage distribution for these cars. You assume that the dealer will not underrate the cars, but you suspect that the mileage might be overrated.

(a) What shall we use for H_0?

We want to see if the dealer's claim $\mu = 47$ can be rejected. Therefore, our null hypothesis is simply that $\mu = 47$. We denote the null hypothesis as

$$H_0: \mu = 47$$

(b) What shall we use for H_1?

From experience with this dealer we have every reason to believe that the advertised mileage is too high. If μ is not 47, we are sure it is less than 47. Therefore, the alternate hypothesis is

$$H_1: \mu < 47$$

GUIDED EXERCISE   1

A company manufactures ball bearings for precision machines. The average diameter of a certain type of ball bearing should be 6.0 mm. To check that the average diameter is correct, the company formulates a statistical test.

(a) What should be used for H_0? (*Hint:* What is the company trying to test?)

⇨ If μ is the mean diameter of the ball bearings, the company wants to test $\mu = 6.0$ mm. Therefore, $H_0\colon \mu = 6.0$.

(b) What should be used for H_1? (*Hint:* An error either way, too small or too large, would be serious.)

⇨ An error either way could occur, and it would be serious. Therefore, $H_1\colon \mu \neq 6.0$ (μ is either smaller than or larger than 6.0).

GUIDED EXERCISE   2

A package delivery service claims it takes an average of 24 hours to send a package from New York to San Francisco. An independent consumer agency is doing a study to test the truth of this claim. Several complaints have led the agency to suspect that the delivery time is longer than 24 hours.

(a) What should be used for the null hypothesis?

⇨ The claim $\mu = 24$ hours is in question, so we take $H_0\colon \mu = 24$.

(b) Assuming that the delivery service does not underrate itself, what should be used for the alternate hypothesis?

⇨ If the delivery service does not underrate itself, then the only reasonable alternate hypothesis is $H_1\colon \mu > 24$.

Types of errors

If we *reject the null hypothesis when it is in fact true,* we have an error that is called a *type I error*. On the other hand, if we *accept the null hypothesis when it is in fact false,* we have made an error that is called a *type II error*. Table 9-1 summarizes these results.

Table 9-1 Type I and Type II Errors

	Our Decision	
Truth of H_0	We do not reject H_0	We reject H_0
H_0 is true	Correct decision; no error	Type I error
H_0 is false	Type II error	Correct decision; no error

In order for tests of hypotheses to be well constructed, they must be designed to minimize possible errors of decision. (Usually we do not know if an error has been made, and therefore, we can only talk about the probability of making an error.) Usually for a given sample size an attempt to reduce the probability of one type of error results in an increase in the probability of the other type of error. In practical applications, one type of error may be more serious than another. In such a case, careful attention is given to the more serious error. If we increase the sample size, it is possible to reduce both types of errors, but increasing the sample size may not be possible.

Level of significance

The probability with which we are willing to risk a type I error is called the *level of significance* of a test. The level of significance is denoted by the Greek letter α (pronounced "alpha"). In good statistical practice, α is specified in advance before any samples are drawn so that results will not influence the choice of the level of significance.

The probability of making a type II error is denoted by the Greek letter β (pronounced "beta"). Methods of hypothesis testing require us to choose α and β values to be as small as possible. In elementary statistical applications we usually choose α first.

Power of a test

The quantity $1 - \beta$ is called the *power* of the test and represents the probability of rejecting H_0 when it is in fact false. For a given level of significance, how much power can we expect from a test? The actual value of the power is usually difficult (and sometimes impossible) to obtain, since it requires us to know the H_1 distribution. However, we can make the following general comments.

1. The power of a statistical test increases as the level of significance α increases. A test performed at the $\alpha = 0.05$ level has more power than one at $\alpha = 0.01$. This means that the less stringent we make our significance level α, the more likely it is that we will reject the null hypothesis when it is false.

2. Using a larger value of α will increase the power, but it also will increase the probability of a type I error. Despite this fact, most business executives, administrators, social scientists, and scientists use *small* α values. This choice reflects the conservative nature of administrators and scientists, who are usually more willing to make an error by failing to reject a claim (i.e., H_0) than to make an error by accepting another claim (i.e., H_1) that is false. Table 9-2 summarizes the probabilities of errors associated with a statistical test.

COMMENT Since the calculation of the probability of a type II error is treated in advanced statistics courses, we will restrict our attention to the probability of a type I error. $\circ$

Meaning of accepting H_0

In most statistical applications, the level of significance is specified to be $\alpha = 0.05$ or $\alpha = 0.01$, although other values can be used. If $\alpha = 0.05$, then we say we are using a 5% level of significance. This means that in 100 similar situations, H_0 will be rejected 5 times, on average, when it should not have been rejected.

Table 9-2 Probabilities Associated with a Statistical Test

	Our Decision	
Truth of H_0	We accept H_0 as true	We reject H_0 as false
H_0 is true	Correct decision, with corresponding probability $1 - \alpha$	Type I error, with corresponding probability α called the *level of significance* of the test
H_0 is false	Type II error, with corresponding probability β	Correct decision, with corresponding probability $1 - \beta$ called the *power* of the test

When we accept (or fail to reject) the null hypothesis, we should understand that we are *not proving the null hypothesis*. We are only saying that the sample evidence (data) is not strong enough to justify rejection of the null hypothesis. The word *accept* sometimes has a stronger meaning in common English usage than we are willing to imply in our application of statistics. Therefore, we often use the expression *fail to reject* H_0 instead of accept H_0. *Fail to reject* the null hypothesis simply means the evidence in favor of rejection was not strong enough (see Table 9-3). Often, in the case in which H_0 cannot be rejected, a confidence interval is used to estimate the parameter in question. The confidence interval gives the statistician a range of possible values for the parameter.

Table 9-3 Meanings of the Terms *Fail to Reject* H_0 and *Reject* H_0

Term	Meaning
Fail to reject H_0	There is not enough evidence in the data (and the test being used) to justify a rejection of H_0. This means that we retain H_0 with the understanding that we have not proved it to be true beyond all doubt.
Reject H_0	There is enough evidence in the data (and the test employed) to justify rejection of H_0. This means that we choose the alternate hypothesis H_1 with the understanding that we have not proved H_1 to be true beyond all doubt.

EXAMPLE 2 ❯ Let's reconsider Example 1 regarding the average gas mileage of a compact car. The hypotheses for the test are

H_0: $\mu = 47$ mpg (manufacturer's claim)

H_1: $\mu < 47$ mpg (consumer researcher's suspicion)

(a) Suppose the level of significance is $\alpha = 0.05$. Describe a type I error and its probability.

SOLUTION: If, based on sample evidence, we reject the manufacturer's claim that $\mu = 47$ mpg when in fact the average number of miles per gallon achieved by the compact car is 47 mpg (or higher), we have committed a type I error. The probability of making such an error is as high as 5% because the level of significance is 5%.

(b) Describe a type II error.

SOLUTION: If, based on sample evidence, we fail to reject the manufacturer's claim of $\mu = 47$ mpg when in fact the alternate hypothesis $H_1: \mu < 47$ mpg is true, we have made a type II error. The probability of such an error is designated by the letter β. (The computation of β is beyond the scope of this text.)

GUIDED EXERCISE   3

Let's reconsider Guided Exercise 1 in which we considered the manufacturing specifications for ball bearings. The hypotheses were

$H_0: \mu = 6.0$ mm (manufacturer's specification)

$H_1: \mu \neq 6.0$ mm (cause for adjusting process)

(a) Suppose the manufacturer requires a 1% level of significance. Describe a type I error, its consequences, and its probability.

A type I error is caused when sample evidence indicates that we should reject H_0 when in fact the average diameter of the ball bearings being produced is 6.0 mm. A type I error will cause a needless adjustment and delay of the manufacturing process. The probability of such an error is 1% because $\alpha = 0.01$.

(b) Discuss a type II error and its consequences.

A type II error occurs if the sample evidence tells us not to reject the null hypothesis $H_0: \mu = 6.0$ mm when in fact the average diameter of the ball bearing is either too large or too small to meet specifications. Such an error would mean that the production process would not be adjusted when it really needed to be adjusted. This could possibly result in a large production of ball bearings that do not meet specifications. Perhaps a control chart would be a useful device to monitor the diameters of the ball bearings (*see* Section 6.1).

Critical regions

We use information from a sample to determine if we should reject or not reject the null hypothesis. However, we know that there will be variability among samples. For instance, if we test drive 40 of the compact automobiles discussed in Examples 1 and 2 and compute the average miles per gallon for the 40 automobiles, we know we are likely to find that $\bar{x}$ differs from the null-hypothesis claim H_0: $\mu = 47$ mpg. A difference is likely to occur regardless of the truth of H_0. The question is how much less than 47 mpg can $\bar{x}$ be before we begin to suspect that the population mean mileage μ is less than 47, as stated in the alternate hypothesis. In other words, how much less can $\bar{x}$ be than 47 mpg before we reject H_0: $\mu = 47$ mpg and accept the alternate hypothesis H_1: $\mu < 47$ mpg?

The answer to the question regarding the relative sizes of $\bar{x}$ and μ as stated in the null hypothesis depends on the sample size, the sampling distribution of $\bar{x}$, the alternate hypothesis H_1, and the level of significance α. If the sample test statistic $\bar{x}$ is sufficiently different from the claim about μ made in the null hypothesis, we reject the null hypothesis. The range of values of $\bar{x}$ for which we reject H_0 is called the critical region of the $\bar{x}$ distribution. Depending on the alternate hypothesis, the critical region is located on the left side, the right side, or both sides of the $\bar{x}$ distribution. If the critical region is on the left side, we call the test a *left-tailed* test. Similarly, if the critical region is on the right side, the test is called a *right-tailed* test. If the critical region is on both sides, the test is called *two-tailed*.

H_1: $\mu < k$	H_1: $\mu > k$	H_1: $\mu \neq k$
Left-tailed test	**Right-tailed test**	**Two-tailed test**

Figure 9-1 shows the relationship of the critical region to the alternate hypothesis and the level of significance α.

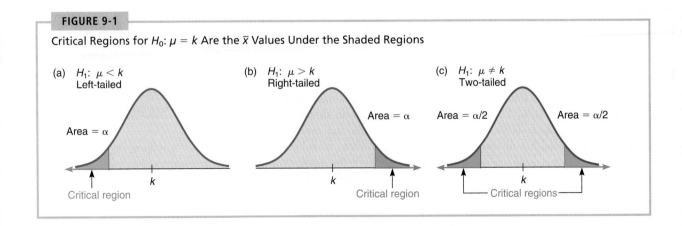

FIGURE 9-1

Critical Regions for H_0: $\mu = k$ Are the $\bar{x}$ Values Under the Shaded Regions

(a) H_1: $\mu < k$
Left-tailed

Area = α

k

Critical region

(b) H_1: $\mu > k$
Right-tailed

Area = α

k

Critical region

(c) H_1: $\mu \neq k$
Two-tailed

Area = $\alpha/2$ Area = $\alpha/2$

k

Critical regions

In the next section we will see how to use the level of significance α and the alternate hypothesis H_1 to compute the boundaries of the critical region(s). Then we will see whether or not the sample test statistic falls in the critical region. If it does, we reject H_0. If it does not, we do not reject H_0.

Summary

In Chapter 8 we used *confidence intervals* to *estimate* the value of a parameter such as the population mean μ or probability of success p. Theory of estimation is very useful and certainly has a valuable place in statistical applications. However, confidence intervals by themselves don't answer every important type of question. For example, there are many situations in real life where the final decision must be a *simple yes or no*. Such a decision is based on the supposed value of a parameter, sample data, and a probability distribution involving the parameter.

Yes-or-no decisions are the essence of *hypothesis testing*, and this has been the main topic of Chapter 9. When we make such yes-or-no decisions about the value of a statistical parameter, we realize that our decision could be wrong. Therefore, we also need to consider the types of errors and associated probabilities for such errors.

In this section we have introduced some of the basic concepts of hypothesis testing. In particular, we have seen the null and alternate hypotheses, the level of significance, the types of errors possible, and the critical region. In the next section we will see how to use information from a sample to conclude the test and determine whether or not to reject the null hypothesis.

V I E W P O I N T

Surf's Up!

How do judges score finalists in the Triple Crown world surfing championship? The Association of Surfing Professionals (ASP) panel of judges uses criteria such as length of ride, wave selection, personal style, and smooth dynamic maneuvers. A 10-point scale is used for each finalist to evaluate the top 5 waves in a 15-wave minimum. The ASP judges enter their scores directly into a computer, where data are stored and overall tallies, averages, and ranks are computed for  each championship finalist. There is a strong element of subjective evaluation in the judges' opinions about personal style and wave selection. However, the *interpretation* of finalists' scores is another example of applied inferential statistics. For more information, see Web site <http://holoholo.org/> and follow the links to Surfing News and the Triple Crown.

SECTION 9.1 PROBLEMS

General: Understanding
Terminology

1. Discuss each of the following topics in class or review the topics on your own. Then write a brief but complete essay in which you answer the following questions.
    (a) What is a null hypothesis H_0?
    (b) What is an alternate hypothesis H_1?
    (c) What is a type I error? What is a type II error?
    (d) What is the level of significance of a test? What is the probability of a type II error?

General: Understanding
Terminology

2. In a statistical test we have a choice of a left-tailed critical region, right-tailed critical region, or two-tailed critical region. Is it the null hypothesis or the alternate hypothesis that is used to determine which type of critical region is used? Explain your answer.

General: Understanding
Terminology

3. If we fail to reject (i.e., "accept") the null hypothesis, does this mean that we have *proved* it to be true beyond *all* doubt? Explain your answer.

General: Understanding
Terminology

4. If we reject the null hypothesis, does this mean that we have *proved* it to be false beyond *all* doubt? Explain your answer.

Veterinary Sciences: Colts

5. The body weight of a healthy 3-month-old colt should be about $\mu = 60$ kg. (Source: *The Merck Veterinary Manual,* a standard reference manual used in most veterinary colleges.)
    (a) If you want to set up a statistical test to challenge the claim that $\mu = 60$ kg, what would you use for the null hypothesis H_0?
    (b) In Nevada there are many herds of wild horses. Suppose that you wanted to test the claim that the average weight of a wild Nevada colt (3 months old) was less than 60 kg. What would you use for the alternate hypothesis H_1?
    (c) Suppose that you wanted to test the claim that the average weight of such a wild colt was greater than 60 kg. What would you use for the alternate hypothesis?
    (d) Suppose that you wanted to test the claim that the average weight of such a wild colt was *different* from 60 kg. What would you use for the alternate hypothesis?
    (e) For each of the tests in parts b, c, and d, would the critical region be on the left, right, or both sides of the mean? Explain your answer in each case.

Business: Honolulu Office
Space

6. Suppose that you are the chief executive officer of an independent real estate business who wants to rent office space in Hawaii. You read the *Wall Street Journal* and find the average cost of "class A" office space in downtown Honolulu is $\mu = \$20.04$ per square foot.
    (a) If you want to set up a statistical test to challenge the claim that $\mu = \$20.04$, what would you use for the null hypothesis?
    (b) Suppose that you are also interested in office space on the Island of Maui. You want to test the claim that office space there has a higher average rent than $20.04 per square foot. What would you use for the alternate hypothesis?
    (c) You are also planning to set up offices on the Island of Molokai. You want to test the claim that the average rent there is less than $20.04 per square foot. What would you use for the alternate hypothesis?

(d) Your company is considering locating an office on the Big Island (Island of Hawaii). You want to test the claim that the average rent there is *different* from $20.04 per square foot. What would you use for the alternate hypothesis?

(e) For each of the tests in parts b, c, and d, would the critical region be on the left, right, or both sides of the mean? Explain your answer in each case.

Pro-football Players: Weight

7. *The Sports Encyclopedia: Pro Football* (11th Edition) indicated that the mean weight for football players in the Western division was approximately 288 pounds in 1992.

(a) Suppose that we want to set up a statistical test that challenges the statement that the mean weight this year is 288 pounds. What would you use for the null hypothesis?

(b) What would you use for the alternate hypothesis if you thought that the average weight was higher, lower, or different (either way)? In each case, would the critical region be in the left, right, or both tails? Explain each answer.

Section 9.2

Tests Involving the Mean μ (Large Samples)

The general procedure for hypothesis testing involves several steps, some of which we examined in the preceding section.

1. Establish the null hypothesis H_0.

2. Establish the alternate hypothesis H_1.

3. Use the level of significance α and the alternate hypothesis to determine the critical region.

In this section we will complete the following steps.

4. Find the critical values that form the boundaries of the critical region(s).

5. Use the sample evidence to draw a conclusion regarding whether or not to reject the null hypothesis H_0.

The basic concepts for hypothesis testing remain the same for all tests of hypotheses, regardless of the parameter in question (such as mean, proportion, difference of means, difference of proportions, or standard deviation). However, the specific details will change depending on the particular parameter being tested and the nature of the sampling distribution of the test statistic.

In this section we will study the particular details of hypothesis tests about a population mean μ when our sample evidence comes from large samples ($n \geq 30$).

Null and alternate hypotheses

We first establish the null and alternate hypotheses. For tests of means, Table 9-4 shows the hypotheses.

To decide whether or not to reject H_0, we look at sample evidence. From the sample, we compute $\bar{x}$. The question is: Is $\bar{x}$ far enough away from the value of μ hypothesized in H_0 to merit rejecting H_0? To answer this question, we look at the $\bar{x}$ distribution *under the assumption that H_0 is true.*

Table 9-4 The Null and Alternate Hypotheses

Null Hypothesis	Alternate Hypotheses and Type of Test		
Claim about μ or historical value of μ	You believe μ is less than value stated in H_0	You believe μ is more than value stated in H_0	You believe μ is different from value stated in H_0
H_0: $\mu = k$	H_1: $\mu < k$	H_1: $\mu > k$	H_1: $\mu \neq k$
	Left-tailed test	Right-tailed test	Two-tailed test

Critical values for testing μ (large samples)

According to the central limit theorem, the $\bar{x}$ distribution based on samples of size n, where n is large ($n \geq 30$), is approximately *normal*. For this reason, our critical values are z values. Figure 9-2 on page 410 gives the critical values for one- and two-tailed tests for two commonly used levels of significance, $\alpha = 0.05$ and $\alpha = 0.01$. Table 4(c) of Appendix I also gives the critical values for a normal distribution.

The critical values given in Figure 9-2 are based on the fact that the *area* above the critical region equals the level of significance. For instance, for a left-tailed test and $\alpha = 0.05$, the area above the critical region is 0.05. Using methods of Chapter 6 and the normal distribution table (Table 4 in Appendix I), we see that $z = -1.645$ is the z value that "cuts off" 5% of the normal curve on the left.

Sample test statistic

To determine if the sample $\bar{x}$ is sufficiently far away from the mean proposed in the null hypothesis to merit rejecting H_0, we convert the sample value $\bar{x}$ to a z value and see if the sample z value falls in the critical region. If so, we reject H_0 and accept H_1. If not, we do not reject H_0. The conversion of $\bar{x}$ to z follows the formula

$$z = \frac{\bar{x} - \mu}{(\sigma/\sqrt{n})}$$

where μ = mean specified in H_0

$\qquad \sigma$ = standard deviation of the x distribution

$\qquad n$ = sample size being used

$\qquad \bar{x}$ = sample test statistic

COMMENT *The methods used in this section will require $\bar{x}$ to be approximately normally distributed*, and we will approximate σ by the sample standard deviation s when necessary. To ensure that our methods yield accurate results, we will *require our sample size to be 30 or larger* ($n \geq 30$). Tests involving small samples ($n < 30$) will be presented in Section 9.4.

FIGURE 9-2

Critical Values z_0 for Tests Involving a Mean (Large Samples)

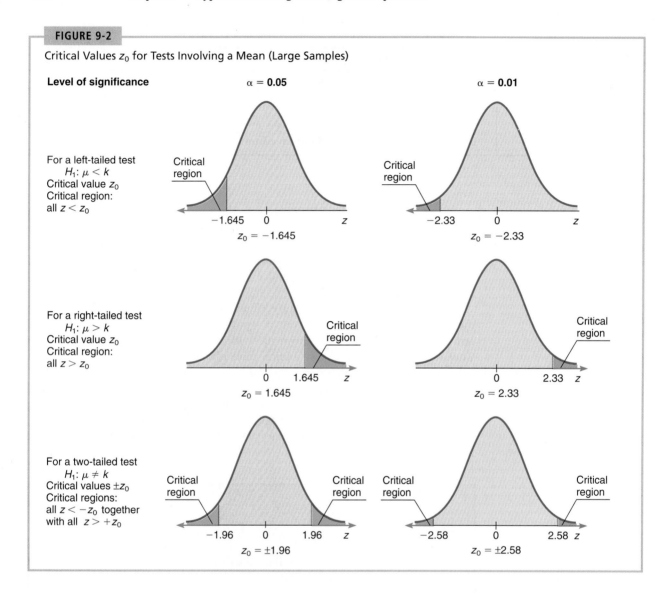

For a left-tailed test
$H_1: \mu < k$
Critical value z_0
Critical region:
all $z < z_0$

For a right-tailed test
$H_1: \mu > k$
Critical value z_0
Critical region:
all $z > z_0$

For a two-tailed test
$H_1: \mu \neq k$
Critical values $\pm z_0$
Critical regions:
all $z < -z_0$ together
with all $z > +z_0$

EXAMPLE 3 ➤ *Statement of Problem:* The St. Louis Zoo wishes to obtain eggs of a rare turtle found near the Mississippi River. The zoo will hatch the eggs and raise the turtles as an exhibit of a rare and endangered species. Carol Wright, the staff biologist at the zoo, has been given the job of finding the eggs to be hatched. Turtles of the area bury their eggs in a nest in sandbanks along the river. Then the nest is abandoned, and the eggs hatch by themselves. A number of different species of turtles live in the region, and eggs from each species look much alike. Past research has shown that lengths of turtle eggs are normally distributed, and lengths of the rare turtle eggs have population mean $\mu = 7.50$ cm with standard deviation $\sigma = 1.5$ cm. The rare turtle egg is the only one with mean length 7.50 cm. The mean lengths of eggs from each of the other species are longer than 7.50 cm. The eggs of all the local turtle species have the same standard deviation $\sigma = 1.5$ cm for length.

After searching for some time, Carol finds a nest with 36 eggs. The way the nest was constructed makes her suspect that it was made by the rare turtle. The mean length of this collection of 36 eggs is $\bar{x} = 7.74$ cm. Since $\bar{x} = 7.74$ cm is longer than the population mean $\mu = 7.50$ cm of the rare turtle, Carol is a little worried that the eggs may come from a species that lays larger eggs.

Let's use a statistical test to help the biologist make a decision. Pay close attention to this example; it contains principal features common to most statistical tests.

I. *Summary of Known Facts:* In any testing problem it is a good idea to make a short summary of known facts before we start to construct the test.

(a) The distribution of lengths of eggs from the rare turtle is *normal*, with population mean $\mu = 7.50$ cm and standard deviation $\sigma = 1.5$ cm.

(b) The nest contains $n = 36$ eggs with an observed mean $\bar{x} = 7.74$ cm.

(c) Since the population is *normal* and μ and σ are known, Theorem 7.1 tells us that $\bar{x}$ is also normally distributed. The mean and standard deviation of the $\bar{x}$ distribution are

$$\mu_{\bar{x}} = \mu = 7.50 \text{ cm}$$

$$\sigma_{\bar{x}} = \frac{s}{\sqrt{n}} = \frac{1.5}{\sqrt{36}} = 0.25 \text{ cm}$$

II. *Establishing H_0 and H_1:* The null hypothesis H_0 is set up for the primary purpose of seeing whether or not it can be rejected. The alternate hypothesis H_1 will be chosen when the null hypothesis must be rejected.

Let μ be the mean length of the population distribution from which our sample of 36 eggs is drawn. Our biologist suspects that the nest contains the rare turtle eggs; therefore, we will use the null hypothesis

$H_0: \mu = 7.50$

since the rare turtle eggs are known to have mean length $\mu = 7.50$ cm. However, all other local species lay eggs with a longer population mean. Therefore, the alternate hypothesis is

$H_1: \mu > 7.50$

III. *Choosing the Level of Significance α:* The null hypothesis says that the eggs in the nest are from the rare turtle. A type I error means that we reject the nest as being formed by the rare turtle when it was in fact formed by the rare turtle. A type II error means that we accept the nest as coming from the rare turtle when it really came from a common species.

A type I error could be serious; we don't want to reject the eggs if they are from the rare turtle. A type II error is not too serious. If the eggs are from a common species, the turtles can be released after they hatch. Naturally, the zoo doesn't want to incubate the wrong eggs, but it would not be too serious if it did.

Although other levels of significance may be used, most researchers use either $\alpha = 0.05$ or $\alpha = 0.01$. Our biologist has agreed to use a level of significance $\alpha = 0.01$. This means that she is willing to risk a type I error with a probability of $\alpha = 0.01$.

IV. *Graphical Model for the Test:* What decision procedure shall we use to test our hypothesis? In a way, the logic of our decision process is similar to that used in a typical courtroom setting, but our methods are mathematical rather than legal. In a courtroom setting, the person charged with a crime is initially considered innocent (null hypothesis). If evidence (data) presented in court can sufficiently discredit the person's innocence, he or she is then judged to be guilty (alternate hypothesis). In a similar way, we will consider the null hypothesis to be *true* until there is enough data (mathematical evidence) to discredit it at the $\alpha = 0.01$ level of significance.

Figure 9-3 shows the graphical model for a test with hypotheses

$$H_0: \mu = 7.50$$
$$H_1: \mu > 7.50$$

and a large sample. Let us examine the details of Figure 9-3.

(a) First, we determine the *sampling distribution.* By the central limit theorem (Theorem 7.2), the test statistic $\bar{x}$ follows a *normal* distribution. We convert the $\bar{x}$ distribution to the standard normal distribution by the formula

$$z = \frac{\bar{x} - \mu}{(\sigma/\sqrt{n})}$$

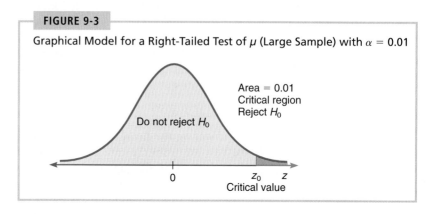

FIGURE 9-3

Graphical Model for a Right-Tailed Test of μ (Large Sample) with $\alpha = 0.01$

where μ is given in H_0, σ is given or estimated from the sample (large sample), and n is the sample size. In our case, $\mu = 7.50$, $\sigma = 1.5$, and $n = 36$. Figure 9-3 is the standard normal distribution.

(b) Next, we find the critical region. Since H_1: $\mu > 7.50$ claims that μ is greater than 7.50, the *critical region* for this problem is a *right tail* of the standard normal distribution. This tail has an area of $\alpha = 0.01$. If the z value corresponding to the sample statistic $\bar{x} = 7.74$ falls in the critical region, we say that there is enough evidence to discredit the null hypothesis, and we *reject* H_0 at the $\alpha = 0.01$ level of significance. If our observed value falls outside the critical region, we conclude that the evidence is not strong enough to reject H_0, so in this case we *fail to reject* H_0. Since H_1: $\mu > 7.50$ claims that μ is greater than 7.50, then an observed sample statistic $\bar{x}$ *far enough to the right* of $\mu = 7.50$ discredits the null hypothesis and supports the alternate hypothesis. In other examples we will encounter left-tailed and two-tailed tests.

(c) The value z_0 is called the *critical value* for the test. It is the separation point for the critical region. If the z value of the sample statistic $\bar{x} = 7.74$ falls to the *left* of z_0, we *fail to reject* H_0. If it falls to the *right* of z_0, we *reject* H_0. Once we know z_0 and convert the sample statistic $\bar{x}$ to z, we can quickly finish the test.

V. *Finding the Critical Value, Critical Region, and z Value for $\bar{x}$*: Figure 9-2 (page 410) gives critical values z_0 to use for tests with sampling distributions that are normal. We find these values as we did in Chapter 6 by using Table 4, "Areas of a Standard Normal Distribution," in Appendix I. In this problem we have a right-tailed test with level of significance $\alpha = 0.01$. We want to find the value z_0 so that 1% of the standard normal curve lies to the right of z_0. Since the areas in Table 4 go from 0 to z, we look up z so that the area from 0 to z is $0.5000 - \alpha = 0.5000 - 0.0100 = 0.4900$. This gives us the value $z_0 = 2.33$. The critical region is all values of z greater than or equal to $z_0 = 2.33$ (see Figure 9-4 on page 414).

The sample "evidence" we have is $\bar{x} = 7.74$ based on a nest of 36 eggs. We convert $\bar{x}$ to z using the formula in part IV:

$$z = \frac{\bar{x} - \mu}{(\sigma/\sqrt{n})} = \frac{7.74 - 7.50}{(1.5/\sqrt{36})} = 0.96$$

Now we have all the information we need to complete the test.

VI. *Conclusion:* Earlier we pointed out a similarity between the logic of a statistical test and proceedings in a court of law. We are now at the point where all the sample evidence has been presented, and we are awaiting a decision. Who makes the decision? In a court, the jury or judge would make the decision, but in our statistical test, it is Mother Nature who makes the final decision. The observed value $\bar{x} = 7.74$ cm for the length of turtle eggs taken from the turtle

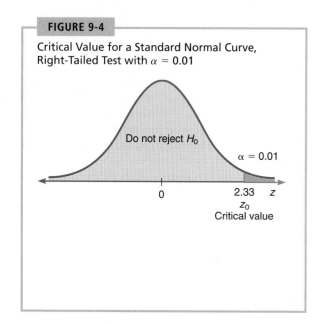

FIGURE 9-4

Critical Value for a Standard Normal Curve, Right-Tailed Test with $\alpha = 0.01$

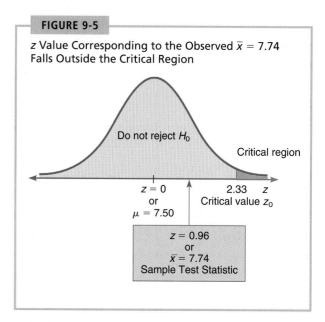

FIGURE 9-5

z Value Corresponding to the Observed $\bar{x} = 7.74$ Falls Outside the Critical Region

nest corresponds to the z value 0.96. This z value does not lie in the critical region (see Figure 9-5).

We conclude that there is not enough evidence to discredit the null hypothesis at the $\alpha = 0.01$ level of significance. Therefore, we do not reject H_0: $\mu = 7.50$, and we still think that the nest could have been made by the rare endangered species of turtle. The biologist should collect the eggs and take them to the zoo for incubation. It is important to remember that we do not claim to have *proved* that the eggs are from the rare turtle; all we say is that there is not enough evidence to reject H_0, and therefore, we "accept" it.

VII. *Meaning of α and β:* Remember that β is the probability of accepting H_0 when it is false. In our setting, β is the probability that the zoo incubates and hatches eggs from a common turtle species. To calculate β requires knowledge of the H_1 distribution. Since there are several species of common turtles in the area and we don't know which species made the nest, we really don't know the H_1 distribution. However, a type II error is not too serious because if the wrong eggs (from the common turtle) are hatched, there is little harm done. The baby turtles simply can be released near the place the eggs were found.

The level of significance α is the probability of rejecting H_0 when it is true. This is the probability of concluding that the eggs are not from the rare turtle species when they in fact are. It would be a serious mistake to reject the nest of the rare turtle. To guard against such a mistake, we have taken a relatively small α value of 0.01.

A research meteorologist has been studying wind patterns over the Pacific Ocean. Based on these studies, a new route is proposed for commercial airlines going from San Francisco to Honolulu. The new route is intended to take advantage of existing wind patterns to reduce flying time. It is known that for the old route the distribution of flying times for a large four-engine jet has mean $\mu = 5.25$ hours with standard deviation $\sigma = 0.6$ hour. Thirty-six flights on the new route have yielded a mean flying time of $\bar{x} = 4.90$ hours. Does this indicate that the average flying time for the new route is less than 5.25 hours? Use a 5% level of significance.

(a) What is H_0?

⇨ The average time for the old route is $\mu = 5.25$ hours, so we will set up the null hypothesis as

$$H_0: \mu = 5.25$$

In words, we are saying that the mean flying time on the new route is the same as that on the old route.

(b) What is H_1?

⇨ We want to see if the average flying time for the new route is less than 5.25 hours, so the alternate hypothesis is

$$H_1: \mu < 5.25$$

(c) What type of critical region must be used?

⇨ Since the < symbol is used in the alternate hypothesis, Figure 9-1 tells us to use a left-tailed test.

(d) What is the critical value?

⇨ By Figure 9-2 (or using Table 4 in Appendix I), the critical value for a left-tailed test with $\alpha = 0.05$ is $z_0 = -1.645$.

(e) What is the z value corresponding to the sample test statistic $\bar{x} = 4.90$?

⇨ We use the formula (from the central limit theorem)

$$z = \frac{\bar{x} - \mu}{(\sigma/\sqrt{n})}$$

with $\mu = 5.25$ (the value specified in H_0), $\sigma = 0.6$, and $n = 36$. Therefore,

$$z = \frac{4.90 - 5.25}{(0.6/\sqrt{36})} = -3.50$$

(f) Do we reject or fail to reject H_0?

⇨ Look at the critical region in Figure 9-6. The sample z value corresponding to the sample test statistic $\bar{x} = 4.90$ falls in the critical region. We reject H_0, and choose H_1 at the $\alpha = 0.05$ level. This means that we conclude that the flying time on the new route is less than the flying time on the old route.

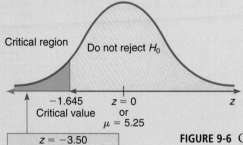

FIGURE 9-6 Critical Region, $\alpha = 0.05$

EXAMPLE 4 ❯   A large company has branch offices in several major cities of the world. From time to time it is necessary for company employees to move their families from one city to another. From long experience, the company knows that its employees move on the average of once every 8.50 years. However, trends for the past few years have led people to think that a change might have occurred. To determine if such a change has occurred, a random sample was taken of 48 employees (from the entire company). The employees were asked to provide either the number of years since the company asked them to move or the number of years employed by the company if they had never been asked to move. For this sample of 48 employees, the mean time was $\bar{x} = 7.91$ years with sample standard deviation 3.62 years.

Let us see if we can reject the hypothesis H_0: $\mu = 8.50$ at the $\alpha = 0.05$ level of significance. Since we have no way of knowing if the average moving time (μ) has increased or decreased, the alternate hypothesis will simply be H_1: $\mu \neq 8.50$.

(a) Do we use a right-tailed, left-tailed, or two-tailed test? Find the critical values.

SOLUTION: Since the alternate hypothesis uses the $\neq$ symbol, we use a two-tailed test. Since the sample size $n = 48$ is large, the sampling distribution $\bar{x}$ is approximately normal by the central limit theorem. Therefore, we can use Figure 9-2 to find the critical values. For $\alpha = 0.05$, Figure 9-2 on page 410 tells us that the critical values are $z_0 = \pm 1.96$.

(b) Convert the sample test statistic $\bar{x} = 7.91$ to a z value. Show the locations of the critical region and sample test statistic on the standard normal distribution.

SOLUTION: We convert $\bar{x} = 7.91$ years to a z value using $\mu = 8.50$ from H_0, $n = 48$, and estimating σ by $s = 3.62$:

$$z = \frac{\bar{x} - \mu}{(\sigma/\sqrt{n})} = \frac{7.91 - 8.50}{(3.62/\sqrt{48})} \approx -1.13$$

FIGURE 9-7

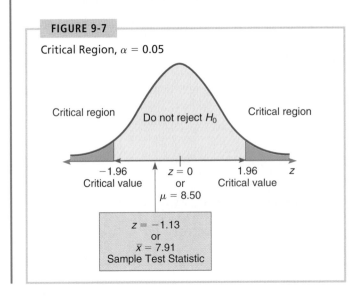

Critical Region, $\alpha = 0.05$

Calculator Note It is best to do the entire calculation for z internally on your calculator. For instance, on the TI-83, key in

$$(7.91 - 8.50) \div (3.62 \div \sqrt{(48)}) \qquad \text{ENTER}$$

Then round z to 2 places after the decimal. If you calculate the denominator $3.62/\sqrt{48}$ separately, be sure to carry at least four places after the decimal. Then divide the quantity $(7.91 - 8.50)$ by that number.

(c) Do we reject or fail to reject H_0 at the 5% level of significance?

> SOLUTION: The sample test statistic does not fall in the critical region (see Figure 9-7 on page 416). Therefore, we *fail to reject* H_0. It appears that employees still move on the average of once every 8.50 years.

GUIDED EXERCISE   5

A machine makes twist-off caps for bottles. The machine is adjusted to make caps of diameter 1.85 cm. Production records show that when the machine is so adjusted, it will make caps with mean diameter 1.85 cm and with standard deviation $\sigma = 0.05$ cm. During production, an inspector checks the diameters of caps to see if the machine has slipped out of adjustment. A random sample of 64 caps is taken. If the mean diameter for this sample is $\bar{x} = 1.87$ cm, does this indicate that the machine has slipped out of adjustment and the average diameter of caps is no longer $\mu = 1.85$ cm? (Use a 1% level of significance.)

(a) What is the null hypothesis:

$H_0: \mu = 0.05, \qquad H_0: \mu = 1.85,$

     or $H_0: \mu = 1.87$?

⇨ $H_0: \mu = 1.85$

Be sure to use the hypothesized parameter, *not* the sample statistic $\bar{x}$ in the null hypothesis.

(b) An error either way would be serious. Therefore, we want to test the null hypothesis against the hypothesis that the mean diameter is *not* 1.85 cm. So what is the alternate hypothesis:

$H_1: \mu < 1.85, \qquad H_1: \mu \neq 1.85,$

     or $H_1: \mu > 1.85$?

⇨ $H_1: \mu \neq 1.85$

(c) Should we use a one- or two-tailed test? What are the critical values?

⇨ Because the alternate hypothesis uses the $\neq$ symbol, we use a two-tailed test. Since the sample size $n = 64$ is large, the central limit theorem tells us that the sampling distribution of $\bar{x}$ is approximately normal. Therefore, we can use Figure 9-2 or Table 4 in Appendix I to find the critical values. For $\alpha = 0.01$, $z_0 = \pm 2.58$.

Exercise continues

Exercise continued

(d) What is the value of the sample test statistic? Convert it to a z value.

⇨ Since we are testing μ, the sample test statistic is $\bar{x}$. The sample of 64 bottle caps yielded a sample mean diameter $\bar{x} = 1.87$ cm. To convert $\bar{x}$ to z, we use $\mu = 1.85$ from H_0, $n = 64$, and $\sigma = 0.05$:

$$z = \frac{\bar{x} - \mu}{(\sigma/\sqrt{n})} = \frac{1.87 - 1.85}{(0.05/\sqrt{64})} = 3.20$$

(e) Show the critical region and the z value of the sample test statistic on the normal curve. Does the sample test statistic fall in the critical region or not? Do we reject or fail to reject H_0? At the 1% level of significance, can we say that the machine needs adjustment?

⇨ **FIGURE 9-8** Critical Region, $\alpha = 0.01$

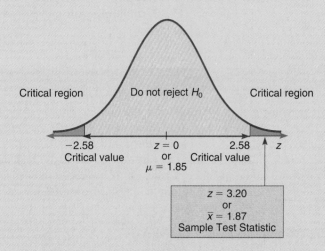

We see that the z value corresponding to the sample test statistic $\bar{x} = 1.87$ falls in the critical region shown in Figure 9-8. This means that we reject H_0 and conclude that at the 1% level of significance the machine needs adjustment.

Calculator Note The TI-83 is one of the calculators that, like computer statistics packages, supports hypothesis testing. For large sample tests of μ, we enter the value μ_0 used in the null hypothesis H_0: $\mu = \mu_0$. Then, to enter the alternate hypothesis, on the line next to μ we select the symbol ($\neq\mu_0$, $<\mu_0$, $>\mu_0$) used in the alternate hypothesis. Also required is an estimate for σ and either the sample statistics $\bar{x}$ and n or the raw data. The output typically provides the z value of the sample test statistic and the corresponding P value (discussed in Section 9.3). The following displays are from the TI-83 using data from Guided Exercise 5 regarding the diameter of twist-off caps.

Press **STAT**, select **TESTS**,
use **Option 1:Z-Test** with **Inpt:Stats**

```
Z-Test
 Inpt:Data Stats
 μ₀:1.85
 σ:.05
 x̄:1.87
 n:64
 μ:≠μ₀ <μ₀ >μ₀
Calculate Draw
```

Results of **Calculate**

```
Z-Test
 μ≠1.85
 z=3.2
 p=.0013744042
 x̄=1.87
 n=64
```

In the output we are given the z value 3.2, which corresponds to the sample mean $\bar{x} = 1.87$. We must compare the given sample z value with the known critical value(s) z_0 for the level of significance for the problem. In Guided Exercise 5, $\alpha = 0.01$, and we know the critical values are $z_0 = \pm 2.58$. The sample statistic $z = 3.2$ falls in the right-most critical region because 3.2 is greater than $z_0 = +2.58$, so we reject H_0.

Statistical significance

In a sense, the sample estimate of the parameter represents the evidence favoring rejection of H_0. For this reason, statisticians say the results of a random sample are *statistically significant* if the estimated parameter falls in the critical region of a test. In Guided Exercise 5 the random sample of 64 caps gave a sample test statistic $\bar{x} = 1.87$. Since the corresponding z value was in the critical region, we say that the results of the sample were statistically significant—that is, there was sufficient evidence to reject H_0 at the specified level of significance.

Statistical Significance

If we reject H_0, we say that the data are *statistically significant*.
If we do not reject H_0, we say that the data are *not statistically significant*.

V I E W P O I N T

Money, Money, Money!

What is the statistical profile of U.S. millionaires? Are they flashy spenders? Are they stingy or generous with friends? Do they drive big cars and wear expensive clothes? Do millionaires work or inherit their wealth? How many millionaires are there in the United States? What's the buying power of a million dollars today compared with 10 or 20 years ago? For a good statistical profile of U.S. millionaires, see *The Millionaire Next Door,* by T. J. Stanley and W. D. Danko. The book points out that most United States millionaires *earned* what they have, behave like ordinary people, and are not flashy or arrogant. A million dollars has a great deal less buying power than it used to have. Anyone who works hard and is willing to take some risks has a good chance of becoming a millionaire. A good understanding of statistics helps.

SECTION 9.2 PROBLEMS

For Problems 1 through 12, please provide the requested information.

(a) What is the null hypothesis? What is the alternate hypothesis? Will we use a left-tailed, right-tailed, or two-tailed test? What is the level of significance?

(b) What sampling distribution will we use? What is the critical value z_0 (or critical values $\pm z_0$)?

(c) Sketch the critical region and show the critical value (or critical values).

(d) Calculate the z value corresponding to the sample statistic $\bar{x}$ and show its location on the sketch in part c.

(e) Based on your answers for parts a to d, shall we reject or fail to reject (i.e., "accept") the null hypothesis at the given level of significance α? Explain your conclusion in the context of the problem.

(f) Are the data statistically significant?

Meteorology: Nor'easter Storms

1. *Weatherwise* is a magazine published in association with the American Meteorological Society. In one issue there is a rating system for classifying nor'easter storms that frequently hit New England states and can cause much damage near the coast. A *severe* storm has an average peak wave height of 16.4 feet for waves hitting the shore. Suppose that a nor'easter is in progress at the severe storm class rating. Peak wave heights are usually measured from land (using binoculars) off fixed cement peers. Suppose that a reading of 36 peak waves showed an average wave height of $\bar{x} = 15.1$ feet with sample standard deviation $s = 3.2$ feet. Does this information indicate that the storm is (perhaps temporarily) retreating from its severe rating? Use $\alpha = 0.01$.

Production: Auto Assembly Time

2. Let x be a random variable that represents assembly time for the Ford Taurus. The *Wall Street Journal* reported that the average assembly time for the Ford Taurus is $\mu = 38$ hours. A modification in the assembly procedure has been made. It is thought that the average assembly time may be reduced because of this modification. A random sample of 47 new Ford Taurus automobiles coming off the assembly line showed the average assembly time to be $\bar{x} = 37.5$ hours with sample standard deviation $s = 1.2$ hours. Does this indicate that the average assembly time has been reduced? Use $\alpha = 0.01$.

Management: Priority List

3. Message mania! According to the *Wall Street Journal*, a professional employee working in a large company receives an average of $\mu = 31.8$ calls per day. Most of the calls are from other employees in the company. Because of the large number of calls, employees find themselves distracted and are unable to concentrate when they return to their tasks. In an effort to reduce distraction caused by such interruptions, one company established a "priority list" that all employees were to use before making phone calls. One month after the new priorities were put into use, a random sample of 63 employees showed they were receiving an average of $\bar{x} = 28.5$ calls per day with sample standard deviation $s = 10.7$ calls per day. Use a 1% level of significance to test the claim that there has been a change (either way) in the average number of calls per day received per employee.

Marketing: Fashion Design

4. Judy Povich is a fashion design artist who designs the display windows in front of a large clothing store in New York City. Electronic counters at the entrances total the number of people entering the store each business day. Before Judy was hired by the store, the mean number of people entering the store each day was 3218. However, since Judy has started working, it is thought that this number has increased. A random sample of 42 business days after Judy began work gave an average $\bar{x} = 3392$ people entering the store each day. The sample standard deviation was $s = 287$ people. Does this indicate that the average number of people entering the store each day has increased? Use a 1% level of significance.

IRS: Tax on Tips

5. Maureen is a cocktail hostess in a very exclusive private club. The Internal Revenue Service is auditing her tax return this year. Maureen claims that her average tip last year was $4.75. To support this claim, she sent the IRS a random sample of 52 credit card receipts showing her bar tips. When the IRS got the receipts, they computed the sample average and found it to be $\bar{x} = \$5.25$ with sample standard deviation $s = \$1.15$. Do these receipts indicate that the average tip Maureen received last year was more than $4.75? Use a 1% level of significance.

Veterinary Science: Alfalfa

6. Let x be a random variable representing the percentage of protein content for early bloom alfalfa hay. The average percentage protein content of such early bloom alfalfa should be $\mu = 17.2\%$. (Source: *The Merck Veterinary Manual*, a manual commonly used in veterinary colleges.) A Colorado rancher is thinking of buying baled hay but suspects that the hay is from a later summer cutting with lower protein content. A small amount of hay was removed from each bale of a random sample of 50 bales. The average protein content from the samples was $\bar{x} = 15.8\%$ with sample standard deviation $s = 5.3\%$. Use a 5% level of significance to test the claim that this hay has lower average protein content than early bloom alfalfa.

Dodge Intrepid:
Acceleration Time

7. *Consumer Reports* indicated that the mean acceleration time (0 to 60 miles per hour) for the Dodge Intrepid was 10.2 seconds. In most tests of this type, regular unleaded gasoline is used. Suppose that 41 such tests were made using premium unleaded gasoline (octane over 91), and the sample mean acceleration time (0 to 60 miles per hour) was $\bar{x} = 9.7$ seconds with sample standard deviation $s = 2.1$ seconds. Does this indicate that premium gasoline tends to reduce average acceleration time (0 to 60 miles per hour)? Use $\alpha = 0.05$.

Mercury Sable: Breaking
Distance

8. *Consumer Reports* also indicated that the mean braking distance (from 60 miles per hour) on wet pavement for the Mercury Sable was 159 feet. Suppose that Sables equipped with tires having a new tread designed to grip the road better on wet pavement were used in 45 tests (braking from 60 miles per hour). The sample mean braking distance was $\bar{x} = 148$ feet with sample standard deviation $s = 23.5$ feet. Does this information indicate that the population mean braking distance for Mercury Sables on wet pavement is reduced for the new tire tread? Use $\alpha = 0.01$.

Health: Vitamins

9. Let x be a random variable that represents milliliters of oxygen per deciliter of whole blood. For healthy adults, the population mean of x is $\mu = 19.0$ milliliters of oxygen per deciliter. (Source: *The Merck Manual*, a commonly used reference in most medical schools and nursing programs.) A company that sells vitamins claims that its multivitamin complex will increase the oxygen capacity of the blood. A random sample of 48 adults took the vitamins for 6 months. After blood tests, it was found that the sample mean was $\bar{x} = 20.7$ milliliters of oxygen per deciliter with sample standard deviation $s = 9.9$. Use a 1% level of significance to test the claim that the average oxygen capacity has been increased.

Racing: Indianapolis 500

10. How fast are the Indianapolis 500 cars? In the history of the race from 1911 to 1997, the fastest average speed of a winner was $\mu = 233.72$ mph (Buddy Lazier, 1996, source: *USA Today*.) Suppose that this year the 35 front-line cars have a qualifying lap-time average speed $\bar{x} = 226.91$ mph with sample standard deviation $s = 16.83$ mph. Use a 1% level of significance to test the claim that the lap-time average speed for the front lineup is different from that of Buddy Lazier.

Health: Arthritis

11. Let x be a random variable that represents the pH of arterial plasma (i.e., the acidity of the blood). For healthy adults, the mean of the x distribution is $\mu = 7.4$ pH. (See source for Problem 9.) A new drug for arthritis has been developed. However, it is thought this drug might change blood pH. A random sample of 33 patients with arthritis took the drug for 3 months. Blood tests showed $\bar{x} = 8.1$ pH with sample standard deviation $s = 0.6$ pH. Use a 5% level of significance to test the claim that the drug has changed (either way) the mean pH of the blood.

Military Contractors: Mess Hall

12. An Air Force base mess hall has received a shipment of 10,000 gallon-size cans of cherries. The supplier claims that the average amount of liquid is 0.25 gallon per can. A government inspector took a random sample of 100 cans and found the average liquid content to be 0.28 gallon per can with a standard deviation of 0.10. Does this indicate that the supplier's claim is too low? (Use a 5% level of significance.)

Comparison: Statistical Test vs U.S. Court System

13. Compare similarities of statistical testing with legal methods used in a U.S. court setting. Then discuss the following topics in class or consider the topics on your own. Please write a brief but complete essay in which you answer the following questions.
    (a) In a court setting, the person charged with a crime is initially considered to be innocent. The claim of innocence is maintained until the jury returns with a decision. Explain how the claim of innocence could be taken to be the null hypothesis. Do we assume that the null hypothesis is true throughout the testing procedure? What would the alternate hypothesis be in a court setting?

(b) The court claims that a person is innocent if the evidence against the person is not adequate to find him or her guilty. This does not mean, however, that the court has necessarily *proved* the person to be innocent. It simply means that the evidence against the person was not adequate for the jury to find him or her guilty. How does this situation compare with a statistical test for which the conclusion is "do not reject" (i.e., accept) the null hypothesis? What would be a type II error in this context?

(c) If the evidence against a person is adequate for the jury to find him or her guilty, then the court claims that the person is guilty. Remember, this does not mean that the court has necessarily *proved* the person to be guilty. It simply means that the evidence against the person was strong enough to find him or her guilty. How does this situation compare with a statistical test for which the conclusion is "reject" the null hypothesis? What would be a type I error in this context?

(d) In a court setting, the final decision as to whether the person charged is innocent or guilty is made at the end of the trial, usually by a jury of impartial people. In hypothesis testing, the final decision to reject or not reject the null hypothesis is made at the end of the test by using information or data from an (impartial) random sample. Discuss these similarities between statistical hypothesis testing and a court setting.

(e) We hope that you are able to use this discussion to increase your understanding of statistical testing by comparing it with something that is a well-known part of our American way of life. However, all analogies have weak points. It is important not to take the analogy between statistical hypothesis testing and legal court methods too far. For instance, the judge does not set a level of significance and tell the jury to determine a verdict that is wrong only 5% or 1% of the time. Discuss some of these weak points in the analogy between the court setting and hypothesis testing.

General: Alternate
Hypothesis

14. Suppose that you did some statistical research on a given topic.
(a) If you had a null hypothesis that you hoped to *reject* based on your sample data, would you be better off using a one-tailed test (either left or right tail as appropriate) or a two-tailed test? Explain your answer.
(b) Answer part a if you had a null hypothesis you hoped *not to reject* based on your sample data. Explain your answer. (*Hint:* Examine Figure 9-2 and the size of the critical values for each test.)
(c) If a report states that certain data were used to reject (or fail to reject) a given hypothesis, would it be a good idea to know what type of test (one-tailed or two-tailed) was used? Explain your answer.

Section **9.3**

The P Value in Hypothesis Testing

The level of significance α in a test of a hypothesis is the probability of making a type I error; that is, α is the probability of rejecting the null hypothesis when it is true.

The experimenter sets α before beginning the statistical test. Then this level of significance is used to determine whether or not H_0 is rejected. However, as we shall see in the next example, the same sample data may lead to two different test conclusions depending on the level of significance used.

EXAMPLE 5 ▷ Viva, a credit card company, lowered its annual interest rate by 1%. Records from before the rate change showed the average outstanding balance on credit card accounts to be $\mu = \$576$. The managers believe that reducing interest rates spurs greater use of credit cards, which results in higher outstanding balances. To test this claim, a random sample of 36 accounts was examined 6 months after the interest-rate reduction. The average outstanding balance was $\bar{x} = \$615$ with standard deviation $s = \$120$. Test the managers' claim at the 0.05 and 0.01 levels of significance.

SOLUTION: For both levels of significance we use the same null and alternate hypotheses:

$$H_0: \mu = 576$$
$$H_1: \mu > 576$$

The sample test statistic $\bar{x} = 615$ is also the same for both cases. Using $\mu = 576$ from the null hypothesis, $n = 36$, and $\sigma \approx s = 120$, we find the z value corresponding to the sample test statistic $\bar{x} = 615$:

$$z = \frac{\bar{x} - \mu}{(\sigma/\sqrt{n})} = \frac{615 - 576}{(120/\sqrt{36})} = 1.95$$

In both cases we have a right-tailed test. Figure 9-9 shows the test conclusion for each of the two levels of significance.

FIGURE 9-9

Test Conclusions

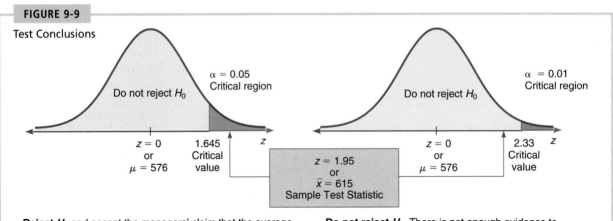

Do not reject H_0

$\alpha = 0.05$
Critical region

$z = 0$
or
$\mu = 576$

1.645
Critical
value

z

$z = 1.95$
or
$\bar{x} = 615$
Sample Test Statistic

Do not reject H_0

$\alpha = 0.01$
Critical region

$z = 0$
or
$\mu = 576$

2.33
Critical
value

z

Reject H_0 and accept the managers' claim that the average outstanding balance is higher at the 5% level of significance.

Do not reject H_0. There is not enough evidence to conclude that the average outstanding balance is higher at the 1% level of significance.

As we see in Example 5, smaller levels of significance move the critical value for the rejection region farther from the mean μ. A natural question arises: What is the smallest level of significance at which the sample data will tell us to reject H_0? The answer is the *P value* associated with the observed sample statistic. The *P* value is also called the *probability of chance* or the *attained level of significance*.

> For the distribution described by the null hypothesis, the *P value* is the smallest level of significance for which the observed sample statistic tells us to reject H_0.

Consequently if

P value $\leq \alpha$, then we reject H_0

P value $> \alpha$, then we do not reject H_0

Finding the *P* value for tests of μ

How do we go about computing the *P* value? We need to express the *P* value in terms of probability. Let's use the model of a right-tailed test of the mean. For a right-tailed test of the mean, the *P* value is simply the probability that the sample mean from any random sample of the same size will be greater than or equal to the observed sample mean $\bar{x}$. In symbols, for right-tailed tests of μ, we have

P value $= P(\bar{x}$ computed from any random sample of size $n \geq$ observed $\bar{x})$

P values are the *areas* in the tail or tails of a probability distribution beyond the observed sample statistic. Figure 9-10 on page 426 shows the *P* values for right-, left-, and two-tailed tests of the mean. The *P* values are the hatched areas of the figure.

EXAMPLE 6 ❯

Compute the *P* value for the hypothesis test of Example 5. Recall that the problem involved the claim that the average outstanding balance on Viva credit cards increased after the annual interest was decreased. We had

$H_0: \mu = 576$

$H_1: \mu > 576$

Observed sample data: $n = 36$, $\bar{x} = \$615$, and $s = \$120$.
For what levels of significance α do we reject H_0?

SOLUTION:

(a) To find the *P* value, we first convert $\bar{x}$ to z using $\mu = 576$ given in the null hypothesis, standard deviation $\sigma \approx s = 120$, and $n = 36$ in the formula

$$z = \frac{\bar{x} - \mu}{(\sigma/\sqrt{n})} = \frac{615 - 576}{(120/\sqrt{36})} = 1.95$$

(b) Sketch the *P* value on a diagram. Because the alternate hypothesis specifies a *right-tailed* test, the *P* value is the area to the *right* of the observed sample statistic (see Figure 9-11 on page 427).

(Example continues on page 427.)

FIGURE 9-10

P Values, Level of Significance α, and Decision Procedure for Different Tests of the Mean

Hypothesis	CASE I: *P* value $\leq \alpha$ This means we reject H_0	CASE II: *P* value $> \alpha$ This means we do not reject H_0

H_0: $\mu = k$
H_1: $\mu < k$
Left-tailed test

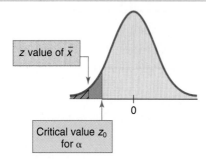

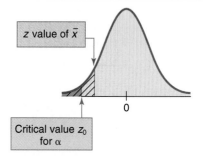

H_0: $\mu = k$
H_1: $\mu > k$
Right-tailed test

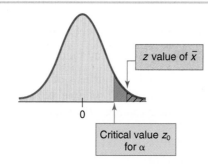

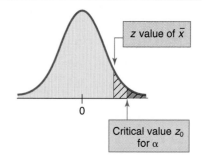

H_0: $\mu = k$
H_1: $\mu \neq k$
Two-tailed test

α = area
P value = area

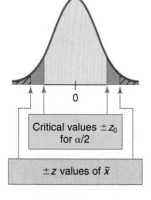

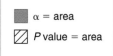

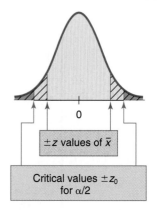

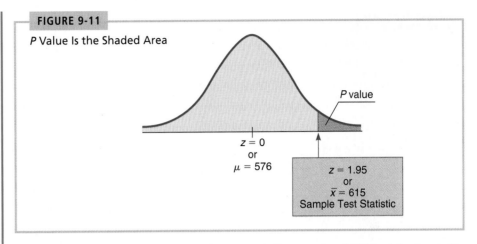

FIGURE 9-11

P Value Is the Shaded Area

P value

$z = 0$
or
$\mu = 576$

$z = 1.95$
or
$\bar{x} = 615$
Sample Test Statistic

(c) Compute the P value. Using techniques of Chapter 6 and Table 4 in Appendix I, we find

$$P \text{ value} = P(\bar{x} \geq 615) = P(z \geq 1.95) = 0.5000 - 0.4744 = 0.0256$$

(d) *Conclusion:* Since the P value is the *smallest* level of significance for which the sample data tell us to reject H_0, we reject H_0 for any $\alpha \geq 0.0256$. For $\alpha < 0.0256$, we fail to reject H_0.

GUIDED EXERCISE   6

Last year the average age of students attending Fremont College was 21.3 years. This semester, in order to meet the needs of older students, more classes were scheduled during evening and weekend hours. Has the average age of the students at Fremont College increased this semester? To answer this question, a random sample of 64 students enrolled this semester was studied. The average age of these students was $\bar{x} = 22.1$ years with sample standard deviation $s = 2.7$ years. Use the P value of the sample statistic $\bar{x}$ to test the hypothesis that the average age of students this semester is higher than it was last year. Use $\alpha = 0.01$.

(a) What are the null and alternate hypotheses?

$H_0: \mu = 21.3$
$H_1: \mu > 21.3$

(b) Convert $\bar{x}$ to z.

Using $\mu = 21.3$ from H_0, $n = 64$, and $\sigma \approx s = 2.7$, we get

$$z = \frac{\bar{x} - \mu}{(\sigma/\sqrt{n})}$$

$$= \frac{22.1 - 21.3}{(2.7/\sqrt{64})}$$

$$\approx 2.37$$

Exercise continues

Exercise continued

(c) Sketch a diagram showing the P value of the sample test statistic $\bar{x} = 22.1$.

➡ **FIGURE 9-12** P Value Is the Shaded Area

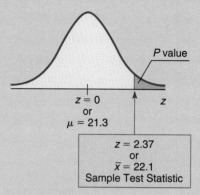

(d) Use Table 4 in Appendix I to compute the P value.

➡ P value $= P(\bar{x} \geq 22.1)$

$= P(z \geq 2.37)$

$= 0.5000 - 0.4911$

$= 0.0089$

(e) Compare the P value with $\alpha = 0.01$. Do we reject H_0 or not? Has the average age at the college increased?

➡ The P value $= 0.0089$ is the smallest level of significance for which we reject H_0. Since

P value $= 0.0089 < 0.01 = \alpha$

we reject H_0 for $\alpha = 0.01$. At the 1% level of significance, it seems that the average age of the students enrolled this semester has increased.

For a distribution described by the null hypothesis, the P value is the probability of obtaining any sample statistic as *far away* from μ as, or even *farther* from μ than, the given sample test statistic $\bar{x}$. In the case of a *two-tailed* test, this means that we must consider *both* tails of the distribution. For instance, the P value associated with a sample z value is the *total* of the area to the *left of* $-z$ added to the area to the *right of* z (see Figure 9-13).

The next example shows how to compute the P value for a *two-tailed* test of the mean (large sample size).

EXAMPLE 7 ➤ Whitehall Construction Company figures the cost of a project based on an average idle time of 68 minutes per worker per 8-hour shift. Idle time does not include lunch breaks but can be caused by delays in material delivery, delays in the completion of a prior stage of the project, worker injury, inadequate crew size due to absenteeism, and so forth. A new work schedule has just been completed by the construction foreman. It is not known whether or not this schedule will change idle time (increase or decrease). To determine if the new schedule *changes* the average

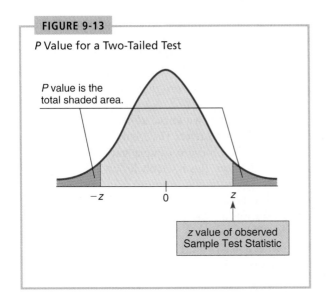

FIGURE 9-13

P Value for a Two-Tailed Test

P value is the total shaded area.

z value of observed Sample Test Statistic

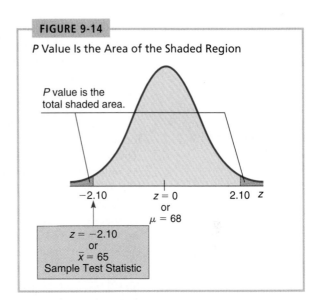

FIGURE 9-14

P Value Is the Area of the Shaded Region

P value is the total shaded area.

$z = -2.10$ or $\bar{x} = 65$ Sample Test Statistic

idle time, a random sample of 49 workers was observed during an 8-hour shift. The average idle time was $\bar{x} = 65$ minutes with standard deviation $s = 10$ minutes. Use a 5% level of significance. Does the new schedule *change* idle time? Use the P value of the sample statistic to make the decision.

SOLUTION:

(a) The null and alternate hypotheses are

$$H_0: \mu = 68 \text{ minutes}$$
$$H_1: \mu \neq 68 \text{ minutes}$$

(b) Next, we convert the sample test statistic $\bar{x} = 65$ to a z value using $n = 49$, $\mu = 68$ from H_0, and $s = 10$ as the approximation for σ:

$$z = \frac{\bar{x} - \mu}{(\sigma/\sqrt{n})} = \frac{65 - 68}{(10/\sqrt{49})} = -2.10$$

(c) Show the P value on a diagram. The alternate hypothesis corresponds to a *two-tailed* test, so the P value is the area to the *left* of $z = -2.10$ plus the area to the *right* of $z = 2.10$ (see Figure 9-14).

(d) Compute the P value. By symmetry, the P value is *twice* the area in the tail to the *left of* -2.10. Using Table 4 in Appendix I, we see that

$$P \text{ value} = 2(0.5000 - 0.4821) = 2(0.0179) = 0.0358$$

(e) Conclude the test. *The P value is the smallest level of significance for which we reject H_0.* This means that we reject H_0 for all $\alpha \geq 0.0358$. In particular, we reject H_0 for $\alpha = 0.05$, since 0.05 is greater than 0.0358. (Note that we fail to reject H_0 for $\alpha = 0.01$ because the P value 0.0358 is greater than 0.01.) ●

Statistical significance based on *P* values

In hypothesis testing, we use sample data to draw a conclusion about the null hypothesis. If the sample data results are quite different from the claim stated in H_0, then we suspect that the difference is due to some effect other than just random chance. In this case, we reject H_0, and we say that the data are *statistically significant*. If we do not reject the null hypothesis H_0, we say that the data are *not statistically significant*.

Interpreting results from *P* values generated on a computer

The advantage of knowing the *P* value is that we know *all* levels of significance for which the observed sample statistic tells us to reject H_0. Many research journals require authors to include the *P* value of the observed sample statistic. Then readers will have more information and will know the test conclusion for any preset level of significance.

Again, let us caution you to establish the level of significance α *before* doing the hypothesis test. The level of significance reflects the probability level at which you are willing to risk a type I error. Also, the accuracy and reliability of your measurement instruments might affect your choice of α. Thus select α first, then use a computer or table to find the *P* value of your test statistic, and finally draw the appropriate conclusion.

Hypothesis Testing Using *P* Values from a Computer

1. Establish the level of significance α of the test.
2. Determine the null and alternate hypotheses, H_0 and H_1.
3. Enter information about the observed sample into the computer and look for the *P* value of the observed sample statistic in this computer output.
4. Compare your level of significance with the *P* value.
    If *P* value $\leq \alpha$, reject H_0.
    If *P* value $> \alpha$, do not reject H_0.

Calculator Note When doing hypothesis testing using a computer statistics package or a calculator similar to the TI-83, the *P* value corresponding to the sample statistic is generally provided. The user compares it with a specified level of significance α to conclude the test. The displays below are from the TI-83. They relate to data in Guided Exercise 6 regarding the average age of students attending Fremont College.

Press **STAT**, select **TESTS**, use **Option 1:Z-Test** with **Inpt:Stats**   Results of **DRAW**

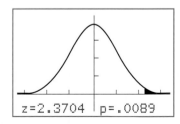

```
Z-Test
 Inpt:Data Stats
 μ₀:21.3
 σ:2.7
 x̄:22.1
 n:64
 μ:≠μ₀ <μ₀ >μ₀
Calculate Draw
```

z=2.3704   p=.0089

To conclude the test, compare the given *P* value of 0.0089 with the level of significance $\alpha = 0.01$ given in the problem. We see that the *P* value is less than α, so we reject H_0 and say that the results are statistically significant.

Summary

The P value (probability of chance) is the area of the sampling distribution that lies beyond the observed sample statistic. It tells us the probability that a sample statistic will be more extreme than the observed sample statistic. We compute the P value as follows:

1. When H_1 indicates a right-tailed test,

    P value = area to the right of the observed sample statistic
    P value = P(sample statistic > observed sample statistic)

2. When H_1 indicates a left-tailed test,

    P value = area to the left of the observed sample statistic
    P value = P(sample statistic < observed sample statistic)

3. When H_1 indicates a two-tailed test,

    P value = sum of the areas in the two tails

    (a) In the case that the observed sample statistic falls in the right half of a symmetric curve, then

    P value = $2P$(sample statistic > observed sample statistic)

    (b) In the case that the observed sample statistic falls in the left half of a symmetric curve, then

    P value = $2P$(sample statistic < observed sample statistic)

To conclude the test, we compare the level of significance α with the P value.

1. If the P value is less than or equal to α, we reject H_0.

2. If the P value is greater than α, we do not reject H_0.

VIEWPOINT

Predator or Prey?

Consider animals such as the arctic fox, gray wolf, desert lion, and South American jaguar. Each animal is a predator. What are the total sleep (hours per day), maximum life span (years), and overall danger index from other animals? Now consider prey such as rabbits, deer, wild horses, and the Brazilian tapir (a wild pig). Are there statistically significant differences in average sleep, life span, and danger index? What about other variables such as the ratio of brain weight to body weight or the sleep exposure index  (sleeping in a well-protected den or out in the open)? How did prehistoric humans fit into this picture? Scientists have collected a lot of data, and a great deal of statistical work has been done regarding such questions. For more information, see Web site <http://lib.stat.cmu.edu/> and follow the links to Datasets and then Sleep.

SECTION 9.3 PROBLEMS

For Problems 1 through 9, please do the following:

(a) Convert the sample mean $\bar{x}$ to a sample z value. (*Hint:* Due to the large sample size, approximate σ with s.)

(b) Sketch a graph of the z distribution. Show the location of the area corresponding to the P value for this test.

(c) Compute the P value for this test.

(d) Are the data significant at the specified level of significance α?

1. Given H_0: $\mu = 5$ and H_1: $\mu > 5$ with observed sample mean $\bar{x} = 6.1$, sample standard deviation $s = 2.5$, and sample size $n = 40$. Use $\alpha = 0.01$.

2. Given H_0: $\mu = 53.1$ and H_1: $\mu < 53.1$ with observed sample mean $\bar{x} = 52.7$, sample standard deviation $s = 4.5$, and sample size $n = 41$. Use $\alpha = 0.01$.

3. Given H_0: $\mu = 21.7$ and H_1: $\mu \neq 21.7$ with observed sample mean $\bar{x} = 20.5$, sample standard deviation $s = 6.8$, and sample size $n = 45$. Use $\alpha = 0.05$.

4. Given H_0: $\mu = 18.7$ and H_1: $\mu \neq 18.7$ with observed sample mean $\bar{x} = 19.1$, sample standard deviation $s = 5.2$, and sample size $n = 32$. Use $\alpha = 0.05$.

Ecology: Coyotes

5. A random sample of 68 adult coyotes in a region of northern Minnesota showed the average age to be $\bar{x} = 2.05$ years with sample standard deviation $s = 0.82$ years (based on information from the book *Coyotes: Biology, Behavior and Management,* by M. Bekoff, Academic Press). However, it is thought that the overall population mean age of coyotes is $\mu = 1.75$. Does the sample data indicate that coyotes in this region of northern Minnesota tend to live longer than an average of 1.75 years? Use $\alpha = 0.01$.

Archaeology: Mesa Verde

6. An archaeology professor claims that the average floor space of Mesa Verde Anasazi kivas is $\mu = 12$ square meters. However, in the book *Architecture of Social Integration in Prehistoric Pueblos* (by W. Lipe and M. Hegmon, Crow Canyon Archaeological Center Press) it is stated that a random sample of 56 Mesa Verde Anasazi kivas had a sample mean area of $\bar{x} = 12.3$ square meters with sample standard deviation $s = 3.4$ square meters. Do these data indicate that the mean floor area of all such Mesa Verde kivas is different from $\mu = 12$ square meters? Use $\alpha = 0.01$.

Leisure: Fishing

7. Pyramid Lake is on the Paiute Indian Reservation in Nevada. The lake is famous for cutthroat trout. Suppose a friend tells you that the average length of a trout caught in Pyramid Lake is $\mu = 19$ inches. However, the January 1995 Creel Survey (published by the Pyramid Lake Paiute Tribe Fisheries Association) reported that of a random sample of 73 fish caught, the mean length was $\bar{x} = 18.7$ inches with estimated standard deviation $s = 3.2$ inches. Do these data indicate that the average length of a trout caught in Pyramid Lake is less than $\mu = 19$ inches? Use $\alpha = 0.05$.

Franchise: Flower and Gift Shops

8. *Franchise and Business Opportunities Annual Report* has over 1500 listings of small business startup opportunities. Each of these involves a startup cost (money needed to get the business going). In most cases, bank financing is readily available if you have a reasonable credit rating. One category of small business is the flower and gift shop business. For this type of business, the mean startup costs (across the entire

United States) amount to $61,400. However, startup costs may vary from one region to the next. Suppose that you are interested in starting a flower and gift shop in San Antonio, Texas. Local newspaper advertising and business sections can be used to obtain a sample of startup costs for this type of business. Suppose that a sample of 34 small flower and gift shops in the San Antonio region gave a sample mean startup cost of $\bar{x} = \$55,200$ with sample standard deviation $s = \$18,800$. Does this indicate that the population average startup cost in this region is lower than the national average? Use $\alpha = 0.05$.

Lifestyle: Student Cars

9. Based on information from *The Statistical Abstract of the United States* (116th Edition), the average daily cost of owning and operating an automobile is $15.35. This includes depreciation, finance charges, general maintenance, gasoline, insurance, and 12,000 miles driven annually. A random sample of 34 college students who own cars found the average cost per day to be $11.85 with sample standard deviation of $6.21. Test the claim that college students' average daily ownership expenses are less than the national average. Use a 5% level of significance.

Autos: Mileage

10. How do the year 2000 automobiles stack up mileage-wise? A random sample of 100 autos from *Consumer Review CAR Buyer's Guide* showed city and highway mileage. The autos tested included sport, luxury, compact, sedan, van, and sport utility models.

(a) Is the city mileage less than 19.5 mpg for the year 2000 cars? Let's look at a Minitab (➤Stat ➤Basic Statistics ➤1-Sample z) display of the sample results.

Z-Test

```
Test of mu = 19.500 vs mu < 19.500
The assumed sigma = 4.48

VARIABLE    N     MEAN    STDEV   SE MEAN    Z       P
City       100   18.750   4.480   0.448    -1.67   0.047
```

Refer to the Minitab display. Write out the null and alternate hypotheses. What is the value of the sample mean? What is the P value? At what levels of significance may we reject the null hypothesis and conclude that the average mileage is less than 19.5 mpg?

(b) Is the city mileage different from 19.5 mpg for the year 2000 cars? Look at the next Minitab display.

Z-Test

```
Test of mu = 19.500 vs mu not = 19.500
The assumed sigma = 4.48

VARIABLE    N     MEAN    STDEV   SE MEAN    Z       P
City       100   18.750   4.480   0.448    -1.67   0.094
```

Refer to the Minitab display. Write out the null and alternate hypotheses. What is the value of the sample mean? What is the P value? At what levels of significance may we reject the null hypothesis and conclude that the average mileage is different than 19.5 mpg?

(c) How does the P value of a one-tailed test compare with the P value of a two-tailed test for the same sample data and same null hypothesis?

Tests Involving the Mean μ (Small Samples)

Sometimes it is not practical or even possible to obtain large samples. Cost, available time, and other factors may require us to work with *small samples*. For our purposes, we will say that a *sample size less than 30 is a small sample*. In Section 8.2 we found that the *t* distribution is suitable for small samples in which the sampled population has a normal distribution. However, it can be shown that our applications of the *t* distribution are still appropriate for populations that are not normal but exhibit a "mound-shaped" and symmetric probability distribution. Because such distributions commonly occur, the *t* distribution is very useful in applied work.

Small sample

When doing hypothesis tests of μ with small samples, we will use essentially the same methods we used for large samples. The main difference is that when σ is not known, the sampling distribution $\bar{x}$ follows a *Student's t distribution with degrees of freedom d.f. = n − 1* instead of a normal distribution.

When we draw a random sample of size *n* from a population that has a normal distribution (or at least a mound-shaped symmetric distribution) with mean μ, then the *t* values—that is,

$$t = \frac{\bar{x} - \mu}{(s/\sqrt{n})}$$

where $\bar{x}$ = sample mean

n = sample size

s = sample standard deviation

follow a Student's *t* distribution with degrees of freedom *d.f. = n − 1*.

Critical values from a t distribution

In Table 5, "Student's *t* Distribution," in Appendix I, you see the column headings *c*, α', and α''. (Up to now we have ignored the α' and α'' headings.) In this table, *c* represents the level of confidence, α' is the significance level for a one-tailed test (α' is the area to the right of *t* or, equivalently, to the left of −*t*), and α'' is the significance level for a two-tailed test (α'' is the area beyond −*t* and *t*, so in fact $\alpha'' = 2\alpha'$). Figure 9-15 illustrates these different values.

To find the critical value(s) t_0, we look in the column headed by the level of significance α' for a *one-tailed test* or α'' for a *two-tailed test*. The critical value t_0 is located in the row headed by degrees of freedom *d.f. = n − 1*.

FIGURE 9-15

Meaning of α' and α'' in Table 5 in Appendix I

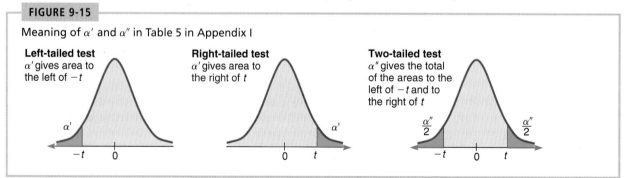

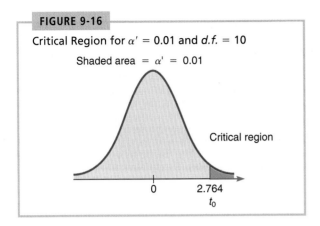

FIGURE 9-16

Critical Region for $\alpha' = 0.01$ and $d.f. = 10$

Shaded area $= \alpha' = 0.01$

Critical region

0   2.764

t_0

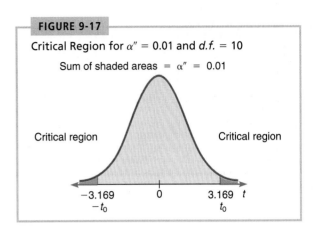

FIGURE 9-17

Critical Region for $\alpha'' = 0.01$ and $d.f. = 10$

Sum of shaded areas $= \alpha'' = 0.01$

Critical region   Critical region

−3.169   0   3.169 t

$-t_0$ t_0

EXAMPLE 8 ▷ Use Table 5 in Appendix I to find the critical value(s) t_0 and critical region for the described tests of μ with sample size $n = 11$.

(a) Suppose that we have a *right-tailed* test of μ with level of significance 0.01.

SOLUTION: Since the test is a *one-tailed* test, we use the column headed by $\alpha' = 0.01$. The degrees of freedom are $d.f. = 11 - 1 = 10$, so we use the row headed by 10. The critical value is $t_0 = 2.764$ (see Figure 9-16).

(b) Suppose that we have a *two-tailed* test of μ with level of significance 0.01.

SOLUTION: The test is a *two-tailed* test, so we look in the column under $\alpha'' = 0.01$ and the row headed by $d.f. = 11 - 1 = 10$. The critical values are $\pm t_0 = \pm 3.169$ (see Figure 9-17). ●

GUIDED EXERCISE 7

Use Table 5 in Appendix I to find the critical value(s) t_0 and critical region for the described test.

(a) A *left-tailed* test of μ with level of significance 0.05 and sample size $n = 8$.

⇨ In this case we use the column headed by $\alpha' = 0.05$ and the row headed by $d.f. = n - 1 = 8 - 1 = 7$. This gives us $t = 1.895$. For a left-tailed test, we use the symmetry of the distribution to get $t_0 = -1.895$ (see Figure 9-18).

(b) A *two-tailed* test of μ with level of significance 0.05 and sample size $n = 8$.

⇨ Use the column headed by $\alpha'' = 0.05$ and the row headed by $d.f. = n - 1 = 7$. By the symmetry of the curve, the critical values are $\pm t_0 = \pm 2.365$ (see Figure 9-19).

Exercise continues

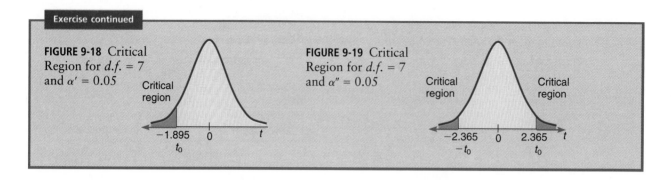

Exercise continued

FIGURE 9-18 Critical Region for $d.f. = 7$ and $\alpha' = 0.05$

FIGURE 9-19 Critical Region for $d.f. = 7$ and $\alpha'' = 0.05$

Sample test statistic

Once we have the critical value(s) t_0 and critical region, we convert the sample test statistic $\bar{x}$ to a t value using the formula

$$t = \frac{\bar{x} - \mu}{(s/\sqrt{n})} \text{ with } d.f. = n - 1$$

where $\bar{x}$ = sample mean and μ is specified in H_0

n = sample size

s = sample standard deviation

To conclude the test, we locate the t value of the sample statistic on a diagram showing the critical region. If the sample t value falls in the critical region, we reject H_0. If the sample t value falls outside the critical region, we fail to reject H_0.

The next example demonstrates the traditional method of using critical regions to conclude a test of a mean using small samples.

EXAMPLE 9 ❯

A company manufactures large rocket engines used to project satellites into space. The government buys the rockets, and the contract specifies that these engines are to use an average of 5500 pounds of rocket fuel the first 15 seconds of operation. The company claims that its engines fit specifications. To test the claim, an inspector randomly selects six such engines from the warehouse. These six engines are fired 15 seconds each, and the fuel consumption for each engine is measured. For all six engines, the mean fuel consumption is $\bar{x}$ = 5690 pounds and the standard deviation is s = 250 pounds. Is the claim justified at the 5% level of significance?

SOLUTION:

(a) We want to see if we can reject the hypothesis μ = 5500 pounds, so the null hypothesis is

$$H_0: \mu = 5500$$

A substantial difference either way from 5500 pounds could be important, so the alternate hypothesis is

$$H_1: \mu \neq 5500$$

(b) Next, we find the critical values. Since our sample is small, we use Table 5 in Appendix I. For a *two-tailed* test with level of significance 0.05, we use the $\alpha'' = 0.05$ column and the row headed by *d.f.* = $n - 1 = 6 - 1 = 5$. This gives us $\pm t_0 = \pm 2.571$.

(c) Using $\mu = 5500$ from H_0, $s = 250$, and $n = 6$, we convert the sample test statistic $\bar{x} = 5690$ to a t value:

$$t = \frac{\bar{x} - \mu}{(s/\sqrt{n})} = \frac{5690 - 5500}{(250/\sqrt{6})} \approx 1.862$$

(d) On a diagram, show the critical regions and the location of the sample test statistic. (See Figure 9-20.) We see that the t value of the sample test statistic falls outside the critical region. Therefore, we cannot reject H_0. The data do not present sufficient evidence to indicate that the average fuel consumption for the first 15 seconds of operation is different from $\mu = 5500$ pounds.

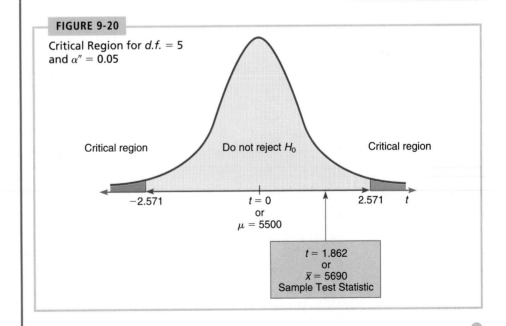

FIGURE 9-20

Critical Region for *d.f.* = 5 and $\alpha'' = 0.05$

Critical region   Do not reject H_0 Critical region

−2.571 $t = 0$ 2.571 t
                  or
                  $\mu = 5500$

$t = 1.862$
or
$\bar{x} = 5690$
Sample Test Statistic

Suppose in Example 9 that a random sample of eight engines was used and the mean fuel consumption was $\bar{x} = 5880$ pounds (for the first 15 seconds). Again, the standard deviation was 250 pounds. Use a 1% level of significance to test the claim that the population average fuel consumption *exceeds* 5500 pounds.

(a) We will use H_0: $\mu = 5500$. What should we use for H_1:

   H_1: $\mu < 5500$, H_1: $\mu \neq 5500$,
      or H_1: $\mu > 5500$?

⇨ H_1: $\mu > 5500$ because we want to test the claim that the average fuel consumption *exceeds* 5500 pounds.

(b) Find the critical value t_0.

⇨ Because the sample size is small, we use Table 5 in Appendix I. For a one-tailed test, we look in the column headed by $\alpha' = 0.01$ and the row headed by *d.f.* = 8 − 1 = 7. The critical value is $t_0 = 2.998$.

(c) Convert the sample test statistic $\bar{x}$ to a t value.

⇨ We use $\mu = 5500$, $n = 8$, and $s = 250$:

$$t = \frac{\bar{x} - \mu}{(s/\sqrt{n})}$$

$$= \frac{5880 - 5500}{(250/\sqrt{8})}$$

$$\approx 4.299$$

(d) Sketch the critical region and show the location of the sample test statistic on the diagram. Do we reject H_0 or not at the 1% level of significance?

⇨ Since the sample test statistic falls in the critical region, we reject H_0. There is evidence that the average fuel consumption exceeds 5500 pounds during the first 15 seconds of operation. (See Figure 9-21.)

FIGURE 9-21 Critical Region for *d.f.* = 7 and $\alpha' = 0.01$

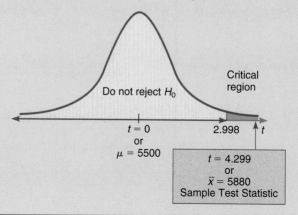

Using *P* values for tests of
μ, small samples

Recall that the *P* value for a statistical test is the probability of getting a sample statistic as far (or even farther) into the tails of the sampling distribution as the observed sample statistic. The smaller the *P* value, the stronger the evidence is to reject H_0.

In the case of small sample tests of μ, the sample distribution $\bar{x}$ follows a Student's *t* distribution. Therefore, we will use Table 5 in Appendix I to estimate the *P* values. However, the areas given in Table 5 are limited. Consequently, when we use Table 5 to estimate *P* values for $\bar{x}$, we usually find an *interval containing the P value* rather than a single number for the *P* value.

NOTE REGARDING TABLE 5 IN APPENDIX I In Table 5, α' represents the area in *one tail* beyond *t*. The α'' values represent the area in the *two tails*. Notice that in each column $\alpha'' = 2\alpha'$. Consequently, we use α' values as endpoints of the *P* value intervals for *one-tailed tests* and α'' values as endpoints of the *P* value intervals for *two-tailed tests* (see Figure 9-22).

Let's find the *P* values associated with the sample test statistic of Example 9 and of Guided Exercise 8.

EXAMPLE 10 >

In Example 9 we tested the rocket manufacturer's claim that its rockets consumed an average of 5500 pounds of fuel in the first 15 seconds of operation. The test was a two-tailed test with sample statistic $\bar{x} = 5690$, $s = 250$, and $n = 6$. Find the *P* value associated with $\bar{x} = 5690$. What does the *P* value tell us?

SOLUTION:

(a) First, we convert $\bar{x} = 5690$ to a *t* value. This was done in Example 9. The corresponding *t* value is 1.862.

FIGURE 9-22

P Values Corresponding to α' and α''

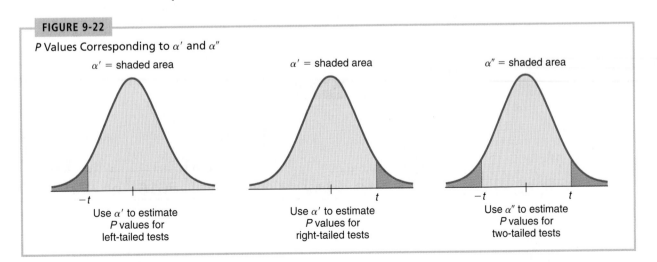

(b) The next step is to find an interval containing the P value corresponding to the sample test statistic $t = 1.862$ using a two-tailed test. Our test is two-tailed, so we will use α'' values. To find the associated P value interval, we look in the row headed by $d.f. = n - 1 = 6 - 1 = 5$. We find that the sample t value $t = 1.862$ falls between 1.699 and 2.015. The corresponding P value then lies between corresponding α'' values 0.150 and 0.100 (see Table 9-5).

Put the adjacent α'' values in increasing order to form the interval

$$0.100 < P \text{ value} < 0.150$$

Table 9-5   Excerpt from Student's t Distribution (Table 5, Appendix I)

One-tailed test P value	α'	...		
✔ Two-tailed test P value	α''	...	0.150	0.100 ...
d.f. 5		...	1.699	2.015 ...

Sample $t = 1.862$

(c) We see that the P value may be almost as large as 0.150. For $\alpha = 0.05$, we see that the P value is greater than α, so we do not reject H_0. This result is consistent with the conclusion of Example 9.

Calculator Note Computer printouts and calculators such as the TI-83 can give more accurate P values from the Student's t distribution than we can obtain from Table 5 in Appendix I. For instance, the results of Example 10 (a two-tailed test regarding rocket fuel consumption) as completed on the TI-83 are displayed.

Press **STAT**, select **TESTS**,   Results of **DRAW**
use **Option 2:T-Test** with **Inpt:Stats**

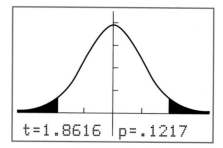

The P value is the area of the shaded region shown. In Example 10 we used Table 5 in Appendix I to find the interval containing the P value, $0.100 < P$ value < 0.150. The calculator gives the P value of 0.1217, which is in the interval. Since the P value is greater than $\alpha = 0.05$, we do not reject H_0.

The next example shows how to find an interval containing the P value for a one-tailed test using the Student's t distribution.

EXAMPLE 11 ❯ In Guided Exercise 8 we again looked at the fuel consumption of rockets. This time the sample mean consumption for the first 15 seconds of flight was $\bar{x} = 5880$ pounds with $s = 250$ pounds and $n = 8$ rockets. We used a right-tailed test with $H_0: \mu = 5500$ and $H_1: \mu > 5500$. Find the P value associated with the sample statistic $\bar{x}$. What does the P value tell you?

SOLUTION:

(a) Again, our first step is to convert the sample statistic $\bar{x} = 5880$ to a t value. We did this in Guided Exercise 8 and got $t = 4.299$.

(b) This test is a one-tailed test, so we use α' values to create the P-value interval. To find the associated P value, we look in the row headed by $d.f. = n - 1 = 8 - 1 = 7$ and find that the sample t statistic $t = 4.299$ falls to the *right* of 3.499 (Table 9-6).

Notice that, reading from left to right, the P values decrease. Therefore, the P value corresponding to the sample t value is smaller than that corresponding to $t = 3.499$. Thus,

P value < 0.005

This relation is shown in Figure 9-23 on page 442.

(c) Since the P value is less than the level of significance $\alpha = 0.01$, it is significant, and we reject H_0. This result is consistent with the conclusion of Guided Exercise 8.

Table 9-6   Excerpt from Student's t Distribution (Table 5, Appendix I)

			0.010	0.005
✔ One-tailed test P value	α'	...	0.010	0.005
Two-tailed test P value	α''	...		
$d.f.$	7	...	2.998	3.499

Sample $t = 4.299$

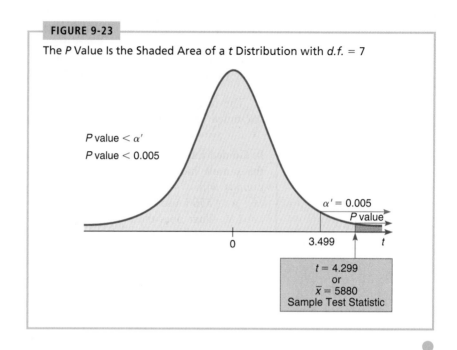

FIGURE 9-23

The *P* Value Is the Shaded Area of a *t* Distribution with *d.f.* = 7

P value $< \alpha'$
P value < 0.005

$\alpha' = 0.005$
P value
0   3.499   *t*

$t = 4.299$
or
$\bar{x} = 5880$
Sample Test Statistic

V I E W P O I N T

Start the Day Right!

Just how healthy is that bowl of cereal? What are the calories, protein, fat, sodium, fiber, and vitamins in a bowl of breakfast cereal? What is the rating compared with other cereals? Each morning millions of Americans pour their favorite cereal into a bowl and believe that they are starting the day right. The breakfast cereal companies spend millions of dollars in advertising trying to convince us their breakfast cereal is better than all the others. What are the facts? Using statistics you already know, you can decide for yourself. See the Cereal Data File, under the subject Food, at Web site <http://lib.stat.cmu.edu/DASL/> or collect your own information at the supermarket by reading the nutrition facts on cereal boxes.

ⓈE C T I O N   9 . 4   P R O B L E M S

In Problems 1–6, use Table 5 in Appendix I to find the critical value(s) t_0 for the described test of the mean (small sample).

1. Sample size $n = 9$, left-tailed test, level of significance 5%

2. Sample size $n = 13$, right-tailed test, level of significance 1%

3. Sample size $n = 24$, two-tailed test, level of significance 1%

4. Sample size $n = 18$, left-tailed test, level of significance 5%

5. Sample size $n = 12$, two-tailed test, level of significance 5%

6. Sample size $n = 29$, right-tailed test, level of significance 1%

For Problems 7–13, please provide the requested information.

(a) What is the null hypothesis? What is the alternate hypothesis? Will we use a left-tailed, right-tailed, or two-tailed test? What is the level of significance?
(b) What sampling distribution will we use? What is the critical value t_0 (or critical values $\pm t_0$)?
(c) Sketch the critical region and show the critical value (or critical values).
(d) Calculate the t value corresponding to the sample statistic $\bar{x}$ and show its location on the sketch in part c.
(e) Find the P value (or interval containing the P value) for your test and explain what the P value means in the context of the problem.
(f) Based on your answers for parts a to e, shall we reject or fail to reject (i.e., "accept") the null hypothesis based on the given level of significance? Explain your conclusion in the context of the problem.

Medical: Red Blood Cell Count

7. Let x be a random variable that represents red blood cell count (RBC) in millions per cubic millimeter of whole blood. Then x has a distribution that is approximately normal, and for the population of healthy female adults, the mean of the x distribution is about 4.8 (based on information from *Diagnostic Tests with Nursing Implications*, Springhouse Corporation). Suppose that a female patient has taken six laboratory blood tests over the past several months and the RBC count data sent to the patient's doctor were

| 3.5 | 4.2 | 4.5 | 4.6 | 3.7 | 3.9 |

(a) Use a calculator with sample mean and sample standard deviation keys to verify that $\bar{x} = 4.07$ and $s = 0.44$.
(b) Do the given data indicate that the population mean RBC count for this patient is lower than 4.8? Use $\alpha = 0.05$.

Medical: Hemoglobin Count

8. Let x be a random variable that represents hemoglobin count (HC) in grams per 100 milliliters of whole blood. Then x has a distribution that is approximately normal with population mean about 14 for healthy adult women (see reference in Problem 7). Suppose that a female patient has taken 12 laboratory blood tests during the past year. The HC data sent to the patient's doctor were

| 19 | 23 | 15 | 21 | 18 | 16 |
| 14 | 20 | 19 | 16 | 18 | 21 |

(a) Use a calculator with sample mean and sample standard deviation keys to verify that $\bar{x} = 18.33$ and $s = 2.71$.
(b) Does this information indicate that the population average HC for this patient is higher than 14? Use $\alpha = 0.01$.

Wild Life: Hummingbirds

9. *Selasphorus platycerus* is the scientific name of the broad-tailed hummingbird that is commonly found throughout the western United States. It is known that the mean incubation time for eggs of this bird is approximately 16.5 days (based on information from *The Hummingbird Book*, by D. Stokes and L. Stokes, Little, Brown and Company). Assume that the incubation times are approximately normally distributed. However, at higher elevations (above 8000 feet), it is thought that the average incubation time might be different from 16.5 days. A number of broad-tailed hummingbird nests were located in mountain country above 8000 feet, and incubation times for 18 eggs gave the following information (in days):

| 15 | 19 | 16 | 16 | 18 | 17 | 21 | 18 | 17 |
| 16 | 17 | 18 | 19 | 16 | 15 | 21 | 23 | 19 |

(a) Use a calculator with mean and standard deviation keys to verify that $\bar{x} = 17.83$ days and $s = 2.20$ days.

(b) Does this information indicate that the population average incubation time above 8000 feet elevation is different from (either more or less than) 16.5 days? Use $\alpha = 0.05$.

Leisure: Fishing

10. Homser Lake, Oregon has an Atlantic salmon catch and release program that has been very successful. The average fisherman's catch has been $\mu = 8.8$ Atlantic salmon per day. (*Source: National Symposium on Catch and Release Fishing*, Humbolt State University.) Suppose that a new quota system restricting the number of fishermen has been put in effect this season. A random sample of fishermen gave the following catch per day:

| 12 | 6 | 11 | 12 | 5 | 0 | 2 |
| 7 | 8 | 7 | 6 | 3 | 12 | 12 |

(a) Use a calculator with mean and sample standard deviation keys to verify that $\bar{x} = 7.36$ and $s = 4.03$.

(b) Assuming that the catch per day has an approximately normal distribution, use a 5% level of significance to test the claim that the population average catch per day is now different from 8.8.

Archaeology: Tree Rings

11. Tree ring dating from archaeological excavation sites is used in conjunction with other chronologic evidence to estimate occupation dates of prehistoric Indian ruins in the southwestern United States. It is thought that Burnt Mesa Pueblo was occupied around 1300 A.D. (based on evidence from potsherds and stone tools). The following data give tree ring dates (A.D.) from adjacent archaeological sites (*Bandelier Archaeological Excavation Project: Summer 1990 Excavations at Burnt Mesa Pueblo*, edited by T. Kohler, Washington State University Department of Anthropology):

| 1189 | 1267 | 1268 | 1275 | 1275 |
| 1271 | 1272 | 1316 | 1317 | 1230 |

(a) Use a calculator with mean and standard deviation keys to verify that $\bar{x} = 1268$ and $s = 37.29$ years.

(b) Assuming that the tree ring dates in this excavation area follow a distribution that is approximately normal, does this information indicate the population mean of tree ring dates in the area is different from (either higher or lower than) 1300 A.D.? Use a 1% level of significance.

Leisure: Car Rental

12. The *Wall Street Journal* (Dow Jones Travel Index) reported that the average cost of renting a car in Orlando, Florida is $\mu = \$38$ per day. A random sample of vacationing families (with rental cars) at Disney World were asked how much they were being charged for their rental cars. The costs per day (in dollars) turned out to be as follows:

29.71	33.16	39.99	35.00	41.28	35.01	38.88
39.42	34.16	37.12	36.79	41.12	33.29	32.15
29.99	32.56					

(a) Use a calculator with mean and sample standard deviation keys to verify that $\bar{x} = \$35.60$ and $s = \$3.76$.
(b) Assuming that daily car rental costs have an approximately normal distribution, use a 5% level of significance to test the claim that the population average car rental cost per day for families visiting Disney World is less than that reported by the *Wall Street Journal*.

Business: Profit per Employee (Food Producers)

13. H. J. Heinz, Campbell Soup, Kellogg, Hershey Foods, Quaker Oats, and other companies are important food producers. How profitable are such food companies? Let x be a random variable that represents annual profit as a percentage of assets for the nation's largest food companies. Assume that x has a distribution that is approximately normal with mean $\mu = 6.6$ (based on information from *Fortune 500*, Vol. 135, No. 8). Suppose that recent financial reports for randomly selected national food companies gave the following x values:

| 6.6 | 9.1 | 3.3 | 2.5 | 8.4 | 5.1 | 4.8 | 3.4 |

(a) Use a calculator with mean and standard deviation keys to verify that $\bar{x} = 5.4$ and $s = 2.4$.
(b) Use a 5% level of significance to test the claim that the population average annual profits as a percentage of assets for large food companies is now less than 6.6.

Veterinary Science: Lions

14. The heart rate of a healthy lion is approximately normally distributed with mean $\mu = 40$ beats per minute. (*Source: The Merck Veterinary Manual,* a reference used in most veterinary colleges.) A heart rate that is too low or too high can indicate a health problem. A veterinarian has removed an abscessed tooth from a young, healthy zoo lion. As the animal slowly starts to come out of the anesthetic, the heart rate (in beats per minute) is taken six times during the half-hour period following the tooth extraction:

| 30 | 37 | 43 | 38 | 35 | 36 |

We want to use the sample information to determine if the heart rate of the lion is different from 40 beats per minute. Look at the following ComputerStat display.

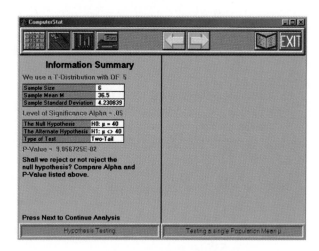

(a) What are the hypotheses of the test?

(b) At the 5% level of significance, what is the test conclusion? Is the population average heart rate of the lion different (either way) from 40 beats per minute?

<table>
</table>

Section **9.5**

Tests Involving a Proportion

Many situations arise that call for tests of proportions or percentages rather than means. For example, a welfare office claims that the proportion of incomplete applications it receives is now 47%. The office is using this claim to justify a request for two more staff members whose main duty will be to help applicants complete the forms properly. The funding agency wants to test the claim that 47% of the applications are incomplete.

How can we make such a test? In this section we will study tests involving proportions (i.e., percentages or proportions). In principle, such tests are the same as those in Sections 9.2 and 9.4. The main difference is that here we are working with distributions of proportions instead of distributions of means.

Tests for a single proportion
Throughout this section we will assume that the situations we are dealing with satisfy the conditions underlying the binomial distribution. In particular, we will let r be a binomial random variable. This means that r is the number of successes out of n independent binomial trials (for the definition of binomial trial, see Section 5.2). We will use $\hat{p} = r/n$ as our estimate for p, the probability of success on each trial. The letter q again represents the probability of failure on each trial, and so $q = 1 - p$. We also assume that the samples are large (i.e., $np > 5$ and $nq > 5$).

For large samples, the distribution of $\hat{p} = r/n$ values is well approximated by a *normal curve* with mean μ and standard deviation σ as follows:

$$\mu = p \qquad \sigma = \sqrt{\frac{pq}{n}}$$

The null and alternate hypotheses for tests of proportions are

| $H_0: p = k$ | $H_0: p = k$ | $H_0: p = k$ |
| $H_1: p < k$ | $H_1: p > k$ | $H_1: p \neq k$ |

depending on what is asked for in the problem. Notice that since p is a probability, the value k must be between 0 and 1.

Critical values for testing p
($np > 5$ and $nq > 5$)

Since the $\hat{p}$ distribution is approximately normal when n is sufficiently large, we will use for our tests the same *critical values* z_0 as those we used for testing the mean with large samples (Section 9.2). These values were shown in Figure 9-2 and in Table 4(c) of Appendix I. Figure 9-24 below repeats the critical values for convenience.

FIGURE 9-24

Critical Values for Tests of Proportions

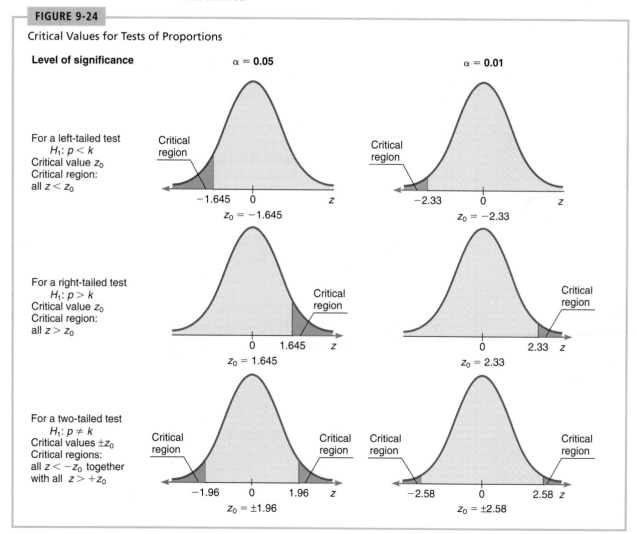

Level of significance

For a left-tailed test
  $H_1: p < k$
Critical value z_0
Critical region:
all $z < z_0$

For a right-tailed test
  $H_1: p > k$
Critical value z_0
Critical region:
all $z > z_0$

For a two-tailed test
  $H_1: p \neq k$
Critical values $\pm z_0$
Critical regions:
all $z < -z_0$ together
with all $z > +z_0$

Sample test statistic

For tests of proportions, we need to convert our sample test statistic $\hat{p}$ to a z value. Then we can compare the sample z value with a specified critical value z_0 to determine whether or not to reject H_0. The $\hat{p}$ distribution is approximately normal with mean p and standard deviation $\sqrt{pq/n}$. Therefore, the conversion of $\hat{p}$ to z follows the formula

$$z = \frac{\hat{p} - p}{\sqrt{\dfrac{pq}{n}}}$$

where $\hat{p} = r/n$ is the sample test statistic; n = number of trials
p = proportion specified in H_0; and $q = 1 - p$

EXAMPLE 12 ▷ A team of eye surgeons has developed a new technique for a risky eye operation to restore the sight of people blinded from a certain disease. Under the old method, it is known that only 30% of the patients who undergo this operation recover their eyesight.

Suppose that surgeons in various hospitals have performed a total of 225 operations using the new method and that 88 have been successful (the patients fully recovered their sight). Can we justify the claim that the new method is better than the old one? (Use a 1% level of significance.)

SOLUTION:

(a) Let p be the probability that a patient fully recovers his or her eyesight. The null hypothesis is that p is still 0.30, even for the new method of operation. Therefore,

H_0: $p = 0.30$

(b) The alternate hypothesis is that the new method has improved the patient's chances for eyesight recovery. Therefore,

H_1: $p > 0.30$

(c) Since the alternate hypothesis is H_1: $p > 0.30$, we use a right-tailed test. For $\alpha = 0.01$, Figure 9-24 on page 447 shows a critical value of $z_0 = 2.33$.

(d) Next, we find the sample test statistic $\hat{p}$ and convert it to a z value:

$$\hat{p} = \frac{r}{n} = \frac{88}{225} \approx 0.39$$

The z value corresponding to $\hat{p}$ is

$$z = \frac{\hat{p} - p}{\sqrt{\dfrac{pq}{n}}} = \frac{0.39 - 0.30}{\sqrt{\dfrac{0.30(0.70)}{225}}} \approx 2.95$$

In the formula, the value for p came from the null hypothesis. H_0 specified that $p = 0.30$, so $q = 1 - 0.30 = 0.70$.

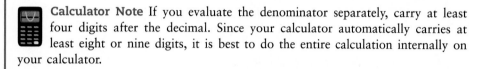 **Calculator Note** If you evaluate the denominator separately, carry at least four digits after the decimal. Since your calculator automatically carries at least eight or nine digits, it is best to do the entire calculation internally on your calculator.

(e) Finally, we draw a normal distribution showing the critical region and the location of the z value corresponding to the sample test statistic $\hat{p}$ (Figure 9-25). Since the sample z value falls inside the critical region, we reject H_0 and conclude that the new method seems to have improved a patient's chances for eyesight recovery.

FIGURE 9-25

Critical Region, with $\alpha = 0.01$

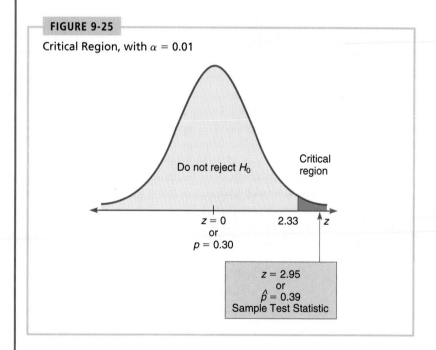

GUIDED EXERCISE   9

A botanist has produced a new variety of hybrid wheat that is better able to withstand drought than other varieties. The botanist knows that for the parent plants the proportion of seeds germinating is 80%. The proportion of seeds germinating for the hybrid variety is unknown, but the botanist claims that it is 80%. To test this claim, 400 seeds from the hybrid plant are tested and it is found that 312 germinated. Use a 5% level of significance to test the claim that the proportion germinating for the hybrid is 80%.

(a) Let p be the proportion of hybrid seeds that will germinate. What is the null hypothesis:

$H_0: p > 0.80,$ $H_0: p = 0.80,$
    or $H_0: p \neq 0.80$?

    ⇨ $H_0: p = 0.80$

(b) Because we have no prior knowledge about germination proportion for the hybrid plant, what would be a good choice for the alternate hypothesis?

    ⇨ $H_1: p \neq 0.80$

(c) Should our test be a one-tailed or a two-tailed test? For $\alpha = 0.05$, what is (are) the critical value(s)?

    ⇨   The test is a two-tailed test. By Figure 9-24, the critical values are $z_0 = \pm 1.96$.

(d) Calculate the sample test statistic $\hat{p}$.

    ⇨   The number of trials is $n = 400$, and the number of successes is $r = 312$. Thus

$$\hat{p} = \frac{r}{n} = \frac{312}{400} = 0.78$$

(e) Next, we convert the sample test statistic $\hat{p} = 0.78$ to a z value. Based on our choice for H_0, what value should we use for p in our formula? Since $q = 1 - p$, what value should we use for q? Using these values for p and q, convert $\hat{p}$ to a z value.

    ⇨   According to H_0, $p = 0.80$. Then $q = 1 - p = 0.20$. Using these values in the formula for z gives

$$z = \frac{\hat{p} - p}{\sqrt{\dfrac{pq}{n}}}$$

$$= \frac{0.78 - 0.80}{\sqrt{\dfrac{0.80(0.20)}{400}}} = -1.00$$

 Calculator Note If you evaluate the denominator separately, be sure to carry at least 4 digits after the decimal.

Exercise continues

(f) On the normal curve, show the critical   ⟹   The critical region is shown in Figure 9-26.
regions and the z value corresponding to   Since the sample statistic does not fall in the
the sample test statistic $\hat{p}$. Do we reject or   critical region, we fail to reject H_0. At the 5%
fail to reject H_0?   level of significance, there is not enough evi-
                                                        dence to say the botanist is wrong.

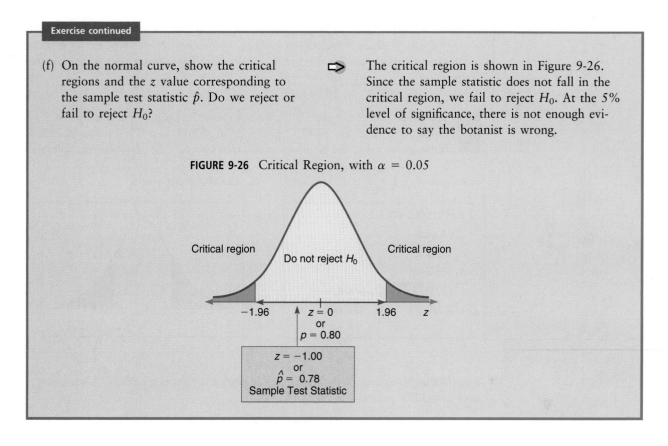

FIGURE 9-26 Critical Region, with $\alpha = 0.05$

P values for tests of   Since the $\hat{p}$ sampling distribution is approximately normal, we use Table 4,
proportions   "Areas of a Standard Normal Distribution," in Appendix I to find P values.

EXAMPLE 13 ⟩   Let's solve Guided Exercise 9 using the P value approach. In that problem, 312 of
400 seeds from a hybrid wheat variety germinated. For the parent plants, the proportion of germinating seeds was 80%. Use a 5% level of significance to test the
claim that the population proportion of germinating seeds from the hybrid wheat
is different from that of the parent plants.

(a) First, we find the sample statistic $\hat{p}$ and the corresponding z value. This was
done in Guided Exercise 9, where we found that $\hat{p} = 0.78$ with corresponding
$z = -1.00$.

(b) Next, we find the P value associated with $z = -1.00$ and a two-tailed test. Using
Table 4 (Appendix I) together with techniques of Chapter 6, we see that
$P(z \leq -1.00) = 0.5000 - 0.3413 = 0.1587$. The test is a two-tailed z test, so
we double the area found in the left tail to get P value $= 0.3174$. Since this P

value is greater than the level of significance, $\alpha = 0.05$, we do not reject the null hypothesis, and we conclude that there is not enough evidence to show that the germination rate for the hybrid wheat is different from that of the parent plants.

 Calculator Note On the TI-83, we can show the results by using **Option 5:1-PropZTest** under the **TESTS** option for the **STAT** key. The symbol p_0 refers to the value set in the null hypothesis H_0: $p = p_0$. The results are as follows.

Input data   Use **DRAW** for results

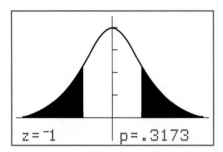

The P value is the area of the shaded region. Since the P value of 0.3173 is greater than $\alpha = 0.05$, we do not reject H_0.

V I E W P O I N T

Who Did What?

Art, music, literature, and science share a common need to classify things: Who painted that picture? Who composed that music? Who wrote that document? Who should get that patent? In statistics, such questions are called *classification problems*. For example, the *Federalist Papers* were published anonymously in 1787–1788 by Alexander Hamilton, John Jay, and James Madison. But who wrote what? That question is addressed by F. Mosteller (Harvard University) and D. Wallace (University of Chicago) in the book *Statistics: A Guide to the Unknown,* edited by J. M. Tanur. Other scholars have studied authorship regarding Plato's *Republic* and Plato's *Dialogues,* including the *Symposium.* For more information on this topic, see the source in Problems 7 and 8 in the exercise set for this section.

SECTION 9.5 PROBLEMS

For each of the following problems, please provide the requested information.

(a) What is the null hypothesis? What is the alternate hypothesis? Will we use a left-tailed, right-tailed, or two-tailed test? What is the level of significance?

(b) What sampling distribution will we use? What is the critical value z_0 (or critical values $\pm z_0$)?

(c) Sketch the critical region and show the critical value (or critical values).

(d) Compute the z value corresponding to the sample statistic $\hat{p}$ and show its location on the sketch in part c.

(e) Find the P value for your test and explain what the P value means in the context of the problem.

(f) Based on your answers for parts a to e, shall we reject or fail to reject (i.e., "accept") the null hypothesis for the given level of significance α? Explain your conclusion in the context of the problem.

Sociology: Crime Rate

1. Is the national crime rate really going down? Some sociologists say yes! They say that the reason for the decline in crime rates for the 1980s and 1990s is demographics. It seems that the population is aging, and older people commit fewer crimes. According to the FBI and the Justice Department, 70% of all arrests are of males aged 15 to 34 years. (Source: *True Odds*, by J. Walsh, Merritt Publishing.) Suppose that you are a sociologist in Rock Springs, Wyoming, and a random sample of police files show that of 32 arrests last month, 24 were of males aged 15 to 34 years. Use a 1% level of significance to test the claim that the population proportion of such arrests in Rock Springs is different from 70%.

Sociology: High School Dropouts

2. *Life in America's Small Cities* is a book by G. S. Thomas. This book reports that in 1990 in Key West, Florida, 22.1% of all 16- to 19-year-olds were high school dropouts. However, in 1995, a random sample of 193 people in this Key West age group showed that 32 were high school dropouts. Does this indicate that the population proportion of high school dropouts (16- to 19-year-olds) in Key West is less than 22.1%? Use a 5% level of significance.

Highway Safety: Intoxication

3. The U.S. Department of Transportation, National Highway Traffic Safety Administration, reported that 77% of all fatally injured automobile drivers were intoxicated. A random sample of 27 records of automobile driver fatalities in Kit Carson County, Colorado showed that 15 involved intoxicated drivers. Do these data indicate that the population proportion of driver fatalities related to alcohol is less than 77% in Kit Carson County? Use $\alpha = 0.01$.

Highway Safety: Air Bags

4. Do drivers with air bags take greater risks? Researchers at Virginia Commonwealth University studied fatal automobile accidents between cars in which one had an air bag and the other did not. They found that in 73% of all such fatal crashes the driver of the air bag vehicle was responsible for the accident (see source in Problem 1). Suppose that you obtain police records of fatal automobile crashes (involving a vehicle with an air bag and one without) near Fargo, North Dakota. A random

sample of 41 such accidents shows that 33 were the fault of the driver of the air bag vehicle. Does this indicate that the population proportion of such accidents in which the driver of the air bag vehicle was at fault is higher than 73% in the Fargo district? Use $\alpha = 0.05$.

Ecology: Wolves

5. This problem is based on information from *The Wolf in the Southwest: The Making of an Endangered Species,* by David E. Brown (University of Arizona Press). Before 1918, the proportion of female wolves in the general population of all southwestern wolves was about 50%. However, after 1918, southwestern cattle ranchers began a widespread effort to destroy wolves. In a recent sample of 34 wolves, there were only 10 females. One theory is that male wolves tend to return sooner than females to their old territory, where their predecessors were exterminated. Do these data indicate that the population proportion of female wolves is now less than 50% in the region? Use $\alpha = 0.01$.

Leisure: Fishing

6. Athabasca Fishing Lodge is located on Lake Athabasca in northern Canada. In one of its recent brochures, the lodge advertises that 75% of its guests catch northern pike weighing over 20 pounds. Suppose that last summer 64 out of a random sample of 83 guests did in fact catch northern pike weighing over 20 pounds. Does this indicate that the population proportion of guests who catch pike over 20 pounds is different from 75% (either higher or lower)? Use $\alpha = 0.05$.

Humanities: Plato's *Republic*

7. Prose rhythm is characterized as the occurrence of five-syllable sequences in long passages of text. This characterization may be used to assess the similarity among passages of text and sometimes the identity of authors. The following information is based on an article by D. Wishart and S. V. Leach appearing in *Computer Studies of the Humanities and Verbal Behavior* (Vol. 3, pp. 90–99). Syllables were categorized as long or short. On analyzing Plato's *Republic,* Wishart and Leach found that about 26.1% of the five-syllable sequences are of the type in which two are short and three are long. Suppose that Greek archaeologists have found an ancient manuscript dating back to Plato's time (about 427–347 B.C.). A random sample of 317 five-syllable sequences from the newly discovered manuscript showed that 61 are of the two-short-and-three-long type. Do the data indicate that the population proportion of this type of five-syllable sequence is different (either way) from the text of Plato's *Republic*? Use $\alpha = 0.01$.

Humanities: Plato's *Symposium*

8. *Symposium* is part of a larger work referred to as Plato's *Dialogues.* Wishart and Leach (see source in Problem 7) found that about 21.4% of five-syllable sequences in *Symposium* are of the type in which four are short and one is long. Suppose that an antiquities store in Athens has a very old manuscript that the owner claims to be part of Plato's *Dialogues.* A random sample of 493 five-syllable sequences from this manuscript showed that 136 were of the four-short-and-one-long type. Do the data indicate that the population proportion of this type of five-syllable sequence is higher than that found in Plato's *Symposium*? Use $\alpha = 0.01$.

Marketing: Brand
Loyalty

9. *USA Today* reported that about 47% of the general consumer population in the United States is loyal to the automobile manufacturer of their choice. Suppose that Chevrolet did a study of a random sample of 1006 Chevrolet owners and found that 490 said that they would buy another Chevrolet. Does this indicate that the population proportion of consumers loyal to Chevrolet is more than 47%? Use $\alpha = 0.01$.

Marketing: Supermarket
Prices

10. *Harper's Index* reported that 80% of all supermarket prices end in the digit 9 or 5. Suppose that you check a random sample of 115 items in a supermarket and find that 88 have prices that end in 9 or 5. Does this indicate that less than 80% of the prices in the store end in 9 or 5? Use $\alpha = 0.05$.

Management: Supervisor
Communications

11. The *Wall Street Journal* reported that 65% of all U.S. office workers decided to take their current jobs because of open communications with their supervisors. Suppose that a random sample of 78 American Express office workers in Los Angeles were asked for the main reason they took their current jobs and 60 responded because of open communications with their supervisors. Does this indicate a difference (either up or down) from the reported 65% in the *Wall Street Journal*? Use $\alpha = 0.05$.

Management: Flex Time
Schedules

12. The *Wall Street Journal* reported that 24% of U.S. office workers prefer "flex time," a system in which the employees choose their own work schedules. Suppose that a random sample of 66 IBM employees in Boston showed that 23 prefer a flex-time work schedule. Does this indicate that the proportion of all IBM employees in Boston that prefer a flex-time schedule is more than 24%? Use $\alpha = 0.01$.

Myers-Briggs: Personality
Types and Student
Government

13. Are most student government leaders extroverts? According to Myers-Briggs estimates, about 82% of college student government leaders are extroverts. (Source: *Myers-Briggs Type Indicator Atlas of Type Tables*.) Suppose that a Myers-Briggs personality preference test was given to a random sample of 73 student government leaders attending a large national leadership conference and that 56 were found to be extroverts. Does this indicate that the population proportion of extroverts among college student government leaders is different (either way) from 82%? Use $\alpha = 0.01$.

Medical: Hypertension

14. This problem is based on information taken from *The Merck Manual* (a reference manual used in most medical and nursing schools). Hypertension is defined as a blood pressure over 140 mm Hg systolic and/or over 90 mm Hg diastolic. Hypertension, if not corrected, can cause long-term health problems. In the college-age population (18–24 years), about 9.2% have hypertension. Suppose that a blood donor program is occurring in a college dormitory this week (final exams week). Before each student gives blood, the nurse takes a blood pressure reading. Of 196 donors, it is found that 29 have hypertension. Do these data indicate that the population proportion of students with hypertension during final exams week is higher than 9.2%? Use the following Minitab (➤**Stat** ➤**Basic Statistics** ➤**1-Prop**) display to draw a conclusion based on a 5% level of significance.

Test and Confidence Interval for One Proportion

```
Test of p = 0.092 vs p > 0.092
Sample   X    N    Sample p        95.0 % CI        Z-Value   P-Value
   1    29   196   0.147959   (0.098252, 0.197667)    2.71     0.003
```

SUMMARY

In this chapter we studied statistical inference methods called *hypothesis testing.* In hypothesis testing, we establish an initial claim about the value of a parameter. This claim is the null hypothesis H_0, which claims the parameter in question equals a certain value. Then we propose an alternate hypothesis H_1, which indicates the parameter is less than, greater than, or different from the value in the null hypothesis. To determine whether or not to reject the null hypothesis, we use the evidence of the sample data and the predetermined level of significance α.

The basic steps we follow in the procedure of hypothesis testing are:

1. State the null and alternate hypotheses H_0 and H_1.

2. Choose α, the level of significance.

3. Find the critical value(s) z_0 or t_0 and the corresponding critical region.

4. Convert the sample test statistic to a z value or t value.

5. If the sample test statistic falls in the critical region, reject H_0. Otherwise, do not reject H_0.

An alternate way to conclude a test of hypotheses is to compare the P value of the sample test statistic with α. The P value of the sample test statistic is the smallest level of significance for which we can reject H_0. Therefore, if P value $\leq \alpha$, we reject H_0, and if P value $> \alpha$, we do not reject H_0. Because statistical computer software packages generate the P value associated with the sample test statistic and the null hypothesis, the method of using P values to conclude tests of hypotheses is widely used and very popular.

In this chapter we used hypothesis testing to conduct tests involving single means μ (large or small sample) and single proportions, p. The methods of hypothesis testing are very general, and we will see them used again in later chapters with other parameters and other probability distributions.

IMPORTANT WORDS & SYMBOLS

Section 9.1
Hypothesis
Hypothesis testing
Null hypothesis H_0
Alternate hypothesis H_1
Type I error
Type II error
α, the level of significance of a test and the probability
    of a type I error
β, the probability of a type II error
Power of a test $(1 - \beta)$
Critical region
Right-tailed tests
Left-tailed tests
Two-tailed tests

Section 9.2
Sample test statistic
Critical value or values for testing means

Section 9.3
P value
Statistical significance

Section 9.4
Degrees of freedom, $d.f.$ for testing μ with small samples
α' and α'' with respect to the t distribution table

Section 9.5
$\hat{p}$ distribution

VIEWPOINT

Will It Rain?

Do cloud-seeding experiments ever work? If you seed the clouds, will it rain? If it did rain, who would benefit? Who would be displeased by the rain? If you seeded the clouds and nothing happened, would taxpayers (who support the effort) complain or rejoice? Maybe this should be studied over a remote island—such as Tasmania (near Australia). Using what you already know about statistical testing, you can conduct your own tests given the appropriate data. Remember, there are sociological questions (pleased/displeased with result) as well as technical questions (number of inches of rain produced). For data regarding cloud-seeding experiments over Tasmania, see Web site <http://lib.stat.cmu.edu/> and look under Datasets for Cloud.

CHAPTER REVIEW PROBLEMS

Before you solve each problem, first categorize it by answering the following questions:
(a) Are we testing a single mean or a single proportion?
(b) Are we using a large sample or a small sample?
(c) Then solve each problem by
    (i)   stating the null and alternate hypotheses.
    (ii) finding the critical values(s) and critical region.
    (iii) sketching the critical region.
    (iv) converting the sample test statistic to a z value or a t value as appropriate and showing its location on the sketch in part iii.
    (v) finding the P value of the sample test statistic.
    (vi) making your decision to reject or fail to reject the null hypothesis.
    (vii) relating your conclusion to the specific question asked in the problem.

Environment: Miles per Vehicle

1. According to the Energy Information Association, the average number of miles driven annually per vehicle in the United States is 11.3 thousand miles. Suppose that a random sample of 36 vehicles owned by residents of Chicago showed that the average mileage driven last year was 11.0 thousand miles with standard deviation 600 miles. Does this indicate that the average miles driven per vehicle in Chicago is different from (higher or lower than) the national average? Use a 5% level of significance.

Business: Leased Cars

2. According to *USA Today*, 25% of cars were leased last year instead of sold. To test if the percentage of leased cars is still the same this year, a random sample of 50 new car deals was examined. Of these, 18 involved car leases instead of car purchases. Does this indicate that the percentage of leased cars is different (either higher or lower) this year? Use a 1% level of significance.

Medical: Waiting Time

3. New patients often have to wait a number of days to get an appointment with a physician. According to *Practice Patterns of General Internal Medicine* (published by the American Medical Association), a new patient must wait an average of 10.8 days for an appointment. Rockwood Clinic did a study of its new patients and found that for a random sample of 20 new patients, the average waiting time for an appointment was 9.2 days with standard deviation 3 days. At the 1% level of significance, do new patients at Rockwood Clinic have an average waiting time for an appointment that is less than 10.8 days?

Lifestyle: Students with Jobs

4. Professor Jennings claims that only 35% of the students at Flora College work while attending school. Dean Renata thinks the professor has underestimated the number of students with part-time or full-time jobs. A random sample of 81 students shows that 39 have jobs. Do the data indicate that more than 35% of the students have jobs? Use a 5% level of significance.

Business: Sales per Employee

5. In the retail business the average annual sales per employee varies from business to business. According to an article in the magazine *American Demographics*, the average value of annual sales per employee in apparel stores is 88.9 thousand dollars. In one large chain of apparel stores it was found that for a random sample of 40

employees, the average value of sales per employee last year was 89.7 thousand dollars with standard deviation 5 thousand dollars. Is this evidence that employees in this chain have higher average annual sales than the apparel industry in general? Use a 5% level of significance.

6. For grocery stores, the average annual sales per employee is 131.4 thousand dollars (*American Demographics* magazine). Great Foods chain did a study of 50 employees from different stores and found the average annual sales per employee to be 135.8 thousand dollars with standard deviation 8 thousand dollars. Does it appear that the average annual sales per Great Foods employee is different from 131.4 thousand? Use a 1% level of significance.

7. Is there a Mr. or Ms. Right? According to a survey conducted by CJ Olson Market Research (reported in *USA Today*), 84.6% of adults between 18 and 24 years of age believe every person has a perfect match. A random sample of 50 people in the 25-to-34 age group were asked the same question. Thirty-six in the older age group said that they believed every person has a perfect match. Use a 1% level of significance to test the hypothesis that the proportion of people in the 25-to-34 age group who believe everyone has a perfect match is less than the proportion in the 18-to-24 age group.

8. A hospital reported that the normal death rate for patients with extensive burns (more than 40% of skin area) has been significantly reduced by the use of new fluid plasma compresses. Before the new treatment the mortality rate for extensive burn patients was about 60%. Using the new compresses, the hospital found that only 40 out of 90 patients with extensive burns died. Use a 1% level of significance to test the claim that the mortality rate has dropped.

9. The Congressional Budget Office reports that 36% of the federal civilian employees have a bachelor's degree or higher (*Wall Street Journal*). A random sample of 120 employees in the private sector showed that 33 have a bachelor's degree or higher. Does this indicate that the percentage of employees who hold bachelor's or higher degrees in the private sector is less than in the federal civilian sector? Use $\alpha = 0.05$.

10. A machine in the student lounge dispenses coffee. The average cup of coffee is supposed to contain 7.0 oz. Eight cups of coffee from this machine show the average content to be 7.3 oz with a standard deviation of 0.5 oz. Do you think the machine has slipped out of adjustment and the average amount of coffee per cup is different from 7 oz? Use a 5% level of significance.

11. The manufacturer of a sports car claims that the fuel injection system lasts 48 months before it needs to be replaced. A consumer group tests this claim by surveying a random sample of 10 owners who had fuel injection systems replaced. The ages of the cars at the time of replacement were (in months):

29	42	49	58	53
46	30	51	42	62

(a) Use your calculator to verify that the mean age of a car when the fuel injection system fails is $\bar{x} = 46.2$ months with standard deviation $s = 10.85$ months.

(b) Test the claim that the fuel injection system lasts less than 48 months before needing replacement. Use a 5% level of significance.

DATA HIGHLIGHTS: GROUP PROJECT

Break into small groups and discuss the following topics. Organize a brief outline in which you summarize the main points of your group discussion.

"With Sampling, There Is Too a Free Lunch" was a headline that appeared in the *Wall Street Journal*. The article is about food product samples available at grocery stores. Giving out food samples is expensive and labor-intensive. It clogs supermarket aisles. It is risky. What if a customer tries an item and spits it out on the floor or says the product is awful? It creates litter. Some customers drop toothpicks or small paper cups on the floor or spill the product. However, the budget that companies are willing to spend to have their products sampled is growing. The director of communications for Bigg's "hypermarket" (a combination grocery and general-merchandise store) says that more than 60% of customers sample products and about 37% of those who sample a product buy it.

(a) Let's test the hypothesis that 60% of customers sample a particular product. What is the null hypothesis? Do you believe that the percentage of customers who sample products is less than, more than, or just different from 60%? What will you use for the alternate hypothesis?

(b) Choose a level of significance α.

(c) Go to a grocery store when special products are being sampled (not just the usual in-house store samples often available at the deli or bakery). Count the number of customers going by the display when a sample is available and the number of customers who try the sample. Be sure the number of customers n is large enough to use the normal distribution to approximate the binomial.

(d) Using your sample data, conclude the hypothesis test. What is your conclusion?

(e) Do you think different food products might have higher or lower percentages of customers trying them? For instance, does a higher percentage of customers try samples of pizza than samples of yogurt? How could you use statistics to justify your answer? (See Chapter 10.)

(f) Do you want to include young children in your sample? Do they pick up items to include in the customer's basket, or do they just munch the samples?

LINKING CONCEPTS: WRITING PROJECTS

Discuss each of the following topics in class or review the topics on your own. Then write a brief but complete essay in which you summarize the main points. Please include formulas and graphs as appropriate.

The most important questions in life usually cannot be answered with absolute certainty. Many important questions are answered by giving an estimate and a measure of confidence in the estimate. This was the focus of Chapter 8. However, sometimes important questions must be answered in a more straightforward manner by a simple *yes* or *no*. Hypothesis testing is the statistical process of answering questions with a straightforward yes or no *and* providing an estimate of the risk in accepting the answer.

1. Review and discuss type I and type II errors associated with hypothesis testing.

2. Review and discuss the level of significance and power of a statistical test.

3. The following statements are very important. Give them some careful thought and discuss them.
   a. When we fail to reject the null hypothesis we do not claim it is absolutely true. We simply claim that at the given level of significance the data were not sufficient to reject the null hypothesis.
   b. When we accept the alternate hypothesis we do not claim the null hypothesis is absolutely false. We do claim that at the given level of significance the data presented enough evidence to reject the null hypothesis.

4. As access to computers becomes more and more prevalent, we see *P* values reported in hypothesis testing more frequently. Review the use of *P* values in hypothesis testing. What is the difference between the level of significance of a test and the *P* value? Considering both the *P* value and level of significance, under what conditions do we reject or fail to reject the null hypothesis?

The problems in this section may be done using statistical computer software or calculators with statistical functions. Displays and suggestions are given for Minitab (Release 12), the TI-83 graphing calculator, ComputerStat, and Excel.

Testing a Mean μ Using Large Samples

Minitab

When testing a single mean using the normal distribution, Minitab requires that the *population* standard deviation be known. If you do not know the population standard deviation and you wish to estimate it by the sample standard deviation s, then you must ask Minitab to compute the sample standard deviation s.

Windows Pull-Down Menu Selection

Enter the data. Then, if you do not know the population standard deviation σ, use the menu selection ➤ **Stat** ➤ **Basic Statistics** ➤ **Descriptive Statistics** to find the sample standard deviation s. Record that value. To test a single mean using a normal distribution, follow the menu selection ➤ **Stat** ➤ **Basic Statistics** ➤ **1-Sample z.** In the dialogue box use the sample estimate s for the value of σ. For Minitab displays showing the output of 1-Sample z test, see Problem 10 in Section 9.3.

TI-83

The TI-83 supports all the tests of hypotheses presented in Chapter 9. For testing μ, large sample, press the **STAT** key, select **TESTS**, and use **Option 1:Z-Test.** For Application 1, select **Inpt:Data**, and put the raw data in list L_1. Use **STAT, Calculate, 1-Var Stats** on L_1 to determine the sample standard deviation s. In the Z-Test, use the value of s as an estimate for σ. The value μ_0 is the value specified in the null hypothesis H_0: $\mu = \mu_0$. Specify the symbol ($\neq$, $<$, $>$) used in the alternate hypothesis. The **Calculate** option produces the value of the sample statistic $\bar{x}$ and the correspond-

ing z value and P value. The **Draw** option shows the area represented by the P value and gives both the sample z value and the corresponding P value. For displays, see the Calculator Notes in Sections 9.2 and 9.3.

ComputerStat

Under the Hypothesis Testing menu, select the program Testing a Single Population Mean. Follow the instructions on the screen.

Excel

Excel has a command ZTEST(array, test value from H_0) that claims to give the P value for a two-tailed test of a mean utilizing the normal distribution. However, as of version Excel 2000, the command does not provide the described P value. Instead, it gives the area of the right tail bounded by the sample statistic $\bar{x}$. You can adjust the given output from ZTEST to provide the P value for a two-tailed test of the mean. If the sample test statistic $\bar{x}$ is greater than the parameter value specified in H_0, double the output provided by ZTEST. If the sample test statistic $\bar{x}$ is less than the parameter value specified in H_0, subtract the ZTEST output from 1 to get the area of the left tail bounded by the sample statistic $\bar{x}$. Then double the result to obtain the P value for the two-tailed test.

APPLICATION

People who do shift work must often adjust their eating habits, sleep habits, exercise habits, family contacts, social lives, and overall lifestyles to accommodate their jobs. Extensive rearrangement of a person's habits and lifestyle can sometimes result in tension, anxiety, and overall health problems. An extensive study of the health consequences of shift work can be found in the following publication. Interested readers are referred to this report:

United States Department of Health, Education, and Welfare, NIOSH Technical Report. Tasto, Colligan, et al., *Health Consequences of Shift Work*. Washington: GPO.

In an effort to study mood levels of nurses working in large hospitals, an opinion scale was used. Opinion ratings ranged from 0 = no feelings of tension and anxiety to a rating of 4 = extensive feelings of tension and anxiety. The scale was continuous, so a nurse could mark any number between 0 and 4. Suppose a random sample of 35 nurses on the day shift of a very large hospital gave the ratings shown in the accompanying table of their feelings of tension and anxiety at the end of the day shift.

(a) On a scale of 0 to 4, a moderate (i.e., medium) level of tension and anxiety is 2. Use the null hypothesis that the population mean tension level for nurses after the day shift is 2. Use the alternate hypothesis that the population mean tension level is different from (either higher or lower than) 2. Use a 5% level of significance. What is the P value? Compare the P value with the level of significance. Do you think we should accept or reject the null hypothesis?

(b) Rerun the program with the same data (in ComputerStat you do not need to reenter the data). Use the same null hypothesis, but use a right-tailed test with level of significance 0.10. Do we accept or reject the null hypothesis? Look at the P value. What is the smallest level of significance that will result in a rejection of the null hypothesis using a right-tailed test?

Data for Day-Shift Nurses

3.50	3.75	2.33	2.16	3.50	0.80	1.25
1.33	2.67	2.50	1.50	0.75	0.00	0.67
4.00	3.75	3.50	3.25	2.40	3.50	2.75
3.50	2.67	2.80	2.33	3.50	3.80	2.75
1.50	1.33	0.00	2.25	1.75	0.50	1.75

Computer Displays

Displays showing Minitab outputs for tests of μ using the normal distribution and for tests of proportion p are presented in Problem 10 in Section 9.3 and in Problem 14 in Section 9.5, respectively.

TI-83 displays for tests of μ using the normal distribution, tests of μ using the Student's t distribution, and tests of proportions are shown in the Calculator Notes in Sections 9.3, 9.4, and 9.5, respectively.

ComputerStat has program options for tests of μ for large and small samples and tests of single proportions under the main menu Hypothesis Testing. The output gives the z or t value of the sample test statistic and compares it with the critical values of the corresponding distribution. In addition, the P values are shown. See Problem 14 in Section 9.4 for a ComputerStat display of a test of μ for small samples.

10 Inferences About Differences

"So what!"

—Anonymous

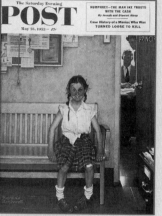

**"Girl with Black Eye"
by Norman Rockwell
(1894–1978)**

Norman Rockwell painted everyday
people and situations. In this cover
for the *Saturday Evening Post* (May
23, 1953), a young lady is about to
have a conference with her school
principal. So what!!

We have all heard the exclamation, "So what!" Philologists (people who study cultural linguistics) tell us that this expression is a shortened version of "So what is the difference!" They also tell us that there are similar popular or slang expressions about differences in all languages and cultures. It is human nature to challenge the claim that something is better, worse, or just simply different. In this chapter we will focus on this very human theme by studying probable differences between population means and probable differences between population proportions. In a way, this is another example of what the great French mathematician Pierre Simon de Laplace (1749–1827) meant when he said, "The theory of probability is at bottom only common sense reduced to calculation."

Section 10.1

Tests Involving Paired Differences (Dependent Samples)

Many statistical applications use *paired data* samples to draw conclusions about the difference between two population means. Data *pairs* occur very naturally in "before and after" situations, where the *same* object or item is measured both before and after a treatment. Applied problems in social science, natural science, and business administration frequently involve studies of matching pairs. Psychological studies of identical twins; biological studies of plant growth on plots of land matched for soil type, moisture, and sun; and business studies on sales of matched inventories are examples of paired data studies.

Creating data pairs

When working with paired data, it is very important to have a definite and uniform method of creating data pairs that clearly utilizes a natural matching of characteristics. The next example and guided exercise demonstrate this feature.

EXAMPLE 1 ❯   A shoe manufacturer claims that among the general population of adults in the United States, the average length of the left foot is greater than that of the right. To compare the average length of the left foot with that of the right, we can take a random sample of 15 U.S. adults and measure the length of the left foot and then the length of the right foot for each person in the sample. Is there a natural way of pairing the measurements? How many pairs will we have?

SOLUTION: In this case, we can pair each left foot measurement with the same person's right foot measurement. The person serves as the "matching link" between the two distributions. We will have 15 pairs of measurements.

GUIDED EXERCISE   1

A psychologist has developed a series of exercises called the Instrumental Enrichment (IE) program, which he claims to be useful in overcoming cognitive deficiencies in mentally retarded children. To test the program, extensive statistical tests are being conducted. In one simple test, a random sample of 10-year-old students with IQ scores below 80 was selected. An IQ test was given to these students before they spent 2 years in an IE program, and an IQ test was given to the same students after the program.

(a) On what basis can you pair the IQ scores?

⇨   Take the "before and after" IQ scores of each individual student.

(b) If there were 20 students in the sample, how many data pairs would you have?

⇨   Twenty data pairs. Note that there would be 40 IQ scores, but only 20 pairs.

To compare two populations we cannot always employ paired data tests, but when we can, what are the advantages? Using matched or paired data can often reduce the danger of introducing extraneous or uncontrollable factors into our sample measurements because the matched or paired data have essentially the *same* characteristics except for the *one* characteristic that is being measured. Furthermore, it can be shown that pairing data has the theoretical effect of reducing measurement variability (i.e., variance), which increases the strength of statistical conclusions.

When we wish to compare the means of two samples, the first item to be determined is whether or not there is a natural pairing between the data in the two samples. Again, data pairs are created from "before and after" situations, or from matching data by using studies of the same object, or by a process of taking measurements of closely matched items.

Testing the differences d

When testing *paired* data, we take the difference *d* of the data pairs *first* and look at the mean difference $\overline{d}$. Then we use a test on $\overline{d}$. Theorem 10.1 provides the basis for our work with paired data.

Theorem 10.1

Consider a random sample of *n* data pairs. Suppose the differences *d* between the first and second members of each data pair are (approximately) normally distributed with population mean μ_d. Then the *t* values

$$t = \frac{\overline{d} - \mu_d}{s_d/\sqrt{n}}$$

where $\overline{d}$ is the sample mean of the *d* values, *n* is the number of data pairs, and

$$s_d = \sqrt{\frac{\Sigma(d - \overline{d})^2}{n - 1}}$$

is the sample standard deviation of the *d* values, follow a Student's *t* distribution with degrees of freedom *d.f.* = *n* − 1.

Hypotheses for testing the mean of the differences

When testing the mean of the differences of paired data values, the null hypothesis is that there is no difference among the pairs. That is, the mean of the differences μ_d is zero.

$$H_0: \mu_d = 0$$

The alternate hypothesis depends on the problem and can be

$$H_1: \mu_d < 0 \text{ (left-tailed)} \quad \text{or} \quad H_1: \mu_d > 0 \text{ (right-tailed)} \quad \text{or}$$
$$H_1: \mu_d \neq 0 \text{ (two-tailed)}$$

Critical values

Since the $\overline{d}$ distribution follows a Student's *t* distribution with degrees of freedom *d.f.* = *n* − 1, our critical values will be found in Table 5, "Student's *t* Distribution," in Appendix I. The methods used for finding the critical values are the same as those in Section 9.4, where we used the Student's *t* distribution to find critical values for tests of the mean with small samples. If the test is a *one-tailed* test, we find the critical *t* value in the column headed by α' = level of significance. If the test is a *two-tailed* test, we find the critical value in the column

Sample test statistic

headed by α'' = level of significance. In each case we use the row containing the degrees of freedom $n - 1$, where n is the *number of data pairs*.

For paired difference tests, we make our decision regarding H_0 according to the evidence of the sample mean $\bar{d}$ of the differences of measurements. By Theorem 10.1, we convert the sample test statistic $\bar{d}$ to a t value using the formula

$$t = \frac{\bar{d} - \mu_d}{(s_d/\sqrt{n})} \text{ with } d.f. = n - 1$$

where s_d = sample standard deviation of the differences d

$\quad n$ = number of data pairs

$\quad \mu_d = 0$ as specified in H_0

EXAMPLE 2 ❯

A team of heart surgeons at Saint Ann's Hospital knows that many patients who undergo corrective heart surgery have a dangerous buildup of anxiety before their scheduled operations. The staff psychiatrist at the hospital has started a new counseling program intended to reduce this anxiety. A test of anxiety is given to patients who know they must undergo heart surgery. Then each patient participates in a series of counseling sessions with the staff psychiatrist. At the end of the counseling sessions, each patient is retested to determine anxiety level. Table 10-1 on page 468 indicates the results for a random sample of nine patients. Higher scores mean higher levels of anxiety.

From the given data, can we conclude that the counseling sessions reduce anxiety? Use a 0.01 level of significance.

SOLUTION: Before we answer this question, let us notice two important points: (1) we have a *random sample* of nine patients, and (2) we have a *pair* of measurements taken on each patient before and after counseling sessions. In our problem, the sample size is $n = 9$ pairs (i.e., patients), and the d values are found in the fourth column of Table 10-1.

(a) First, we need to find the sample mean $\bar{d}$ and sample standard deviation s_d of the d values. Using the formulas for $\bar{d}$ and for s_d or a calculator with mean and standard deviation keys, we find

$$\bar{d} \approx 33.33 \qquad \text{and} \qquad s_d \approx 22.92$$

(b) Next, we set the hypotheses. In our problem we want to test the claim that the counseling sessions reduce anxiety. This means that the anxiety level B before counseling is expected to be higher than the anxiety level A after counseling. In symbols, $d = B - A$ should tend to be positive, and the population mean of differences μ_d also should be positive. Therefore, we have the hypotheses

$$H_0: \mu_d = 0$$
$$H_1: \mu_d > 0$$

(c) Find the critical values. Critical values are found using the Student's t distribution. We have a right-tailed test, with level of significance 0.01, so we look in the column headed by $\alpha' = 0.01$. The degrees of freedom are $d.f. = n - 1 = 9 - 1 = 8$. In Table 5 in Appendix I, we find the critical value is $t_0 = 2.896$.

(d) Convert the sample test statistic $\bar{d}$ to a t value. Using $n = 9$, $\bar{d} = 33.33$, $s_d = 22.92$, and $\mu_d = 0$ from H_0, we get

$$t = \frac{\bar{d} - \mu_d}{(s_d/\sqrt{n})} = \frac{33.33 - 0}{(22.92/\sqrt{9})} \approx 4.363$$

(e) Finally, sketch the critical region, show the sample statistic on the sketch, and decide whether or not to reject H_0. Figure 10-1 shows a Student's t distribution with 8 degrees of freedom. We see that the t value corresponding to the observed sample statistic $\bar{d} \approx 33.33$ falls in the critical region. We reject H_0, select H_1, and conclude that the counseling sessions do reduce anxiety (at the 0.01 level of significance).

P-value approach

(f) To find an interval containing the P value associated with the sample test statistic $t = 4.363$, we use techniques shown in Section 9.4. In Table 5 (Appendix I), use the row with $d.f. = 8$ to find the location of $t = 4.363$. Then, since this is a one-tailed test, use α' values to create the interval containing the corresponding P value (see Table 10-2). Reading from left to right, we notice that as the t values increase, the P values decrease. Therefore,

$$P \text{ value} < 0.005$$

Table 10-1

Patient	B Score before Counseling	A Score after Counseling	$d = B - A$ Difference
Jan	121	76	45
Tom	93	93	0
Diane	105	64	41
Barbara	115	117	−2
Mike	130	82	48
Bill	98	80	18
Frank	142	79	63
Carol	118	67	51
Alice	125	89	36

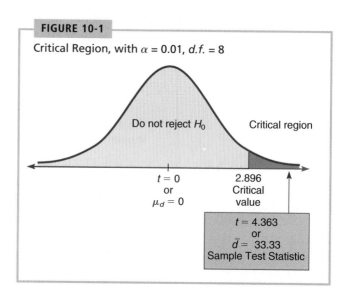

FIGURE 10-1

Critical Region, with $\alpha = 0.01$, $d.f. = 8$

Table 10-2   Excerpt from Student's t Distribution Table   (Table 5, Appendix I)

✔ P value for one-tailed test α'	...	0.010	0.005
d.f. 8	...	2.896	3.355

Sample $t = 4.363$

This means that we reject H_0 for all levels of significance α greater than or equal to 0.005. In particular, we reject H_0 for $\alpha = 0.01$.

GUIDED EXERCISE   2

Do educational toys make a difference in the age at which a child learns to read? To study this question, researchers designed an experiment in which one group of preschool children spent 2 hours each day (for 6 months) in a room well supplied with "educational" toys such as alphabet blocks, puzzles, ABC readers, coloring books featuring letters, and so forth. A control group of children spent 2 hours a day for 6 months in a "noneducational" toy room. It was anticipated that IQ differences and home environment might be uncontrollable factors unless identical twins could be used.

Therefore, six pairs of identical twins of preschool age were randomly selected. From each pair, one member was randomly selected to participate in the experimental (i.e., educational toy room) group and the other in the control (i.e., noneducational toy room) group. For each twin the data item recorded is the age in months when the child began reading at the primary level (Table 10-3).

Table 10-3   Reading Ages for Identical Twins in Months

Twin Pair	Experimental Group B = Reading Age	Control Group A = Reading Age	Difference $d = B - A$
1	58	60	
2	61	64	
3	53	52	
4	60	65	
5	71	75	
6	62	63	

(a) Compute the entries in the $d = B - A$ column of Table 10-3.

⇨

Pair	$d = B - A$
1	−2
2	−3
3	1
4	−5
5	−4
6	−1

(b) Using formulas for the mean and sample standard deviation or a calculator with mean and sample standard deviation keys, compute $\bar{d}$ and s_d.

⇨ $\bar{d} \approx -2.33$
$s_d \approx 2.16$

Exercise continues

(c) What is the null hypothesis?

➡ $H_0: \mu_d = 0$

(d) To test the claim that the experimental group learned to read at a *different age* (either younger or older), what should the alternate hypothesis be?

➡ $H_1: \mu_d \neq 0$

(e) What is the value of the degrees of freedom (*d.f.*)? For a 5% level of significance, find the critical values.

➡ $d.f. = n - 1 = 6 - 1 = 5$
Use Table 5 in Appendix I. Since we have a two-tailed test, use the column headed by $\alpha'' = 0.05$. The critical values are $\pm t_0 = \pm 2.571$.

(f) Convert the sample test statistic $\overline{d}$ to a *t* value.

➡ Using $\mu_d = 0$ from H_0, $\overline{d} = -2.33$, $n = 6$, and $s_d = 2.16$, we get

$$t = \frac{\overline{d} - \mu_d}{(s_d/\sqrt{n})}$$

$$= \frac{-2.33 - 0}{(2.16/\sqrt{6})}$$

$$\approx -2.642$$

(g) Sketch the critical regions and place the *t* value corresponding to the sample test statistic $\overline{d}$ on the sketch. Do we reject or fail to reject H_0 at the 5% level of significance?

➡ See Figure 10-2. Since the sample statistic falls in the critical region, we reject H_0 at the 5% level of significance.

(h) Do the results of this experiment indicate that there is a difference in reading age when a preschool child is exposed to educational toys?

➡ This experiment indicates that at the 5% level of significance, educational toys make a difference.

(i) Find an interval containing the *P* value of the sample test statistic, and conclude the test using the *P* value.

➡ In Table 5 in Appendix I, use the row with *d.f.* = 5 and find the location of the positive value corresponding to the sample test statistic $t = -2.642$ (see Table 10-4). Because this is a two-tailed test, use α'' values to form the *P*-value interval.

FIGURE 10-2 Critical Region, with $\alpha = 0.05$, *d.f.* = 5

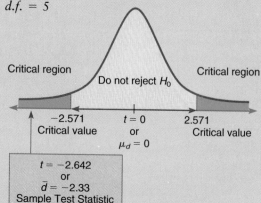

Critical region

Do not reject H_0

Critical region

−2.571
Critical value

$t = 0$
or
$\mu_d = 0$

2.571
Critical value

$t = -2.642$
or
$\overline{d} = -2.33$
Sample Test Statistic

Table 10-4 Excerpt from Table 5 (Appendix I)

α''	0.050	0.020
d.f. 5	2.571	3.365

Sample $t = 2.642$

The corresponding *P* value is in the interval

0.020 < *P* value < 0.050

This says that we reject H_0 for all levels of significance greater than or equal to 0.05. In particular, we reject H_0 for $\alpha = 0.05$.

Example 2 regarding counseling heart patients was a paired difference problem of the "before and after" type. Guided Exercise 2 regarding children spending time in a room with educational toys demonstrates a paired difference problem of the "matched pair" type.

V I E W P O I N T

DUI

DUI usually means "driving under the influence" of *alcohol*, but driving under the influence of *sleep loss* can be just as dangerous. Researchers in Australia have found that after staying awake for 24 hours straight, a person will be about as impaired as if he had enough alcohol to be legally drunk in most U.S. states (*Source: Rocky Mountain News*). Using driver simulation exams and statistical tests (paired difference tests) found in this section, it is possible to show that the null hypothesis $H_0: \mu_d = 0$ cannot be rejected. Or put another way, the average level of impairment for a given individual from alcohol (at the DUI level) is about the same as the average level of impairment for sleep loss (24 hours without sleep).

SECTION 10.1 PROBLEMS

For Problems 1–13, please provide the requested information.
(a) What is the null hypothesis? What is the alternate hypothesis? Will we use a left-tailed, right-tailed, or two-tailed test? What is the level of significance?
(b) What sampling distribution will we use? What is the critical value t_0 (or critical values $\pm t_0$)?
(c) Sketch the critical region and show the critical value (or critical values).
(d) Calculate the t value corresponding to the sample statistic $\bar{d}$ and show its location on the sketch in part c.
(e) Find the P value (or interval containing the P value) for your test and explain what the P value means in the context of the problem.
(f) Based on your answers for parts a to e, shall we reject or fail to reject (i.e., "accept") the null hypothesis for the given level of significance α? Explain your conclusion in the context of the problem.

Business: CEO Salaries

1. Are America's top chief executive officers (CEOs) really worth all that money? One way to answer this question is to look at row *B*, the annual company percentage increase in revenue, versus row *A*, the CEO's annual percentage salary increase in that same company. (*Source: Forbes*, Vol. 159, No. 10.) A random sample of companies such as John Deere & Co., General Electric, Union Carbide, Dow Chemical, and so forth yields the following data.

B: Percent for company	24	23	25	18	6	4	21	37
A: Percent for CEO	21	25	20	14	−4	19	15	30

Do these data indicate that the population mean percentage increase in corporate revenue (row *B*) is different from the population mean percentage increase in CEO's salary? Use a 5% level of significance.

Leisure: Fishing

2. Is fishing better from a boat or from the shore? Pyramid Lake is on the Paiute Indian Reservation in Nevada. Presidents, movie stars, and people who just want to catch fish go to Pyramid Lake for really large cutthroat trout. Let row *B* represent hours per fish fishing from the shore, and let row *A* represent hours per fish using a boat. The following data are paired by month from October through April. (*Source: Pyramid Lake Fisheries*, Paiute Reservation, Nevada.)

	Oct.	Nov.	Dec.	Jan.	Feb.	March	April
B: Shore	1.6	1.8	2.0	3.2	3.9	3.6	3.3
A: Boat	1.5	1.4	1.6	2.2	3.3	3.0	3.8

Use a 1% level of significance to test if there is a difference in the population mean hours per fish using a boat compared with fishing from the shore.

Ecology: Rocky Mountain National Park

3. This problem is based on information taken from *Winter Wind Studies in Rocky Mountain National Park*, by D. E. Glidden (Rocky Mountain Nature Association). At five weather stations on Trail Ridge Road in Rocky Mountain National Park, the peak wind gusts (miles per hour) in January and April are recorded below.

Weather Station	1	2	3	4	5
January	139	122	126	64	78
April	104	113	100	88	61

Does this information indicate that the peak wind gusts are higher in January than in April? Use $\alpha = 0.01$.

Wild Life: Highways and Deer

4. The western United States has a number of four-lane interstate highways that cut through long tracts of wilderness. To prevent car accidents with wild animals, the highways are bordered on both sides with 12-foot-high woven wire fences. Although the fences prevent accidents, they also disturb the winter migration patterns of many animals. To compensate for this disturbance, the highways have frequent wilderness underpasses designed for exclusive use by deer, elk, and other animals.

In Colorado, there is a large group of deer that spend their summer months in a region on one side of a highway and survive the winter months in a lower region on the other side. To determine if the highway has disturbed deer migration to the winter feeding area, the following data were gathered on a random sample of 10 wilderness districts in the winter feeding area. Row *B* represents the average January deer count in a 5-year period before the highway was built, and row *A* represents the average January deer count for a 5-year period after the highway was built. The highway department claims that the January population has not changed. Test this against the claim that the January population has dropped. Use a 1% level of significance. Units used in the table are hundreds of deer.

Wilderness District	1	2	3	4	5	6	7	8	9	10
B: Before highway	10.3	7.2	12.9	5.8	17.4	9.9	20.5	16.2	18.9	11.6
A: After highway	9.1	8.4	11.6	5.8	16.1	10.2	22.7	14.3	21.8	9.9

Climate: Buffalo vs.
Grand Rapids

5. The following information is taken from the U.S. Department of Commerce Environmental Data Service. In the table, the months January to December are listed along with the average temperatures in degrees Fahrenheit for Buffalo, New York and Grand Rapids, Michigan.

	Jan.	Feb.	March	April	May	June
Buffalo	25.1	24.5	32.3	43.3	54.6	64.7
Grand Rapids	24.4	24.4	33.9	46.5	57.9	68.0

	July	August	Sept.	Oct.	Nov.	Dec.
Buffalo	70.3	68.9	62.6	51.8	40.0	29.5
Grand Rapids	72.6	70.6	63.2	52.1	39.3	28.5

Do these data indicate that the average temperature in Buffalo is different from (either higher or lower than) that in Grand Rapids? Use a 0.05 level of significance.

Climate: Miami vs. Honolulu

6. The following information is taken from the U.S. Department of Commerce Environmental Data Service. In the table, the months January to December are listed along with the average temperatures in degrees Fahrenheit for Miami, Florida and Honolulu, Hawaii.

	Jan.	Feb.	March	April	May	June
Miami	67.5	68.0	71.3	74.9	78.0	80.9
Honolulu	74.4	72.6	73.3	74.7	76.2	78.0

	July	August	Sept.	Oct.	Nov.	Dec.
Miami	82.2	82.7	81.6	77.8	72.3	68.5
Honolulu	79.1	79.8	79.5	78.4	76.1	73.7

Do these data indicate that the average temperature in Miami is different from (either higher or lower than) that in Honolulu? Use $\alpha = 0.01$.

Navajo Reservations: Hogans

7. The following data are based on information taken from the book *Navajo Architecture: Forms, History, Distributions,* by S. C. Jett and V. E. Spencer (University of Arizona Press). A survey of houses and traditional hogans was made in a number of different regions in the modern Navajo Indian Reservation. The following table is the result of a random sample of eight regions on the Navajo Reservation.

Area on Navajo Reservation	Number of Inhabited Houses	Number of Inhabited Hogans
Bitter Springs	18	13
Rainbow Lodge	16	14
Kayenta	68	46
Red Mesa	9	32
Black Mesa	11	15
Canyon de Chelly	28	47
Cedar Point	50	17
Burnt Water	50	18

Does this information indicate that the population mean number of inhabited houses is greater than that of hogans on the Navajo Reservation? Use a 5% level of significance.

Archaeology: Stone Tools

8. This problem is based on information taken from *Bandelier Archaeological Excavation Project: Summer 1990 Excavations at Burnt Mesa Pueblo and Casa del Rito,* edited by T. A. Kohler (Washington State University Department of Anthropology). The artifact frequency for an excavation of a kiva in Bandelier National Monument gave the following information:

Stratum	Flaked Stone Tools	Nonflaked Stone Tools
1	7	3
2	3	2
3	10	12
4	1	8
5	4	18
6	38	33
7	51	38

Does this information indicate that there tend to be more flaked stone tools than nonflaked stone tools at this excavation site? Use a 5% level of significance.

Archaeology: Pottery

9. On the same stratum of an excavated block of rooms at Bandelier National Monument, the following information was obtained about sherds of service ware in two different subareas (see reference in Problem 8).

Service Ware	Subarea 1	Subarea 2
Socorro black on white	10	4
Santa Fe black on white	42	39
Galisteo black on white	15	21
Puerco black on red	6	9
Wingate black on red	11	6

Does this information indicate that the population mean number of service ware sherds in subarea 1 is different from (either higher or lower than) that in subarea 2? Use a 5% level of significance.

Psychology: Parent Response

10. Many mothers say that they can recognize the cries of their own babies and that they can distinguish between a cry of pain and a hunger cry. A psychologist studied this phenomenon using the following experiment. A random sample of mothers listened to tape-recorded sets of five cries from different babies, one of which was their own. They had to decide which was their baby. Each mother heard 20 such sets, in which 10 were cries of hungry babies and 10 were cries produced by a slight pin prick on a foot. The results are shown in the following table, where row *B* is the correct number of identifications (out of 10) for a hunger cry, and row *A* is the correct number of identifications (out of 10) for a pain cry. The psychologist claims

that the mothers are more successful in picking out their own babies when a hunger cry is involved, since the mothers have more experience with that situation. Test this claim at the 5% level of significance.

Mother	1	2	3	4	5	6	7	8
B: Hunger cry	6	6	6	5	3	7	9	2
A: Pain cry	5	4	7	3	2	6	4	3

Medical: Heart Rate

11. This problem is based on information taken from *Diagnostic Tests with Nursing Implications* (Springhouse). A patient who had undergone a triple coronary artery bypass was tested in the following way. The patient took his pulse at rest and then exercised on a treadmill at a 10% grade at 1.7 miles per hour for 2 minutes. This procedure was repeated on six different days. The pulse rate 6 minutes after the test and the pulse rate before the test are shown in the following table (beats per minute).

Test	1	2	3	4	5	6
Pulse before	69	72	75	73	70	74
Pulse after	85	79	83	84	87	78

Do these data indicate that the population mean heart rate 6 minutes after the test is higher than before the test? Use a 5% level of significance.

Medical: Blood Pressure

12. The patient described in Problem 11 also had his systolic blood pressure measured before and after the treadmill tests. The results for the tests are shown in the following table.

Test	1	2	3	4	5	6
Before	135	120	138	127	122	133
After	146	132	144	122	121	130

Does this indicate that the population mean systolic blood pressure before the test is different from (either higher or lower than) that 6 minutes after the test? Use a 5% level of significance.

Pro Golf: Scores

13. Do professional golfers play better in the first round of a tournament? Let row B represent the score in the fourth (and final) round, and let row A represent the score in the first round of a professional golf tournament. A random sample of finalists in the British Open gave the following data for their first and last rounds in the tournament. (*Source: Golf Almanac.*)

B: Last	73	68	73	71	71	72	68	68	74
A: First	66	70	64	71	65	71	71	71	71

Do these data indicate that the population mean score on the last round is higher than that on the first? Use a 5% level of significance.

Academic: Professors

14. The following Excel output shows data based on information taken from *Academe: Bulletin of the American Association of University Professors* (Vol. 79, No. 2). The data are from a random sample of nine small and medium-sized colleges and

universities in the western United States and give the numbers of male and female assistant professors at the selected colleges. The Excel menu choices ➤**Tools** ➤**Data Analysis** ➤**t-Test: Paired Two-Sample for Means** produced the following results.

t-Test: Paired Two-Sample for Means

Male	Female			Male	Female
17	16	Mean		25.33333333	25.22222222
59	55	Variance		244.75	225.1944444
7	8	Observations		9	9
41	36	Pearson Correlation		0.941529561	
22	30	Hypothesized Mean Difference		0	
25	34	df		8	
20	15	t Stat		0.063150898	
23	22	P(T<=t) one-tail		0.475597855	
14	11	t Critical one-tail		1.85954832	
		P(T<=t) two-tail		0.95119571	
		t Critical two-tail		2.306005626	

In the dialogue box for t-Test, we used Male as array 1 and Female as array 2, so the results are based on differences $d =$ #Male $-$ #Female professors.

(a) Looking at the output, what is the null hypothesis H_0?

(b) The entry t Stat gives the t value of the sample test statistic $\bar{d}$. The sign of t Stat is positive. Does this indicate that the value of $\bar{d}$ is positive or negative?

(c) For a one-tailed test of $H_1: \mu_d > 0$, the P value is given by P(T<=t). Does this P value mean that we reject or fail to reject H_0 for a one-tailed test?

(d) In the dialogue box for the t-Test, we specified a level of significance $\alpha = 0.05$. Compare the t Stat value with the critical value for a one-tailed test. Based on critical values, do we reject H_0 or not? Is this result consistent with the result from part c?

(e) State your conclusion from part d in the context of the question: do we have statistical evidence that in small and medium-sized colleges in the western United States the population mean number of male assistant professors is greater than that of female assistant professors?

(f) The output P(T<=t) two-tail gives the P value of the sample test statistic for a two-tailed test while t Critical two-tail gives the positive critical value for a 5% level of significance. State H_1 for a two-tailed test. Then complete parts c through e for a two-tailed test.

Section **10.2**

Inferences About the Difference of Two Means (Large, Independent Samples)

Many practical applications of statistics involve comparison of two population means or two population proportions. In Section 10.1 we considered tests of differ-

Independent samples

ences of means for *dependent samples*. With dependent samples, we could pair the data and then consider the difference of data measurements *d*. In this section we will turn our attention to tests of differences of means for *independent samples*. We will see new techniques for testing the differences of means for *independent samples*.

First, let's consider independent samples.

> **Definition**
>
> We say that two sampling distributions are *independent* if there is no relation whatsoever between specific values of the two distributions.

EXAMPLE 3 ▷

A teacher wishes to compare the effectiveness of two teaching methods. Students are randomly divided into two groups: the first group is taught by method 1; the second group by method 2. At the end of the course, a comprehensive exam is given to all students, and the mean score $\bar{x}_1$ for group 1 is compared with the mean score $\bar{x}_2$ for group 2. Are the samples independent or dependent?

SOLUTION: Because the students are *randomly* divided into two groups, it is reasonable to say that the $\bar{x}_1$ distribution is independent of the $\bar{x}_2$ distribution.   ●

EXAMPLE 4 ▷

In Section 10.1 we considered a situation in which a shoe manufacturer claims that for the general population of adult U.S. citizens, the average length of the left foot is greater than the average length of the right foot. To study this claim, the manufacturer gathers data in the following fashion. Sixty adult U.S. citizens are drawn at random, and for these 60 people, both their left and right feet are measured. Let $\bar{x}_1$ be the mean length of the left feet and $\bar{x}_2$ be the mean length of the right feet.

Are the $\bar{x}_1$ and $\bar{x}_2$ distributions independent for this method of collecting data?

SOLUTION: In this method, there is only *one* random sample of people drawn, and both the left and right feet are measured from this sample. The length of a person's left foot is usually related to the length of the right foot, so in this case the $\bar{x}_1$ and $\bar{x}_2$ distributions are *not* independent. In fact, we could pair the data and consider the distribution of the differences, left foot length minus right foot length. Then we would use the techniques of paired difference tests as discussed in Section 10.1.   ●

Suppose the shoe manufacturer in Example 4 gathers data in the following way. Sixty adult U.S. citizens are drawn at random and their left feet are measured; then another 60 adult U.S. citizens are drawn at random and their right feet are measured. Again, $\bar{x}_1$ is the mean of the left foot measurements and $\bar{x}_2$ is the mean of the right foot measurements.

Are the $\bar{x}_1$ and $\bar{x}_2$ distributions independent for this method of collecting data?

For this method of gathering data, two random samples are drawn: one for the left foot measurements and one for the right foot measurements. The first sample is not related to the second sample. The $\bar{x}_1$ and $\bar{x}_2$ distributions are independent.

Testing Difference of Means for Large, Independent Samples

Properties of $\bar{x}_1 - \bar{x}_2$ distribution, large sample size

In this subsection we will use distributions that arise from a difference of means from independent samples. How do we obtain such distributions? If we have two statistical variables x_1 and x_2, each with its own distribution, we take independent random samples of size n_1 from the x_1 distribution and size n_2 from the x_2 distribution. Then we can compute the respective means $\bar{x}_1$ and $\bar{x}_2$. Consider the difference $\bar{x}_1 - \bar{x}_2$. This represents a difference of means. If we repeat the sampling process over and over, we will come up with lots of $\bar{x}_1 - \bar{x}_2$ values. These values can be arranged in a frequency table, and we can make a histogram for the distribution of $\bar{x}_1 - \bar{x}_2$ values. This will give us an experimental idea of the theoretical distribution of $\bar{x}_1 - \bar{x}_2$.

Fortunately, it is not necessary to carry out this lengthy process for each example. The results have already been worked out mathematically. The next theorem presents the main results for large, independent samples.

Theorem 10.2

Let x_1 have a normal distribution with mean μ_1 and standard deviation σ_1. Let x_2 have a normal distribution with mean μ_2 and standard deviation σ_2. If we take independent random samples of size n_1 from the x_1 distribution and of size n_2 from the x_2 distribution, then the variable $\bar{x}_1 - \bar{x}_2$ has the following characteristics.

1. A normal distribution

2. Mean $\mu_1 - \mu_2$

3. Standard deviation

$$\sqrt{\frac{\sigma_1^2}{n_1} + \frac{\sigma_2^2}{n_2}}$$

COMMENT Theorem 10.2 requires that x_1 and x_2 have normal distributions. However, if both n_1 and n_2 are 30 or larger, then for most practical applications, the central limit theorem assures us that $\bar{x}_1$ and $\bar{x}_2$ are approximately normally distributed. In this case the conclusions of the theorem are again valid even if the original x_1 and x_2 distributions were not normal.   ○

Hypotheses for testing difference of means

When testing the difference of means, it is customary to use the null hypothesis

$$H_0: \mu_1 - \mu_2 = 0 \text{ or, equivalently, } H_0: \mu_1 = \mu_2$$

As mentioned in Section 9.1, the null hypothesis is set up to see if it can be rejected. When testing the difference of means, we first set up the hypothesis H_0 that there is no difference. The alternate hypothesis could then be any of the ones listed in Table 10-5. The alternate hypothesis and consequent type of test used depend on the particular problem. Note that μ_1 is always listed first.

Table 10-5   Alternate Hypotheses and Type of Test: Difference of Two Means

H_1			Type of Test
$H_1: \mu_1 - \mu_2 < 0$	or equivalently	$H_1: \mu_1 < \mu_2$	Left-tailed test
$H_1: \mu_1 - \mu_2 > 0$	or equivalently	$H_1: \mu_1 > \mu_2$	Right-tailed test
$H_1: \mu_1 - \mu_2 \neq 0$	or equivalently	$H_1: \mu_1 \neq \mu_2$	Two-tailed test

Critical values

Since the $\bar{x}_1 - \bar{x}_2$ distribution is *normal*, we use the usual critical values z_0 for normal distributions given in Figure 9-2 in Section 9.2. For convenience, we list the values again in Table 10-6.

Table 10-6   Critical Values for Difference of Means Tests, Independent Samples, Large Sample Size

Level of Significance	$\alpha = 0.05$	$\alpha = 0.01$
Critical value z_0 for a left-tailed test	−1.645	−2.33
Critical value z_0 for a right-tailed test	1.645	2.33
Critical values $\pm z_0$ for a two-tailed test	±1.96	±2.58

Sample test statistic

For tests of difference of means $\mu_1 - \mu_2$, independent samples, we make a decision regarding H_0 based on the sample evidence $\bar{x}_1 - \bar{x}_2$. For large samples, Theorem 10.2 tells us that the $\bar{x}_1 - \bar{x}_2$ distribution is normal with mean $\mu_1 - \mu_2$ and standard deviation $\sqrt{\sigma_1^2/n_1 + \sigma_2^2/n_2}$. We use this information to convert the sample $\bar{x}_1 - \bar{x}_2$ value to a standard z value using the following formula.

$$z = \frac{(\bar{x}_1 - \bar{x}_2) - (\mu_1 - \mu_2)}{\sqrt{\dfrac{\sigma_1^2}{n_1} + \dfrac{\sigma_2^2}{n_2}}}$$

where $\mu_1 - \mu_2 = 0$, as stated in the null hypothesis

$\bar{x}_1$ = sample mean of x_1 data

$\bar{x}_2$ = sample mean of x_2 data

σ_1 = standard deviation of the x_1 distribution

σ_2 = standard deviation of the x_2 distribution

n_1 = sample size from the x_1 distribution

n_2 = sample size from the x_2 distribution

EXAMPLE 5 >

A consumer group is testing camp stoves. To test the heating capacity of a stove, the group measures the time required to bring 2 quarts of water from 50°F to boiling (at sea level).

Two competing models are under consideration. Thirty-six stoves of each model are tested and the following results are obtained.

Model 1: mean time $\bar{x}_1 = 11.4$ min; standard deviation $s_1 = 2.5$ min

Model 2: mean time $\bar{x}_2 = 9.9$ min; standard deviation $s_2 = 3.0$ min

Is there any difference between the performances of these two models? (Use a 5% level of significance.) Also find the P value for the sample test statistic.

SOLUTION:

(a) The problem is whether or not the observed difference between the sample means is significant at the 0.05 level. Let μ_1 and μ_2 be the means of the distributions of times for models 1 and 2, respectively. We set up the null hypothesis to say that there is no difference:

$H_0: \mu_1 = \mu_2$ or $H_0: \mu_1 - \mu_2 = 0$

The alternate hypothesis says that there is a difference:

$H_1: \mu_1 \neq \mu_2$ or $H_1: \mu_1 - \mu_2 \neq 0$

(b) Since the alternate hypothesis is $H_1: \mu_1 \neq \mu_2$, we use a two-tailed test. By Table 10-6 we see that the critical values are $\pm z_0 = \pm 1.96$ for $\alpha = 0.05$.

(c) The next step is to compute the sample test statistic $\bar{x}_1 - \bar{x}_2$ and then convert it to a z value.

We are given the values $\bar{x}_1 = 11.4$ and $\bar{x}_2 = 9.9$. Therefore, the sample test statistic is $\bar{x}_1 - \bar{x}_2 = 11.4 - 9.9 = 1.5$. To convert this value to z, we use the values $\sigma_1 \approx s_1 = 2.5$, $\sigma_2 \approx s_2 = 3.0$, $n_1 = 36$, and $n_2 = 36$ in the formula. From the null hypothesis, we use $\mu_1 - \mu_2 = 0$ as well. Then

$$z = \frac{(\bar{x}_1 - \bar{x}_2) - (\mu_1 - \mu_2)}{\sqrt{\dfrac{\sigma_1^2}{n_1} + \dfrac{\sigma_2^2}{n_2}}} = \frac{1.5 - 0}{\sqrt{\dfrac{2.5^2}{36} + \dfrac{3.0^2}{36}}} \approx 2.30$$

(d) Figure 10-3 below shows the critical region and the location of the sample test statistic. We see that the sample test statistic $\bar{x}_1 - \bar{x}_2$ falls in the critical region. Therefore, we reject H_0 at the 5% level of significance. The mean times for the two stoves to boil water are statistically different.

(e) To find the P value of the sample test statistic $\bar{x}_1 - \bar{x}_2 = 1.5$, we first find the corresponding z value. By part c, the z value is 2.30. Since this is a *two-tailed* test, the P value represents the total of the areas to the right of $z = 2.30$ and to the left of $z = -2.30$ (see Figure 10-4 below). We use Table 4 in Appendix I to find the P value. Since we are using a *two-tailed* test, we *double* the area for one tail found in Table 4:

$$P \text{ value} = 2(0.5000 - 0.4893) = 0.0214$$

This means that we reject H_0 for any $\alpha \geq 0.0214$. In particular, we reject H_0 for $\alpha = 0.05$, but we fail to reject H_0 for $\alpha = 0.01$ because 0.01 is not greater than 0.0214.

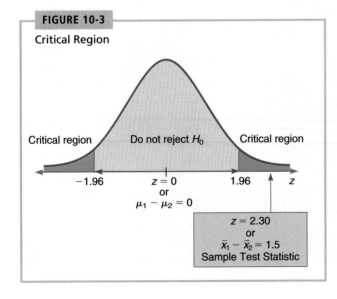

FIGURE 10-3

Critical Region

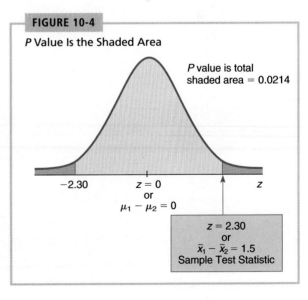

FIGURE 10-4

P Value Is the Shaded Area

GUIDED EXERCISE   4

Let us return to Example 3 at the beginning of this section. A teacher wishes to compare the effectiveness of two teaching methods. Students are randomly divided into two groups. The first group is taught by method 1; the second group by method 2. At the end of the course, a comprehensive exam is given to all students.

The first group consists of $n_1 = 49$ students with a mean score of $\bar{x}_1 = 74.8$ points and standard deviation $s_1 = 14$ points. The second group has $n_2 = 50$ students with a mean score of $\bar{x}_2 = 81.3$ points and standard deviation $s_2 = 15$ points. The teacher claims that the second method will increase the mean score on the comprehensive exam. Is this claim justified at the 5% level of significance?

Let μ_1 and μ_2 be the mean scores of the distributions of all scores using method 1 and method 2, respectively.

(a) Which is the null hypothesis:

$H_0\colon \mu_1 = \mu_2$, $H_0\colon \mu_1 \neq \mu_2$,
$H_0\colon \mu_1 < \mu_2$, or $H_0\colon \mu_1 > \mu_2$?

⇨ $H_0\colon \mu_1 = \mu_2$ or $H_0\colon \mu_1 - \mu_2 = 0$

(b) To examine the validity of the teacher's claim, what will we use for the alternate hypothesis:

$H_1\colon \mu_1 \neq \mu_2$, $H_1\colon \mu_1 > \mu_2$, or $H_1\colon \mu_1 < \mu_2$?

⇨ $H_1\colon \mu_1 < \mu_2$ (the second method gives a higher average score) or $H_1\colon \mu_1 - \mu_2 < 0$.

(c) Find the critical value z_0.

⇨ By Table 10-6, we see that for a left-tailed test where $\alpha = 0.05$, $z_0 = -1.645$.

(d) Compute the sample test statistic $\bar{x}_1 - \bar{x}_2$.

⇨ $\bar{x}_1 - \bar{x}_2 = 74.8 - 81.3 = -6.5$

(e) Convert $\bar{x}_1 - \bar{x}_2 = -6.5$ to a z value.

⇨ By the null hypothesis, $\mu_1 - \mu_2 = 0$. From the problem, we have $s_1 = 14$ and $s_2 = 15$. Since the samples are both large, we can estimate σ_1 and σ_2 by these values, respectively. Also, $n_1 = 49$ and $n_2 = 50$. Putting all these values in the formula gives

$$z = \frac{(\bar{x}_1 - \bar{x}_2) - (\mu_1 - \mu_2)}{\sqrt{\dfrac{\sigma_1^2}{n_1} + \dfrac{\sigma_2^2}{n_2}}}$$

$$= \frac{-6.5 - 0}{\sqrt{\dfrac{14^2}{49} + \dfrac{15^2}{50}}} \approx -2.23$$

Exercise continues

(f) Show the z value of the sample test statistic and the critical region on a diagram, and then conclude the test. Is the claim that method 2 is better justified at the 5% level of significance?

⇨ See Figure 10-5. Since the z value of the sample test statistic falls in the critical region, we reject H_0, select H_1, and conclude that method 2 is better at the 5% level of significance.

(g) Find the P value for the sample test statistic $\bar{x}_1 - \bar{x}_2$. What does the P value tell us?

⇨ First, we convert $\bar{x}_1 - \bar{x}_2 = -6.5$ to a z value. In part e we found that the corresponding z value is -2.23. Since the test is a *left-tailed* test, the P value is the area to the left of $z = -2.23$ (see Figure 10-6). From Table 4 in Appendix I we have

P value $= 0.5000 - 0.4871 = 0.0129$

This tells us to reject H_0 for any $\alpha \geq 0.0129$.

FIGURE 10-5 Critical Region

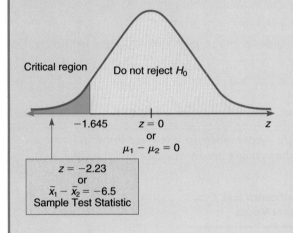

FIGURE 10-6 P Value Is the Shaded Area

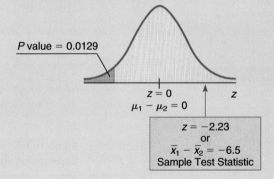

Confidence Intervals for $\mu_1 - \mu_2$ (Large, Independent Samples)

Another way to study the difference in means of two populations is to look at the confidence interval for $\mu_1 - \mu_2$ where μ_1 and μ_2 are the means of independent populations. Techniques similar to those discussed in Section 8.1 tell us to use the value $\bar{x}_1 - \bar{x}_2$ as the point estimate for $\mu_1 - \mu_2$. The error of estimate involves the standard deviation of the $\mu_1 - \mu_2$ distribution (described in Theorem 10.2) and the critical value z_c that depends on the confidence level c. The formula for the $c\%$ confidence interval follows.

---

c Confidence Interval for $\mu_1 - \mu_2$ (Large, Independent Samples)

A c confidence interval for $\mu_1 - \mu_2$ is

$$(\bar{x}_1 - \bar{x}_2) - E < \mu_1 - \mu_2 < (\bar{x}_1 - \bar{x}_2) + E$$

μ_1 = mean of population 1

μ_2 = mean of population 2

where $\bar{x}_1$ = sample mean for population 1

$\bar{x}_2$ = sample mean for population 2

$$E = z_c \sqrt{\frac{s_1^2}{n_1} + \frac{s_2^2}{n_2}}$$

s_1 = sample standard deviation for population 1

n_1 = sample size from population 1 ($n_1 \geq 30$)

s_2 = sample standard deviation for population 2

n_2 = sample size from population 2 ($n_2 \geq 30$)

z_c = critical value for confidence level c (See Table 10-7 for frequently used values.)

c = confidence level ($0 < c < 1$)

Critical values for frequently used confidence levels are listed in Table 8-2 in Section 8.1. For convenience, these values are listed again in Table 10-7.

Table 10-7 Some Levels of Confidence and Their Corresponding Critical Values

Level of Confidence c	Critical Value z_c
0.75	1.15
0.80	1.28
0.85	1.44
0.90	1.645
0.95	1.96
0.99	2.58

EXAMPLE 6 ❯

In the summer of 1988, Yellowstone National Park had some major fires that destroyed large tracts of old timber near many famous trout streams. Fishermen were concerned about the long-term effects of the fires on these streams. However, biologists claimed the new meadows that would spring up under dead trees would produce a lot more insects, which would in turn mean better fishing in the years ahead. Guide services registered with the park provided data about the daily catch for fishermen over many years. Ranger checks on the streams also provided data about the daily number of fish caught by fishermen. *Yellowstone Today*, 1993 (a national park publication) indicated that the biologists' claim had been basically correct and that Yellowstone anglers were delighted by their average increased catch.

Suppose that you are a biologist studying data from Yellowstone streams. A random sample of $n_1 = 167$ fishing reports in the years 1983 to 1988 (before the fires) showed that the average catch per day was $\bar{x}_1 = 5.2$ trout with sample standard deviation $s_1 = 1.9$. Then another random sample of $n_2 = 125$ fishing reports in the years 1990 to 1993 (after the fires) showed that the average catch per day was $\bar{x}_2 = 6.8$ trout with sample standard deviation $s_2 = 2.3$.

(a) For each sample, what is the population? Are the samples dependent or independent? Explain.

SOLUTION: The population for the first sample is the average number of trout caught per day by fishermen before the fires. The population of the second sample is the average number of trout caught per day after the fires. Both samples were random samples taken in their respective time periods. There was no effort to pair individual data values. Therefore, the samples can be thought of as independent samples.

(b) Compute a 95% confidence interval for $\mu_1 - \mu_2$, the difference of population means.

SOLUTION: Since $n_1 = 167, \bar{x}_1 = 5.2, s_1 = 1.9, n_2 = 125, \bar{x}_2 = 6.8, s_2 = 2.3$, and $z_{0.95} = 1.96$ (see Table 10-7), then

$$E = 1.96\sqrt{\frac{(1.9)^2}{167} + \frac{(2.3)^2}{125}} \approx 1.96\sqrt{0.0639} \approx 0.4955 \approx 0.50$$

The 95% confidence interval is

$$(\bar{x}_1 - \bar{x}_2) - E < \mu_1 - \mu_2 < (\bar{x}_1 - \bar{x}_2) + E$$

$$(5.2 - 6.8) - 0.50 < \mu_1 - \mu_2 < (5.2 - 6.8) + 0.50$$

$$-2.10 < \mu_1 - \mu_2 < -1.10$$

(c) Explain the meaning of the confidence interval you computed in part b.

SOLUTION: Since μ_1 represents the population average daily catch before the fires and μ_2 represents the population average daily catch after the fires, we are 95% confident the difference $\mu_1 - \mu_2$ is between -2.10 and -1.10 fish per day. Put another way, since the confidence interval contains only *negative values*, we can be 95% sure that $\mu_1 - \mu_2 < 0$. This means that we are 95% sure that $\mu_1 < \mu_2$. In words, we are 95% sure that the average catch before the fires is less than the average catch after the fires.

Meaning of confidence intervals for $\mu_1 - \mu_2$

In Example 6 we saw that all the numbers contained in the confidence interval for $\mu_1 - \mu_2$ were negative. Two other cases can occur. The next display summarizes the interpretations for all three cases.

Interpretation of Confidence Intervals for $\mu_1 - \mu_2$

Suppose that we construct a $c\%$ confidence interval for $\mu_1 - \mu_2$. Then three cases arise:

1. The $c\%$ confidence interval contains only *negative values*. In this case, we conclude that $\mu_1 - \mu_2 < 0$, and we are therefore $c\%$ confident that $\mu_1 < \mu_2$.

2. The $c\%$ confidence interval contains only *positive values*. In this case, we conclude that $\mu_1 - \mu_2 > 0$, and we can be $c\%$ confident that $\mu_1 > \mu_2$.

3. The $c\%$ confidence interval contains *both positive and negative values*. In this case, we cannot at the $c\%$ confidence level conclude that either μ_1 or μ_2 is larger. However, if we *reduce* the confidence level c to a *smaller value*, then the confidence interval will, in general, be shorter (explain why). A shorter confidence interval *might* put us back into case 1 or case 2 above (again, explain why).

GUIDED EXERCISE   5

(a) A study reported a 90% confidence interval for the difference of means to be

$$10 < \mu_1 - \mu_2 < 20$$

For this interval, what can you conclude about the respective values of μ_1 and μ_2?

⟹   At a 90% level of confidence, we can say that the difference $\mu_1 - \mu_2$ is positive, so $\mu_1 - \mu_2 > 0$ and $\mu_1 > \mu_2$.

(b) A study reported a 95% confidence interval for the difference of means to be

$$-5.32 < \mu_1 - \mu_2 < 3.16$$

From this interval, what can you conclude about the respective values of μ_1 and μ_2?

⟹   At the 95% confidence level, we see that the difference of means ranges from negative to positive values. We cannot tell from this interval whether μ_1 is greater than μ_2 or μ_1 is less than μ_2.

Calculator Note The TI-83 calculator, like most statistical computer packages, supports testing the difference of means for large samples. Press the **STAT** key and select **TESTS**. Option **3:2-SampZTest...** conducts hypothesis tests for the difference of means, large independent samples. As in many tests and confidence intervals, you have a choice of using raw data with the Data input option or summary statistics with the Stats option in input. The results for Example 5 regarding comparison of time to boil water for two different models of camp stoves are as follows.

Select **Inpt:Stats**, enter $\sigma 1 = 2.5$, $\sigma 2 = 3$, $\bar{x}1 = 11.4$, **n1** = 36, $\bar{x}2 = 9.9$, **n2** = 36, select $\neq \mu 2$ for a two-tailed test, and finally select **DRAW**.

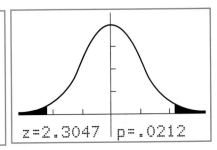

We see that the z-value for the sample test statistic $\bar{x}_1 - \bar{x}_2 = 2.3047$ with P value = 0.0212. Differences from results in Example 5 are due to rounding.

 Calculator Note The TI-83 also supports confidence intervals for difference of means for large, independent samples. Press the **STAT** key and select the **TESTS** option. Then use option **9: 2-SampZInt...**. Again, you have a choice of using raw data or summary statistics.

ⓈECTION 10.2 PROBLEMS

For Problems 1–8, please provide the requested information.

(a) What is the null hypothesis? What is the alternate hypothesis? Will we use a left-tailed, right-tailed, or two-tailed test? What is the level of significance?

(b) What sampling distribution will we use? Find the appropriate critical value z_0 (or critical values $\pm z_0$).

(c) Sketch the critical region and show the critical value (or critical values).

(d) Calculate the z value as appropriate that corresponds to the sample statistic. Show its location on the sketch in part c.

(e) Find the P value for your test and explain what the P value means in the context of the problem.

(f) Based on your answers for parts a to e, shall we reject or fail to reject (i.e., "accept") the null hypothesis? Explain your conclusion in simple, nontechnical terms.

Medical: Sleep

1. This problem is based on information from the book *Secrets of Sleep*, by Alexander Borbely. Professor Borbely is director of the Sleep Laboratory at the University of Zurich Medical School. REM sleep is rapid eye movement sleep, during which most dreams occur. Each night a person has both REM and non-REM sleep. However, it is thought that children have more REM sleep than older adults. A sample of $n_1 = 33$ children 10 years old showed they had an average REM sleep of $\bar{x}_1 = 2.6$ hours per night with sample standard deviation $s_1 = 0.5$ hour. Another random sample of $n_2 = 32$ people 35 years old showed they had an average of $\bar{x}_2 = 1.9$ hours of REM sleep per night with sample standard deviation $s_2 = 0.8$ hour. Does this indicate that, on average, 10-year-old children tend to have more REM sleep than 35-year-old adults? Use a 1% level of significance.

Franchise: Startup Cost

2. *Franchise and Business Opportunities 1994 Annual Report* has listed over 1500 business startup opportunities. Based on information from this source, a random sample of $n_1 = 62$ small clothing stores had a sample mean startup cost $\bar{x}_1 = \$83,000$ with sample standard deviation $s_1 = \$17,000$. Another random sample of $n_2 = 51$ small bakeries had a sample mean startup cost of $\bar{x}_2 = \$91,000$ with sample standard deviation $s_2 = \$22,000$. Does this indicate that there is a difference (either higher or lower) in the mean startup costs of small clothing stores compared with small bakeries? Use a 1% level of significance.

Environment: Air Pollution

3. Based on information from the *Rocky Mountain News*, a random sample of $n_1 = 45$ winter days in Denver gave a sample mean pollution index $\bar{x}_1 = 43$ with sample standard deviation $s_1 = 22$. For Englewood (a suburb of Denver), a random sample of $n_2 = 47$ winter days gave a sample mean pollution index $\bar{x}_2 = 30$ with sample standard deviation $s_2 = 12$. Does this indicate that the mean pollution index for Englewood is less than that for Denver in the winter? Use a 1% level of significance.

Leisure: Outdoor Activities

4. A Michigan study concerning preferences for outdoor activities used a questionnaire with a 6-point Likert-type response in which 1 designated "not important" and 6

designated "extremely important." A random sample of $n_1 = 201$ adults were asked about fishing as an outdoor activity. The mean response was $\bar{x}_1 = 4.7$ with sample standard deviation $s_1 = 1.1$. Another random sample of $n_2 = 135$ adults were asked about camping as an outdoor activity. For this group, the mean response was $\bar{x}_2 = 4.2$ with $s_2 = 1.4$. (Based on information taken from *National Symposium on Catch and Release Fishing*, Humboldt State University.) Does this indicate a difference (either way) regarding preference for camping or preference for fishing as an outdoor activity? Use a 1% level of significance.

Psychology: Peer Tutoring

5. In the journal *Mental Retardation* (April 1985), an article reported the results of a peer tutoring program to help mildly mentally retarded children learn to read. In the experiment to assess the value of peer tutoring, the mildly retarded children were divided into two groups: one, the experimental group, received peer tutoring along with regular instruction, and the other, the control group, received regular instruction with no peer tutoring. There were 30 children in each group. The Gates-MacGintie Reading Test was given to both groups before instruction began. For the experimental group, the mean score on the vocabulary portion of the test was 344.5 with standard deviation 49.1. For the control group, the mean score on the same test was 345.9 with standard deviation 50.9. Use a 5% level of significance to test the hypothesis that there was no difference in the vocabulary scores of the two groups before the instruction began.

Psychology: Peer Tutoring

6. In the same article cited in Problem 5, the results of the following experiment were reported. Form 2 of the Gates-MacGintie Reading Test was administered to both a control group and an experimental group after 6 weeks of instruction during which the experimental group received peer tutoring and the control group did not. The average score for the 30 subjects in the control group on the vocabulary portion of the test was 349.2 with standard deviation 56.6. For the experimental group of 30 children, the mean score on the same portion of the test was 368.4 with standard deviation 39.5. Use a 1% level of significance to test the claim that the experimental group performed better than the control group.

Pro Athletes: Heights

7. Is there a difference in average heights of professional football players and professional basketball players? The following Excel output utilizes a random sample of

z-Test: Two Sample for Means

		HtFt	HtBk
var 1 = 0.134028	Mean	6.178888872	6.453249979
var 2 = 0.098735	Known Variance	0.134	0.0987
	Observations	45	40
	Hypothesized Mean Difference	0	
	z	−3.718026896	
	P(Z<=z) one-tail	0.000100422	
	z Critical one-tail	1.644853	
	P(Z<=z) two-tail	0.000200843	
	z Critical two-tail	1.959961082	

45 pro football players and a random sample of 40 pro basketball players (*sources: Sports Encyclopedia Pro Football* and *The Official NBA Basketball Encyclopedia*, respectively). Excel requires the user to input estimates for the variance of each population. We used the menu choice ➤**paste function** f_x ➤**VAR** to compute the sample variance for each data set. Then we used the menu choices ➤**Tools** ➤**Data Analysis** ➤**z-Test Two Sample for Means** to generate the Excel output. A 5% level of significance was chosen.

(a) The hypothesized Mean Difference 0 implies that the null hypothesis is H_0: $\mu_1 - \mu_2 = 0$. Look at the printout. What is the z value of the sample test statistic?

(b) What is the alternate hypothesis for a two-tailed test? Look at the printout. What is the P value of the sample test statistic for a two-tailed test [see P(Z<= z) two-tail]? For a 5% level of significance, what is the test conclusion regarding football player heights (HtFt) versus basketball player heights (HtBk)?

Botany: Iris

8. Species of iris differ in many ways. A study by E. Anderson in the *Bulletin of the American Iris Society* shows a random sample of petal lengths for *Iris virginica* and a random sample of petal lengths for *Iris setosa*. The data are included on the data disk for this text and in the software package ComputerStat. (See Data Set 16 in Appendix II.) ComputerStat generates confidence intervals for the difference of means $\mu_1 - \mu_2$. The following printout shows the results.

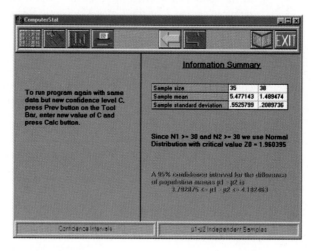

(a) Based on the printout, what level of confidence is used for the confidence interval?

(b) Examine the confidence interval and explain what it means in this context. The confidence interval contains all positive numbers. What does this tell you about the relationship between the average petal length μ_1 for *Iris virginica* and the average petal length μ_2 for *Iris setosa*?

Nursing: Parent Sensitivity

9. "Parental Sensitivity to Infant Cues: Similarities and Differences Between Mothers and Fathers," by M. V. Graham (*Journal of Pediatric Nursing*, Vol. 8, No. 6), reports a study of parental empathy for sensitivity cues and baby temperament (higher scores mean more empathy). Let x_1 be a random variable that represents the score of a mother on an empathy test (as regards her baby). Let x_2 be the empathy score of a father. A random sample of 32 mothers gave a sample mean $\bar{x}_1 = 69.44$ with sample standard deviation $s_1 = 11.69$. A random sample of 32 fathers gave $\bar{x}_2 = 59$ with $s_2 = 11.60$.
   (a) Let μ_1 be the population mean of x_1 and μ_2 be the population mean of x_2. Find a 99% confidence interval for $\mu_1 - \mu_2$.
   (b) Examine the confidence interval and explain what it means in this context. Does the confidence interval contain all positive, all negative, or both positive and negative numbers? What does this tell you about the relationship between average empathy scores for mothers and those for fathers at the 99% confidence level?

U.S. Geological Survey:
Old Faithful Geyser

10. The U.S. Geological Survey compiled historical data about Old Faithful Geyser (Yellowstone National Park) from 1870 to 1987. Some of these data are published in the book, *The Story of Old Faithful*, by G. D. Marler (Yellowstone Association Press). Let x_1 be a random variable that represents the time interval (minutes) between Old Faithful eruptions in the years 1948 to 1952. Based on 9340 observations, the sample mean interval was $\bar{x}_1 = 63.3$ minutes with sample standard deviation $s_1 = 9.17$ minutes. Let x_2 be a random variable that represents the time interval in minutes between Old Faithful eruptions in the years 1983 to 1987. Based on 25,111 observations, the sample mean time interval was $\bar{x}_2 = 72.1$ minutes with sample standard deviation $s_2 = 12.67$ minutes. Let μ_1 be the population mean of x_1 and μ_2 be the population mean of x_2.
   (a) Compute a 99% confidence interval for $\mu_1 - \mu_2$.
   (b) Comment on the meaning of the confidence interval in the context of this problem. Does the interval consist of positive numbers only, negative numbers only, or a mix of positive and negative numbers? Does it appear (at the 99% confidence level) that a change in the interval length between eruptions has occurred? Many geologic experts believe that the distribution of eruption times of Old Faithful changed after the major earthquake that occurred in 1959.

Wild Life: Deer

11. Mule deer in Colorado have been studied extensively by G. W. Mierau and J. L. Schmidt. In their book, *The Mule Deer of Mesa Verde National Park* (Mesa Verde Museum Association), they give the following information about weights of adult male mule deer in two Colorado regions. In the Cache la Poudre Region, a random sample of $n_1 = 51$ deer weighed an average of $\bar{x}_1 = 74.04$ kg with sample standard deviation $s_1 = 17.19$ kg. In the Mesa Verde Region, a random sample of $n_2 = 36$ deer gave a mean weight of $\bar{x}_2 = 94.53$ kg with sample standard deviation $s_2 = 19.66$ kg.

(a) Let μ_1 be the population mean weight of all bucks in the Cache la Poudre Region. Let μ_2 be the population mean weight of all bucks in the Mesa Verde Region. Find a 95% confidence interval for $\mu_1 - \mu_2$.

(b) Examine the confidence interval and comment on its meaning. Does it include numbers that are all positive, all negative, or mixed? What conclusion can you draw (at the 95% level) about the average weight of bucks in the Cache la Poudre Region compared with those in the Mesa Verde Region? The bucks in Mesa Verde National Park are not hunted (and thus tend to be older), and the browse is more abundant. How might these conditions account for the results shown by the confidence interval?

Life Insurance:
Male vs. Female

12. What does life insurance cost? Does it matter if you are male or female? This problem is based on information from *Consumer Reports*. For similar benefits (male and female), the annual premiums paid by a person 45 years old for a $250,000 annual renewable term life insurance policy were as follows:

Males: x_1 = annual premium for described life insurance; sample size
  $n_1 = 47$; sample mean $\bar{x}_1 = \$483.43$; sample standard deviation
  $s_1 = \$126.62$
Females: x_2 = annual premium for described life insurance; sample size
  $n_2 = 51$; sample mean $\bar{x}_2 = \$414.43$; sample standard deviation
  $s_2 = \$105.99$

(a) Assume that the preceding values are representative of the premiums paid for the described life insurance by all males and females (45 years old). Let μ_1 be the population mean annual premium for the described insurance for a male who is 45 years old. Let μ_2 be the population mean annual premium for the described insurance for a female who is 45 years old. Find a 90% confidence interval for $\mu_1 - \mu_2$.

(b) Explain the meaning of the confidence interval found in part a in the context of the problem. Does the interval contain numbers that are all positive, all negative, or both positive and negative? Can you conclude that the population mean premium for males is greater than the population mean premium for females at the 90% confidence level?

General: Interpretation

13. Suppose that a 95% confidence interval for the difference of means (large sample) contains both positive and negative numbers. Will a 99% confidence interval based on the same data necessarily contain both positive and negative numbers? Explain. What about a 90% confidence interval? Explain.

Section 10.3 — Inferences About the Difference of Two Means (Small, Independent Samples)

Testing Difference of Means for Small Samples

Statistical methods involving small samples and the difference between two means are much like the methods for large samples. However, for small samples, we will use the Student's t distribution for critical values instead of the normal distribution.

Independent random samples of sizes n_1 and n_2, respectively, are drawn from two populations that have means μ_1 and μ_2. *We assume that the parent populations have normal distributions, and we also assume that the standard deviations σ_1 and σ_2 for the two populations are equal.* The condition $\sigma_1 = \sigma_2$ may seem quite restrictive. However, in a great many practical applications this condition is satisfied. Furthermore, our methods still apply even if the standard deviations are known to be only approximately equal.

Suppose we draw two independent random samples, one from the x_1 population and one from the x_2 population. Say the sample from the x_1 population is of size n_1 and has sample standard deviation s_1. Likewise for the x_2 population, the sample size is n_2 and the sample standard deviation is s_2. We estimate the common standard deviation for the two populations by using a *pooled variance* of the s_1^2 and s_2^2 values. Research shows that the best estimate of the common variance of the x_1 and x_2 populations is given by the formula

$$s^2 = \frac{(n_1 - 1)s_1^2 + (n_2 - 1)s_2^2}{n_1 + n_2 - 2}$$

Pooled standard deviation   The best estimate of the common or pooled standard deviation is then

$$s = \sqrt{\frac{(n_1 - 1)s_1^2 + (n_2 - 1)s_2^2}{n_1 + n_2 - 2}}$$

Null hypothesis   Just as with tests of difference of means $\mu_1 - \mu_2$ for large independent samples, the null hypothesis we use for testing the difference of means for small samples is that the population means do not differ. In symbols, we have

$$H_0: \mu_1 - \mu_2 = 0 \quad \text{or, equivalently,} \quad H_0: \mu_1 = \mu_2$$

Alternate hypotheses   For tests of difference of means $\mu_1 - \mu_2$ for small independent samples, we use the same alternate hypotheses as we did for tests of difference of means for large samples. Table 10-8 on page 494 lists the possibilities for the alternate hypothesis. This table is simply a repetition of Table 10-5 in Section 10.2.

The next theorem presents the main results necessary to conduct tests of hypotheses regarding $\mu_1 - \mu_2$ for small samples.

Table 10-8 Alternate Hypotheses and Type of Test: Difference of Two Means

	H_1			Type of Test
$H_1: \mu_1 - \mu_2 < 0$	or equivalently	$H_1: \mu_1 < \mu_2$		Left-tailed test
$H_1: \mu_1 - \mu_2 > 0$	or equivalently	$H_1: \mu_1 > \mu_2$		Right-tailed test
$H_1: \mu_1 - \mu_2 \neq 0$	or equivalently	$H_1: \mu_1 \neq \mu_2$		Two-tailed test

THEOREM 10.3 Let x_1 and x_2 have normal (or approximately normal) distributions with means μ_1 and μ_2 and standard deviations σ_1 and σ_2, respectively. We assume that σ_1 and σ_2 are (approximately) equal. Suppose that we take independent random samples of sizes n_1 and n_2 from the x_1 and x_2 populations. Let $\bar{x}_1$ and $\bar{x}_2$ and s_1 and s_2 be the sample means and standard deviations. If $n_1 < 30$ and/or $n_2 < 30$, then

$$t = \frac{(\bar{x}_1 - \bar{x}_2) - (\mu_1 - \mu_2)}{s\sqrt{\dfrac{1}{n_1} + \dfrac{1}{n_2}}}$$

has a Student's t distribution with degrees of freedom

$$d.f. = n_1 + n_2 - 2 .$$

Critical values

Since the sample statistic $\bar{x}_1 - \bar{x}_2$ follows a Student's t distribution, we find the critical value(s) t_0 and critical regions by using Table 5 in Appendix I. The critical value(s) t_0 is (are) found in the row headed by $d.f. = n_1 + n_2 - 2$. For *one-tailed* tests, we use the column headed by $\alpha' =$ level of significance. For *two-tailed tests*, we use the column headed by $\alpha'' =$ level of significance.

Sample test statistic

Using Theorem 10.3 with the null hypothesis $H_0: \mu_1 - \mu_2 = 0$, we see that we can use the following formulas to convert the sample test statistic $\bar{x}_1 - \bar{x}_2$ to a t value.

$$t = \frac{\bar{x}_1 - \bar{x}_2}{s\sqrt{\dfrac{1}{n_1} + \dfrac{1}{n_2}}} \text{ with } d.f. = n_1 + n_2 - 2$$

where the pooled standard deviation s is

$$s = \sqrt{\frac{(n_1 - 1)s_1^2 + (n_2 - 1)s_2^2}{n_1 + n_2 - 2}}$$

Then we compare the sample t statistic with the critical value(s) t_0 to conclude the test.

EXAMPLE 7 ▷   Two competing headache remedies claim to give fast-acting relief. An experiment was performed to compare the mean lengths of time required for bodily absorption of brand A and brand B headache remedies.

Twelve people were randomly selected and given an oral dose of brand A. Another 12 were randomly selected and given an equal dose of brand B. The length of time in minutes for the drugs to reach a specified level in the bloodstream was recorded. The means, standard deviations, and sizes of the two samples were as follows.

$$Brand\ A: \bar{x}_1 = 20.1\ \text{min.;} \qquad s_1 = 8.7\ \text{min.;} \qquad n_1 = 12$$

$$Brand\ B: \bar{x}_2 = 18.9\ \text{min.;} \qquad s_2 = 7.5\ \text{min.;} \qquad n_2 = 12$$

Past experience with the drug compositions of the two remedies permits researchers to assume that the standard deviations of the two time distributions are approximately equal. Let us use a 5% level of significance to test the claim that there is no difference in the mean times required for bodily absorption. Also find the P value of the sample test statistic.

SOLUTION:

(a) The null hypothesis is

$$H_0: \mu_1 = \mu_2 \qquad \text{or} \qquad H_0: \mu_1 - \mu_2 = 0$$

Since we have no prior knowledge about which brand is faster, the alternate hypothesis will be simply

$$H_1: \mu_1 \neq \mu_2 \qquad \text{or} \qquad H_1: \mu_1 - \mu_2 \neq 0$$

(b) Since the alternate hypothesis is $H_1: \mu_1 \neq \mu_2$, we use a two-tailed test. The critical values are found in Table 5 in Appendix I using the row headed by $d.f. = n_1 + n_2 - 2 = 12 + 12 - 2 = 22$ and the column headed by $\alpha'' = 0.05$. The critical values are $\pm t_0 = \pm 2.074$.

(c) Next, we compute the sample test statistic $\bar{x}_1 - \bar{x}_2$ and convert it to a t value. Since $\bar{x}_1 = 20.1$ and $\bar{x}_2 = 18.9$, their difference is

$$\bar{x}_1 - \bar{x}_2 = 20.1 - 18.9 = 1.2$$

To find the t value corresponding to 1.2, we first find the *pooled standard deviation s*.

$$s = \sqrt{\frac{(n_1 - 1)s_1^2 + (n_2 - 1)s_2^2}{n_1 + n_2 - 2}}$$

$$= \sqrt{\frac{(12 - 1)(8.7)^2 + (12 - 1)(7.5)^2}{12 + 12 - 2}}$$

$$= \sqrt{\frac{(11)(75.69) + (11)(56.25)}{22}}$$

$$= \sqrt{\frac{832.59 + 618.75}{22}}$$

$$= \sqrt{65.97}$$

$$\approx 8.12$$

Then, using this value for s and $n_1 = 12$ and $n_2 = 12$, we find the t value corresponding to the sample statistic $\bar{x}_1 - \bar{x}_2 = 1.2$ to be

$$t = \frac{\bar{x}_1 - \bar{x}_2}{s\sqrt{\dfrac{1}{n_1} + \dfrac{1}{n_2}}} = \frac{1.2}{8.12\sqrt{\dfrac{1}{12} + \dfrac{1}{12}}} \approx 0.362$$

(d) Figure 10-7 shows the critical region and the location of the sample test statistic. We see that the sample test statistic falls outside the critical region. Therefore, we fail to reject H_0. Our evidence is not strong enough to conclude that there is a difference of mean times.

(e) To find a P-value interval for the sample test statistic, we use the Student's t distribution (Table 5 of Appendix I). Since the test is a two-tailed test, we use α'' values to form the interval. The first step is to convert the sample test statistic $\bar{x}_1 - \bar{x}_2 = 1.2$ to a t value. By part c the corresponding t value is 0.362. In the Student's t distribution table we look in the row headed by $d.f. = n_1 + n_2 - 2 = 22$ (see Table 10-9).

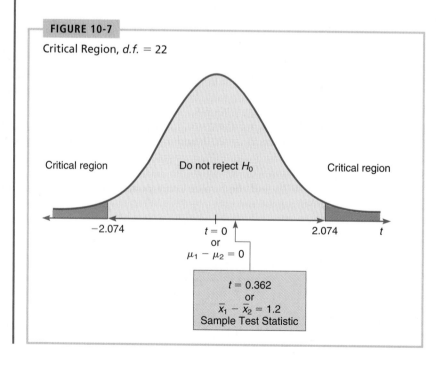

FIGURE 10-7

Critical Region, *d.f.* = 22

Critical region   Do not reject H_0 Critical region

−2.074 $t = 0$
                or
        $\mu_1 - \mu_2 = 0$ 2.074 t

$t = 0.362$
or
$\bar{x}_1 - \bar{x}_2 = 1.2$
Sample Test Statistic

Table 10-9 Excerpt from Student's t Distribution Table (Table 5, Appendix I)

P value for two-tailed test	α''	...	0.250	0.200
d.f.	22	...	1.182	1.321

Sample t value $= 0.362$

Since the sample t value 0.362 is smaller than 1.182, the corresponding P value of 0.362 is larger than $\alpha'' = 0.250$. Therefore, we have the interval

$$P \text{ value} > 0.250$$

This means that we fail to reject H_0 for any significance level α smaller than 0.250. In particular, we fail to reject H_0 for $\alpha = 0.05$. This result is consistent with the conclusion in part d.

GUIDED EXERCISE   6

Suppose the experiment for measuring the time in minutes for the headache remedies to enter the bloodstream (Example 7) yielded sample means, standard deviations, and sample sizes as follows:

Brand A: $\bar{x}_1 = 20.1$ min.; $s_1 = 8.7$ min.; $n_1 = 12$
Brand B: $\bar{x}_2 = 11.2$ min.; $s_2 = 7.5$ min.; $n_2 = 12$

Brand B claims to be faster. Is this claim justified at the 1% level of significance? (Use the following steps to obtain the answer.)

(a) Choose H_0 from the following:

$H_0: \mu_1 = \mu_2$ $H_0: \mu_1 \neq \mu_2$
$H_0: \mu_1 < \mu_2$ $H_0: \mu_1 > \mu_2$

⟹ $H_0: \mu_1 = \mu_2$ or $H_0: \mu_1 - \mu_2 = 0$

(b) What would we use for H_1:

$H_1: \mu_1 = \mu_2$, $H_1: \mu_1 < \mu_2$,
$H_1: \mu_1 > \mu_2$, or $H_1: \mu_1 \neq \mu_2$?

⟹ $H_1: \mu_1 > \mu_2$ (or $H_1: \mu_1 - \mu_2 > 0$). This says that the mean time for brand B is less than the mean time for brand A.

(c) Find the critical value t_0.

⟹ The test is a right-tailed test, so we look in the column headed by $\alpha' = 0.01$. The degrees of freedom are $d.f. = n_1 + n_2 - 2 = 12 + 12 - 2 = 22$. Table 5 in Appendix I gives $t_0 = 2.508$.

Exercise continues

(d) Compute the sample test statistic $\bar{x}_1 - \bar{x}_2$ and convert it to a t value. (*Hint:* You will need values for n_1, n_2, and the pooled standard deviation s.)

➡ $\bar{x}_1 - \bar{x}_2 = 20.1 - 11.2 = 8.9$. Since n_1, n_2, s_1, and s_2 have the same values as in Example 7, the value of the pooled standard deviation will again be $s \approx 8.12$. The sample sizes are $n_1 = 12$ and $n_2 = 12$. Using these values, we find t as follows:

$$t = \frac{\bar{x}_1 - \bar{x}_2}{s\sqrt{\dfrac{1}{n_1} + \dfrac{1}{n_2}}}$$

$$= \frac{8.9}{8.12\sqrt{\dfrac{1}{12} + \dfrac{1}{12}}}$$

$$\approx 2.685$$

(e) Sketch the critical region and show the sample test statistic on the diagram. Do we reject or fail to reject H_0? Does brand B seem to work faster?

➡ See Figure 10-8. Note that the sample statistic falls in the critical region, so at the 1% level of significance, we reject H_0. Brand B seems to work faster.

(f) Find the P value corresponding to the sample test statistic $\bar{x}_1 - \bar{x}_2$. What does the P value tell us?

➡ We look in the row of Table 5 in Appendix I headed by $d.f. = n_1 + n_2 - 2 = 22$. In this row we see that the sample t value of 2.685 falls between t values 2.508 and 2.819 (see Table 10-10). Because the test is a one-tailed test we use α' values to estimate the P value.

FIGURE 10-8 Critical region with $d.f. = 22$

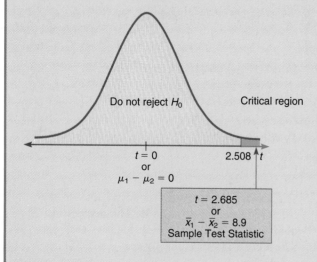

Table 10-10   Excerpt from Table 5 (Appendix I)

P value for one-tailed test	α'	...	0.010	0.005
	$d.f.$ 22	...	2.508	2.819
			↑ Sample t value = 2.685	

From the table, we see that

$0.005 < P$ value < 0.010

This tells us to reject H_0 for any level of significance greater than or equal to 0.010. In particular, we reject H_0 at the 0.01 level of significance. This result is consistent with the conclusion in part e.

Confidence Intervals for $\mu_1 - \mu_2$ (Small, Independent Samples)

As we have seen with large samples, another way to study the difference in means of two populations is to look at the confidence interval for $\mu_1 - \mu_2$, where μ_1 and μ_2 are the means of independent populations. Techniques similar to those discussed in Section 10.2 tell us to use the value $\bar{x}_1 - \bar{x}_2$ as the point estimate for $\mu_1 - \mu_2$. For small samples from approximately normal populations with equal standard deviations σ_1 and σ_2, Theorem 10.3 leads to the following result.

---

c Confidence Interval for $\mu_1 - \mu_2$ (Small, Independent Samples)

$$(\bar{x}_1 - \bar{x}_2) - E < \mu_1 - \mu_2 < (\bar{x}_1 - \bar{x}_2) + E$$

where the samples are independent, and the standard deviations are approximately equal.

Population 1	Population 2
$n_1 \, (< 30)$ = sample size	$n_2 \, (< 30)$ = sample size
$\bar{x}_1$ = sample mean	$\bar{x}_2$ = sample mean
s_1 = sample standard deviation	s_2 = sample standard deviation

$$s = \sqrt{\frac{(n_1 - 1)s_1^2 + (n_2 - 1)s_2^2}{n_1 + n_2 - 2}}$$

$$E = t_c \, s \sqrt{\frac{1}{n_1} + \frac{1}{n_2}}$$

c = confidence level, $0 < c < 1$

t_c = critical value for confidence level c
and degrees of freedom $d.f. = n_1 + n_2 - 2$
(See Table 5 in Appendix I.)

EXAMPLE 8 ➤   Alexander Borbely is a professor at the Medical School of the University of Zurich, where he is director of the Sleep Laboratory. Dr. Borbely and his colleagues are experts on sleep, dreams, and sleep disorders. In his book *Secrets of Sleep*, Dr. Borbely discusses brain waves, which are measured in hertz, the number of oscillations per second. Rapid brain waves (wakefulness) are in the range of 16 to 25 hertz. Slow brain waves (sleep) are in the range of 4 to 8 hertz. During normal sleep, a person goes through several cycles (each cycle is about 90 minutes) of rapid to slow and back to rapid brain waves. During deep sleep, brain waves are at their lowest.

In his book, Professor Borbely comments that alcohol is a *poor* sleep aid. In one study, a number of subjects were given 1/2 liter of red wine before they went to sleep. The subjects fell asleep quickly but did not remain asleep the entire night. Toward morning, between 4 and 6 A.M., they tended to wake up and have trouble going back to sleep.

Suppose that a random sample of 29 college students were randomly divided into two groups. The first group of $n_1 = 15$ people were given 1/2 liter of red wine before going to sleep. The second group of $n_2 = 14$ people were given no alcohol before going to sleep. Everyone in both groups went to sleep at 11 P.M. The average brain wave activity from 4 to 6 A.M. was determined for each individual in each group. The results follow:

Group 1 (x_1 values): $n_1 = 15$ (with alcohol)
Average brain wave activity in the hours 4 to 6 A.M.

16.0	19.6	19.9	20.9	20.3	20.1	16.4	20.6
20.1	22.3	18.8	19.1	17.4	21.1	22.1	

For group 1, we have the sample mean and standard deviation of

$$\bar{x}_1 = 19.65 \quad \text{and} \quad s_1 = 1.86$$

Group 2 (x_2 values): $n_2 = 14$ (no alcohol)
Average brain wave activity in the hours 4 to 6 A.M.

8.2	5.4	6.8	6.5	4.7	5.9	2.9
7.6	10.2	6.4	8.8	5.4	8.3	5.1

For group 2, we have the sample mean and standard deviation of

$$\bar{x}_2 = 6.59 \quad \text{and} \quad s_2 = 1.91$$

(a) Do you think that the samples are independent or dependent? Explain.

SOLUTION: Since the original random sample of 29 students was randomly divided into two groups, it is reasonable to say that the samples are independent.

(b) What assumptions are we making about the data?

SOLUTION: We are assuming that the populations of x_1 and x_2 values are each approximately normally distributed with approximately the same population standard deviations.

(c) Compute a 90% confidence interval for $\mu_1 - \mu_2$, the difference of population means.

SOLUTION: First, we need to find s, the pooled standard deviation:

$$s = \sqrt{\frac{(n_1 - 1)s_1^2 + (n_2 - 1)s_2^2}{n_1 + n_2 - 2}}$$

$$= \sqrt{\frac{(15 - 1)1.86^2 + (14 - 1)1.91^2}{15 + 14 - 2}}$$

$$= \sqrt{3.55} \approx 1.88$$

Next, we find the $t_{0.90}$ value and compute E. Since $d.f. = n_1 + n_2 - 2 = 15 + 14 - 2 = 27$, Table 5 in Appendix I gives $t_{0.90} = 1.703$. Then

$$E = t_c s \sqrt{\frac{1}{n_1} + \frac{1}{n_2}}$$

$$= 1.703(1.88)\sqrt{\frac{1}{15} + \frac{1}{14}}$$

$$= 1.1898 \approx 1.19$$

The c confidence interval is

$$(\overline{x}_1 - \overline{x}_2) - 1.19 < \mu_1 - \mu_2 < (\overline{x}_1 - \overline{x}_2) + 1.19$$
$$(19.65 - 6.59) - 1.19 < \mu_1 - \mu_2 < (19.65 - 6.59) + 1.19$$
$$11.87 < \mu_1 - \mu_2 < 14.25$$

Therefore, after further rounding, we have

11.9 hertz $< \mu_1 - \mu_2 <$ 14.3 hertz

(d) Explain the meaning of the confidence interval you computed in part c.

SOLUTION: μ_1 represents the population average brain wave activity for people who drink 1/2 liter of wine before sleeping. μ_2 represents the population average brain wave activity for people who take no alcohol before sleeping. Both periods of measurement are from 4 to 6 A.M. We are 90% confident the difference $\mu_1 - \mu_2$ is between 11.9 and 14.3 hertz (rounded values). It would seem reasonable to conclude that people who drink before sleeping might wake up in the early morning and have trouble going back to sleep. Since the confidence interval from 11.9 to 14.3 contains only *positive values*, we could express this by saying that we are 90% confident that $\mu_1 - \mu_2$ is *positive*. This means that $\mu_1 - \mu_2 > 0$. Thus we are 90% confident that $\mu_1 > \mu_2$ (that is, average brain wave activity from 4 to 6 A.M. for the group drinking wine is more than average brain wave activity for the group not drinking).

Calculator Note The TI-83 supports hypothesis testing and confidence intervals for difference of means, small sample. Press the STAT key and select TESTS. Option **4:2-SampTTest...** does hypothesis testing. Option **0:2-SampTInt...** gives confidence intervals. For both options you have a choice of inputting data in lists (**Inpt: Data**) or using summary statistics (**Inpt: Stat**). In addition, you indicate whether or not to use the pooled standard deviation. For problems in this text, respond **Pooled: Yes.** For example, the results follow for Example 8, in which we examined the average brain wave activity for subjects who had alcohol before going to sleep compared with those who did not.

Select **Inpt: Stats**, enter $\bar{x}_1 = 19.65$, $S_{x1} = 1.86$, $n_1 = 15$, $\bar{x}_2 = 6.59$, $S_{x2} = 1.91$, $n_2 = 14$, **C-Level** = 90; **Pooled: yes**.

```
2-SampTInt
 Inpt:Data Stats
 x̄1:19.65
 Sx1:1.86
 n1:15
 x̄2:6.59
 Sx2:1.91
↓n2:14
```

```
2-SampTInt
↑n1:15
 x̄2:6.59
 Sx2:1.91
 n2:14
 C-Level:90
 Pooled:No Yes
 Calculate
```

The results are

```
2-SampTInt
 (11.867,14.253)
 df=27
 X̄1=19.65
 X̄2=6.59
 Sx1=1.86
↓SX2=1.91
■
```

```
2-SampTInt
 (11.867,14.253)
↑Sx1=1.86
 Sx2=1.91
 Sxp=1.8842397
 n1=15
 n2=14
```

The confidence interval is $11.867 < \mu_1 - \mu_2 < 14.253$. Notice that the pooled standard deviation is given as $S_{xp} \approx 1.88$.

SECTION 10.3 PROBLEMS

For Problems 1–6, please provide the requested information.
   (a) What is the null hypothesis? What is the alternate hypothesis? Will we use a left-tailed, right-tailed, or two-tailed test? What is the level of significance?
   (b) What sampling distribution will we use? Find the appropriate critical value t_0 (or critical values $\pm t_0$).
   (c) Sketch the critical region and show the critical value (or critical values).
   (d) Calculate the t value that corresponds to the sample statistic. Show its location on the sketch in part c.
   (e) Find the P value for your test and explain what the P value means in the context of the problem.
   (f) Based on your answers for parts a to e, shall we reject or fail to reject (i.e., "accept") the null hypothesis for the given level of significance α? Explain your conclusion in the context of the problem.

Wild Life: Fox Rabies

1. A study of fox rabies in southern Germany gave the following information about different regions and the occurrence of rabies in each region (B. Sayers et al., "A Pattern Analysis Study of a Wild Life Rabies Epizootic," *Medical Informatics* 2:11–34). Based on information from this article, a random sample of $n_1 = 16$ locations in region I gave the following information about the number of cases of fox rabies near that location.

   x_1 data: 1   8   8   8   7   8   8   1
              3   3   3   2   5   1   4   6

   A second random sample of $n_2 = 15$ locations in region II gave the following information about the number of cases of fox rabies near that location.

   x_2 data: 1   1   3   1   4   8   5   4
              4   4   2   2   5   6   9

   (a) Use a calculator with sample mean and sample standard deviation keys to verify that $\bar{x}_1 = 4.75$ and $s_1 = 2.82$ in region I and that $\bar{x}_2 = 3.93$ and $s_2 = 2.43$ in region II.
   (b) Does this information indicate that there is a difference in the mean number of cases of fox rabies between the two regions? Use a 5% level of significance.

Marketing: Restaurants

2. *Life in America's Small Cities* is a book by G. S. Thomas. The following problem is based on information taken from this book. In Key West, Florida, a random sample of nine adult residents who regularly eat at local restaurants were asked to record the amount of money each person spent in a given month at local restaurants. The dollar amounts for this group were

   Key West x_1
   73   63   88   55   81   90   44   52   97

   The same question was asked of a random sample of 11 adult residents (who regularly eat in local restaurants) in Fredericksburg, Virginia. The dollar amounts for this group were

Fredericksburg x_2

46	52	75	51	44	28	93	45	66	50	47

(a) Use a calculator with sample mean and sample standard deviation keys to verify that $\bar{x}_1 = \$71.44$, $s_1 = \$18.84$, $\bar{x}_2 = \$54.27$, and $s_2 = \$17.62$.

(b) In his book, Thomas points out that per capita spending at restaurants in Key West is higher than in most other small cities. Do the data indicate that the mean spending in Key West is higher than in Fredericksburg? Use a 5% level of significance.

Management: *Red Ink Behaviors* 3. In her book *Red Ink Behaviors*, Jean Hollands reports the assessment of leading Silicon Valley companies regarding a manager's lost time due to inappropriate behavior of employees. Consider the following independent random variables—manager's hours per week lost due to hot tempers, flaming e-mails, and general unproductive tensions, x_1.

1	5	8	4	2	4	10

Manager's hours per week lost due to disputes regarding technical workers' superior attitudes that their colleagues are "dumb and dispensable," x_2:

10	5	4	0	0	4	10

(a) Use a calculator with sample mean and standard deviation keys to verify that $\bar{x}_1 = 4.86$, $s_1 = 3.18$, $\bar{x}_2 = 4.71$, and $s_2 = 4.11$.

(b) Does this indicate that the population mean time lost for hot tempers is different from population mean time lost due to disputes arising from technical workers' superior attitudes? Use $\alpha = 0.05$.

Management: *Red Ink Behaviors* 4. This problem is based on information regarding productivity in leading Silicon Valley companies. (See reference in Problem 3.) In large corporations, an "intimidator" is an employee who tries to stop communication and sometimes sabotages others but, above all, likes to listen to himself or herself talk. Let x_1 be a random variable representing productive hours per week lost by peer employees of an intimidator.

8	3	6	2	2	5	2

A "stressor" is an employee with a hot temper that leads to unproductive tantrums in corporate society. Let x_2 be a random variable representing productive hours per week lost by peer employees of a stressor.

3	3	10	3	4	2	5

(a) Use a calculator with mean and standard deviation keys to verify that $\bar{x}_1 = 4.00$, $s_1 = 2.38$, $\bar{x}_2 = 4.29$, and $s_2 = 2.69$.

(b) Assuming that the variables x_1 and x_2 are independent, do the data indicate that the population mean time lost due to stressors is greater than the population mean time lost due to intimidators? Use a 5% level of significance.

Environment: Electric Power Plant 5. A large electric power plant uses ocean water for its cooling system (and returns the water to the ocean). A random sample of 10 temperature readings showed the changes in water temperature to be (in °F):

6	8	4	5	10	3	9	11	7	9

(a) For these temperatures, verify that the mean is $\bar{x}_1 = 7.2°F$ and the sample standard deviation is $s_1 = 2.7°F$.

A new generator was added to the plant, and environmentalists fear that the average change in water temperature has increased. A random sample of 12 temperature readings showed the changes in water temperature now to be (in °F):

9	11	15	12	7	12
10	13	8	11	14	8

(b) For these temperatures, verify that the sample mean is $\bar{x}_2 = 10.8°F$ with sample standard deviation $s_2 = 2.5°F$.

(c) Use a 5% level of significance to test the claim that the mean change in water temperature has increased since the addition of the new generator.

Psychology: Perceptual Illusions   6. Irv is doing a psychology study of susceptibility of people to perceptual illusions. He claims airplane pilots are less susceptible than the general population. To test the claim, a random sample of 11 airplane pilots judged the length of a line in a large illusionary figure. For each judgment, the magnitude of the deviation from the actual length was recorded. The mean of these magnitudes was $\bar{x}_1 = 76$ mm with sample standard deviation $s_1 = 9$ mm. A random sample of 15 nonpilots were then asked to judge the length, and the magnitudes of their deviations were recorded. For the nonpilots the mean of the magnitudes was $\bar{x}_2 = 82$ mm with standard deviation $s_2 = 7$ mm. Do the data support the claim at the 1% level of significance?

Business: Profit Sectors   7. Many companies take in a lot of money. How much of that money can they keep as profit? Do some industry groups have a better performance than others? One way to answer such questions is to study profit as a percentage of revenue. Annual profit as a percentage of revenue is shown below for a group of companies that manufacture electronic equipment. Names such as General Electric, Motorola, Whirlpool, Maytag, and so on are included.

8.1	5.1	5.3	6.2	6.8	4.8	5.5	5.4
6.0	4.9	4.5	5.1	5.2	4.7	6.3	

For a group of food and drug stores, annual profit as a percentage of revenue is shown next. Names such as Kroger, Safeway, Albertson's, Walgreen, and so on are included.

1.5	2.7	3.6	2.0	3.2	2.5
2.9	1.9	1.5	2.1	2.0	4.5

(*Source: Fortune 500*, Vol. 135, No. 8.)

(a) Assume that the preceding data are representative of each given industry group. Let x_1 represent annual profit as percentage of revenue for the electronic companies, and let x_2 represent profit as percentage of revenue for food and drug companies. Use a calculator with mean and sample standard deviation keys to verify that $\bar{x}_1 \approx 5.6$, $s_1 \approx 1.0$, $\bar{x}_2 \approx 2.5$, and $s_2 \approx 0.9$.

(b) Let μ_1 be the population mean for x_1 and let μ_2 be the population mean for x_2. Find a 99% confidence interval for $\mu_1 - \mu_2$.

(c) Examine the confidence interval and explain what it means in this context. Does the interval consist of numbers that are all positive, all negative, or of different signs? At the 99% level of confidence, does it appear that one industry group has a higher profit as a percentage of revenue?

Business: Profit Sectors

8. **It's what you keep that matters.** What percentage of a company's revenue is profit? Let's consider the insurance industry and the health care industry. Annual profit as a percentage of revenue is shown below for a group of companies in the insurance industry. Names such as New York Life, Prudential, John Hancock, Mutual of Omaha, and so on are included.

2.5	3.3	6.8	5.1	3.1	3.6
2.5	3.3	4.5	5.9	6.6	5.9

For a group of health care organizations, annual profit as a percentage of revenue is shown next. Names such as Humana, Columbia Health Care, Manor Care, United Health Care, and so on are included.

7.6	3.5	6.3	4.8	4.2	3.3	3.8	5.3	3.2	4.2

(*Source: Fortune 500,* Vol. 135, No. 8.)

(a) Assume that the preceding data are representative of each given industry group. Let x_1 represent annual profit as a percentage of revenue for the insurance companies, and let x_2 represent profit as a percentage of revenue for the health care companies. Use a calculator with mean and sample standard deviation keys to verify that $\bar{x}_1 \approx 4.4$, $s_1 \approx 1.6$, $\bar{x}_2 \approx 4.6$, and $s_2 \approx 1.4$.

(b) Let μ_1 be the population mean for x_1 and let μ_2 be the population mean for x_2. Find a 90% confidence interval for $\mu_1 - \mu_2$.

(c) Examine the confidence interval and explain what it means in this context. Does the interval consist of numbers that are all positive, all negative, or of different signs? At the 90% level of confidence, does it appear that one industry group has a higher profit as a percentage of revenue?

Psychology: Self-esteem

9. Female undergraduates in randomized groups of 15 took part in a self-esteem study ("There's More to Self-Esteem than Whether It Is High or Low: The Importance of Stability of Self-Esteem," by M. H. Kernis et al., *Journal of Personality and Social Psychology,* Vol. 65, No. 6). The study measured an index of self-esteem from the point of view of competence, social acceptance, and physical attractiveness. Let x_1, x_2, and x_3 be random variables representing the measures of self-esteem through competence, social acceptance, and attractiveness, respectively. Higher index values mean a more positive influence on self-esteem.

Variable	Sample Size	$\bar{x}$ Mean	s Standard Deviation	Population Mean
x_1	15	19.84	3.07	μ_1
x_2	15	19.32	3.62	μ_2
x_3	15	17.88	3.74	μ_3

(a) Find an 85% confidence interval for $\mu_1 - \mu_2$.
(b) Find an 85% confidence interval for $\mu_1 - \mu_3$.
(c) Find an 85% confidence interval for $\mu_2 - \mu_3$.
(d) Comment on the meaning of each of the confidence intervals found in parts a, b, and c. At the 85% confidence level, what can you say about the average differences in influence on self-esteem between competence and social acceptance, between competence and attractiveness, and between social acceptance and attractiveness?

Wild Life: Wolf

10. David E. Brown is an expert on wildlife conservation. In his book *The Wolf in the Southwest: The Making of an Endangered Species* (University of Arizona Press), he records the following weights of adult gray wolves from two regions in Mexico.

Chihuahua region: x_1 variable in pounds

86	75	91	70	79
80	68	71	74	64

Durango region: x_2 variable in pounds

68	72	79	68	77	89	62	55	68
68	59	63	66	58	54	71	59	67

(a) Use a calculator with mean and standard deviation keys to verify that $\bar{x}_1 = 75.80$ pounds, $s_1 = 8.32$ pounds, $\bar{x}_2 = 66.83$ pounds, and $s_2 = 8.87$ pounds.
(b) Let μ_1 be the mean weight of the population of all gray wolves in the Chihuahua region. Let μ_2 be the mean weight of the population of all gray wolves in the Durango region. Find an 85% confidence interval for $\mu_1 - \mu_2$.
(c) Examine the confidence interval and explain what it means in this context. Does the interval consist of numbers that are all positive, all negative, or of different signs? At the 85% level of confidence, what can you say about the comparison of the average weight of gray wolves in the Chihuahua region with the average weight of those in the Durango region?

Sociology: Income vs. Family Size

11. Does level of family income make a difference in family size? A study for the 1941 Canada census involved family size in rural Ontario. The following Minitab (►**Stat** ►**Basic Statistics** ►**2-Sample t**, with equal variances checked) printout shows the family sizes for a random sample of low-income families and for a random sample of high-income families.

Two Sample T-Test and Confidence Interval

```
Two sample T for Low Inc vs High Inc
                N     MEAN    STDEV   SEMEAN
Low Inc        14     4.86    2.77     0.74
High Inc       15     4.13    2.56     0.66
95% CI for mu Low Inc − mu High Inc: (−1.31, 2.75)
T-Test mu Low Inc = mu High Inc (vs not =): T = 0.73 P = 0.47 DF = 27
Both use Pooled StDev = 2.66
```

(a) Does the 95% confidence interval contain all positive, all negative, or different signs? At the 95% level of confidence, does there seem to be a difference in the average number of children for low- and high-income families?

(b) In this printout we also have the results for a two-tailed hypothesis test of $\mu_1 = \mu_2$. What is the P value for this test? Should we reject or fail to reject the null hypothesis H_0: $\mu_1 = \mu_2$? Does the test indicate there is a difference in family size according to income level? How does this result compare to your conclusion in part a?

Section 10.4   Inferences About the Difference of Two Proportions

Tests for Difference of Proportions

The construction of critical regions for tests involving a difference of proportions is essentially the same as we used for the difference of means for large samples. The main change is that we use a probability distribution of proportions rather than one of means.

Suppose that we draw independent random samples of sizes n_1 and n_2, respectively, from two binomial distributions. Let r_1 be the number of successes in the sample of size n_1 and let r_2 be the number of successes in the sample of size n_2. Let the probability of success on each trial be p_1 for the first binomial distribution and let it be p_2 for the second binomial distribution.

For *large* values of n_1 and n_2, the distribution of the sample differences

$$\hat{p}_1 - \hat{p}_2 = \frac{r_1}{n_1} - \frac{r_2}{n_2}$$

is closely approximated by a *normal distribution* with mean μ and standard deviation σ, as follows

$$\mu = p_1 - p_2$$

$$\sigma = \sqrt{\frac{p_1 q_1}{n_1} + \frac{p_2 q_2}{n_2}}$$

where $q_1 = 1 - p_1$ and $q_2 = 1 - p_2$.

Pooled estimate $\hat{p}$

For most practical problems involving a comparison of two binomial populations, the experimenters will want to test the null hypothesis that $p_1 = p_2$. Consequently, this is the only type of test we shall consider. Since the values of p_1 and p_2 are unknown, and since specific values are not assumed under the hypothesis $p_1 = p_2$, they must be approximated by sample estimates. Under the condition $p_1 = p_2$, the best estimate for their common value is the total number of successes $(r_1 + r_2)$ divided by the total number of trials $(n_1 + n_2)$. If we denote this pooled estimate by $\hat{p}$ (read "p hat"), then

$$\hat{p} = \frac{r_1 + r_2}{n_1 + n_2}$$

Critical values

This formula gives the best sample estimate $\hat{p}$ for p_1 and p_2 under the assumption that $p_1 = p_2$. Also, $\hat{q} = 1 - \hat{p}$.

Since the $\hat{p}_1 - \hat{p}_2$ distribution is approximately normal, the critical values z_0 will be the same ones we have been using for normal distributions (see Table 10-11).

Table 10-11   Critical Values for Tests Involving a Difference of Two Proportions (Large Samples)

Level of Significance	$\alpha = 0.05$	$\alpha = 0.01$
Critical value z_0 for a left-tailed test	-1.645	-2.33
Critical value z_0 for a right-tailed test	1.645	2.33
Critical values $\pm z_0$ for a two-tailed test	± 1.96	± 2.58

COMMENT For most practical applications, the sample sizes n_1 and n_2 will be considered large samples if the four quantities

$$n_1 \hat{p} \qquad n_1 \hat{q} \qquad n_2 \hat{p} \qquad n_2 \hat{q}$$

are larger than 5 (see Section 6.4).

Sample test statistic

As stated earlier, the sample statistic $\hat{p}_1 - \hat{p}_2$ follows a normal distribution with mean $\mu = p_1 - p_2$ and standard deviation $\sigma = \sqrt{p_1 q_1/n_1 + p_2 q_2/n_2}$. Under the null hypothesis, we assume that $p_1 = p_2$ and then use the pooled estimate $\hat{p}$ in place of each p. Using all this information, we find that the sample test statistic is

$$z = \frac{\hat{p}_1 - \hat{p}_2}{\sqrt{\dfrac{\hat{p}\hat{q}}{n_1} + \dfrac{\hat{p}\hat{q}}{n_2}}}$$

$$\text{where } \hat{p} = \frac{r_1 + r_2}{n_1 + n_2}$$

$$\text{and } \hat{q} = 1 - \hat{p}$$

$$\hat{p}_1 = \frac{r_1}{n_1}$$

$$\hat{p}_2 = \frac{r_2}{n_2}$$

EXAMPLE 9 ▶ The Macek County Clerk wishes to increase voter registration. One method under consideration is to send reminders in the mail to all citizens in the county who are eligible to register. As a pilot study to determine if this method will actually increase voter registration, a random sample of 1250 potential voters was taken. Then this sample was randomly divided into two groups.

Group 1: There were 625 people in this group. No reminders to register were sent to them. The number of potential voters from this group who registered was 295.

Group 2: This group also contained 625 people. Reminders were sent in the mail to each member in the group, and the number who registered to vote was 350.

The county clerk claims that the proportion of people who registered was significantly greater in group 2. On the basis of this claim the clerk recommends that the project be funded for the entire population of Macek County. Use a 5% level of significance to test the claim that the proportion of potential voters who registered was greater in group 2, the group that received reminders. Also find the *P* value of the sample test statistic.

SOLUTION:

(a) Let p_1 be the proportion of voters who registered from group 1, and let p_2 be the proportion registered from group 2. The null hypothesis is that there is no difference in proportions, so

$$H_0\colon p_1 = p_2 \quad \text{or} \quad H_0\colon p_1 - p_2 = 0$$

The alternate hypothesis is that the proportion of voters was greater in the group that received reminders.

$$H_1\colon p_1 < p_2 \quad \text{or} \quad H_1\colon p_1 - p_2 < 0$$

(b) Because $H_1\colon p_1 < p_2$, we use a left-tailed test. For $\alpha = 0.05$, Table 10-11 gives the critical value $z_0 = -1.645$.

(c) Next we compute the sample statistic $\hat{p}_1 - \hat{p}_2$ and convert it to a z value.

 Calculator Note Carry the values for $\hat{p}_1, \hat{p}_2, \hat{q}_1, \hat{q}_2$, and the pooled estimates $\hat{p}$ and $\hat{q}$ out to at least three places after the decimal. Then round the z value of the corresponding test statistic to two places after the decimal.

For the first group, the number of successes is $r_1 = 295$ out of $n_1 = 625$ trials. For the second group, there are $r_2 = 350$ successes out of $n_2 = 625$ trials. Since

$$\hat{p}_1 = \frac{r_1}{n_1} = \frac{295}{625} = 0.472 \quad \text{and} \quad \hat{p}_2 = \frac{r_2}{n_2} = \frac{350}{625} = 0.560$$

then

$$\hat{p}_1 - \hat{p}_2 = 0.472 - 0.560 = -0.088$$

To convert this $\hat{p}_1 - \hat{p}_2$ value to a z value, we need to find the *pooled estimate* $\hat{p}$ for the common values of p_1 and p_2 and the corresponding value for $\hat{q}$:

$$\hat{p} = \frac{r_1 + r_2}{n_1 + n_2} = \frac{295 + 350}{625 + 625} = 0.516$$

$$\hat{q} = 1 - \hat{p} = 0.484$$

Using these values, we find

$$z = \frac{\hat{p}_1 - \hat{p}_2}{\sqrt{\dfrac{\hat{p}\hat{q}}{n_1} + \dfrac{\hat{p}\hat{q}}{n_2}}} = \frac{-0.088}{\sqrt{\dfrac{(0.516)(0.484)}{625} + \dfrac{(0.516)(0.484)}{625}}} \approx -3.11$$

(d) Next, we show the critical region and the location of the sample test statistic with respect to the critical region. From this information, we can conclude the test (see Figure 10-9). We see that the z value corresponding to the sample test statistic $\hat{p}_1 - \hat{p}_2 = -0.088$ falls in the critical region. Therefore, we reject H_0 and conclude that the county clerk's claim is valid at the 0.05 level of significance.

(e) Since this is a *left-tailed* test, the P value corresponding to the test statistic is the area to the *left* of $z = -3.11$ (see Figure 10-10). By Table 4 in Appendix I, we see that

$$P \text{ value} = 0.5000 - 0.4991 = 0.0009$$

This means that we reject H_0 for any $\alpha \geq 0.0009$. In particular, we reject H_0 for $\alpha = 0.05$ (and also for $\alpha = 0.01$). This is consistent with the conclusion reached in part d.

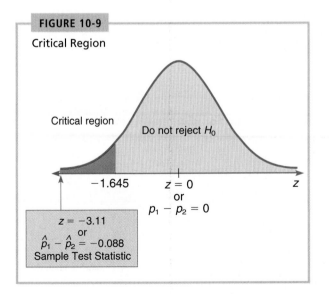

FIGURE 10-9

Critical Region

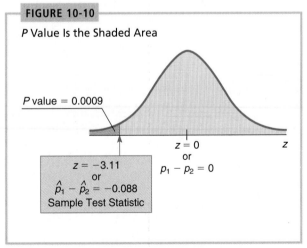

FIGURE 10-10

P Value Is the Shaded Area

In Example 9 about voter registration, suppose that a random sample of 1100 potential voters was randomly divided into two groups.

Group 1: 500 potential voters; no registration reminders sent; 248 registered to vote

Group 2: 600 potential voters; registration reminders sent; 332 registered to vote

Do these data support the claim that the proportion of voters who registered was greater in the group that received reminders than in the group that did not? Use a 1% level of significance.

(a) What should we use for H_0 and H_1? Should we use a left-, right- or two-tailed test?

⇨ As before,

$$H_0: p_1 = p_2 \quad \text{and} \quad H_1: p_1 < p_2$$

We use a left-tailed test.

(b) What is $\hat{p}$ for this problem? What is $\hat{q}$?

⇨ $n_1 = 500$ and $r_1 = 248$; $n_2 = 600$ and $r_2 = 332$. Thus

$$\hat{p} = \frac{r_1 + r_2}{n_1 + n_2} = \frac{248 + 332}{500 + 600} \approx 0.527$$

$$\hat{q} = 1 - \hat{p} \approx 1 - 0.527 \approx 0.473$$

(c) Find the critical value z_0.

⇨ Since this is a left-tailed test with $\alpha = 0.01$, the critical value is $z_0 = -2.33$.

(d) What is the value of the sample test statistic $\hat{p}_1 - \hat{p}_2$?

⇨ $\hat{p}_1 = \frac{r_1}{n_1} = \frac{248}{500} = 0.496$

$$\hat{p}_2 = \frac{r_2}{n_2} = \frac{332}{600} \approx 0.553$$

$$\hat{p}_1 - \hat{p}_2 = -0.057$$

(e) Convert the sample test statistic $\hat{p}_1 - \hat{p}_2 = -0.057$ to a z value.

⇨ $z = \dfrac{\hat{p}_1 - \hat{p}_2}{\sqrt{\dfrac{\hat{p}\hat{q}}{n_1} + \dfrac{\hat{p}\hat{q}}{n_2}}}$

$$= \frac{-0.057}{\sqrt{\dfrac{(0.527)(0.473)}{500} + \dfrac{(0.527)(0.473)}{600}}}$$

$$\approx -1.89$$

Exercise continues

(f) Show the critical region and the sample test statistic on a diagram. Do we reject or fail to reject H_0?

⇨ Since the sample test statistic falls outside the critical region, we fail to reject H_0 (see Figure 10-11). At the 1% level of significance, the data do not support the claim that reminders increase the proportion of registered voters.

(g) Find the P value for the sample test statistic. What does this tell you?

⇨ The z value corresponding to the sample test statistic is -1.89. We are using a *left-tailed* test, so the P value is the area to the left of $z = -1.89$ (see Figure 10-12). By Table 5 in Appendix I,

$$P \text{ value} = 0.5000 - 0.4706$$
$$= 0.0294$$

This tells us to reject H_0 for all $\alpha \geq 0.0294$. We cannot reject H_0 for $\alpha = 0.01$.

FIGURE 10-11 Critical Region

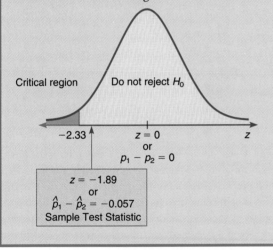

FIGURE 10-12 P Value Is the Shaded Area

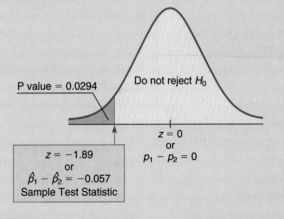

Confidence Intervals for $p_1 - p_2$

Estimating the difference of proportions $p_1 - p_2$

We conclude this section with a discussion of confidence intervals $p_1 - p_2$, the difference of two proportions from binomial probability distributions. Based on information similar to that used for hypothesis testing, we can use the following results.

> ### c Confidence Interval for $p_1 - p_2$ (Large Samples)
>
> $$(\hat{p}_1 - \hat{p}_2) - E < p_1 - p_2 < (\hat{p}_1 - \hat{p}_2) + E$$
>
> where
>
> $$E = z_c \hat{\sigma} = z_c \sqrt{\frac{\hat{p}_1 \hat{q}_1}{n_1} + \frac{\hat{p}_2 \hat{q}_2}{n_2}}$$
>
> c = confidence level, $0 < c < 1$
>
> z_c = critical value for confidence level c (See Table 10-7 on page 484 or Table 4(b) in Appendix I for frequently used values.)
>
> n_1 = number of trials in binomial experiment 1
>
> r_1 = number of successes in binomial experiment 1
>
> n_2 = number of trials in binomial experiment 2
>
> r_2 = number of successes in binomial experiment 2
>
> $$\hat{p}_1 = \frac{r_1}{n_1} \quad \text{and} \quad \hat{q}_1 = 1 - \frac{r_1}{n_1}$$
>
> $$\hat{p}_2 = \frac{r_2}{n_2} \quad \text{and} \quad \hat{q}_2 = 1 - \frac{r_2}{n_2}$$
>
> We assume that the four quantities
>
> $$n_1 \hat{p}_1 \qquad n_1 \hat{q}_1 \qquad n_2 \hat{p}_2 \qquad n_2 \hat{q}_2$$
>
> are all greater than 5.

EXAMPLE 10 ▷ In his book *Secrets of Sleep,* Professor Alexander Borbely describes research on dreams in the Sleep Laboratory at the University of Zurich Medical School. During normal sleep, there is a phase known as REM (rapid eye movement). For most people, REM sleep occurs about every 90 minutes or so, and it is thought that dreams occur just before or during the REM phase. Using electronic equipment in the sleep laboratory, it is possible to detect the REM phase in a sleeping person. If a person is wakened immediately after the REM phase, he or she usually can describe a dream that has just taken place. Based on a study of over 650 people in the Zurich Sleep Laboratory, it was found that about one-third of all dream reports contain feelings of fear, anxiety, or aggression. There is a conjecture that if a person is in a good mood when going to sleep, the proportion of "bad" dreams (fear, anxiety, aggression) might be reduced.

Suppose that two groups of subjects were randomly chosen for a sleep study. In group I, before going to sleep, the subjects spent 1 hour watching a comedy movie. In this group, there were a total of $n_1 = 175$ dreams recorded, of which $r_1 = 49$ were dreams with feelings of anxiety, fear, or aggression. In group II, the subjects did not watch a movie but simply went to sleep. In this group, there were a total of $n_2 = 180$ dreams recorded, of which $r_2 = 63$ were dreams with feelings of anxiety, fear, or aggression.

(a) Why could groups I and II be considered independent binomial distributions? Why do we have a "large sample" situation?

SOLUTION: Since the two groups were chosen randomly, it is reasonable to assume that neither group's response would be related to the other. In both groups, each recorded dream could be thought of as a trial, with success being a dream with feelings of fear, anxiety, or aggression.

$$\hat{p}_1 = \frac{r_1}{n_1} = \frac{49}{175} = 0.28 \qquad \text{and} \qquad \hat{q}_1 = 1 - \hat{p}_1 = 0.72$$

$$\hat{p}_2 = \frac{r_2}{n_2} = \frac{63}{180} = 0.35 \qquad \text{and} \qquad \hat{q}_2 = 1 - \hat{p}_2 = 0.65$$

Since

$$n_1\hat{p}_1 = 49 > 5 \qquad n_1\hat{q}_1 = 126 > 5$$
$$n_2\hat{p}_2 = 63 > 5 \qquad n_2\hat{q}_2 = 117 > 5$$

then large-sample theory is appropriate.

(b) What is $p_1 - p_2$? Compute a 95% confidence interval for $p_1 - p_2$.

SOLUTION: p_1 is the population proportion of success (bad dreams) for all people who watch comedy movies before bed. Thus p_1 can be thought of as the percentage of bad dreams for all people who are in a "good mood" when they go to bed. Likewise, p_2 is the percentage of bad dreams for the population of all people who just go to bed (no movie). The difference $p_1 - p_2$ is the population difference.

To find a confidence interval for $p_1 - p_2$, we need the values of z_c, $\hat{\sigma}$, and then E. From Table 10-7 on page 484, we see that $z_{0.95} = 1.96$, so

$$\hat{\sigma} = \sqrt{\frac{\hat{p}_1\hat{q}_1}{n_1} + \frac{\hat{p}_2\hat{q}_2}{n_2}} = \sqrt{\frac{(0.28)(0.72)}{175} + \frac{(0.35)(0.65)}{180}}$$

$$= \sqrt{0.0024} \approx 0.0492$$

$$E = z_c\,\hat{\sigma} = 1.96(0.0492) \approx 0.096$$

$$(\hat{p}_1 - \hat{p}_2) - E < p_1 - p_2 < (\hat{p}_1 - \hat{p}_2) + E$$

$$(0.28 - 0.35) - 0.096 < p_1 - p_2 < (0.28 - 0.35) + 0.096$$

$$-0.166 < p_1 - p_2 < 0.026$$

(c) Explain the meaning of the confidence interval that you constructed in part c.

SOLUTION: We are 95% sure that the percentage difference of "bad" dreams for group I and group II is between -16.6% and 2.6%. Since the interval -0.166 to 0.026 is not all negative (nor all positive), we cannot say that $p_1 - p_2 < 0$

(or $p_1 - p_2 > 0$). Thus, at the 95% confidence level, we *cannot* conclude that $p_1 < p_2$ or $p_1 > p_2$. The comedy movies before bed help some people reduce the percentage of "bad" dreams, but at the 95% confidence level, we cannot say the *population difference* is reduced.

Interpreting the meaning of a confidence interval for $p_1 - p_2$

In Example 10 the 95% confidence interval for the difference between proportions of "bad" dreams in groups I and II contained both positive and negative numbers. Because the confidence interval contained both positive and negative numbers, we could not conclude that there was a difference in proportions at the 95% confidence level. The following summary shows how to interpret the results when the confidence interval contains only negative numbers or when the confidence interval contains only positive numbers.

Interpreting a $c\%$ Confidence Interval for $p_1 - p_1$

Suppose that we construct a $c\%$ confidence interval for $p_1 - p_2$. Then one of the following three cases arise:

1. The $c\%$ confidence interval contains only *negative values*. In this case, we conclude that $p_1 - p_2 < 0$, and we are therefore $c\%$ confident that $p_1 < p_2$.

2. The $c\%$ confidence interval contains only *positive values*. In this case, we conclude that $p_1 - p_2 > 0$, and we can be $c\%$ confident that $p_1 > p_2$.

3. The $c\%$ confidence interval contains *both positive and negative values*. In this case, we cannot at the $c\%$ confidence level conclude that either p_1 or p_2 is larger. However, if we *reduce* the confidence level c to a *smaller value*, then the confidence interval will, in general, be shorter. A shorter confidence interval *might* put us back into case 1 or case 2 above.

GUIDED EXERCISE   8

In Example 10 we created a 95% confidence interval for the difference of proportions of "bad" dreams experienced by members of group I (those who watched a comedy movie before going to bed) and group II (those who did not watch a comedy) before retiring. In group I, 49 out of 175 participants had "bad" dreams, so $\hat{p}_1 = 0.28$. In group II, 63 out of 180 participants had "bad" dreams, so $\hat{p}_2 = 0.35$.

Exercise continues

Exercise continued

(a) Construct an 80% confidence interval for $p_1 - p_2$.

⇨ From Example 10 we have

$\hat{\sigma} = 0.0492$. Also $z_{80\%} = 1.28$ (see Table 10-7), so the maximal error of estimate $E = 1.28(0.0492) \approx 0.063$. The confidence interval is

$$(\hat{p}_1 - \hat{p}_2) - E < p_1 - p_2 < (\hat{p}_1 - \hat{p}_2) + E$$
$$(0.28 - 0.35) - 0.063 < p_1 - p_2 <$$
$$(0.28 - 0.35) + 0.063$$

or

$$-0.133 < p_1 - p_2 < -0.007$$

(b) Interpret the meaning of the 80% confidence interval.

⇨ Since the 80% confidence interval contains only negative numbers, we conclude at the 80% confidence level that p_2 is greater than p_1. In other words, at the 80% confidence level, it seems that group II participants had a greater proportion of "bad" dreams.

Calculator Note The TI-83 supports both testing and confidence intervals for the difference of proportions. Both functions can be found by pressing the **STAT** key and selecting **TESTS**. Testing is option **6:2_PropZTest...** while confidence intervals is option **B:2PropZInt...** When we input data, the number of successes r_1 is entered as the value of x_1 while the number of successes r_2 is entered as the value of x_2. Input respective sample sizes for n_1 and n_2. Using the data of Example 10 regarding dreams of anxiety, we see the following results for testing that there is no difference, and for a 95% confidence interval.

Result for Testing

```
2-PropZTest
 P1≠P2
 z=-1.41891162
 P=.1559248802
 P̂1=.28
 P̂2=.35
↓P̂=.3154929577
```

95% Confidence Interval

```
2-PropZInt
 (-.1663,.02634)
 P̂1=.28
 P̂2=.35
 n1=175
 n2=180
```

Notice that when looking at the confidence interval we see both positive and negative numbers, so we cannot conclude that the proportion of dreams with anxiety is less after watching comedies. Likewise, the hypothesis test gives a P value of 0.156 for a two-tailed test. We cannot reject H_0 at the 5% level of significance.

VIEWPOINT

Yukon News

On May 16, 1997, the *Yukon News* (see Web site <http://www.yukonweb.com/>, and link to the *Yukon News*) featured an article entitled "Resurgence of the Dreaded White Plague," about the resurgence of tuberculosis (TB) in the far north. TB, also known as the white plague, has been present in Canada since it was brought in by European immigrants in the 17th century. Although antibiotics are widely used today, the disease has never been eradicated. Canadian National

Health data suggest that TB is spreading faster in the Yukon than elsewhere in Canada. Because of this, the Canadian government has established many new TB clinics in remote Yukon villages. Using what you have learned in this section and Canadian National Health data, can you think of a way to use a test of difference of proportions to determine if the rate of TB in the population of villages with the clinics is different from the rate of TB in the independent population of villages without the clinics?

SECTION 10.4 PROBLEMS

For Problems 1–6, please provide the requested information.

(a) What is the null hypothesis? What is the alternate hypothesis? Will we use a left-tailed, right-tailed, or two-tailed test? What is the level of significance?

(b) What sampling distribution will we use? Find the appropriate critical value z_0 (or critical values $\pm z_0$).

(c) Sketch the critical region and show the critical value (or critical values).

(d) Calculate the z value as appropriate that corresponds to the sample statistic. Show its location on the sketch in part c.

(e) Find the P value for your test and explain what the P value means in the context of the problem.

(f) Based on your answers for parts a to e, shall we reject or fail to reject (i.e., "accept") the null hypothesis for the given level of significance α? Explain your conclusion in the context of the problem.

Sociology: High School Dropouts

1. This problem is based on information taken from *Life in America's Fifty States*, by G. S. Thomas. A random sample of $n_1 = 153$ people ages 16 to 19 was taken from the Island of Oahu, Hawaii, and 12 were found to be high school dropouts. Another random sample of $n_2 = 128$ people ages 16 to 19 was taken from Sweetwater

County, Wyoming, and 7 were found to be high school dropouts. Do these data indicate that the population proportion of high school dropouts on Oahu is different (either way) from that of Sweetwater County? Use a 1% level of significance.

Political Science: Voter Turnout

2. A random sample of $n_1 = 288$ voters registered in the state of California showed that 141 voted in the last general election. A random sample of $n_2 = 216$ registered voters in the state of Colorado showed that 125 voted in the most recent general election. (See reference in Problem 1.) Do these data indicate that the population proportion of voter turnout in Colorado is higher than that in California? Use a 5% level of significance.

Opinion Poll: Extraterrestrials

3. Based on information from *Harper's Index*, 37 out of a random sample of 100 adult Americans who did not attend college believe in extraterrestrials. However, out of a random sample of 100 adult Americans who did attend college, 47 claim that they believe in extraterrestrials. Does this indicate that the proportion of people who attended college who believe in extraterrestrials is higher than the proportion who did not attend college? Use $\alpha = 0.01$.

Opinion Poll: Organ Donors

4. Based on information from *Harper's Index*, 78 out of a random sample of 100 American adults claim that they would donate a loved one's organs after death. However, out of a random sample of 100 adult Americans, only 20 claim that they would donate their own organs after death. Does this information indicate that the proportion of adult Americans who would donate a loved one's organs is higher? Use $\alpha = 0.01$.

History: Billy the Kid

5. In 1881, Lincoln County, New Mexico Territory, was part of the wild, wild west. Cattle rustling, horse stealing, and stagecoach holdups were common events. Famous people such as Jesse Chisum, Kit Carson, and Billy the Kid all paid visits to the Lincoln County Courthouse. Because the courthouse is so famous, archaeological excavations have been conducted at different sites near the courthouse. One categorization of artifacts is arms, indulgences, and entertainment. Such artifacts include guns, cartridges, liquor bottles, tobacco cans, playing cards, and so on. At one excavation site, $n_1 = 444$ artifacts were found, of which 10 were of the preceding classification. At a second site, $n_2 = 326$ artifacts were found, and 8 were of the given categorization. (*Source: Archaeological Testing at Three Historic Sites at Lincoln State Monument,* by Y. R. Oakes, Museum of New Mexico Research Section.) Do these data indicate that the population proportion of arms, indulgences, and entertainment artifacts in the general region around one site is different (either way) from the other? Use a 5% level of significance.

History: Billy the Kid

6. For the Lincoln County Courthouse excavations (see Problem 5), another classification of artifacts was domestics, food items, and stable items. Such artifacts include zinc cans, glass jars, cups, plates, horseshoes, harness rings, spurs, and so on. At one site, $n_1 = 338$ artifacts were found, and 18 were of the preceding classification. At another site, $n_2 = 329$ artifacts were found, of which 9 were of the described classification. Do these data indicate that the population proportion of such artifacts is higher in the general region around the first site? Use a 5% level of significance.

Myers-Briggs: Married
Couples

7. Isabel Myers was a pioneer in the study of personality types. She identified four basic personality preferences which are described at length in the book *A Guide to the Development and Use of the Myers-Briggs Type Indicator*, by Myers and McCaulley (Consulting Psychologists Press, Inc., 1990). Marriage counselors know that couples who have none of the four preferences in common may have a stormy marriage. Myers took a random sample of 375 married couples and found that 289 had two or more personality preferences in common. In another random sample of 571 married couples it was found that only 23 had no preferences in common. Let p_1 be the population proportion of all married couples who have two or more personality preferences in common. Let p_2 be the population proportion of all married couples who have no personality preferences in common.
   (a) Find a 99% confidence interval for $p_1 - p_2$.
   (b) Explain the meaning of the confidence interval in part a in the context of this problem. Does the confidence interval contain all positive, all negative, or both positive and negative numbers? What does this tell you (at the 99% confidence level) about the proportion of married couples with two or more personality preferences in common compared with the proportion of married couples sharing no personality preferences in common?

Myers-Briggs: Married
Couples

8. Most married couples have two or three personality preferences in common (see reference in Problem 7). Myers used a random sample of 375 married couples and found that 132 had three preferences in common. Another random sample of 571 couples showed that 217 had two personality preferences in common. Let p_1 be the population proportion of all married couples who have three personality preferences in common. Let p_2 be the population proportion of all married couples who have two personality preferences in common.
   (a) Find a 90% confidence interval for $p_1 - p_2$.
   (b) Examine the confidence interval in part a and explain what it means in this context. Does the confidence interval contain all positive, all negative, or both positive and negative numbers? What does this tell you about the proportion of married couples with three personality preferences in common compared with the proportion of couples with two preferences in common (at the 90% confidence level)?

Navajo Reservation:
Permanent Dwellings

9. S. C. Jett is a professor of geography at the University of California, Davis. He and a colleague, V. E. Spencer, are experts on modern Navajo culture and geography. The following information is taken from their book, *Navajo Architecture: Forms, History, Distribution* (University of Arizona Press). On the Navajo Reservation, a random sample of 210 permanent dwellings in the Fort Defiance Region showed that 65 were traditional Navajo hogans. In the Indian Wells Region, a random sample of 152 permanent dwellings showed that 18 were traditional hogans. Let p_1 be the population proportion of all traditional hogans in the Fort Defiance Region, and let p_2 be the population proportion of all traditional hogans in the Indian Wells Region.
   (a) Find a 99% confidence interval for $p_1 - p_2$.
   (b) Examine the confidence interval and comment on its meaning. Does it include numbers that are all positive, all negative, or mixed? What if it is hypothesized that Navajo who follow the traditional culture of their people tend to live in hogans? Comment on the confidence interval for $p_1 - p_2$ in this context.

Anthropology: Cultural
Affiliation

10. "Unknown cultural affiliations and loss of identity at high elevations." These are
words used to propose the hypothesis that archaeological sites tend to lose their
identity as altitude extremes are reached. This idea is based on the notion that prehistoric people tended *not* to take trade wares to temporary settings and/or isolated
areas (reference: *Prehistoric New Mexico: Background for Survey*, by D. E. Stuart
and R. P. Gauthier, University of New Mexico Press). As elevation zones of prehistoric people (in what is now the state of New Mexico) increased, there seemed to
be a loss of artifact identification. Consider the following information.

Elevation Zone	Number of Artifacts	Number Unidentified
7000–7500 ft	112	69
5000–5500 ft	140	26

Let p_1 be the population proportion of unidentified archaeological artifacts at the
elevation zone 7000–7500 ft in the given archaeological area. Let p_2 be the population proportion of unidentified archaeological artifacts at the elevation zone
5000–5500 ft in the given archaeological area.
(a) Find a 99% confidence interval for $p_1 - p_2$.
(b) Explain the meaning of the confidence interval in part a in the context of this
problem. Does the confidence interval contain all positive numbers, all negative
numbers, or both positive and negative numbers? What does this tell you (at
the 99% confidence level) about the comparison of the population proportion
of unidentified artifacts at high elevations (7000–7500 ft) with the population
proportion of unidentified artifacts at lower elevations (5000–5500 ft)? How
does this relate to the stated hypothesis?

Ecology: Wood Ducks

11. The National Wildlife Federation published an article entitled "The Trouble with
Wood Ducks" (*National Wildlife*, Vol. 31, No. 5). Wood ducks became in danger of
extinction and in 1918 a federal ban on hunting wood ducks helped save the species.
Now they are in trouble because of loss of nesting habitat. One solution is the use
of artificial nesting boxes. However, the placement of the nesting boxes is crucial. An
experiment was conducted using different types of nesting box placement. In group
I, the boxes were well separated from each other and well hidden by available brush.
In group II, the boxes were closely grouped together and highly visible. In group I
boxes, there were a total of 474 eggs, of which a field count showed that about 270
hatched. In group II boxes, there were a total of 805 eggs, of which a field count
showed that, again, about 270 hatched. Does group I placement produce a significantly greater hatch proportion than group II? To answer this question, let's look at
a Minitab (**►Stat ►Basic Statistics ►2 proportions**, with summarized data and use
of pooled estimate for p checked) printout (see next page) of the results.

(a) Does a 95% confidence interval contain all positive, all negative, or both positive and negative numbers? From the confidence interval, does it appear that
group I placement of nesting boxes produces a higher hatch proportion?
(b) Consider the P value of the hypothesis test $H_0: p_1 - p_2 = 0$ against $H_1: p_1 -
p_2 > 0$. Do we reject or fail to reject H_0? Is this result consistent with the infor-
mation from the 95% confidence interval?

Test and Confidence Interval for Two Proportions

Sample	X	N	Sample p
1	270	474	0.569620
2	270	805	0.335404

Estimate for p(1) − p(2): 0.234217

95% CI for p(1) − p(2): (0.178985, 0.289448)

Test for p(1) − p(2) = 0 (vs > 0) : Z = 8.19 P-Value = 0.000

SUMMARY

In this chapter we continued our discussion of hypothesis testing. We applied the techniques to the difference of means μ_d formed from paired data of dependent samples. For large and small, independent samples we considered the difference of means $\mu_1 - \mu_2$, where μ_1 is the mean of the first population and μ_2 is the mean of the second population. We used hypothesis testing to determine if the sample difference $\bar{x}_1 - \bar{x}_2$ was significant. We also constructed confidence intervals for $\mu_1 - \mu_2$.

Finally, we looked at the difference of proportions $p_1 - p_2$ from large, independent samples. The sample statistic $\hat{p}_1 - \hat{p}_2$ was used to determine if the difference in proportions was significant. The same sample statistic $\hat{p}_1 - \hat{p}_2$ was used as a point estimate for $p_1 - p_2$ in the construction of confidence intervals.

IMPORTANT WORDS & SYMBOLS

Section 10.1
Dependent samples
Data pairs
μ_d, difference of means from data pairs
$d.f.$ for testing μ_d

Section 10.2
Independent samples

Section 10.3
Pooled variance

Section 10.4
Pooled estimate of proportions, $\hat{p}$

VIEWPOINT

Who Watches Cable TV?

Consider the following claim: Average cable TV viewers are, generally speaking, as affluent as newspaper readers and better off than radio or magazine audiences. How do we know that this claim is true? One way to answer such a question is to construct several individual tests of difference of means. One test would compare cable TV viewers with newspaper readers, a second test would compare cable TV viewers with radio audiences, and a third test would compare cable TV viewers with magazine audiences. Another way to handle all three tests at once would be to use techniques of ANOVA (analysis of variance)—a subject often introduced in a second course in statistics. For more information and data, see *American Demographics* (Vol. 17, No. 6).

CHAPTER REVIEW PROBLEMS

For each of these problems, please do the following:
(a) Determine whether the samples are independent or dependent.
(b) Determine whether the sampling distribution for the difference follows a Student's *t* distribution or a normal distribution.
(c) In hypothesis testing problems, please provide the following information:
    (i) The null and alternate hypotheses.
    (ii) The appropriate critical value t_0 to z_0 (or critical values $\pm t_0$ or $\pm z_0$).
    (iii) A sketch of the sampling distribution showing the critical region(s).
    (iv) Calculate the *t* or *z* value as appropriate that corresponds to the sample statistic. Show this value on the sketch in part (iii).
    (v) Find the *P* value for the test.
    (vi) Based on your answers for parts i–v, shall we reject or fail to reject the null hypothesis? Explain your conclusion in the context of the problem.

Super Bowl Football: Super Ads

1. The broadcast of the Super Bowl on commercial television offers a stage for "super ads." *USA Today* asks readers to watch and rate the ads aired during the Super Bowl on a scale of 1 to 10 (with 10 the highest score). How do ratings on 30-second ads compare with ratings on 60-second ads? Ratings for samples of ads aired during past Super Bowls follow. Assume that these data are representative of all 30- and 60-second ads.

30-Second Ads: x_1 = rating; sample size $n_1 = 142$; $\bar{x}_1 = 6.03$; estimated sample standard deviation $s_1 = 1.2$

60-Second Ads: x_2 = rating; sample size $n_2 = 47$; $\bar{x}_2 = 7.09$; estimated sample standard deviation $s_2 = 1.3$

Let μ_1 represent the population of x_1 values and let μ_2 represent the population of x_2 values.

(a) Use a 5% level of significance to test the claim that there is no difference in the average ratings for 30-second ads and 60-second ads.

(b) Find a 95% confidence interval. Comment on the meaning of the confidence interval. Does the confidence interval contain values that are all positive, all negative, or mixed? At the 95% confidence level, what does the interval tell you about the average rating of 30-second ads compared with the average rating of 60-second ads?

Psychology: Rats in a Maze

2. The following data are based on information from the Regis University Psychology Department. In an effort to determine if rats perform certain tasks more quickly when offered larger rewards, the following experiment was performed. On day 1, a group of 3 rats were given a reward of one food pellet each time they ran a maze. A second group of 3 rats were given a reward of five food pellets each time they ran the maze. On day 2, the groups were reversed, so the first group now got five food pellets for running the maze and the second group got only one pellet for running the same maze. The average times in seconds for each rat to run the maze 30 times are shown in the following table.

Rat	A	B	C	D	E	F
Time with one food pellet	3.6	4.2	2.9	3.1	3.5	3.9
Time with five food pellets	3.0	3.7	3.0	3.3	2.8	3.0

Do these data indicate that rats receiving larger rewards tend to run the maze in less time? Use a 5% level of significance.

Psychology: Rats Climbing Ladders

3. The same experimental design discussed in Problem 2 also was used to test rats trained to climb a sequence of short ladders. Times in seconds for eight rats to do this task are shown in the following table.

Rat	A	B	C	D	E	F	G	H
Time 1 pellet	12.5	13.7	11.4	12.1	11.0	10.4	14.6	12.3
Time 5 pellets	11.1	12.0	12.2	10.6	11.5	10.5	12.9	11.0

Do these data indicate that rats receiving larger rewards tend to perform the ladder climb in less time? Use a 5% level of significance.

Transportation: Two Bus Lines

4. A comparison is made between two bus lines to determine if arrival times of their regular buses from Denver to Durango are off schedule by the same amount of time. For 81 randomly selected runs, bus line A was observed to be off schedule an average time of 53 min with standard deviation 19 min. For 100 randomly selected runs, bus line B was observed to be off schedule an average of 62 min with standard deviation 15 min. Do the data indicate a significant difference in off-schedule times? Use a 5% level of significance.

Surveys: Telephone vs. Face to Face

5. The book *Survey Responses: An Evaluation of Their Validity,* by E. J. Wentland and K. Smith (Academic Press, 1993), includes studies reporting accuracy of answers to questions from surveys. A study by Locander et al. (1976) considered the question, "Are you a registered voter?" Accuracy of response was confirmed by a check of city voting records. Two methods of survey were used: a face-to-face interview and a tele-

phone interview. A random sample of 93 people were asked the voter registration question face to face. Seventy-nine respondents gave accurate answers (as verified by city records). Another random sample of 83 people were asked the same question during telephone interviews. Seventy-four respondents gave accurate answers. Assume that the samples are representative of the general population. Let p_1 be the population proportion of all people who answer the voter registration question accurately during face-to-face interviews. Let p_2 be the population proportion of all people who answer the question accurately during telephone interviews.

(a) Use a 1% level of significance to test the hypothesis that the proportion of respondents who answer accurately during face-to-face interviews is different from the proportion who answer accurately during telephone interviews.

(b) Find a 99% confidence interval for $p_1 - p_2$. Comment on the meaning of the confidence interval in the context of this problem. At the 99% level, do you detect any difference in the proportion of accurate responses from the face-to-face interviews compared with the proportion of accurate responses from the phone interviews?

Surveys: Sensitive Material

6. Locander et al. (1976) also studied the accuracy of responses on questions involving material more sensitive than voter registration. From public records, individuals were identified as having been charged with drunken driving not less than 6 months or more than 12 months from the starting date of the study. Two random samples from this group were studied. In the first sample of 30 individuals, the respondents were asked in face-to-face interviews if they had been charged with drunken driving in the last 12 months. Of these 30 people interviewed face to face, 16 answered the question accurately. The second random sample consisted of 46 people who had been charged with drunken driving. During telephone interviews, 25 of these responded accurately to the question asking if they had been charged with drunken driving during the past 12 months. Assume that the samples are representative of all people recently charged with drunken driving. Let p_1 represent the population proportion of all people with recent charges of drunken driving who respond accurately to face-to-face interviews asking if they have been charged with drunken driving during the past 12 months. Let p_2 represent the population proportion of people who respond accurately to the question when it is asked in telephone interviews.

(a) Use a 5% level of significance to test the hypothesis that the proportion of respondents who answer accurately during face-to-face interviews is different from the proportion who answer accurately during telephone interviews.

(b) Find a 95% confidence interval for $p_1 - p_2$. Comment on the meaning of the confidence interval in the context of this problem. At the 95% level, do you detect any difference in the proportion of accurate responses from the face-to-face interviews compared with the proportion of accurate responses from the phone interviews?

Hamburger Stands: Service

7. An independent rating service is trying to determine which of two hamburger stands has quicker service. Over a period of 16 randomly selected times, the average waiting period at Burger Queen is 4.8 min with standard deviation 2.0 min. The average waiting period at McGregor's over a period of 14 randomly selected times is 5.2 min with standard deviation 1.8 min.

(a) Find a 90% confidence interval for the difference of population mean waiting times.

(b) Does the confidence interval contain all positive numbers, all negative numbers, or some positive and some negative numbers? What conclusion can you draw regarding the difference in mean waiting time for these two hamburger stands?

Highway Maintenance: Reflecting Paint

8. The highway department tested two types of reflecting paint for concrete bridge end pillars. The two kinds of paint were alike in every respect except that one was red and the other was yellow. The red paint was applied to 12 bridges, and the yellow paint was applied to 12 bridges. After a period of 1 year, reflectometer readings were made on all these bridge end pillars. (A higher reading meant better visibility.) For the red paint, the mean reflectometer reading was $\bar{x}_1 = 9.4$ with standard deviation $s_1 = 2.1$. For the yellow paint, the mean was $\bar{x}_2 = 6.8$ with standard deviation $s_2 = 2.0$. Based on these data, can we conclude that the yellow paint had less visibility after 1 year? (Use a 1% level of significance.)

Archaeology: Projectile Points

9. The Wind Mountain archaeological site is in southwestern New Mexico. Prehistoric Native Americans called Anasazi once lived and hunted small game in this region. A stemmed projectile point is an arrowhead that has a notch on each side of the base. Both stemmed and stemless projectile points were found at the Wind Mountain site. A random sample of $n_1 = 55$ stemmed projectile points showed the mean length to be $\bar{x}_1 = 3.0$ cm with sample standard deviation $s_1 = 0.8$ cm. Another random sample of $n_2 = 52$ stemless projectile points showed the mean length to be $\bar{x}_2 = 2.7$ cm with $s_2 = 0.9$ cm. (*Source: Mimbres Mogollon Archaeology*, by A. I. Woosley and A. J. McIntyre, University of New Mexico Press.) Do these data indicate a difference (either way) for the population mean length of the two types of projectile points? Use a 5% level of significance.

10. Given H_0: $\mu = 5$, H_1: $\mu \neq 5$, and P value 0.0213,
    (a) do we reject or fail to reject H_0 at a 1% level of significance?
    (b) do we reject or fail to reject H_0 at a 5% level of significance?

11. Given H_0: $p = 0.5$, H_1: $p > 0.5$, and P value 0.0023,
    (a) do we reject or fail to reject H_0 at a 1% level of significance?
    (b) do we reject or fail to reject H_0 at a 5% level of significance?

DATA HIGHLIGHTS: GROUP PROJECT

Break into small groups and discuss the following topic. Organize a brief outline in which you summarize the main points of your group discussion.

"Sweets May Not Be Culprit in Hyper Kids" was a *USA Today* (February 3, 1994) headline reporting results of a study that appeared in the *New England Journal of Medicine*. In this study, the subjects were 25 normal preschoolers aged 3 to 5, and 23 kids aged 6 to 10, who had been described as "sensitive to sugar." The kids and their families were put on three different diets for 3 weeks each. One diet was high in sugar, one was low in sugar and contained aspartame, and one was low in sugar and contained saccharin. The diets were all free of additives, artificial food coloring, preservatives, and chocolate. All food in the household was removed, and then meals were delivered to the families. Researchers gathered information about the kids' behavior from parents, babysitters, and teachers. In addition, researchers tested the kids for memory, concentration, reading, and math skills. The result: "We couldn't find any difference in terms of their behavior or their learning on any of the three diets," said Mark Wolraich, professor of pediatrics at Vanderbilt University Medical Center, who oversaw the project. In another interview, Dr. Wolraich was quoted as saying, "Our study would say there is no evidence sugar has an adverse effect on children's behavior."

(a) This research involved comparison of several means, not just two. However, let us take a simplified view of the problem and consider the difference of behavior when children consumed the diet with sugar compared with their behavior when they consumed the diet with aspartame and low sugar. List some variables that might be measured to reflect the behavior of the children.

(b) Let's assume that the general null hypothesis was that there is no difference in children's behavior when they have a diet high in sugar. Was the evidence sufficient to allow the researchers to reject the null hypothesis and conclude that there are differences in children's behavior when they have a diet high in sugar? When we cannot reject H_0, have we *proved* that H_0 is true? In your own words, paraphrase the comments made by Dr. Wolraich.

⒧INKING CONCEPTS: **WRITING PROJECTS**

Discuss each of the following topics in class or review the topics on your own. Then write a brief but complete essay in which you summarize the main points. Please include formulas and graphs as appropriate.

Is there a relationship between confidence intervals and two-tailed hypothesis tests? The answer is yes. Let c be the level of confidence used to construct a confidence interval from sample data. Let α be the level of significance for a two-tailed hypothesis test. The following statement applies to hypothesis tests of the mean.

> For a two-tailed hypothesis test with level of significance α and null hypothesis H_0: $\mu = k$, we *reject* H_0 whenever k falls *outside* the $c = 1 - \alpha$ confidence interval for μ based on the sample data. When k falls within the $c = 1 - \alpha$ confidence interval, we do not reject H_0.

A corresponding relationship between confidence intervals and two-tailed hypothesis tests is also valid for other parameters such as p, $\mu_1 - \mu_2$, and $p_1 - p_2$.

(a) Consider the hypotheses H_0: $\mu_1 - \mu_2 = 0$ and H_1: $\mu_1 - \mu_2 \neq 0$. Suppose a 95% confidence interval for $p_1 - p_2$ contains only positive numbers. Should you reject the null hypothesis when $\alpha = 0.05$? Why or why not?

(b) Consider the hypothesis H_0: $p_1 - p_2 = 0$ and H_1: $p_1 - p_2 \neq 0$. Suppose a 99% confidence interval for $p_1 - p_2$ contains both positive and negative numbers. Should you reject the null hypothesis when $\alpha = 0.01$? Why or why not?

USING TECHNOLOGY

The problems in this section may be done using statistical computer software or calculators with statistical functions. Displays and suggestions are given for Minitab (Release 12), the TI-83 graphing calculator, ComputerStat, and Excel.

Inferences for $\mu_1 - \mu_2$ (Large, Independent Samples)

Minitab

When testing the difference of means from independent samples, Minitab always uses the t distribution regardless of sample size. The P value of the sample test statistic will be slightly larger than the P value generated using the normal distribution. Enter the data in two columns. Then use menu selections ➤Stat ➤Basic Statistics ➤2-Sample t. Do not check equal variances for large samples. The output gives an option for a confidence interval as well as for dotplots or boxplots of the two variables. The output displays the mean, the standard deviation, and the standard error of the mean for each of the variables. The t value of the sample test statistic and the P value are based on the Student's t distribution.

TI-83

You have the option of using raw data entered into two separate lists or summary statistics. Press the STAT key, select the TESTS option, and then use option 3:2-SampZTest. You will need to enter an estimate for the population standard deviation for each variable. The output gives the z value of the sample test statistic and the P value for the test. Under the Draw option, you can produce a graph showing the region of the P value. See the Calculator Note in Section 10.2 for a display of the output. For confidence intervals, use option 9:2-SampZInt under the Tests option.

ComputerStat

Under the Hypothesis Testing menu, select the program Testing a Difference of Means (Independent Sample). Then follow the instructions on the screen. For confidence intervals, use the Confidence Interval menu and select the program Confidence Intervals for MU1 = MU2. An example of the confidence interval output is given in Problem 8 in Section 10.2.

Excel

Enter the data in two columns. Then use the menu selection ➤Tools ➤Data Analysis ➤z-Test Two Sample for Means. Excel requires estimates of the variance for each variable. The output provides the mean and known variance for each variable, the z value of the sample test statistic, the P values for a one-tailed test and for a two-tailed test, and the critical z_0 values for a one-tailed test and for a two-tailed test. For an example of the output, see Problem 7 in Section 10.2.

APPLICATION 1

People who do shift work must often adjust their eating habits, sleep habits, exercise habits, family contacts, social lives, and overall lifestyles to accommodate their jobs. Extensive rearrangement of a person's habits and lifestyle can sometimes result in tension, anxiety, and overall health problems. An extensive study of the health consequences of shift work can be found in the following publication. Interested readers are referred to this report:

United States Department of Health, Education, and Welfare, NIOSH Technical Report. Tasto, Colligan, et al., *Health Consequences of Shift Work*. Washington GPO, 1978.

In an effort to study mood levels of nurses working in large hospitals, an opinion scale was used. Opinion ratings ranged from 0 = no feelings of tension and anxiety to a rating of 4 = extensive feelings of tension and anxiety. The scale was continuous, so a nurse could mark any number between 0 and 4. Suppose a random sample of 35 nurses on the day shift of a very large hospital gave the ratings shown in the accompanying table of their feelings of tension and anxiety at the end of the day shift.

Data for Day-Shift Nurses

3.50	3.75	2.33	2.16	3.50	0.80	1.25
1.33	2.67	2.50	1.50	0.75	0.00	0.67
4.00	3.75	3.50	3.25	2.40	3.50	2.75
3.50	2.67	2.80	2.33	3.50	3.80	2.75
1.50	1.33	0.00	2.25	1.75	0.50	1.75

Suppose a random sample of 33 nurses on the night shift also were asked to give their opinions about feelings of tension and anxiety at the end of the night shift. They used the same rating scale as described for the day shift. The ratings they gave are as follows.

Data for Night-Shift Nurses

3.50	3.75	3.50	3.10	3.20	3.33	1.75
2.75	2.50	2.75	3.20	3.75	4.00	2.00
1.00	0.00	1.80	2.50	3.50	3.00	2.60
3.10	2.75	4.00	2.90	1.75	2.20	3.50
1.00	2.50	0.80	3.70	2.60		

(a) Explain why the samples of day-shift and night-shift nurses are independent, or at least why it is reasonable to assume they are independent. We want to compare mean tension levels of day- and night-shift nurses.

(b) In this problem let MU1 = population mean tension level for the day-shift nurses and let MU2 = population mean tension level for the night-shift nurses. What is the null hypothesis?

(c) If we want to test the claim that there is a difference (either way) between tension levels of day-shift and night-shift nurses, which alternate hypothesis do we use? Use a 10% level of significance. What is the P value? Compare the P value with the level of significance. Shall we accept or reject the null hypothesis? What is the smallest level of significance that will result in a rejection of the null hypothesis in this two-tailed test?

(d) Rerun the program with the same data (you do not need to reenter the data). Use the same null hypothesis, but use a left-tailed test with a 10% level of significance. Do we accept or reject the null hypothesis? What is the smallest level of significance at which we can say the night-shift nurses show a higher average level of tension?

Paired Difference Test

Minitab

Enter the data in two columns. Then use the menu selections ➤Stats ➤Basic Statistics ➤Paired-t. The output includes an option for a confidence interval. Other items include the number of data pairs n, the mean, the standard deviation, and the standard error of the mean for each data set as well as for the differences. The t value of the sample test statistic is given as well as the P value. There is also an option to generate a histogram, dotplot, or boxplot for the differences.

For Problem 14 in Section 10.1 regarding the paired difference of male and female faculty in small colleges, the Minitab output is

Paired T-Test and Confidence Interval

```
Paired T for Male - Female

              N     MEAN   STDEV   SE MEAN
Male          9    25.33   15.64   5.21
Female        9    25.22   15.01   5.00
Difference    9     0.11    5.28   1.76

95% CI for mean difference: (-3.95, 4.17)
T-Test of mean difference = 0 (vs not = 0):
T-Value = 0.06  P-Value = 0.951
```

TI-83

Enter the first number of each data pair in the list L_1 and the corresponding value in the second list L_2. Create the list L_3 by subtracting L_2 from L_1. List L_3 contains the differences d. Our next step is to conduct a t test on the differences in L_3. Press **STAT**, select **TESTS**, and choose **Option 2:T-Test.** Use **Inpt:Data.** The value for μ_0 is 0 because we are testing the hypothesis $H_0: \mu_0 = 0$. Indicate that the data are in list L_3 with **Freq: 1.** Select the appropriate symbol ($\neq$, $<$, or $>$) for the alternate hypothesis, and then choose **Calculate,** which gives the value of the sample test statistic and its t value and P value, or **Draw,** which shows the area corresponding to the P value of the sample test statistic and gives the t value of the sample test statistic as well as the P value. See the displays in Section 9.4.

ComputerStat

Under the hypothesis testing menu, select Testing $\mu_1 - \mu_2$ (Dependent Samples). Then follow the instructions on the screens.

Excel

Enter the data in two columns. Then use the menu choices ►**Tools** ►**Data Analysis** ►**t-Test: Paired Two-Sample for Means.** The output contains the mean and variance for each data column, the degrees of freedom for the test, the sample t statistic, the P value for a one-tailed test, the P value for a two-tailed test, and the critical values t_0 for the specified level of significance for both a one-tailed test and a two-tailed test. See Problem 14 in Section 10.1 for a display of the output.

APPLICATION 2

Suppose a random sample of eight nurses was changed from the day shift to a rotating shift of some night work and some day work. For each of these nurses, the information shown in the table below was recorded about feelings of tension and anxiety at the end of a day shift and also at the end of a night shift.

Data for Nurses Working Both Shifts

Nurse	1	2	3	4	5	6	7	8
Day shift (B)	1.5	3	2	3	2	2	1	2
Night shift (A)	3.5	2	4	4	3.5	2	3.5	3

(a) Explain why the sample data for the day shift cannot be thought of as independent of the sample data for the night shift.

(b) Let us say that A is the random variable representing tension levels of night nurses and B is the random variable representing tension levels of day nurses. If we want to test the claim that nurses have a higher level of tension after the night shift, what would we use for the null hypothesis? What would we use for the alternate hypothesis? Choose the appropriate hypotheses and enter your choices on the computer. Use a 2% level of significance.

Shall we accept or reject the claim that after a night shift nurses express more feelings of tension on the average than they do after a day shift?

(c) What is the smallest level of significance at which these data will allow us to accept the claim that after a night shift nurses express more feelings of tension?

Computer Displays

Inferences for $\mu_1 - \mu_2$ (Small, Independent Samples)

Minitab

Enter the data in two columns and then use the menu selections ➤Stat ➤Basic Statistics ➤2-Sample t. Check the box for equal variances. An example of the output is shown in Problem 11 in Section 10.3.

TI-83

You may use raw data entered in two lists or summary statistics. Press the STAT key and arrow to the TESTS option. The option 4:2-SampTTest gives the P value and sample test statistic for hypothesis testing. The option 0:2-SampTInt gives confidence intervals for the difference of means, small samples. See the Calculator Note in Section 10.3 for an example of the TI-83 displays.

ComputerStat

Under the menu choice Hypothesis Testing, select Testing MU1 − MU2 (Independent Samples) and follow the screen instructions. The output provides summary statistics, the t value of the sample test statistic, and the P value. The user is then asked to draw a conclusion, and can check the results by pressing NEXT. The conclusion for a specified level of signifi-

cance and the critical values for that level are provided. For confidence intervals, use Confidence Intervals main menu and select Confidence Intervals for MU1 − MU2.

Excel

Enter the data in two columns. Then use the menu selections ➤Tools ➤Data Analysis ➤t-Test:Two-Sample Assuming Equal Variances. The output includes the means and variances of the two variables, the pooled variance, the t value of the sample test statistic, the P values for a one-tailed test and for a two-tailed test, and the critical values t_0 for the specified level of significance for both a one-tailed and a two-tailed test. The display is similar in style to that for Paired Difference Tests as shown in Problem 14, Section 10.1.

Inferences for $p_1 - p_2$

Minitab

Use the menu selections ➤Stat ➤Basic Statistics ➤2-Proportions. Enter the data as summarized data. Under options, select the alternate hypothesis, and include confidence intervals if you wish. A sample of the outcome display is shown in Problem 11 in Section 10.4.

TI-83

Press the STAT key. Under the TEST option, select item 6:2-PropZTest for testing the difference of two proportions and select item B:2-PropZInt for confidence intervals. See the Calculator Note in Section 10.4 for an example of the displays.

ComputerStat

Under the Hypothesis Testing main menu, select Testing $p_1 - p_2$. For confidence intervals, use the main menu choice Confidence Intervals and select Confidence Intervals for $p_1 - p_2$. The following display gives the results of testing to see if there is a difference in the proportion of urban residents compared with suburban residents subscribing to a magazine.

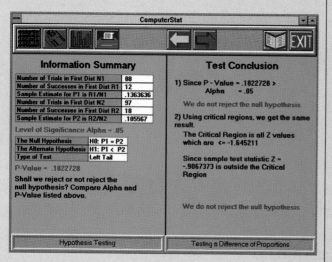

Excel

Excel does not have menu options for testing the difference of proportions or creating confidence intervals for the difference of two proportions.

11 Additional Topics Using Inference

Truth does not happen, it just is.

—Traditional Hopi
Proverb

If we wonder often, the gift of knowledge will come.

—Traditional Arapahoe
Proverb

U.S. Senator Ben Nighthorse Campbell (Northern Cheyenne Tribe). Senator Campbell, from Colorado, initiated successful legislation to establish the National Museum of the American Indian at the Smithsonian Institute.

The Arapahoe and Northern Cheyenne are Western Plains Indians. The Hopi are pueblo dwelling Native Americans who still retain their cultural roots today while residing on beautiful mesas in Northern Arizona. Today you can visit the Hopi Nation, enjoy their cultural center, and stay in a modern hotel, or camp if you like. See Web site *http//www.psv.com/hopi.html.*

In our study of statistics, it is important to remember that our methods are *not* intended to make truth "happen"! Instead, we make every effort to investigate and find out what truth "is". Good statistical practice should *not* be used in an effort to impose a preset (possibly hidden) agenda. Statistical work used to buttress up a preset agenda is an effort to "make truth happen" rather than discover what truth "is". In a sense the wisdom of the Hopi proverb encourages us to be *honest* statisticians working to discover the truth independent of other agendas. The Arapahoe proverb encourages us to wonder often (and perhaps study or investigate) with the assurance that the gift of knowledge will indeed come.

In this chapter, we investigate more applications of statistical inference. In Part I we look at some applications of the chi-square distribution. In Part II we look at some inferences relating to linear regression.

533

PART I: HYPOTHESIS TESTS USING THE CHI-SQUARE DISTRIBUTION
Overview of the Chi-Square Distribution

So far we have used several probability distributions for hypothesis testing and confidence intervals, with the most frequently used being the normal distribution and the Student's t distribution. In this chapter we will use another probability distribution—namely, the chi-square (where *chi* is pronounced like the first two letters in the word *kite*). In Part I we will see applications of the chi-square distribution.

Chi is a Greek letter denoted by the symbol χ, so chi-square is denoted by the symbol χ^2. Because the distribution is of chi-*square* values, the χ^2 values begin at 0 and then are all positive. The graph of the χ^2 distribution is not symmetrical, and like the Student's t distribution, it depends on the number of degrees of freedom. Figure 11-1 shows the χ^2 distribution for several degrees of freedom (*d.f.*).

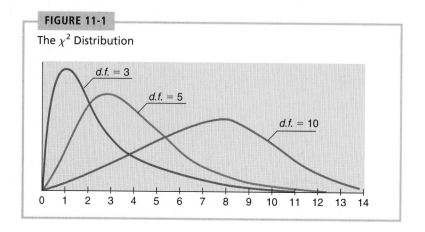

FIGURE 11-1

The χ^2 Distribution

As the degrees of freedom increase, the graph of the chi-square distribution becomes more bell-like and begins to look more and more symmetric. Notice that the mode, or high point, of the graph with n degrees of freedom occurs over $n - 2$ (for $n \geq 3$).

We use Table 6 in Appendix I to find critical values of chi-square distributions for which a designated area α falls to the right of the critical value. Table 11-1 gives an excerpt from Table 6. Notice that the row headers are degrees of freedom and the column headers are areas α in the *right tail* of the distribution. For instance, according to the table, for a χ^2 distribution with 3 degrees of freedom, $\alpha = 0.995$ is the area falling to the right of $\chi^2 = 0.072$. For a χ^2 distribution with 4 degrees of freedom, $\alpha = 0.010$ is the area falling to the right of $\chi^2 = 13.28$.

In the next three sections we will see how to apply the chi-square distribution to different applications.

Table 11-1   Excerpt from Table 6 (Appendix I): The χ^2 Distribution

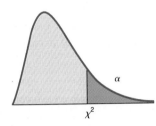

d.f. \ α	0.995	0.990	0.975	. . .	0.010	0.005
:	:	:	:		:	:
3	0.072	0.115	0.352		11.34	12.84
4	0.207	0.297	0.484		13.28	14.86

Section **11.1**

Chi Square: Tests of Independence

Innovative Machines Incorporated has developed two new letter arrangements for computer keyboards. They wish to see if there is any relationship between the arrangement of letters on the keyboard and the number of hours it takes a new typing student to learn to type at 20 words per minute—or, from another point of view, if the time it takes a student to learn to type is *independent* of the arrangement of the letters on the keyboard.

To answer questions of this type, we test the hypotheses

H_0: Keyboard arrangement and learning times *are independent.*

H_1: Keyboard arrangement and learning times *are not independent.*

Chi-square distribution

In problems of this sort we are testing the *independence* of two factors. The probability distribution we use to make the decision is the *chi-square distribution.* Recall from the overview of the chi-square distribution that *chi* is pronounced like the first two letters of the word *kite* and that chi is a Greek letter denoted by the symbol χ, so chi square is denoted by χ^2.

The first task for Innovative Machines is to gather data. Suppose that the company took a random sample of 300 beginning typing students and randomly assigned them to learn to type on one of three keyboards. The learning times for this sample are shown in Table 11-2 below.

Table 11-2   Keyboard Versus Time to Learn to Type at 20 wpm

Keyboard	21–40 h	41–60 h	61–80 h	Row Total
A	#1 25	#2 30	#3 25	80
B	#4 30	#5 71	#6 19	120
Standard	#7 35	#8 49	#9 16	100
Column total	90	150	60	300 Sample size

Contingency table

Table 11-2 is called a *contingency table*. The *shaded boxes* that contain observed frequencies are called *cells*. The row and column totals are not considered to be cells. This contingency table is of size 3 × 3 (read, "three-by-three"), since there are three rows of cells and three columns. When giving the size of a contingency table, we always list the number of *rows first*.

GUIDED EXERCISE   1

Give the size of the contingency tables in Figures 11-2 and 11-3. Also count the number of cells in each table. (Remember, each cell is a shaded box.)

(a) **FIGURE 11-2** Contingency Table

⇨ There are two rows and four columns, so this is a 2 × 4 table. There are eight cells.

(b) **FIGURE 11-3** Contingency Table

⇨ Here we have three rows and two columns, so this is a 3 × 2 table with six cells.

Null hypothesis

We are testing the *null hypothesis* that the keyboard arrangement and the time it takes a student to learn to type are *independent*.

H_0: Variables are independent
H_1: Variables are not independent

Expected frequency

We use the null hypothesis to determine the *expected frequency* of each cell.

For instance, to compute the expected frequency of cell 1 in Table 11-2, we observe that cell 1 consists of all the students in the sample who learned to type on keyboard A and who mastered the skill at the 20 word per minute level in 21 to 40 hours. By the assumption (null hypothesis) that the two events are independent, we use the multiplication law to obtain the probability that a student is in cell 1.

$$P(\text{cell 1}) = P(\text{keyboard A } and \text{ skill in 21–40 h})$$
$$= P(\text{keyboard A}) \cdot P(\text{skill in 21–40 h})$$

Since there are 300 students in the sample and 80 used keyboard A,

$$P(\text{keyboard A}) = \frac{80}{300}$$

Also, 90 of the 300 students learned to type in 21–40 hours, so

$$P(\text{skill in 21–40 h}) = \frac{90}{300}$$

Using these two probabilities and the assumption of independence,

$$P(\text{keyboard A } and \text{ skill in 21–40 h}) = \frac{80}{300} \cdot \frac{90}{300}$$

Finally, since there are 300 students in the sample, we have the *expected frequency* E for cell 1.

$$E = P(\text{student in cell 1}) \cdot (\text{no. of students in sample})$$

$$= \frac{80}{300} \cdot \frac{90}{300} \cdot 300$$

$$= \frac{80 \cdot 90}{300} = 24$$

We can repeat this process for each cell. However, the last step yields an easier formula for the expected frequency E.

Formula for Expected Frequency E

$$E = \frac{(\text{row total})(\text{column total})}{\text{sample size}}$$

Note: If the expected value is not a whole number, do *not* round it to the nearest whole number.

Let's use this formula in Example 1 to find the expected frequency for cell 2.

EXAMPLE 1 ▷ Find the expected frequency for cell 2 in contingency Table 11-2 on page 535.

SOLUTION: Cell 2 is in row 1 and column 2. The row total is 80, and the column total is 150. The size of the sample is still 300.

$$E = \frac{(\text{row total})(\text{column total})}{\text{sample size}} = \frac{(80)(150)}{300} = 40$$

GUIDED EXERCISE 2

Table 11-3 contains the *observed frequencies* O and *expected frequencies* E for the contingency table giving keyboard arrangement and number of hours it takes a student to learn to type at 20 words per minute. Fill in the missing expected frequencies.

Table 11-3 Complete Contingency Table of Keyboard Arrangement and Time to Learn to Type

Keyboard	21–40 h	41–60 h	61–80 h	Row Total
A	#1 O = 25 E = 24	#2 O = 30 E = 40	#3 O = 25 E = __	80
B	#4 O = 30 E = 36	#5 O = 71 E = __	#6 O = 19 E = __	120
Standard	#7 O = 35 E = __	#8 O = 49 E = 50	#9 O = 16 E = 20	100
Column total	90	150	60	300 Sample size

For cell 3 we have

$$E = \frac{(80)(60)}{300} = 16$$

For cell 5 we have

$$E = \frac{(120)(150)}{300} = 60$$

For cell 6 we have

$$E = \frac{(120)(60)}{300} = 24$$

For cell 7 we have

$$E = \frac{(100)(90)}{300} = 30$$

Computing the sample statistic χ^2

Now we are in a position to compute the sample statistic χ^2 for the sample of typing students. The χ^2 value is a measure of the sum of differences between observed frequency O and expected frequency E in each cell. These differences are listed in Table 11-4 on page 539 for the nine cells.

As we see, if we sum the differences between the observed frequencies and the expected frequencies of the cells, we get the value zero. This total certainly does not reflect the fact that there were differences between the observed and expected frequencies. To obtain a measure whose sum does reflect the magnitude of the differences, we square the differences and work with the quantities $(O - E)^2$. But

Table 11-4   Difference Between the Observed and Expected Frequencies

Cell	Observed O	Expected E	Difference $(O - E)$
1	25	24	1
2	30	40	-10
3	25	16	9
4	30	36	-6
5	71	60	11
6	19	24	-5
7	35	30	5
8	49	50	-1
9	16	20	-4
			$\Sigma(O - E) = 0$

instead of using the terms $(O - E)^2$, we use the values $(O - E)^2/E$. The reason we use this expression is that a small difference between the observed and expected frequency is not nearly as important if the expected frequency is large as it is if the expected frequency is small. For instance, for both cells 1 and 8, the squared difference $(O - E)^2$ is 1. However, this difference is more meaningful in cell 1, where the expected frequency is 24, than it is in cell 8, where the expected frequency is 50. When we divide the quantity $(O - E)^2$ by E, we take the size of the difference with respect to the size of the expected value. We use the sum of these values to form the sample statistic χ^2:

$$\chi^2 = \Sigma \frac{(O - E)^2}{E}$$

where the sum is over all cells in the contingency table.

COMMENT If you look up the word *irony* in a dictionary, you will find one of its meanings is described as "the difference between actual (or observed) results and expected results." Since irony is so prevalent in much of our human experience, it is not surprising that statisticians have incorporated a related chi-square distribution into their work. As we will soon see, the chi-square distribution has many applications in social science, business administration, and natural science.

GUIDED EXERCISE   3

(a) Complete Table 11-5 from the data of Table 11-4.

Table 11-5

Cell	O	E	O − E	(O − E)²	(O − E)²/E
1	25	24	1	1	0.04
2	30	40	−10	100	2.50
3	25	16	9	81	5.06
4	30	36	−6	36	1.00
5	71	60	11	121	2.02
6	19	24	−5	25	1.04
7	35	30	5	25	0.83
8	49	50	___	—	___
9	16	20	___	—	___

$$\Sigma \frac{(O - E)^2}{E} = \underline{\quad}$$

(b) Compute the statistic χ^2 for this sample.

⟹ The last two rows of Table 11-5 are

Cell	O	E	O − E	(O − E)²	(O − E)²/E
8	49	50	−1	1	0.02
9	16	20	−4	16	0.80

$$\Sigma \frac{(O - E)^2}{E} = \text{total of last column} = 13.31$$

⟹ Since $\chi^2 = \Sigma \dfrac{(O - E)^2}{E}$, then $\chi^2 = 13.31$.

Critical value χ_α^2

Notice that when the observed frequency and the expected frequency are very close, the quantity $(O - E)^2$ is close to zero, and so the statistic χ^2 is near zero. As the difference increases, the statistic χ^2 also increases. To determine how large the statistic can be before we must reject the null hypothesis of independence, we find a *critical value* χ_α^2 in Table 6 in Appendix I for the specified level of significance α and the number of degrees of freedom in the sample.

As we saw in the chi-square overview at the beginning of this chapter, the chi-square distribution changes as the degrees of freedom change. To find a critical value χ_α^2, we need to know the degrees of freedom in the sample as well as the level of significance designated. To test independence, the degrees of freedom *d.f.* of a sample are determined by the following formula.

> **Degrees of Freedom for Test of Independence**
>
> Degrees of freedom = (number of rows − 1) · (number of columns − 1)
>
> or $\quad\quad\quad\quad d.f. = (R - 1)(C - 1)$
>
> where R = number of cell rows
>
> $\quad\quad\quad C$ = number of cell columns

| GUIDED EXERCISE | 4 |

Determine the number of degrees of freedom in the example of keyboard arrangements (see Table 11-2). Recall that the contingency table had three rows and three columns.

$$d.f. = (R - 1)(C - 1)$$
$$= (3 - 1)(3 - 1) = (2)(2) = 4$$

To test the hypothesis that the letter arrangement on a keyboard and the time it takes to learn to type at 20 words per minute are independent at the $\alpha = 0.05$ level of significance, we look up the critical value $\chi^2_{0.05}$ in Table 6 in Appendix I. For $d.f. = 4$ and $\alpha = 0.05$, we see $\chi^2_{0.05} = 9.49$. When we compare the sample statistic $\chi^2 = 13.31$ with the critical value $\chi^2_{0.05}$, we see that the sample statistic is larger. Since it is larger, we *reject* the null hypothesis of independence and conclude that keyboard arrangement and learning time are *not* independent (Figure 11-4).

Test conclusion

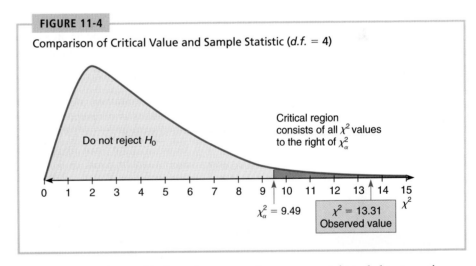

FIGURE 11-4

Comparison of Critical Value and Sample Statistic (*d.f.* = 4)

Do not reject H_0

Critical region consists of all χ^2 values to the right of χ^2_α

$\chi^2_\alpha = 9.49$

$\chi^2 = 13.31$
Observed value

COMMENT For tests of independence, we always use a *right-tailed* test on the chi-square distribution. This is because we are testing to see if the χ^2 measure of the difference between the observed and expected frequencies is too large to be due to chance alone.

Summary

Let's summarize how we use the chi-square distribution to test the independence of two variables.

Step 1: Set up the hypotheses:

H_0: The variables *are* independent.

H_1: The variables *are not* independent.

Step 2: Compute the expected frequency for each cell in the contingency table by use of the formula

$$E = \text{expected frequency} = \frac{(\text{row total})(\text{column total})}{\text{sample size}}$$

Step 3: Compute the statistic χ^2 for the sample:

$$\chi^2 = \Sigma \frac{(O - E)^2}{E}$$

where O is the observed frequency, E is the expected frequency, and the sum Σ is over all cells.

Step 4: Find the critical value χ_α^2 in Table 6 in Appendix I. Use the level of significance α and the number of degrees of freedom *d.f.* to find the critical value.

$$d.f. = (R - 1)(C - 1)$$

where R is the number of rows and C is the number of columns of cells in the contingency table. The critical region consists of all values of χ^2 to the *right* of the critical value χ_α^2.

Step 5: Compare the sample statistic χ^2 in step 3 with the critical value χ_α^2 in step 4. If the sample statistic is *larger,* reject the null hypothesis of independence. Otherwise, do not reject the null hypothesis.

NOTE We compare the sample statistic χ^2 with the critical value χ_α^2. But the distribution of sample statistics is only approximately the same as the theoretical distribution whose critical values χ_α^2 are found in Table 6 in Appendix I. In order to safely use critical values χ_α^2, we must be sure that all the cells have an *expected frequency* larger than or equal to 5. If this condition is not met, the sample size should be increased.

GUIDED EXERCISE   5

Super Vending Machines Company is to install soda pop machines in elementary schools and high schools. The market analysts wish to know if flavor preference and school level are independent. A random sample of 200 students was taken. Their school levels and soda pop preferences are given in Table 11-6. Is independence indicated at the $\alpha = 0.01$ level of significance?

Step 1: State the null and alternate hypotheses. H_0: School level and soda pop preference are independent.

H_1: School level and soda pop preference are not independent.

Exercise continues

Exercise continued

Step 2: Complete the contingency Table 11-6 by filling in the required expected frequencies.

Table 11-6 School Level and Soda Pop Preference

Soda Pop	High School	Elementary School	Row Total
Kula Kola	$O = 33$ #1 $E = 36$	$O = 57$ #2 $E = 54$	90
Mountain Mist	$O = 30$ #3 $E = 20$	$O = 20$ #4 $E = 30$	50
Jungle Grape	$O = 5$ #5 $E = \underline{\quad}$	$O = 35$ #6 $E = \underline{\quad}$	40
Diet Pop	$O = 12$ #7 $E = \underline{\quad}$	$O = 8$ #8 $E = \underline{\quad}$	20
Column Total	80	120	200 Sample size

➡ The expected frequency

for cell 5 is $\dfrac{(40)(80)}{200} = 16$

for cell 6 is $\dfrac{(40)(120)}{200} = 24$

for cell 7 is $\dfrac{(20)(80)}{200} = 8$

for cell 8 is $\dfrac{(20)(120)}{200} = 12$

Note: In this example the expected frequencies are all whole numbers. If the expected frequency has a decimal part such as 8.45, do *not* round the value to the nearest whole number; rather, give the expected frequency as the decimal number.

Step 3: Fill in Table 11-7 and use the table to find the sample statistic χ^2.

Table 11-7 Computational Table for χ^2

Cell	O	E	$O - E$	$(O - E)^2$	$(O - E)^2/E$
1	33	36	-3	9	0.25
2	57	54	3	9	0.17
3	30	20	10	100	5.00
4	20	30	-10	100	3.33
5	5	16	-11	121	7.56
6	35	24	11	____	____
7	12	8	____	____	____
8	8	12	____	____	____

➡ The last three rows of Table 11-7 read as follows:

Cell	O	E	$O - E$	$(O - E)^2$	$(O - E)^2/E$
6	35	24	11	121	5.04
7	12	8	4	16	2.00
8	8	12	-4	16	1.33

χ^2 = total of last column

$$= \Sigma \frac{(O - E)^2}{E} = 24.68$$

Step 4: What is the size of the contingency table? Use the number of rows and columns to determine the number of degrees of freedom. For $\alpha = 0.01$, use Table 6 in Appendix I to find the critical value $\chi^2_{0.01}$.

➡ The contingency table is of size 4×2. Since there are four rows and two columns,

$d.f. = (4 - 1)(2 - 1) = 3$

For $\alpha = 0.01$ the critical value χ^2_α is 11.34.

Step 5: Do we reject or fail to reject the null hypothesis that school level and soda pop flavor preference are independent?

➡ Since the statistic χ^2 is larger than the critical value χ^2_α, we reject the null hypothesis of independence and conclude that school level and soda pop preference are dependent.

P values

COMMENT *P* values for tests of independence can be estimated from Table 6 in Appendix I. For instance, to estimate the *P* value for the sample statistic $\chi^2 = 24.68$ found in Guided Exercise 5, we look in the row headed by the degrees of freedom for the contingency table, *d.f.* $= 3$. We see that the sample statistic $\chi^2 = 24.68$ falls to the right of the row entry 12.84. This means that the *P* value is smaller than the column header 0.005 corresponding to the entry 12.84. Therefore,

$$P \text{ value} < 0.005$$

Consequently, we reject H_0 for all $\alpha \geq 0.005$. In particular, we reject H_0 for $\alpha = 0.01$. This result is consistent with the conclusion stated in Guided Exercise 5.

Calculator Note The TI-83 performs chi-square tests of independence. The user enters the original table of observed values (as a matrix) and sets the dimension of matrix [B] to match the dimension of the matrix of observed values. Then the calculator computes the sample χ^2 statistic and its corresponding *P* value. On the TI-83, press **STAT**, select **TESTS**, and choose option **C: χ^2-Test....** Use the **MATRIX** key to edit the matrix of observed values and set the dimension of matrix [B], the matrix in which expected values are placed.

V I E W P O I N T

Loyalty: Going, Going, Gone!

Was there a time in the past when people worked for the same company all their lives, regularly purchased the same brand names, always voted for candidates from the same political party, and loyally cheered for the same sports team? One way to look at this question is to consider tests of *statistical independence*. Is customer loyalty independent of company profits? Can a company maintain its productivity independent of loyal workers? Can politicians do whatever they please independent of the voters back home? Americans may be ready to act on a pent-up desire to restore a sense of loyalty in their lives. For more information, see *American Demographics*, Vol. 19, No. 9.

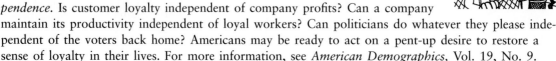

SECTION 11.1 PROBLEMS

For Problems 1–9, please do the following:
(a) State the null and alternate hypotheses.
(b) Find the value of the chi-square statistic from the sample.
(c) Find the degrees of freedom and the appropriate critical chi-square value.

(d) Sketch the critical region and locate your sample chi-square value and critical chi-square value on the sketch.

(e) Decide whether you should reject or fail to reject the null hypothesis using the given level of significance α.

Use the expected values E to the hundredths place.

Myers-Briggs: Clergy, M.D., Lawyer

1. The following table shows the Myers-Briggs personality preference and professions for a random sample of 2408 people in the listed professions (*Atlas of Type Tables*, by Macdaid, McCaulley, and Kainz). E refers to extroverted, and I refers to introverted.

| | Personality Preference Type | | |
Occupation	E	I	Row Total
Clergy (all denominations)	308	226	534
M.D.	667	936	1603
Lawyer	112	159	271
Column total	1087	1321	2408

Use the chi-square test to determine if the listed occupations and personality preferences are independent at the 0.01 level of significance.

Myers-Briggs: Clergy, M.D., Lawyer

2. The following table shows the Myers-Briggs personality preference and professions for a random sample of 2408 people in the listed professions (*Atlas of Type Tables*, by Macdaid, McCaulley, and Kainz). T refers to thinking, and F refers to feeling.

| | Personality Preference Type | | |
Occupation	T	F	Row Total
Clergy (all denominations)	114	420	534
M.D.	785	818	1603
Lawyer	176	95	271
Column total	1075	1333	2408

Use the chi-square test to determine if the listed occupations and personality preferences are independent at the 0.01 level of significance.

Archaeology: Pottery

3. The following table shows site type and type of pottery for a random sample of 628 sherds at a location in Sand Canyon Archaeological Project, Colorado (*The Sand Canyon Archaeological Project*, edited by Lipe).

| | Pottery Type | | | |
Site Type	Mesa Verde Black on White	McElmo Black on White	Mancos Black on White	Row Total
Mesa Top	75	61	53	189
Cliff-Talus	81	70	62	213
Canyon Bench	92	68	66	226
Column total	248	199	181	628

Use a chi-square test to determine if site type and pottery type are independent at the 0.05 level of significance.

Prehistoric Pueblos:
Ceremonial Rankings

4. The following table shows ceremonial ranking and type of pottery sherds for a random sample of 1404 sherds at a location in the Sand Canyon Archaeological Project, Colorado (*The Architecture of Social Integration in Prehistoric Pueblos*, edited by Lipe and Hegmon).

Ceremonial Ranking	Cooking Jar Sherds	Decorated Jar Sherds (Noncooking)	Row Total
A	242	26	268
B	658	45	703
C	371	62	433
Column total	1271	133	1404

Use a chi-square test to determine if ceremonial ranking and pottery type are independent at the 0.05 level of significance.

Yellowstone National Park:
Buffalo

5. The following table shows age distribution and location of a random sample of 166 buffalo in Yellowstone National Park (based on information from *The Bison of Yellowstone National Park*, National Park Service Scientific Monograph Series).

Age	Lamar District	Nos Perce District	Firehole District	Row Total
Calf	13	13	15	41
Yearling	10	11	12	33
Adult	34	28	30	92
Column total	57	52	57	166

Use a chi-square test to determine if age distribution and location are independent at the 0.05 level of significance.

Myers-Briggs: College Majors

6. The following table shows the Myers-Briggs personality preference and area of study for a random sample of 519 college students (*Applications of the Myers-Briggs Type Indicator in Higher Education,* edited by Provost and Anchors). In the table, IN refers to introvert, intuitive; EN refers to extrovert, intuitive; IS refers to introvert, sensing; and ES refers to extrovert, sensing.

Myers-Briggs Preference	Arts & Science	Business	Allied Health	Row Total
IN	64	15	17	96
EN	82	42	30	154
IS	68	35	12	115
ES	75	42	37	154
Column total	289	134	96	519

Use a chi-square test to determine if Myers-Briggs preference type is independent of area of study at the 0.01 level of significance.

Sociology: Movies

7. Mr. Acosta, a sociologist, is doing a study to see if there is a relationship between the age of a young adult (18 to 35 years old) and the type of movie preferred. A random sample of 93 adults revealed the following data. Test if age and type of movie preferred are independent at the 0.05 level.

Movie	Person's Age			Row Total
	18–23 yr	24–29 yr	30–35 yr	
Musical	8	15	11	34
Science fiction	12	10	8	30
Comedy	9	8	12	29
Column total	29	33	31	93

Library Funds: Ethnic Groups

8. After a large fund drive to help the Boston City Library, the following information was obtained from a random sample of contributors to the library fund. Using a 1% level of significance, test the claim that the amount contributed to the library fund is independent of ethnic group.

Ethnic Group	Number of People Making Contributions					Row Total
	$1–50	$51–100	$101–150	$151–200	Over $200	
A	310	715	201	105	42	1373
B	619	511	312	97	22	1561
C	402	624	217	88	35	1366
D	544	571	309	79	29	1532
Column total	1875	2421	1039	369	128	5832

Movie Theaters: Billings

9. Blue Bird Consolidated Theaters has more than 600 theaters located across the country. Each theater has four separate screens, and a customer can choose from one of four different movies. The president of Blue Bird Consolidated wanted to know if a variety of shows (spy, comedy, horror, children's) or a coordinated bill (all spy, all comedy, all horror, all children's) would have any effect on the total ticket sales at a theater. The president randomly assigned 47 theaters to use a variety of shows and 53 other theaters to use a coordinated bill of shows. For all theaters, total ticket sales for one week were recorded.

We used Minitab (enter the data in columns, then select ▶Stat ▶Tables ▶Chisquare Test) to compute the χ^2 value of the sample test statistic. The output follows, where 1 represents variety as the type of billing and 2 represents coordinated. In the output, 1–2tho represents 1000 to 2000 ticket sales, while 2–3tho represents 2001 to 3000 ticket sales.

Chi-Square Test

```
Expected counts are printed below observed counts
          <1000    1–2tho    2–3tho    >3000     Total
1            10       12       18        7         47
           7.52     13.16    18.80     7.52
2             6       16       22        9         53
           8.48     14.84    21.20     8.48
Total        16       28       40       16        100
Chi-Sq = 0.818 + 0.102 + 0.034 + 0.036 +
         0.725 + 0.091 + 0.030 + 0.032 = 1.868
DF  = 3,  P-Value  = 0.600
```

(a) What are the understood hypotheses H_0 and H_1?
(b) Look at the Minitab output. For variety billing (row 1 header), what is the expected number of theaters with ticket sales of more than 3000?
(c) What is the χ^2 value of the sample statistic?
(d) Look at the P value for the test. Should we reject or fail to reject the null hypothesis at the 5% level of significance?

Section **Chi Square: Goodness of Fit**

Last year the labor union bargaining agents listed five items and asked each employee to mark the *one* most important to her or him. The items and corresponding percentages of favorable responses are shown in Table 11-8. The bargaining agents need to determine if the distribution of responses *now* "fits" last year's distribution or if it is different.

Hypotheses

In questions of this type, we are asking if a population follows a specified distribution. In other words, we are testing the hypotheses

> H_0: The population fits the given distribution.
>
> H_1: The population has a different distribution.

Table 11-8 Bargaining Items (last year)

Item	Percentage of Favorable Responses
Vacation time	4%
Salary	65%
Safety regulations	13%
Health and retirement benefits	12%
Overtime policy and pay	6%

Computing sample χ^2

We use the chi-square distribution to test "goodness-of-fit" hypotheses. Just as with tests of independence, we compute the sample statistic:

$$\chi^2 = \Sigma \frac{(O - E)^2}{E} \text{ with degrees of freedom} = n - 1$$

where E = expected frequency

O = observed frequency

$\frac{(O - E)^2}{E}$ is summed for each item in the distribution

n = number of items in the distribution

Then we compare it with an appropriate critical value χ^2_α from Table 6 in Appendix I. In the case of a *goodness-of-fit test,* we use the null hypothesis to compute the expected values. Let's look at the bargaining item problem to see how this is done.

In the bargaining item problem, the two hypotheses are

H_0: The present distribution of responses is the same as last year's.

H_1: The present distribution of responses is different.

The null hypothesis tells us that the expected frequencies of the present response distribution should follow the percentages indicated in last year's survey. To test this hypothesis, a random sample of 500 employees was taken. If the null hypothesis is true, then there should be 4%, or 20 responses, out of the 500 rating vacation time as the most important bargaining issue. Table 11-9 on page 550 gives the other expected values and all the information necessary to compute the sample statistic χ^2. We see that the sample statistic is

$$\chi^2 = \Sigma \frac{(O - E)^2}{E} = 14.15$$

Table 11-9 Observed and Expected Frequencies for Bargaining Items

Item	O	E	$(O - E)^2$	$(O - E)^2/E$
Vacation time	30	4% of 500 = 20	100	5.00
Salary	290	65% of 500 = 325	1225	3.77
Safety	70	13% of 500 = 65	25	0.38
Health and retirement	70	12% of 500 = 60	100	1.67
Overtime	40	6% of 500 = 30	100	3.33
	$\Sigma O = 500$	$\Sigma E = 500$		$\Sigma \dfrac{(O - E^2)}{E} = 14.15$

Critical value χ_α^2

Again, larger values of the sample statistic χ^2 indicate greater differences between the proposed probability distribution and the one followed by the sample. The critical value χ_α^2 tells us how large the sample statistic can be before we reject the null hypothesis that the population does follow the distribution proposed in that hypothesis.

To find the critical value χ_α^2, we need to know the level of significance α and the number of degrees of freedom $d.f.$ In the case of a goodness-of-fit test, the degrees of freedom are found by the following formula.

Degrees of Freedom for Goodness-of-Fit Test

$d.f.$ = (number of E entries) $- 1$

Notice that when we compute the expected values E, we must use the null hypothesis to compute all but the last one. To compute the last one, we can subtract the previous expected values from the sample size. For instance, for the bargaining issues, we could have found the number of responses for overtime policy by adding up the other expected values and subtracting that sum from the sample size 500. We would again get an expected value of 30 responses. The degrees of freedom, then, is the number of E values that *must* be computed by using the null hypothesis.

For the bargaining issues, we have

$$d.f. = 5 - 1 = 4$$

To test the hypothesis at the 0.05 level of significance, we find the critical value $\chi_{0.05}^2$ for four degrees of freedom in Table 6 in Appendix I. The critical value, 9.49, and the sample statistic, 14.15, are shown in Figure 11-5. As shown in the figure, the sample statistic χ^2 is in the critical region. Note that the critical region is always

FIGURE 11-5

Critical Value and Test Statistic (*d.f.* = 4)

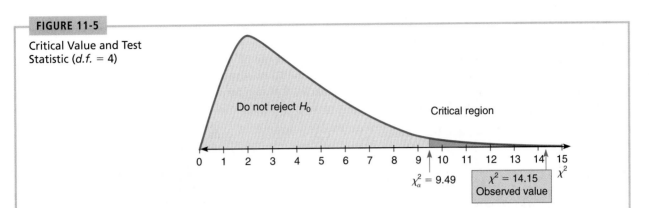

to the right of the critical value χ^2_α. Since the sample statistic is in the critical region, we reject the null hypothesis and conclude that the distribution of responses to the bargaining issues now is different from the distribution of last year.

One important application of goodness-of-fit tests is to genetics theories. Such an application is shown in Guided Exercise 6.

GUIDED EXERCISE   6

According to genetics theory, red-green colorblindness in humans is a recessive sex-linked characteristic. In this case, the gene is carried on the X chromosome only. We will denote an X chromosome with the gene by X_c and one without the gene by X_n. Women have two X chromosomes, and they will be red-green colorblind only if both chromosomes have the gene, designated $X_c X_c$. A woman can have normal vision but still carry the colorblind gene if only one of the chromosomes has the gene, designated $X_c X_n$. A man carries an X chromosome and a Y chromosome; if the X chromosome carries the colorblind gene ($X_c Y$), the man is colorblind.

According to genetics theory, if a man with normal vision ($X_n Y$) and a woman carrier ($X_c X_n$) have a child, the probabilities that the child will have red-green colorblindness, have normal vision and not carry the gene, and have normal vision and carry the gene are given by the *equally likely* events in Table 11-10.

$P(\text{child has normal vision and is not a carrier}) = P(X_n Y) + P(X_n X_n) = \dfrac{1}{2}$

$P(\text{child has normal vision and is a carrier}) = P(X_c X_n) = \dfrac{1}{4}$

$P(\text{child is red-green colorblind}) = P(X_c Y) = \dfrac{1}{4}$

**Table 11-10
Red-Green Colorblindness**

Mother	Father	
	X_n	Y
X_c	$X_c X_n$	$X_c Y$
X_n	$X_n X_n$	$X_n Y$

Exercise continues

To test this genetics theory, Genetics Labs took a random sample of 200 children whose mothers were carriers of the colorblind gene and whose fathers had normal vision. The results are shown in Table 11-11. We wish to test the hypothesis that the population follows the distribution predicted by the genetics theory (see Table 11-10).

(a) State the null and alternate hypotheses.

⇨ H_0: The population fits the distribution predicted by genetics theory.

H_1: The population does not fit the distribution predicted by genetics theory.

(b) Fill in the rest of Table 11-11 and use the table to compute the sample statistic χ^2.

Table 11-11 Colorblindness Sample

Event	O	E	$(O - E)^2$	$(O - E)^2/E$
Red-green colorblind	35	50	225	4.50
Normal vision, noncarrier	105	__	____	____
Normal vision, carrier	60	__	____	____

⇨ **Table 11-12 Completion of Table 11-11**

Event	O	E	$(O - E)^2$	$(O - E)^2/E$
Red-green colorblind	35	50	225	4.50
Normal vision, noncarrier	105	100	25	0.25
Normal vision, carrier	60	50	100	2.00

The sample statistic is $\chi^2 = \Sigma \dfrac{(O - E)^2}{E} = 6.75$

(c) There are three expected frequencies listed in Table 11-11. Use this information to compute the degrees of freedom.

⇨ $d.f. = $ (no. of E values) $- 1 = 3 - 1 = 2$

(d) Find the critical value $\chi^2_{0.01}$ for a 0.01 level of significance. Do we reject or fail to reject the hypothesis that the population follows the distribution predicted by genetics theory?

⇨ From Table 6 in Appendix I, we see that for $d.f. = 2$ and level of significance 0.01, the critical value is $\chi^2_{0.01} = 9.21$. Since the sample statistic $\chi^2 = 6.75$ is less than the critical value, we do not reject the null hypothesis that the population follows the distribution predicted by genetics theory.

P values

COMMENT To estimate *P* values for goodness-of-fit tests, we use Table 6 in Appendix I. As an example, let's find an interval containing the *P* value of the sample chi-square statistic $\chi^2 = 6.75$ in Guided Exercise 6. The degrees of freedom for this test are $d.f. = 2$. Therefore, we look in the row headed by $d.f. = 2$ in Table 6. Notice that the sample statistic $\chi^2 = 6.75$ falls between the row entries 5.99 and 7.38. Therefore, the *P* value falls between the corresponding column headers 0.050 and 0.025.

$$0.025 < P \text{ value} < 0.050$$

This means that we reject H_0 for all $\alpha \geq 0.050$. In particular, we reject H_0 for $\alpha = 0.05$, but we fail to reject H_0 for $\alpha = 0.01$. This result is consistent with the test conclusion in Guided Exercise 6.

VIEWPOINT

Run! Run! Run!

What description would you use for marathon runners? How about age distribution? Body weight? Length of stride? Heart rate? Blood pressure? What countries do these runners come from? What are their best running times? Make your own estimated distribution for these variables, and then consider a goodness-of-fit test for your distribution compared with available data. For a good start, see these Web sites: site <http://www.honolulumarathon.org/> and follow the links to the results of the most recent races, and site <http://www.runnersworld.com/stats/> and look up progression records for men and women.

SECTION 11.2 PROBLEMS

For each of the problems, please do the following:
(a) State the null and alternate hypotheses.
(b) Find the value of the chi-square statistic from the sample.
(c) Find the degrees of freedom and the appropriate critical chi-square value.
(d) Sketch the critical region and locate your sample chi-square value and critical chi-square value on the sketch.
(e) Decide whether you should reject or fail to reject the null hypothesis.

Census: Age Distribution

1. The age distribution of the Canadian population and the age distribution of a random sample of 455 residents in the Indian community of Red Lake (Northwest Territories) are shown on page 554 (based on *U.S. Bureau of the Census, International Data Base*).

Age (years)	Percent of Canadian Population	Observed Number in Red Lake Village
Under 5	7.2%	47
5 to 14	13.6%	75
15 to 64	67.1%	288
65 and older	12.1%	45

Use a 5% level of significance to test the claim that the age distribution of the general Canadian population fits the age distribution of Red Lake Village.

Sociology: Distribution of Households

2. The distribution of types of households for the U.S. population and a random sample of 411 households in the community of Dove Creek, Montana are shown below (based on *Statistical Abstract of the United States*).

Type of Household	Percent of U.S. Households	Observed Number of Households in Dove Creek
Married, with children	26%	102
Married, no children	29%	112
Single parent	9%	33
One person	25%	96
Other (e.g., roommates, siblings)	11%	68

Use a 5% level of significance to test the claim that the distribution of U.S. households fits the Dove Creek distribution.

Archaeology: Stone Tools

3. The types of raw material used to construct stone tools found at the archaeological site Casa del Rito are shown below (*Bandelier Archaeological Excavation Project*, edited by Kohler and Root). A random sample of 1486 stone tools was obtained from a current excavation site.

Raw Material	Regional Percent of Stone Tools	Observed Number of Tools at Current Excavation Site
Basalt	61.3%	906
Obsidian	10.6%	162
Welded tuff	11.4%	168
Pedernal chert	13.1%	197
Other	3.6%	53

Use a 1% level of significance to test the claim that the regional distribution of raw materials fits the distribution at the current excavation site.

Ecology: Deer Browse

4. The types of browse favored by deer are shown in the table below (*The Mule Deer of Mesa Verde National Park*, edited by Mierau and Schmidt). Using binoculars, volunteers observed feeding habits of a random sample of 320 deer.

Type of Browse	Plant Composition in Study Area	Observed Number of Deer Feeding on This Plant
Sage brush	32%	102
Rabbit brush	38.7%	125
Salt brush	12%	43
Service berry	9.3%	27
Other	8%	23

Use a 5% level of significance to test the claim that the natural distribution of browse fits the deer feeding pattern.

Climate: Normal Distribution

5. This problem is based on information from the *National Oceanic and Atmospheric Administration (NOAA) Environmental Data Service*. Let x be a random variable that represents the average daily temperature (degrees Fahrenheit) in July in the town of Kit Carson, Colorado. The x distribution has a mean μ of approximately 75°F and a standard deviation σ of approximately 8°F. A 20-year study (620 July days) gave the entries in the right column of the following table.

I Region under Normal Curve	II x°F	III Expected % from Normal Curve	IV Observed Number of Days in 20 Years
$\mu - 3\sigma \leq x < \mu - 2\sigma$	$51 \leq x < 59$	2.15%	16
$\mu - 2\sigma \leq x < \mu - \sigma$	$59 \leq x < 67$	13.6%	78
$\mu - \sigma \leq x < \mu$	$67 \leq x < 75$	34.1%	212
$\mu \leq x < \mu + \sigma$	$75 \leq x < 83$	34.1%	221
$\mu + \sigma \leq x < \mu + 2\sigma$	$83 \leq x < 91$	13.6%	81
$\mu + 2\sigma \leq x < \mu + 3\sigma$	$91 \leq x < 99$	2.15%	12

(a) Remember that $\mu = 75$ and $\sigma = 8$. Examine Figure 6-5 in Chapter 6 (page 261). Then write a brief explanation for columns I, II, and III in the context of this problem. Note that because of rounding, column III does not total 100%.

(b) Use a 1% level of significance to test the claim that the average daily July temperature follows a normal distribution with $\mu = 75$ and $\sigma = 8$.

Climate: Normal Distribution

6. Let x be a random variable that represents the average daily temperature (degrees Fahrenheit) in January at the town of Hana, Maui (Hawaii). The x variable has a mean μ of approximately 68°F and a standard deviation σ of approximately 4°F (see reference in Problem 5). A 20-year study (620 January days) gave the entries in the right column of the table on page 556.

I	II	III	IV
Region under Normal Curve	$x°F$	Expected % from Normal Curve	Observed Number of Days in 20 Years
$\mu - 3\sigma \leq x < \mu - 2\sigma$	$56 \leq x < 60$	2.15%	14
$\mu - 2\sigma \leq x < \mu - \sigma$	$60 \leq x < 64$	13.6%	86
$\mu - \sigma \leq x < \mu$	$64 \leq x < 68$	34.1%	207
$\mu \leq x < \mu + \sigma$	$68 \leq x < 72$	34.1%	215
$\mu + \sigma \leq x < \mu + 2\sigma$	$72 \leq x < 76$	13.6%	83
$\mu + 2\sigma \leq x < \mu + 3\sigma$	$76 \leq x < 80$	2.15%	15

(a) Remember that $\mu = 68$ and $\sigma = 4$. Examine Figure 6-5 in Chapter 6. Then write a brief explanation for columns I, II, and III in the context of this problem. Note that because of rounding, column III does not total 100%.

(b) Use a 1% level of significance to test the claim that the average daily January temperature follows a normal distribution with $\mu = 68$ and $\sigma = 4$.

Game and Fish Department: Population Distribution

7. The Fish and Game Department stocked Lake Lulu with fish in the following proportions: 30% catfish, 15% bass, 40% bluegill, and 15% pike. Five years later they sampled the lake to see if the distribution of fish had changed. They found the 500 fish in the sample were distributed as follows:

Catfish	Bass	Bluegill	Pike
120	85	220	75

In the 5-year interval, did the distribution of fish change at the 0.05 level?

Library: Book Distribution by Subject

8. The director of library services at Fairmont College did a survey of types of books (by subject) in the circulation library. Then she used library records to take a random sample of 4217 books checked out last term and classified the books in the sample by subject. The results are shown below.

Subject Area	Percent of Books in Circulation Library on This Subject	Number of Books in Sample on This Subject
Business	32%	1210
Humanities	25%	956
Natural science	20%	940
Social science	15%	814
All other subjects	8%	297

Using a 5% level of significance, test the claim that the subject distribution of books in the library fits the distribution of books checked out by students.

Census: Ethnic Origin

9. The accuracy of a census report on a city in southern California was questioned by some government officials. A random sample of 1215 people living in the city was used to check the report, and the results are shown here.

Ethnic Origin	Census Percent	Sample Result
Black	10%	127
Asian	3%	40
Anglo	38%	480
Spanish American	41%	502
Native American	6%	56
All others	2%	10

Using a 1% level of significance, test the claim that census distribution and sample distribution agree.

Market Survey: Age of Customer Distribution

10. Snoop Incorporated is a firm that does market surveys. The Rollum Sound Company hired Snoop to study the age distribution of people who buy compact discs. To check the Snoop report, Rollum used a random sample of 519 customers and obtained the following data.

Customer Age (years)	Percent of Customers from Snoop Report	Number of Customers in Sample
Less than 14	12%	88
14–18	29%	135
19–23	11%	52
24–28	10%	40
29–33	14%	76
More than 33	24%	128

Using a 1% level of significance, test the claim that the distribution of customer ages in the Snoop report agrees with the sample report.

Section **11.3**

Testing σ^2

Testing a Single Variance or Standard Deviation

Many problems arise that require us to make decisions about variability. In this section we will test hypotheses about the variance (or standard deviation) of a population. It is customary to talk about variance instead of standard deviation because our techniques employ the sample variance rather than the standard deviation. Of course, the standard deviation is just the square root of the variance, so any discussion about variance is easily converted to a similar discussion about standard deviation.

Let us consider a specific example in which we might wish to test a hypothesis about the variance. Almost everyone has had to wait in line. In a grocery store, bank, post office, or registration center, there are usually several check-out or service areas. Frequently, each service area has its own independent line. However, many businesses and government offices are adopting a "single-line" procedure.

In a single-line procedure there is only one waiting line for everyone. As any service area becomes available, the next person in line gets served. The old independent-lines procedure has a line at each service center. An incoming customer simply picks the shortest line and hopes it will move quickly. In either procedure, the number of clerks and the rate at which they work is the same, so the average waiting time is the *same*. What is the advantage of the single-line procedure? The difference is in the *attitudes* of people who wait in the lines. A lengthy waiting line will be more acceptable if the variability of waiting times is smaller, even though the average waiting time is the same. When the variability is small, the inconvenience of waiting (although it might not be reduced) does become more predictable. This means impatience is reduced and people are happier.

To test the hypothesis that variability is less in a single-line process, we use the chi-square distribution. The next theorem tells us how to use the sample and population variance to compute values of χ^2.

Theorem 11.1

If we have a normal population with variance σ^2 and a random sample of n measurements is taken from this population with sample variance s^2, then

$$\chi^2 = \frac{(n-1)s^2}{\sigma^2}$$

has a chi-square distribution with degrees of freedom $d.f. = n - 1$.

Finding Critical Values

Recall that the chi-square distribution is *not* symmetrical and that there are different chi-square distributions for different degrees of freedom. Table 6 in Appendix I gives chi-square values for which the area α is to the *right* of the given chi-square value.

EXAMPLE 2 ❯ (a) Find the χ^2 value such that the area to the right of χ^2 is 0.05 when $d.f. = 10$.

SOLUTION: Since the area to the *right* of χ^2 is to be 0.05, we look in the $\alpha = 0.050$ column and the row with $d.f. = 10$. We find $\chi^2 = 18.31$ (see Figure 11-6a).

(b) Find the χ^2 value such that the area to the *left* of χ^2 is 0.05 when $d.f. = 10$.

SOLUTION: When the area to the left of χ^2 is 0.05, the corresponding area to the *right* is $1 - 0.05 = 0.95$, so we look in the $\alpha = 0.950$ column and the row with $d.f. = 10$. We find $\chi^2 = 3.94$ (see Figure 11-6b). ●

FIGURE 11-6

χ^2 Distribution with $d.f. = 10$

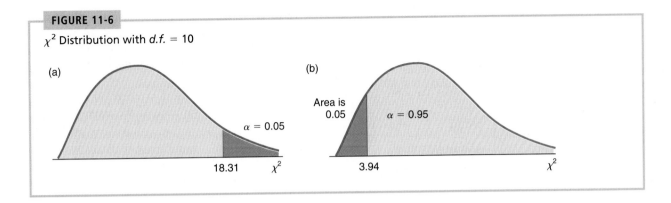

(a) $\alpha = 0.05$ at 18.31 χ^2

(b) Area is 0.05, $\alpha = 0.95$, 3.94 χ^2

Now let's use Theorem 11.1 and our knowledge of the chi-square distribution to determine if a single-line procedure has less variance of waiting times than independent lines.

EXAMPLE 3 ❯

A large discount hardware store in San Antonio had been using the independent-lines procedure to check out customers. After long observation, the manager knew that the standard deviation of waiting times was 7 minutes. The manager decided to introduce the single-line procedure on a trial basis to see if a reduction in waiting time variability would occur. A random sample of 25 customers was monitored, and their waiting times for check-out were determined. The sample standard deviation was $s = 4$ minutes. We will use a 5% level of significance to test the claim that the variance of waiting times has been reduced.

As a null hypothesis, we assume that the variance in waiting times is the same as that of the former independent-lines procedure. The alternate hypothesis is that the variance for the single-line procedure is less than that for the independent lines. If we let σ be the standard deviation of waiting times for the single-line procedure, then σ^2 is the variance, and we have

$H_0: \sigma^2 = 49$ (use $7^2 = 49$)

$H_1: \sigma^2 < 49$

We use the chi-square distribution to test the hypotheses. Assuming that the waiting times are normally distributed, we compute our observed value of χ^2 by using Theorem 11.1. Since

$n = 25$

$s = 4$ so $s^2 = 16$ (observed)

$\sigma = 7$ so $\sigma^2 = 49$ (from $H_0: \sigma^2 = 49$)

$\chi^2 = \dfrac{(n-1)s^2}{\sigma^2} = \dfrac{(25-1)16}{49} \approx 7.8$ (by Theorem 11.1)

The critical value is obtained from Table 6 in Appendix I using the degrees of freedom and level of significance. By the alternate hypothesis H_1: $\sigma^2 < 49$, we want a *left* tail with area 0.05. Therefore, the area in the corresponding right tail is $\alpha = 1 - 0.05 = 0.95$. Since $d.f. = n - 1 = 25 - 1 = 24$, the desired critical value is $\chi^2_{0.95} = 13.85$ (see Figure 11-7).

Since the observed value of $\chi^2 = 7.8$ is in the critical or rejection region, we reject H_0: $\sigma^2 = 49$ and accept H_1: $\sigma^2 < 49$. The variance of the single-line procedure is less than 49.

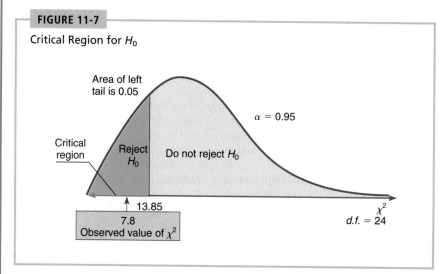

FIGURE 11-7

Critical Region for H_0

GUIDED EXERCISE   7

Certain industrial machines require overhaul when wear on their parts introduces too much variability to pass inspection. A government official is visiting a dentist's office to inspect the operation of an x-ray machine. If the machine emits too little radiation, clear photographs cannot be obtained. However, too much radiation can be harmful to the patient. Government regulations specify an average emission of 60 millirads with standard deviation σ of 12 millirads, and the machine has been set for these readings. After examining the machine, the inspector is satisfied that the average emission is still 60 millirads. However, there is wear on certain mechanical parts. To test variability, the inspector takes a random sample of 30 x-ray emissions and finds the sample standard deviation to be $s = 15$ millirads. Does this support the claim that the variance is too high (i.e., the machine should be overhauled)? Use a 1% level of significance.

Let σ be the (population) standard deviation of emissions (in millirads) of the machine in its present condition.

Exercise continues

Exercise continued

(a) Which of the following shall we use for the null hypothesis? Explain.

$H_0: \sigma^2 = 12$ $H_0: \sigma^2 = 144$ $H_0: \sigma^2 > 144$

⇨ $H_0': \sigma^2 = 144$. We use $\sigma = 12$ and the initial claim that the variance is still what it should be according to specifications.

(b) Which of the following shall we use for the alternate hypothesis? Explain.

$H_1: \sigma^2 > 12$ $H_1: \sigma^2 \neq 144$ $H_1: \sigma^2 > 144$

⇨ $H_1: \sigma^2 > 144$. We want to test the claim that the variance is too large.

(c) What is the observed value of χ^2?

⇨ $\chi^2 = \dfrac{(n-1)s^2}{\sigma^2} = \dfrac{(30-1)15^2}{144} \approx 45.3$

(since $n = 30$, $s = 15$ and by H_0, $\sigma^2 = 144$)

(d) What are the degrees of freedom? Are we using a left-, right-, or two-tailed test? Use Table 6 in Appendix I to find the chi-square critical value.

⇨ $d.f. = n - 1 = 30 - 1 = 29$. Since H_1 is $\sigma^2 > 144$, we use a right-tailed test. The problem calls for $\alpha = 0.01$, and an area of 0.01 is to the right of $\chi^2 = 49.59$ when $d.f. = 29$. The critical value is $\chi^2_{0.01} = 49.59$.

(e) Sketch the critical region on a chi-square curve. Locate the observed chi-square value on the sketch.

⇨ **FIGURE 11-8** Critical Region with $d.f. = 29$

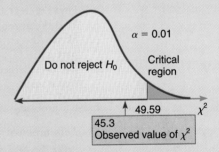

(f) Do we reject or fail to reject H_0? Should the inspector recommend that the machine be overhauled?

⇨ Since the observed chi-square value 45.3 is not in the critical region, we fail to reject H_0 and conclude that the machine does not need an overhaul at this time.

VIEWPOINT

Adoption—A Good Choice!

Cuckoos are birds that are known to lay their eggs in the nests of other (host) birds. The host birds then hatch the eggs and adopt the cuckoo chicks as their own. Birds such as the meadow pipit, tree pipit, hedge sparrow, robin, and wren have all played host to cuckoo eggs and adopted their chicks. L. H. C. Tippett (1902–1985) was a pioneer in the field of statistical quality control who collected data on cuckoo eggs found in the nests of other birds. For the data and possible tests of variances using the data, see Web site <http://lib.stat.cmu.edu/DASL/>. Find Biology under Data Subjects, and then select the Cuckoo Egg Length Data file.

SECTION 11.3 PROBLEMS

In each of the problems, please do the following:
(a) State the null and alternate hypotheses.
(b) Find the degrees of freedom and appropriate critical value or critical values.
(c) Find the appropriate chi-square value using the sample standard deviation.
(d) Sketch the critical region and show the critical chi-square value and the value in part c.
(e) Decide whether to reject or fail to reject the null hypothesis at the given level of significance.

In each problem, assume a normal population distribution.

Archaeology:
Significant Discoveries

1. This problem is based on information from *Archaeological Surveys of Chaco Canyon, New Mexico*, by A. Hayes, D. Brugge, and W. Judge, University of New Mexico Press. A *transect* is an archaeological study area that is 1/5 mile wide and 1 mile long. A *site* in a transect is the location of a significant archaeological find. Let x represent the number of sites per transect. In a section of Chaco Canyon, a large number of transects showed that x has a population variance $\sigma^2 = 42.3$. In a different section of Chaco Canyon, a random sample of 23 transects gave a sample variance $s^2 = 46.1$ for the number of sites per transect. Use a 5% level of significance to test the claim that the variance in the new section is greater than 42.3.

Sociology:
Age at First Marriage

2. This problem is based on information from an article by N. Keyfitz in *The American Journal of Sociology* (Vol. 53, pp. 470–480). Let x be the age in years of a rural Quebec woman at the time of her first marriage. In the year 1941, the population variance of x was approximately $\sigma^2 = 5.1$. Suppose that a recent study of age of first marriage for a random sample of 41 women in rural Quebec gave a

sample variance $s^2 = 3.3$. Use a 5% level of significance to test the claim that the current variance is less than 5.1.

Mountaineering: Accidents

3. This problem is based on information taken from *Accidents in North American Mountaineering* (jointly published by The American Alpine Club and The Alpine Club of Canada). Let x represent the number of mountain climbers killed each year. The long-term variance of x is approximately $\sigma^2 = 136.2$. Suppose that for the past 8 years the variance has been $s^2 = 115.1$. Use a 1% level of significance to test the claim that the recent variance for number of mountain-climber deaths is less than 136.2.

Academics: Salaries

4. This problem is based on information taken from *Academe, Bulletin of the American Association of University Professors*. Let x represent the average annual salary of college and university professors (in thousands of dollars) in the United States. For all colleges and universities in the United States, the population variance of x is approximately $\sigma^2 = 47.1$. However, a random sample of 15 colleges and universities in Kansas showed that x has a sample variance $s^2 = 83.2$. Use a 5% level of significance to test the claim that the variance for colleges and universities in Kansas is greater than 47.1.

Health: Typhoid Shots

5. A new kind of typhoid shot is being developed by a medical research team. The old typhoid shot was known to protect the population for a mean of 36 months with a standard deviation of 3 months. To test the variability of the new shot, a random sample of 24 people were given the new shot. Regular blood tests showed that the sample standard deviation of protection times was 1.9 months. Using a 0.05 level of significance, test the claim that the new typhoid shot has a smaller variance of protection times.

Veterinarian: Tranquilizer Shots

6. Jim Mead is a veterinarian who visits a Vermont farm to examine prize bulls. In order to examine a bull, Jim first gives the animal a tranquilizer shot. The effect of the shot is supposed to last an average of 65 minutes, and it usually does. However, Jim sometimes gets chased out of the pasture by a bull that recovers too soon, and at other times he becomes worried about prize bulls that take too long to recover. By reading his journals, Jim found that the tranquilizer should have a mean duration of 65 minutes with standard deviation of 15 minutes. A random sample of 10 of Jim's bulls had a mean tranquilized duration time of close to 65 minutes but a standard deviation of 24 minutes. At the 0.01 level of significance, is Jim justified in the claim that the variance is larger than that stated in his journal?

Jet Engines: Safety Regulations

7. The fan blades on commercial jet engines must be replaced when wear on these parts indicates too much variability to pass inspection. If a single fan blade broke during operation, it could severely endanger a flight. A large engine contains thousands of fan blades, and safety regulations require that variability measurements on the population of all blades not exceed $\sigma^2 = 0.15$ mm^2. An engine inspector took a random sample of 61 fan blades from an engine. She measured each blade and found a sample variance of 0.27 mm^2. Using a 0.01 level of significance, is the inspector justified in claiming that all the engine fan blades must be replaced?

PART II: INFERENCES RELATING TO LINEAR REGRESSION

Section **11.4** **Linear Regression: Confidence Intervals for Predictions**

In Chapter 3 we studied linear regression models. Given a collection of ordered pairs (x, y) we found the equation of the line that best fit the data using the "least-squares" criterion. Then we used the equation of the line to "predict" y values for specific x values.

Review of linear regression

In the next example we'll review the process we used to generate the equation for the least-squares line. Recall the formulas from Chapter 3.

Least-Squares Line

The equation of the least-squares line is

$$y = a + bx$$

where $b = \dfrac{SS_{xy}}{SS_x}$ is the slope and $a = \bar{y} - b\bar{x}$ is the y-intercept

$\bar{y}$ = mean of y values in the data set

$\bar{x}$ = mean of x values in the data set

$$SS_{xy} = \Sigma xy - \frac{(\Sigma x)(\Sigma y)}{n}$$

$$SS_x = \Sigma x^2 - \frac{(\Sigma x)^2}{n}$$

n = number of points in the data set

EXAMPLE 4 ❯

Suppose you want to buy a new pickup truck. To be specific, let us say you are considering a Ford Ranger and the model you are interested in has a list price of $14,000. As the salesperson at the dealership approaches, you wonder how far you can bargain the price down without any consideration of a possible trade-in. What is a reasonable estimate for the "best price" you can get? To answer this question, we will look at data from *Consumers Digest*. For a random sample of seven Ford Rangers, the list price is given together with the "best" price negotiated by staff members of *Consumers Digest*. The prices are given in units of one thousand dollars. Therefore, the data value 9.8 represents $9800.

x: List price	9.8	11.6	12.3	13.7	16.2	17.7	18.7
y: Best price	8.4	10.4	10.9	12.6	14.4	15.1	16.5

We will generate the least-squares line for these data, and then use the line to predict the best price for the model you want.

(a) Draw a scatter diagram for the data. Figure 11-9 on page 566 shows the scatter diagram as well as the least-squares line (figure generated by Minitab).

(b) Next we determine the equation of the least-squares line. If you are using a calculator that supports two-variable statistics, you may enter the data and generate the equation of the least-squares line directly. Otherwise, use the formulas and the sums shown in the computation table (Table 11-13 on page 566). The column showing y^2 and the sum Σy^2 will be used later in this section.

Using the sums of the computation table, we find

$$\bar{x} = \frac{100}{7} \approx 14.286 \text{ and } \bar{y} = \frac{88.3}{7} \approx 12.614$$

$$SS_{xy} = \Sigma xy - \frac{(\Sigma x)(\Sigma y)}{n} = 1318.75 - \frac{(100)(88.3)}{7} \approx 57.321$$

$$SS_x = \Sigma x^2 - \frac{(\Sigma x)^2}{n} = 1495 - \frac{100^2}{7} \approx 66.429$$

$$b = \frac{SS_{xy}}{SS_x} \approx \frac{57.321}{66.429} \approx 0.863$$

$$a = \bar{y} - b\bar{x} \approx 12.614 - 0.863(14.286) \approx 0.285$$

The equation of the least-squares line is

$$y = 0.285 + 0.863x$$

Note: Because of rounding, the equation for the least-squares line that we obtain by using formulas and a computation table might differ slightly from the least-squares equation produced by a computer or calculator.

(c) Now use the regression equation to produce the "best" price for a Ford Ranger with list price of 14 thousand dollars.

We use 14 in place of x in the regression formula. Recall that we use the symbol y_p to represent the predicted value.

$$y = 0.285 + 0.863x$$

$$y_p = 0.285 + 0.863(14) \approx 12.4 \text{ thousand dollars}$$

If your negotiation skills are good, your best price for the truck is predicted to be 12.4 thousand dollars.

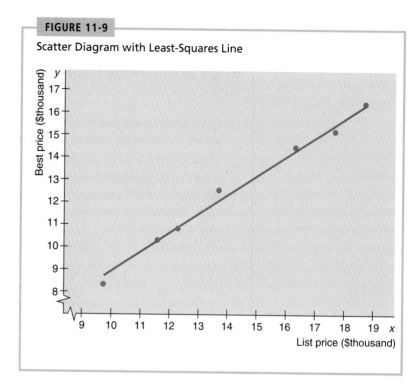

FIGURE 11-9

Scatter Diagram with Least-Squares Line

Best price ($thousand) vs List price ($thousand)

Table 11-13 Computation Table

Pair	x	y	x^2	xy	y^2
1	9.8	8.4	96.04	82.32	70.56
2	11.6	10.4	134.56	120.64	108.16
3	12.3	10.9	151.29	134.07	118.81
4	13.7	12.6	187.69	172.62	158.76
5	16.2	14.4	262.44	233.28	207.36
6	17.7	15.1	313.29	267.27	228.01
7	18.7	16.5	349.69	308.55	272.25
	$\Sigma x = 100$	$\Sigma y = 88.3$	$\Sigma x^2 = 1495$	$\Sigma xy = 1318.75$	$\Sigma y^2 = 1163.91$

Standard error of estimate

Recall from Section 3.2 that the linear-regression equation satisfies the *least-squares criteria*. That is, the equation of the line is such that the sum of the squares of the vertical distances from the data points to the line is as small as possible. Looking at Figure 11-10, we see that the vertical distance from a data point (x, y) to the line is

$$y - y_p$$

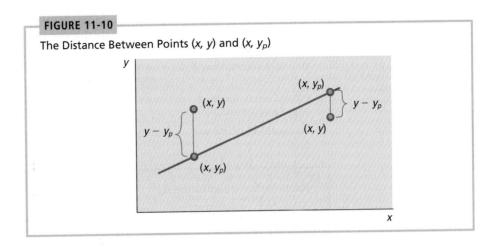

FIGURE 11-10

The Distance Between Points (x, y) and (x, y_p)

where y_p is the value produced by using the same x value in the equation of the least-squares line. The value $y - y_p$ is known as the *residual.*

Residual

The least-squares line gives more reliable predictions when the data points lie "close" to the least-squares line—that is, when the residuals are small. The *standard error of estimate* S_e is a measurement that incorporates all the residuals. Predictions based on the least-squares line are better when S_e is smaller.

Let's look at the formula for the standard error of estimate S_e.

$$\text{Standard error of estimate} = S_e = \sqrt{\frac{\Sigma(y - y_p)^2}{n - 2}} \tag{1}$$

where $n \geq 3$

Note: To compute the standard error of estimate, we require that there be at least three points on the scatter diagram. If we had only two points, the line would be a perfect fit, since two points determine a line. In such a case, there would be no need to compute S_e.

The nearer the scatter points lie to the least-squares line, the smaller S_e will be. In fact, if $S_e = 0$, it follows that each $y - y_p$ is also zero. This means that all the scatter points lie *on* the least-squares line if $S_e = 0$. The larger S_e becomes, the more scattered the points are.

The formula for the standard error of estimate is reminiscent of the formula for the standard deviation. It too is a measure of dispersion. However, the standard deviation involves differences of data values from a mean, whereas the standard error of estimate involves the differences between experimental and predicted y values for a given x.

The actual computation of S_e using Equation (1) is quite long because the formula requires us to use the least-square line equation to compute a predicted value y_p for *each* x value in the data pairs. There is a computation formula that we strongly recommend that you use. However, as with all the computation formulas, be careful about rounding. This formula is sensitive to rounding, and you should carry as many digits as seems reasonable for your problem. Answers will vary depending on rounding used. We give the formula here and follow it with an example of its use.

Formula for Calculating S_e

$$S_e = \sqrt{\frac{SS_y - b SS_{xy}}{n - 2}} \tag{2}$$

where

$$SS_y = \Sigma y^2 - \frac{(\Sigma y)^2}{n}$$

$$SS_{xy} = \Sigma xy - \frac{(\Sigma x)(\Sigma y)}{n}$$

$$SS_x = \Sigma x^2 - \frac{(\Sigma x)^2}{n}$$

$$b = \frac{SS_{xy}}{SS_x}$$

n = number of points in scatter diagram

Use caution in rounding.

With a considerable amount of algebra, Equations (1) and (2) can be shown to be mathematically equivalent. Equation (1) shows the strong similarity between the standard error of estimate and standard deviation. Equation (2) is a shortcut calculation formula because it involves few subtractions and uses quantities (SS_x, b, and SS_{xy}) that are also used to determine the equation for the least-squares line.

In the next example we show you how to compute the standard error of estimate using the computation formula. Then, in the following example and guided exercise, we will show you how to use S_e to create confidence intervals for the y value corresponding to a given x value.

EXAMPLE 5 ▷ Let's find the standard error of estimate for the list price and "best" price of a Ford Ranger truck using the data in Example 4.

SOLUTION: The computation formula for S_e uses the values of b, SS_{xy}, and SS_y. In Example 4 we found that $b \approx 0.863$ and $SS_{xy} \approx 57.321$. We need to compute the value of SS_y. From Table 11-13 in Example 4, we see that $\Sigma y = 88.3$ and $\Sigma y^2 = 1163.91$. Therefore

$$SS_y = \Sigma y^2 - \frac{(\Sigma y)^2}{n} = 1163.91 - \frac{88.3^2}{7} \approx 50.069$$

Next we use these values to compute S_e.

$$S_e = \sqrt{\frac{SS_y - bSS_{xy}}{n-2}}$$

$$= \sqrt{\frac{50.069 - (0.863)(57.321)}{5}}$$

$$\approx 0.347$$

Confidence intervals for *y*

The least-squares line gives us a predicted value y_p for a specified x value. However, we used sample data to get the equation of the line. The line derived from the population of all data pairs is likely to have a slightly different slope, which we designate by the symbol β for population slope, and a slightly different y intercept, which we designate by the symbol α for population intercept. In addition, there is some random error ϵ, so the true y value would be

$$y = \alpha + \beta x + \epsilon$$

Because of the random variable ϵ, for each x value there is a corresponding distribution of y values. The methods of linear regression were developed so that the distribution of y values for a given x could be centered on the population regression line. Furthermore, the distributions of y values corresponding to the various x values all have the same standard deviation, which we estimate by the standard error of estimate S_e.

Using all this background, the theory tells us that for a specific x, a c confidence interval for y is given by the next formula.

---

c Confidence Interval for y

$$y_p - E \le y \le y_p + E$$

where $E = t_c S_e \sqrt{1 + \frac{1}{n} + \frac{(x - \bar{x})^2}{SS_x}}$

y_p = the predicted value of y from the least-squares line for the specified x value

t_c = the value from the Student's t distribution for a c confidence level using $n - 2$ degrees of freedom

S_e = the standard error of estimate [see Equation (2)]

$SS_x = \Sigma x^2 - \frac{(\Sigma x)^2}{n}$

n = number of data pairs

The formulas involved in the computation of a c confidence interval look complicated. However, they involve quantities we have already computed or values we can easily look up in tables. The next example illustrates this point.

EXAMPLE 6 ▸

Using the data in Table 11-13, find an 85% confidence interval for the "best" price of a Ford Ranger having a list price of 14 thousand dollars.

SOLUTION: First we need to find y_p. This was done in Example 4 part c, where we used the value 14 in place of x in the equation of the least-squares line. We obtained

$$y_p = 0.285 + 0.863x \approx 0.285 + 0.863(14) \approx 12.4$$

An 85% confidence interval for the predicted y value is then

$$12.4 - E \le y \le 12.4 + E$$

where $E = t_{0.85}S_e\sqrt{1 + \dfrac{1}{n} + \dfrac{(x - \bar{x})^2}{SS_x}}$

Using $n - 2 = 7 - 2 = 5$ degrees of freedom, we find from Table 5 in Appendix I that $t_{0.85} = 1.699$. We computed S_e in Example 5 and SS_x and $\bar{x}$ in Example 4. Using all these values, we find

$$E = (1.699)(0.347)\sqrt{1 + \dfrac{1}{7} + \dfrac{(14 - 14.286)^2}{66.429}}$$

$$\approx 0.6 \text{ (thousand)}$$

An 85% confidence interval for y is

$$12.4 - 0.6 \le y \le 12.4 + 0.6$$
$$11.8 \le y \le 13.0 \text{ (in thousands)}$$

This means that we are 85% sure that the actual amount we will pay (assuming very good negotiation skills) for the Ford Ranger with list price of 14 thousand dollars will be between 11.8 and 13 thousand dollars.

GUIDED EXERCISE **8**

Let's use the same Ford Ranger data to compute an 85% confidence interval for the "best" price of a Ford Ranger with extra features and a list price of 18 thousand dollars.

(a) From Example 4 we have

$$y = 0.285 + 0.863x$$

Compute y_p for $x = 18$.

$\Rightarrow$

$$y_p = 0.285 + 0.863x$$

$$= 0.285 + 0.863(18)$$

$$\approx 15.8$$

Exercise continues

Exercise continued

(b) The bound E on the error of estimate is

$$E = t_c S_e \sqrt{1 + \frac{1}{n} + \frac{(x - \bar{x})^2}{SS_x}}$$

From Examples 4 and 5 we know that
$S_e = 0.347$, $SS_x = 66.429$, and $\bar{x} = 14.286$.
Look up $t_{0.85}$ and then evaluate E.

⇨

$t_{0.85} = 1.699$

$$E = (1.699)(0.347)\sqrt{1 + \frac{1}{7} + \frac{(18 - 14.286)^2}{66.429}}$$

$$\approx 0.7$$

(c) Find the 85% confidence interval

$$y_p - E \leq y \leq y_p + E$$

⇨

The confidence interval is

$$15.8 - 0.7 \leq y \leq 15.8 + 0.7$$

$$15.1 \leq y \leq 16.5$$

Caution about using x
values beyond the range
of sample x values

Using a computer to find
confidence intervals for
predictions

As we compare the results of Example 6 and Guided Exercise 8, we notice that the 85% confidence interval of y values for $x = 14$ thousand dollars is 0.6 unit above and below the least-squares line whereas the 85% confidence interval of y values for $x = 18$ thousand dollars is 0.7 unit above and below the least-squares line. This comparison reflects the general property that confidence intervals for y are narrower the nearer we are to the mean of the x values. As we move near the extremes of the x distribution, the confidence intervals for y become wider.

If we were to compute an 85% confidence interval for all x values in the range of the sample x values, the confidence-interval band would curve away from the least-squares line, as shown in Figure 11-11 on page 572.

The fact that the confidence-interval band curves away from the least-squares line as x values become farther away from $\bar{x}$ is another reason that we should *not use* the least-squares line to predict y values for x values beyond the data extremes of the sample x distribution. (Recall that in Section 3.2 we cautioned about *extrapolating* predictions for x values beyond the sample x values found in the data used for linear regression.) Both ComputerStat and Minitab software print a warning when you use an x value beyond the range of x values in the original sample data pairs.

The computations involved in finding confidence intervals for predictions are lengthy. Most statistical software packages will find the confidence intervals for you. Figure 11-12 shows the Minitab output for the list and best prices of Ford Rangers. We asked for a prediction (designated Fit in the Minitab output) for $x = 14$.

In the output we see the equation of the least-squares line, the value of the standard error of estimate (designated s), and an 85% prediction interval (under 85.0% P.I.) based on an x value of 14. The predicted value y_p is under Fit. Values differ slightly from those shown in Examples 4, 5, and 6 because of rounding.

FIGURE 11-11

Confidence Bands
for Predicted Values y_p

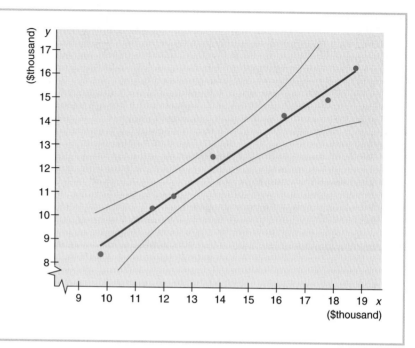

FIGURE 11-12

Minitab Output for
Examples 4, 5, and 6

Enter your data in columns. Then use the menu selections
➤ Stat ➤ Regression. Under Options, use 14 for prediction, and
select the confidence level.

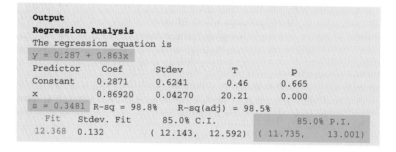

```
Output
Regression Analysis
The regression equation is
y = 0.287 + 0.863x
Predictor    Coef      Stdev       T        p
Constant    0.2871    0.6241     0.46     0.665
x           0.86920   0.04270   20.21     0.000
s = 0.3481  R-sq = 98.8%    R-sq(adj) = 98.5%
   Fit    Stdev. Fit    85.0% C.I.              85.0% P.I.
12.368    0.132      ( 12.143,  12.592)    ( 11.735,    13.001)
```

Calculator Note Although most calculators do not provide the value of the standard error of estimate S_e directly, many do provide all the sums needed to evaluate Equation (2). For example, the **2-Var Stats** on the TI-83 produces the screens shown on page 573 for the data in Example 4. Compare these results with those shown in Table 11-13. Using such results from a calculator simplifies the work of computing S_e. On the TI-83, the value for S_e is given as s under **TEST**, option **E:Lin Reg TTest**.

```
2-Var Stats
  x̄=14.28571429
  Σx=100
  Σx²=1495
  Sx=3.27375628
  σx=3.080551236
↓n=7
```

```
2-Var Stats
  ↑ȳ=12.61428571
  Σy=88.3
  Σy²=1163.91
  Sy=2.88730154
  σy=2.67444444
↓Σxy=1318.75
```

VIEWPOINT

Synoptic Climatology

Synoptic means "giving a summary from the same basic point of view." In this case, the point of view is Niwot Ridge, high above timberline in the Rocky Mountains. Vegetation, water, temperature, and wind all affect the delicate balance of this alpine environment. How do these elements of nature interact to sustain life in such a harsh land? One answer can be found by collecting data at the location and using regression to study the relationship between variables. A study of multiple regression would enable us to see the interaction of several variables. For more information, see Web site <http://culter.colorado.edu:1030/Niwot?>.

SECTION 11.4 PROBLEMS

These problems were originally presented in Section 3.2. In that section, you were asked to make a scatter diagram and find the equation of the least-squares line for each problem. Now please complete the following steps for Problems 1–8.

(a) Look at the scatter diagram again (or make a new one).

(b) Verify the given values for SS_{xy}, b, and $\bar{x}$, and the equation of the least-squares line. Remember that rounding may make your answers differ from those in the answer section.

(c) Compute SS_y and then find the value of the standard error of estimate S_e. Remember that rounding may make your answers differ from those in the answer section.

Basketball: Number of Fouls

1. Data for this problem are based on information from *STATS Basketball Scoreboard*. It is thought that basketball teams that make too many fouls in a game tend to lose the game even if they otherwise play well. Let x = number of fouls more than (i.e., over and above) the opposing team. Let y = percentage of times the team with the larger number of fouls wins the game.

x	0	2	5	6
y	50	45	33	26

$SS_{xy} \approx -89.5$ $SS_x \approx 22.75$ $b \approx -3.934$ $\bar{x} \approx 3.25$ $\bar{y} \approx 38.5$
$y = 51.286 - 3.934x$

Complete parts a through c.
(d) If a team had x = 4 fouls over and above the opposing team, what does the least-squares equation forecast for y?
(e) Find an 80% confidence interval for the forecast in part d.

Veterinary Science: Calves

2. You are the foreman of the Bar-S cattle ranch in Colorado. A neighboring ranch has calves for sale, and you are going to buy some calves to add to the Bar-S ranch herd. How much should a healthy calf weigh? Let x = age of calf (weeks) and let y = weight of calf (kg). The following information is based on data taken from *The Merck Veterinary Manual* (a reference used by many ranchers).

x	1	3	10	16	26	36
y	42	50	75	100	150	200

$SS_{xy} \approx 4181.33$ $SS_x \approx 927.33$ $b \approx 4.509$ $\bar{x} \approx 15.333$ $\bar{y} \approx 102.83$
$y = 33.696 + 4.509x$

Complete parts a through c.
(d) The calves you want to buy are 12 weeks old. What does the least-squares line predict for a healthy weight?
(e) Find a 90% confidence interval for the forecast y value in part d.

Environment: Gas Consumption

3. Do heavier cars really use more gasoline? Suppose a car is chosen at random. Let x = weight of car (hundreds of pounds), and let y = miles per gallon (mpg). The following information is based on data taken from *Consumer Reports* (Vol. 62, No. 4).

x	27	44	32	47	23	40	34	52
y	30	19	24	13	29	17	21	14

$SS_{xy} \approx -427.625$ $SS_x \approx 711.875$ $b \approx -0.601$ $\bar{x} \approx 37.375$ $\bar{y} \approx 20.875$
$y = 43.326 - 0.601x$

Complete parts a through c.
(d) Suppose a car weighs x = 38 (hundred pounds). What does the least-squares line forecast for y = miles per gallon?
(e) Find an 80% confidence interval for the forecast in part d.

Psychology: Irrelevant Responses

4. A child psychiatrist is studying the mental development of children. A random sample of nine children were asked a standard set of questions appropriate to the age

of each child. The number of irrelevant responses to the questions was recorded for each child. In the following data, x = age of child in years and y = number of irrelevant responses.

x	2	3	4	5	7	9	10	11	12
y	15	15	12	13	11	10	8	6	5

$SS_{xy} \approx -104$ $SS_x \approx 108$ $b \approx -0.963$ $\bar{x} \approx 7$ $\bar{y} \approx 10.556$
$y = 17.296 - 0.963x$

Complete parts a through c.
(d) If a child is 9.5 years old, what does the least-squares line predict for the number of irrelevant responses?
(e) Find a 99% confidence interval for the number of irrelevant responses for a child who is 9.5 years old.

Sociology: College vs. Income   5. The following data are based on information from the book *Life in America's Small Cities* (by G. S. Thomas, Prometheus Books). Let x = percentage of those 25 years or older with 4 or more years of college. Let y = per capita income in thousands of dollars. Five small cities in South Carolina (Greenwood, Hilton Head Island, Myrtle Beach, Orangeburg, and Sumpter) reported the following information regarding the x and y variables.

x	13.8	21.9	12.5	12.7	11.5
y	9.0	10.8	8.8	6.9	7.2

$SS_{xy} \approx 22.854$ $SS_x \approx 71.488$ $b \approx 0.320$ $\bar{x} \approx 14.48$ $\bar{y} \approx 8.54$
$y = 3.91 + 0.320x$

Complete parts a through c.
(d) For a small city in South Carolina where x = 20 percent of the population 25 years or older who have had 4 or more years of college, what would the least-squares equation forecast for y = per capita income (in thousands of dollars)?
(e) Find an 80% confidence interval for your forecast y value in part d.

Sociology:
High School Dropouts
vs. Income
6. Five small cities in California (El Centro, Eureka, Hanford, Madera, and San Luis Obispo–Atascadero) reported the following information. Let x = percentage of 16- to 19-year-olds not in school and not high school graduates. Let y = per capita income in thousands of dollars. The following information was obtained (see reference in Problem 5).

x	16.2	9.9	19.5	19.7	9.8
y	7.2	8.8	7.9	8.1	10.3

$SS_{xy} \approx -17.026$ $SS_x \approx 96.828$ $b \approx -0.176$ $\bar{x} \approx 15.02$ $\bar{y} \approx 8.46$
$y = 11.1 - 0.176x$

Complete parts a through c.
(d) For a small city in California where x = 17, what would the least-squares equation forecast for y = per capita income (in thousands of dollars)?
(e) Find a 75% confidence interval for the forecast y value in part d.

Archaeology:
Cultural Affiliation

7. Data for this problem are based on information taken from *Prehistoric New Mexico: Background for Survey* (by D. E. Stuart and R. P. Gauthier, University of New Mexico Press). It is thought that prehistoric Indians did not take their best tools, pottery, and household items when they visited higher elevations for their summer camps. It is hypothesized that archaeological sites tend to lose their cultural identity and specific cultural affiliation as the elevation of the site increases. Let x = elevation (in thousands of feet) for an archaeological site in the southwestern United States. Let y = percentage of unidentified artifacts (no specific cultural affiliation) at a given elevation. The following data were obtained for a collection of archaeological sites in New Mexico.

x	5.25	5.75	6.25	6.75	7.25
y	19	13	33	37	62

$SS_{xy} \approx 55$ $SS_x \approx 2.5$ $b \approx 22$ $\bar{x} \approx 6.25$ $\bar{y} \approx 32.8$
$y = -104.7 + 22x$

Complete parts a through c.
(d) At an archaeological site with elevation $x = 6.5$ (thousand feet), what does the least-squares equation forecast for y = percentage of culturally unidentified artifacts?
(e) Find a 75% confidence interval for the forecast y value in part d.

Climate: Frost-free Days

8. Data for this problem are from *Climatology Report No. 77-3* (by J. F. Benci and T. B. McKee, Department of Atmospheric Science, Colorado State University). Let x = elevation (in thousands of feet) and let y = average number of frost-free days in a year. For Denver, Gunnison, Aspen, Crested Butte, and Dillon, Colorado, the following data were obtained.

x	5.3	7.7	7.9	8.9	9.8
y	162	63	73	49	21

$SS_{xy} \approx -352.260$ $SS_x \approx 11.408$ $b \approx -30.878$ $\bar{x} \approx 7.92$ $\bar{y} \approx 73.6$
$y = 318.156 - 30.878x$

Complete parts a through c.
(d) Colorado Springs is at an elevation of $x = 6$ (thousand feet). What does the least-squares equation forecast for the average number of frost-free days per year in Colorado Springs?
(e) Find an 85% confidence interval for the forecast y value in part d.

General Discussion:
Residual Plots

9. There are several ways to assess how well a least-squares line serves as a model for the data. One method is a graphic tool called a *residual plot*. Let x be the given explanatory variable and y the corresponding response variable of the data. Let the symbol y_p represent the predicted value for x determined by the least-squares line. To make a residual plot, we put the x values in order on the horizontal axis and plot the corresponding residuals $y - y_p$ in the vertical direction. Since for a least-squares model the mean of the residuals is always zero, we dash in a horizontal line at zero. The accompanying figure shows a residual plot for the data of Guided Exercise 3 on page 135 of Section 3.2, in which the relationship between the number of ads run per week and the number of cars sold that week was explored. To make

the residual plot, first compute all the residuals. Remember that x and y are the given data values and y_p is computed from the least-squares line $y_p \approx 6.56 + 1.01x$.

Residual

x	y	y_p	$y - y_p$
6	15	12.6	2.4
20	31	26.8	4.2
0	10	6.6	3.4
14	16	20.7	−4.7
25	28	31.8	−3.8

Residual

x	y	y_p	$y - y_p$
16	20	22.7	−2.7
28	40	34.8	5.2
18	25	24.7	0.3
10	12	16.7	−4.7
8	15	14.6	0.4

Residual Plot (Produced by Minitab)

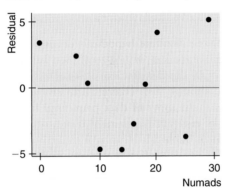

(a) If the least-squares line provides a reasonable model for the data, the pattern of points in the plot will seem random and unstructured around the horizontal line at 0. Is this the case for the residual plot?

(b) If a point on the residual plot seems far outside the pattern of other points, it might reflect an unusual data point (x, y), called an *outlier*. Such points may have quite an influence on the least-squares model. Do there appear to be any outliers in the data for the residual plot?

10. Consider the data in Problem 3.
    (a) Make a residual plot for the least-squares model.
    (b) Use the residual plot to comment about the appropriateness of the least-squares model for these data. See Problem 9.

Section **11.5**

Testing the Correlation Coefficient

A basic assumption in the study of economics is that people will spend more if they earn more. Economists claim that there is a high positive linear correlation between x = amount earned and y = amount spent. How could you test this claim? One way would be to obtain all possible (x, y) pairs for all people in the United States with an income. If you did this impossible task, you would have the *population* of all possible (x, y) pairs. You could then compute the *population correlation coefficient*, which we call ρ (the lowercase Greek letter *rho*, pronounced like "row"). If $\rho = 1$, you have a perfect positive linear correlation. If $\rho = 0$, you have no linear correlation.

The population correlation coefficient ρ

Most people would not even attempt to take the entire population of all incomes and corresponding amounts spent. Usually, we take a random sample and compute the correlation coefficient of the sample. We call this the *sample correlation coefficient r*. Recall that the sample correlation coefficient r is discussed in Section 3.3. If r is near 1, we have evidence that ρ, the population coefficient, is near 1, or at least greater than 0.

Different random samples will give different values of r. We need a test to decide when a sample value of r is far enough from zero to indicate correlation in the population.

Testing ρ

For simplicity, we will *assume that both the x and y variables are normally distributed*. To test if the (x, y) values are correlated *in the population*, we will set up the null hypothesis that they are not correlated.

> H_0: x and y are not correlated, so $\rho = 0$.

The choice of the alternate hypothesis depends on the belief that the correlation is positive, negative, or simply not zero. (See Table 11-14.)

When $\rho = 0$, the distribution of sample correlation coefficients (r values) will be symmetric about $r = 0$. Figure 11-13 shows the distribution for some values of n, where n is the number of data pairs used to compute r.

The type of test (left-tailed, right-tailed, or two-tailed) depends on the choice of H_1, as shown in Table 11-14.

To find the critical values for a test, we use Table 7 in Appendix I. The entries in Table 7 are critical values of r corresponding to given n, the number of data points, and α, the level of significance. Each critical value is listed without a sign. The choice of sign, $+$ or $-$, depends on the type of test used.

Meaning of the term "significant"

Whenever we reject H_0: $\rho = 0$, we say that our r value is *significant*. If we can not reject H_0, we say that our sample r value is *not significant*.

Table 11-14   Alternate Hypotheses

If You Think	Then Use	Type of Test
$\rho > 0$	H_0: $\rho = 0$ H_1: $\rho > 0$	Right-tailed test
$\rho < 0$	H_0: $\rho = 0$ H_1: $\rho < 0$	Left-tailed test
$\rho \neq 0$	H_0: $\rho = 0$ H_1: $\rho \neq 0$	Two-tailed test

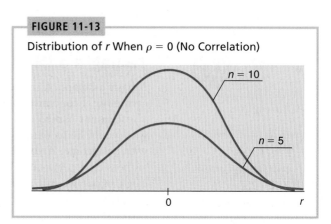

FIGURE 11-13

Distribution of r When $\rho = 0$ (No Correlation)

EXAMPLE 7 ❯ For her sociology class, Zelma interviewed a random sample of $n = 30$ married couples. For each couple, Zelma found x = number of years of formal education for the man and y = number of years of formal education for the woman. Assume that both x and y are normally distributed. After collecting the data, Zelma worked out the correlation coefficient to be $r = 0.28$. Determine whether r is significant at the 5% level of significance.

SOLUTION:

(a) Our null hypothesis is H_0: $\rho = 0$. Since Zelma had no reason to believe that ρ is either positive or negative, the alternate hypothesis is H_1: $\rho \neq 0$. Therefore, we are testing

$$H_0: \rho = 0$$
$$H_1: \rho \neq 0$$

(b) The alternate hypothesis H_1: $\rho \neq 0$ indicates a two-tailed test. Since $\alpha = 0.05$ and $n = 30$, Table 7 in Appendix I gives the critical value of 0.36. The critical region is shown in Figure 11-14.

(c) The value of the sample correlation coefficient is $r = 0.28$. This value does not fall in the critical region. Therefore, we cannot reject H_0: $\rho = 0$ (there is no correlation), and we conclude that $r = 0.28$ *is not significant* at the 0.05 level.

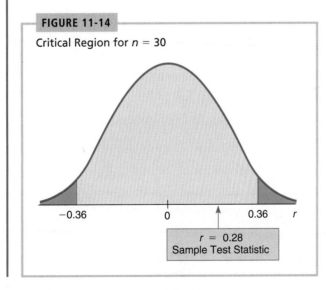

FIGURE 11-14

Critical Region for $n = 30$

-0.36 0 0.36 r

$r = 0.28$
Sample Test Statistic

Even though a significance test indicates the existence of a correlation between x and y in the population, *it does not signify a cause-and-effect relationship.* For instance, a positive correlation between x = annual income and y = age at death does *not* imply that higher income causes later death. Even after a significant correlation between variables has been established, the cause of the correlation must be identified. In addition, we also must decide if the correlation is high enough to be of practical value for the particular application in which it is to be used.

GUIDED EXERCISE 9

Big Rock Insurance Company has reason to believe that among white-collar workers, people who earn more money tend to file fewer insurance claims. The Big Rock Company took a random sample of $n = 30$ white-collar workers. For each person, x = income in dollars per year and y = number of claims filed in the last three years. The sample correlation coefficient was found to be $r = -0.33$. Determine whether r is significant at the 0.05 level of significance.

(a) Which of the following is the null hypothesis?

$H_0: \rho > 0$ $H_0: \rho = 0$
$H_0: \rho < 0$ $H_0: \rho \neq 0$

⇨ $H_0: \rho = 0$; that is, there is no correlation.

(b) Which of the following alternate hypotheses reflects the statement that people who earn more tend to file fewer insurance claims?

$H_1: \rho \neq 0$ $H_1: \rho > 0$ $H_1: \rho < 0$

⇨ The statement indicates a negative correlation between salary and number of claims filed. So we want the alternate hypothesis to claim there is a negative correlation. The appropriate alternate hypothesis is $H_1: \rho < 0$.

(c) Using the alternate hypothesis selected in part b, what kind of test should we use: right-tailed, left-tailed, or two-tailed?

⇨ Use a left-tailed test.

(d) What critical value should be used? Sketch the critical region.

⇨ The critical value is -0.31. See Figure 11-15.

FIGURE 11-15
Critical Region

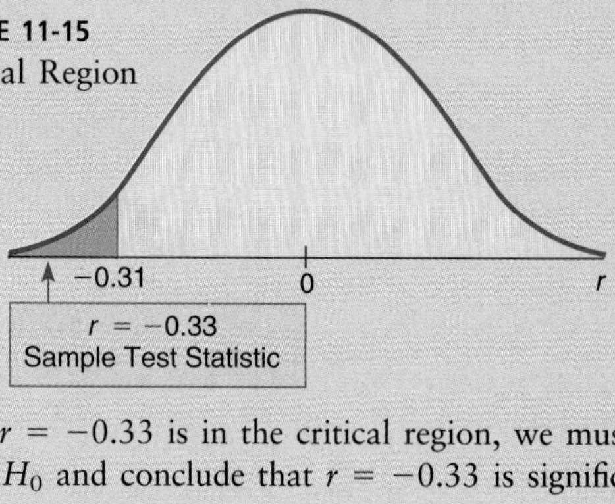

-0.31 0 r

$r = -0.33$
Sample Test Statistic

(e) Is $r = -0.33$ significant or not?

⇨ Since $r = -0.33$ is in the critical region, we must reject H_0 and conclude that $r = -0.33$ is significant.

(f) What assumptions were made about the x distribution and the y distribution?

⇨ We assumed that both x and y had normal distributions.

Problems that require calculation and testing of the correlation coefficient must use a *random sample* of data points. This statement means that both x and y are statistical variables whose numerical values are obtained only after the random sample has been drawn from the population of all possible data pairs. This procedure is somewhat different from our preceding regression (e.g., least-squares line) problems, where the x values can be chosen in advance. In this case, the y values may depend heavily on the choice of x values. The least-squares methods can be applied to random sample data or to data in which the x values are specified in advance. But the interpretation of r as a measure of the linear correlation between x and y *requires* us to use a *random sample* of data points in which both x and y are statistical variables.

P values

COMMENT *P* values for testing the correlation coefficient can be estimated from Table 7 in Appendix I. For instance, in Guided Exercise 9, the sample value of r is -0.33. The test is a left-tailed test with sample size $n = 30$. We look in the $n = 30$ row and find that the r values indicated for one-tailed (left-tailed) tests are -0.31 and -0.42. Since the sample statistic $r = -0.33$ falls between these values, the corresponding *P* value falls between the column headers $\alpha = 0.05$ and $\alpha = 0.01$.

$$0.01 < P \text{ value} < 0.05$$

Remember, we reject the null hypothesis when the *P* value is less than (or equal to) the level of significance α. Consequently, the sample r value is significant when $\alpha = 0.05$ but not significant when $\alpha = 0.01$.

V I E W P O I N T

Hazardous to Your Health!

The Federal Trade Commission (FTC) evaluates domestic cigarettes according to their tar, nicotine, and carbon monoxide contents. The United States Surgeon General considers each of these substances to be a health hazard. If a cigarette has a "richer" taste, with more tar or nicotine, does it also tend to produce more carbon monoxide? Is there a significant positive correlation? For more information, see Problem 9 in this section. Also see Web site <http://www.stat.ncsu.edu/info/jse/>, and look in the data archive for cigarette data.

SECTION 11.5 PROBLEMS

For each problem, please do the following:
(a) State the null and alternate hypotheses.
(b) Find the critical value or values.
(c) Sketch the critical region.

(d) Locate the sample correlation coefficient r on the sketch in part c.

(e) Decide whether you should reject or not reject the null hypothesis at the given level of significance.

(f) In nontechnical terms, summarize the results of the test. If the sample r value *is not significant,* what does this mean about the x and y variables in the context of this problem? If the sample r value *is significant,* what can you say about the x and y variables? In some cases do you think there might be other underlying variables that may help explain a *significant* correlation? Explain your answer.

In the following problems, assume that both x and y are approximately normally distributed.

Movies: Revenues

1. *The New York Times* did a study of the top 10 movies. Let x be the number of weeks a movie has been showing in major theaters. Let y be the total revenues in millions of dollars for that movie. The sample correlation coefficient computed from 10 data pairs printed in the *Times* was $r = 0.384$. Use a 5% level of significance to test the claim that there is a positive population correlation between the x and y variables.

Stocks: Percentage Change

2. It is sometimes claimed that if stocks show a good performance in January, they will show a good performance during the remainder of the year. For major Standard and Poor's stock groups, *USA Today* listed $x =$ January percentage change and $y =$ annual percentage change for that same major stock group. A random sample of 15 of these groups showed the sample correlation coefficient for the data pairs to be $r = 0.646$. Use a 5% level of significance to test the claim that there is a positive population correlation between the x and y variables.

Economics: Wages vs CPI

3. *The Economist* did a study of seven major free-enterprise countries as regards the variables $x =$ percentage change in wages and $y =$ percentage change in consumer prices. It was found that the sample correlation coefficient was $r = 0.918$. Use a 5% level of significance to test the claim that there is a positive population correlation between the x and y variables.

Sociology: High School Dropouts vs Death Rate

4. *Life in America's Small Cities* is a book by G. S. Thomas. In this book, 10 small cities in New York State were studied as regards the variables $x =$ percentage of high school dropouts and $y =$ death rate per 1000 residents. It was found that the sample correlation coefficient was $r = 0.737$. Use a 5% level of significance to test the claim that there is a positive population correlation between the x and y variables.

Sociology: College Grads vs Death Rate

5. In the same reference as in Problem 4, the same 10 cities in New York State were studied as regards the variables $x =$ percentage of population 25 years old and older who have had 4 or more years of college and $y =$ death rate per 1000 residents. It was found that the sample correlation coefficient was $r = -0.766$. Use a 5% level of significance to test the claim that there is a negative population correlation between the x and y variables.

Climate: Nor'easter Storms

6. Nor'easter storms have been known to cause much damage to New England coastal regions. *Weatherwise* is a magazine published in association with the American

Meteorological Society. In *Weatherwise* (Vol. 46, No. 6), a study was reported of the variables x = average peak height of ocean waves (feet) and y = average storm duration (hours). A study of five classifications of storms showed that the sample correlation coefficient for x and y data pairs was $r = 0.999$. Use a 1% level of significance to test the claim that there is a positive population correlation between the x and y variables.

Ecology: Yellowstone

7. *Yellowstone Vegetation* is a book by D. G. Despain, who is a research biologist at Yellowstone National Park. In this book, the following variables are studied: x = elevation in thousands of feet and y = April 1 water content of snowpack in inches. A sample of 12 east-side snow courses showed that the sample correlation coefficient between x and y values was $r = 0.945$. Use $\alpha = 0.01$ to test the claim that there is a positive population correlation between the x and y variables.

Ecology: Yellowstone

8. Let x be the total available water in inches (annual precipitation), and let y be the root depth in inches. A sample of seven different soil types in Yellowstone National Park showed that the sample correlation coefficient between x and y values was $r = 0.412$ (see reference in Problem 7). Use a 1% level of significance to test the claim that the population correlation between the x and y variables is different from zero.

Health: Cigarette Smoking

9. The Federal Trade Commission has published data concerning health issues and cigarette smoking. See Web site <http://www.stat.ncsu.edu/info/jse/>, and then look in the data archive for cigarette data. Let x be the tar content (in milligrams) and let y be the carbon monoxide content (in milligrams) obtained from a burning cigarette. A random sample of 12 popular brand-name cigarettes showed that the sample correlation coefficient between x and y values was $r = 0.970$. Use a 1% level of significance to test the claim that the population correlation coefficient for x and y values is positive.

---

⑤UMMARY

In this chapter we saw more applications of inferential statistics. In Part I we looked at three applications of the chi-square distribution. We used the distribution to test for independence and goodness of fit. In each of these cases we computed expected values and then used the differences between the expected values and the observed values to compute a sample χ^2 statistic. We then used the chi-square distribution table to determine if the sample χ^2 value was significant. Finally, we used the chi-square distribution to test the value of the variance of a normal distribution.

In Part II we returned to the linear-regression model introduced in Chapter 3. First we computed confidence intervals for predicted values. Then we tested the sample correlation coefficient r to see if it was significant.

MPORTANT WORDS & SYMBOLS

Section 11.1
Test of independence
Chi-square distribution, χ^2
Degrees of freedom, $d.f. = (R - 1)(C - 1)$, for χ^2
    distribution and tests of independence
Contingency table with cells
Expected frequency of a cell, E
Observed frequency of a cell, O
Critical chi-square values, χ_α^2
Row total
Column total

Section 11.2
Test of goodness-of-fit

Degrees of freedom $d.f.$ for χ^2 distribution and
    goodness-of-fit tests

Section 11.3
Hypothesis test about σ^2

Section 11.4
Standard error of estimate S_e
Confidence interval for y

Section 11.5
Population correlation coefficient ρ
Significant sample correlation
    coefficient r

VIEWPOINT

Movies and Money!

Young adults are the movie industry's best customers. However, going to the movies is expensive, which may explain why attendance rates increase with household income. Using what you have learned in this chapter, you can create appropriate chi-square tests to determine the goodness of the fit between national percentage rates of attendance by household income and attendance rates in your demographic area. For more information and national data, see *American Demographics* (Vol. 18, No. 12).

CHAPTER REVIEW PROBLEMS

Part I: Hypothesis Tests Using the Chi-Square Distribution

For each of the problems, please do the following:
(a) State the null and alternate hypotheses.
(b) Find the value of the chi-square statistic from the sample.
(c) Find the degrees of freedom and the appropriate critical chi-square value.
(d) Sketch the critical region and locate your sample chi-square value and critical chi-square value on the sketch.
(e) Decide whether you should reject or fail to reject the null hypothesis.

Academic: Grades

1. Professor Fair believes extra time does not improve grades on exams. He randomly divided a group of 300 students into two groups and gave them all the same test.

One group had exactly 1 hour in which to finish the test, and the other group could stay as long as desired. The results follow. Test at the 0.01 level of significance that time to do a test and test results are independent.

Time	Exam Grades				Row Total
	A	B	C	F	
1 h	23	42	65	12	142
Unlimited	17	48	85	8	158
Column total	40	90	150	20	300

Academic: Grades

2. Professor Stone complains that student teacher ratings depend on the grades the students received. In other words, according to Professor Stone, a teacher who gives good grades gets good ratings, and a teacher who gives bad grades gets bad ratings. To test this claim, the Student Assembly took a random sample of 300 teacher ratings on which the students' grades for the course were also indicated. The results follow. Test the hypothesis that teacher ratings and student grades are independent at the 0.01 level of significance.

Rating	Student Grades				Row Total
	A	B	C	F (or withdrawal)	
Excellent	14	18	15	3	50
Average	25	35	75	15	150
Poor	21	27	40	12	100
Column total	60	80	130	30	300

Sociology: Age Distribution

3. A sociologist is studying the age of the population in Blue Valley. Ten years ago the population was such that 20% were under 20 years old, 15% were in the 20–35-year-old bracket, 30% were between 36 and 50, 25% were between 51 and 65, and 10% were over 65. A study done this year used a random sample of 210 residents. The results follow.

Under 20	20–35	36–50	51–65	Over 65
15	25	70	80	20

At the 0.01 level of significance, has the age distribution of the population of Blue Valley changed?

Marketing: New Cars

4. In a study done 10 years ago, *Market Trends* magazine discovered that 20% of the new car buyers planned to keep their new cars more than 5 years, 30% planned to keep them between 2 and 5 years, and 50% intended to sell the cars in less than 2 years. This year the magazine did a similar study. A random sample of 200 new car buyers was taken. In this sample 48 people planned to keep their cars more than 5 years, 75 said they would keep them between 2 and 5 years, and 77 indicated that they planned to sell the cars in less than 2 years. Test the hypothesis that the present buyers plan to keep their cars the same length of time as buyers 10 years ago. Use a 0.01 level of significance.

Sciences: Solar Batteries

5. A set of solar batteries is used in a research satellite. The satellite can run on only one battery, but it runs best if more than one battery is used. The variance σ^2 of lifetimes of these batteries affects the useful lifetime of the satellite before it goes dead. If the variance is too small, all the batteries will tend to die at once. Why? If the variance is too large, the batteries are simply not dependable. Why? Engineers have determined a variance of $\sigma^2 = 15$ months (squared) is most desirable for these batteries. A random sample of 22 batteries gave a sample variance of 14.3 months (squared). Using a 0.05 level of significance, test the claim that $\sigma^2 = 15$ against the claim that σ^2 is different from 15.

Production: Corn Flakes

6. A machine that puts corn flakes in boxes is adjusted to put an average of 15 oz in each box with standard deviation of 0.25 oz. If a random sample of 12 boxes gave a sample standard deviation of 0.38 oz, do these data support the claim that the variance has increased and the machine needs to be brought back into adjustment? (Use a 0.05 level of significance.)

Part II: Inferences Relating to Linear Regression

For these problems, make a scatter plot, verify the given values, and find the requested values.

Marketing: Income vs. Sales

7. Five small cities in Kentucky (Bowling Green, Madisonville, Paducah, Radcliff-Elizabethtown, and Richmond) reported the following information about the random variables x = per capita income and y = per capita retail sales (both in thousands of dollars). (Data are based on information from *Life in America's Small Cities*, by G. S. Thomas.)

x	9.0	8.9	9.9	8.2	7.6
y	5.1	4.5	6.2	3.7	4.2

$SS_{xy} \approx 2.926$ $SS_x \approx 3.028$ $b \approx 0.966$ $\bar{x} \approx 8.720$ $\bar{y} \approx 4.74$
$y = -3.682 + 0.966x$

Note that your answers might vary slightly because of rounding.

(a) Suppose that you plan to open a retail store in a small city in Kentucky where the per capita income is $x = 9.5$ (thousand dollars). What does the least-squares equation forecast for y = per capita retail sales (in thousands of dollars)?

(b) Find the value of the standard error of estimate S_e. Find an 80% confidence interval for the forecast y value in part a.

(c) Verify that the sample correlation coefficient $r = 0.875$. Use a 5% level of significance to test if ρ is positive.

Economics: Wages vs. CPI

8. The following data are based on information taken from *The Economist* magazine. Let x = percentage change in wages and y = percentage change in consumer prices for the past year in Australia, Austria, Canada, France, Italy, Spain, and the United States.

x	3.3	4.1	1.9	2.6	3.6	6.3	2.5
y	2.2	3.5	1.9	2.1	4.0	4.9	2.7

$SS_{xy} \approx 8.759$ $SS_x \approx 12.614$ $b \approx 0.694$ $\bar{x} \approx 3.47$ $\bar{y} \approx 3.04$
$y = 0.633 + 0.694x$

Note that your answers might vary slightly because of rounding.

(a) Suppose that the percentage change in wages is $x = 5$. What does the least-squares equation forecast for y, the corresponding percentage change in consumer prices?

(b) Find the value of S_e. Find an 80% confidence interval for your forecast y value in part a.

(c) Verify that the correlation coefficient $r = 0.895$. Use a 1% level of significance to test if ρ is different from 0.

DATA HIGHLIGHTS: GROUP PROJECTS

Break into small groups and discuss the following topics. Organize a brief outline in which you summarize the main points of your group discussion.

The Statistical Abstract of the United States reported information about the percentage of arrests of all drunk drivers according to age group. In the following table, the entry 3.7 in the first row means that in the entire United States about 3.7% of all people arrested for drunk driving were in the age group 16 to 17 years. The Freemont County Sheriff's Office obtained data about the number of drunk drivers arrested in each age group over the past several years. In the following table, the entry 8 in the first row means that eight people in the age group 16 to 17 years were arrested for drunk driving in Freemont County.

Distribution of Drunk Driver Arrests by Age

Age	National Percentage	Number in Freemont County
16–17	3.7	8
18–24	18.9	35
25–29	12.9	23
30–34	10.3	19
35–39	8.5	12
40–44	7.9	14
45–49	8.0	16
50–54	7.9	13
55–59	6.8	10
60–64	5.7	9
65 and over	9.4	15
	100%	174

Use a chi-square test with 5% level of significance to test the claim that the age distribution of drunk drivers arrested in Freemont County is the same as the national age distribution of drunk drivers arrested.

(a) State the null and alternate hypotheses.
(b) Find the value of the chi-square statistic from the sample.
(c) Find the degrees of freedom and the appropriate chi-square critical value.
(d) Sketch the critical region and locate your sample chi-square value and critical chi-square value on the sketch.
(e) Decide whether you should reject or not reject the null hypothesis. State your conclusion in the context of the problem.

ⓁINKING CONCEPTS: WRITING PROJECTS

Discuss each of the following topics in class or review the topics on your own. Then write a brief but complete essay in which you summarize the main points. Please include formulas and graphs as appropriate.

Consider the results of a study conducted by the makers of Advil. A random sample of patients were assigned to one of two groups: one received ibuprofen and the other received a placebo. Then patients were asked if they experienced stomach upset (yes or no). The results of the study follow.

Group	Stomach Upset		Row Total
	Yes	No	
Ibuprofen	8	664	672
Placebo	6	645	651
Column total	14	1309	1323

1. In Chapter 4 on probability (Section 4.2) we considered contingency tables and the probabilities of events described by different combinations of the rows and columns. Use these techniques to compute the probability that a patient had stomach upset, *given* that the patient was in the ibuprofen group. Compute the probability that a patient had stomach upset, *given* that the patient was in the placebo group.

2. Are the proportions of patients having stomach upset in the two different treatment groups the same or not? We treated questions of this type in Section 10.4 where we applied the normal distribution to test the difference of proportions. Use such a test to determine whether or not the proportions are different at the 5% level of significance.

3. Now consider the 2×2 contingency table. We can apply techniques of tests of independence to see if the proportions of patients experiencing stomach upset in the two treatment groups are the same or not. When we apply these techniques to tests of proportions, we assume that the sample size for each treatment group is *assigned* ahead of time and is not itself a random variable. In such a case we apply a test of independence, but we change the hypotheses to H_0: the proportions are equal and H_1: the proportions are not equal. Perform such a test.

4. Compare the results of Problems 2 and 3. Do the test conclusions agree? When testing two proportions you can use either the normal distribution test of Section 10.4 or a chi-square test. Can you think of any advantages of one choice over the other?

The application below may be done using statistical computer software or calculators with statistical functions. Displays and suggestions are given for Minitab (Release 12), the TI-83 graphing calculator, ComputerStat, and Excel.

Tests of Independence

Minitab

In Minitab, enter data for tests of independence in columns. Be sure to enter data only and not row or column totals. Then use the menu choices ►Stat ►Tables ►Chisquare Test. An example of a typical Minitab display for tests of independence is given in Problem 9 in Section 11.1.

TI-83

Press the **Matrix** key and enter the observed values in a matrix. Name the matrix [A]. Then dimension matrix [B] to the same dimensions. Matrix [B] will hold the expected values. Press the **STAT** key and under the **TESTS** menu, select option **C: χ^2-Test. Calculate** will give the sample χ^2 value and its P value. **Draw** will show the sample test statistic on the χ^2 distribution with the P value shaded.

ComputerStat

Under the main menu item Hypothesis Testing, select Chi-Square Test for Independence. Follow the instructions on the screen to enter data and generate the test results.

Excel

There are no built-in commands for executing a chi-square test.

A study involving people who are food processors gave the following information about work shift and number of sick days.

| Shift | Number of Sick Days | | | | |
	0	1	2	3	4 or more
Day	134	44	24	10	61
Aft/ev	90	39	23	18	99
Night	107	37	21	20	82
Rotating	56	20	14	17	92

Source: United States Department of Health, Education, and Welfare, NIOSH Technical Report. Tasto, Colligan, *et al. Health Consequences of Shift Work.* Washington: GPO, 1978, 29. (*Note:* This table was adapted from Table 7 on page 25 of the source.)

Use a 1% level of significance to test the null hypothesis that work shift and number of sick days are independent against the alternate hypothesis that they are not independent. What is the sample chi-square value? What is the test conclusion?

Computer Display

ComputerStat Windows Display: Test of Goodness of Fit for
Chapter Review Problem 3

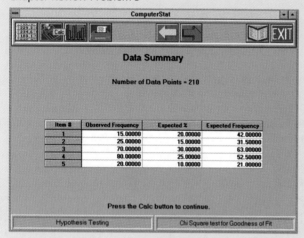

ComputerStat

Data Summary

Number of Data Points = 210

Item #	Observed Frequency	Expected %	Expected Frequency
1	15.00000	20.00000	42.00000
2	25.00000	15.00000	31.50000
3	70.00000	30.00000	63.00000
4	80.00000	25.00000	52.50000
5	20.00000	10.00000	21.00000

Press the Calc button to continue.

Hypothesis Testing	Chi Square test for Goodness of Fit

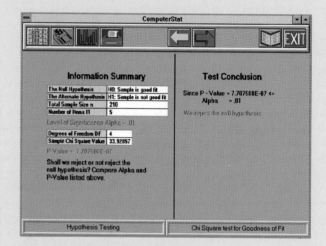

ComputerStat

Information Summary

The Null Hypothesis	H0: Sample is good fit
The Alternate Hypothesis	H1: Sample is not good fit
Total Sample Size n	210
Number of Items I1	5

Level of Significance Alpha = .01

Degrees of Freedom DF	4
Sample Chi Square Value	33.92857

P-Value = 7.707500E-07

Shall we reject or not reject the
null hypothesis? Compare Alpha and
P-Value listed above.

Test Conclusion

Since P - Value = 7.707500E-07 <=
Alpha = .01

We reject the null hypothesis

Hypothesis Testing	Chi Square test for Goodness of Fit

Appendix I Tables

1. Random Numbers
2. Binomial Coefficients $C_{n,r}$
3. Binomial Probability Distribution $C_{n,r}p^r q^{n-r}$
4. Table 4 Areas of a Standard Normal Distribution
   (a) Table of Areas from 0 to z
   (b) Confidence Interval Critical Values z_c
   (c) Hypothesis Testing Critical Values z_0

5. Student's t Distribution
6. The χ^2 Distribution
7. Critical Values of Pearson Product-Moment Correlation, r

Table 1 Random Numbers

92630	78240	19267	95457	53497	23894	37708	79862	76471	66418
79445	78735	71549	44843	26104	67318	00701	34986	66751	99723
59654	71966	27386	50004	05358	94031	29281	18544	52429	06080
31524	49587	76612	39789	13537	48086	59483	60680	84675	53014
06348	76938	90379	51392	55887	71015	09209	79157	24440	30244
28703	51709	94456	48396	73780	06436	86641	69239	57662	80181
68108	89266	94730	95761	75023	48464	65544	96583	18911	16391
99938	90704	93621	66330	33393	95261	95349	51769	91616	33238
91543	73196	34449	63513	83834	99411	58826	40456	69268	48562
42103	02781	73920	56297	72678	12249	25270	36678	21313	75767
17138	27584	25296	28387	51350	61664	37893	05363	44143	42677
28297	14280	54524	21618	95320	38174	60579	08089	94999	78460
09331	56712	51333	06289	75345	08811	82711	57392	25252	30333
31295	04204	93712	51287	05754	79396	87399	51773	33075	97061
36146	15560	27592	42089	99281	59640	15221	96079	09961	05371
29553	18432	13630	05529	02791	81017	49027	79031	50912	09399
23501	22642	63081	08191	89420	67800	55137	54707	32945	64522
57888	85846	67967	07835	11314	01545	48535	17142	08552	67457
55336	71264	88472	04334	63919	36394	11196	92470	70543	29776
10087	10072	55980	64688	68239	20461	89381	93809	00796	95945
34101	81277	66090	88872	37818	72142	67140	50785	21380	16703
53362	44940	60430	22834	14130	96593	23298	56203	92671	15925
82975	66158	84731	19436	55790	69229	28661	13675	99318	76873
54827	84673	22898	08094	14326	87038	42892	21127	30712	48489
25464	59098	27436	89421	80754	89924	19097	67737	80368	08795
67609	60214	41475	84950	40133	02546	09570	45682	50165	15609
44921	70924	61295	51137	47596	86735	35561	76649	18217	63446
33170	30972	98130	95828	49786	13301	36081	80761	33985	68621
84687	85445	06208	17654	51333	02878	35010	67578	61574	20749
71886	56450	36567	09395	96951	35507	17555	35212	69106	01679

Source: Reprinted from *A Million Random Digits with 100,000 Normal Deviates* by the Rand Corporation (New York: The Free Press, 1955). Copyright 1955 and 1983 by the Rand Corporation. Used by permission.

Table 2 Binomial Coefficients $C_{n,r}$

n \ r	0	1	2	3	4	5	6	7	8	9	10
1	1	1									
2	1	2	1								
3	1	3	3	1							
4	1	4	6	4	1						
5	1	5	10	10	5	1					
6	1	6	15	20	15	6	1				
7	1	7	21	35	35	21	7	1			
8	1	8	28	56	70	56	28	8	1		
9	1	9	36	84	126	126	84	36	9	1	
10	1	10	45	120	210	252	210	120	45	10	1
11	1	11	55	165	330	462	462	330	165	55	11
12	1	12	66	220	495	792	924	792	495	220	66
13	1	13	78	286	715	1,287	1,716	1,716	1,287	715	286
14	1	14	91	364	1,001	2,002	3,003	3,432	3,003	2,002	1,001
15	1	15	105	455	1,365	3,003	5,005	6,435	6,435	5,005	3,003
16	1	16	120	560	1,820	4,368	8,008	11,440	12,870	11,440	8,008
17	1	17	136	680	2,380	6,188	12,376	19,448	24,310	24,310	19,448
18	1	18	153	816	3,060	8,568	18,564	31,824	43,758	48,620	43,758
19	1	19	171	969	3,876	11,628	27,132	50,388	75,582	92,378	92,378
20	1	20	190	1,140	4,845	15,504	38,760	77,520	125,970	167,960	184,756

Table 3 Binomial Probability Distribution $C_{n,r}\,p^r q^{n-r}$

This table shows the probability of r successes in n independent trials, each with probability of success p.

											p									
n	r	.01	.05	.10	.15	.20	.25	.30	.35	.40	.45	.50	.55	.60	.65	.70	.75	.80	.85	.90	.95
2	0	.980	.902	.810	.723	.640	.563	.490	.423	.360	.303	.250	.203	.160	.123	.090	.063	.040	.023	.010	.002
	1	.020	.095	.180	.255	.320	.375	.420	.455	.480	.495	.500	.495	.480	.455	.420	.375	.320	.255	.180	.095
	2	.000	.002	.010	.023	.040	.063	.090	.123	.160	.203	.250	.303	.360	.423	.490	.563	.640	.723	.810	.902
3	0	.970	.857	.729	.614	.512	.422	.343	.275	.216	.166	.125	.091	.064	.043	.027	.016	.008	.003	.001	.000
	1	.029	.135	.243	.325	.384	.422	.441	.444	.432	.408	.375	.334	.288	.239	.189	.141	.096	.057	.027	.007
	2	.000	.007	.027	.057	.096	.141	.189	.239	.288	.334	.375	.408	.432	.444	.441	.422	.384	.325	.243	.135
	3	.000	.000	.001	.003	.008	.016	.027	.043	.064	.091	.125	.166	.216	.275	.343	.422	.512	.614	.729	.857
4	0	.961	.815	.656	.522	.410	.316	.240	.179	.130	.092	.062	.041	.026	.015	.008	.004	.002	.001	.000	.000
	1	.039	.171	.292	.368	.410	.422	.412	.384	.346	.300	.250	.200	.154	.112	.076	.047	.026	.011	.004	.000
	2	.001	.014	.049	.098	.154	.211	.265	.311	.346	.368	.375	.368	.346	.311	.265	.211	.154	.098	.049	.014
	3	.000	.000	.004	.011	.026	.047	.076	.112	.154	.200	.250	.300	.346	.384	.412	.422	.410	.368	.292	.171
	4	.000	.000	.000	.001	.002	.004	.008	.015	.026	.041	.062	.092	.130	.179	.240	.316	.410	.522	.656	.815
5	0	.951	.774	.590	.444	.328	.237	.168	.116	.078	.050	.031	.019	.010	.005	.002	.001	.000	.000	.000	.000
	1	.048	.204	.328	.392	.410	.396	.360	.312	.259	.206	.156	.113	.077	.049	.028	.015	.006	.002	.000	.000
	2	.001	.021	.073	.138	.205	.264	.309	.336	.346	.337	.312	.276	.230	.181	.132	.088	.051	.024	.008	.001
	3	.000	.001	.008	.024	.051	.088	.132	.181	.230	.276	.312	.337	.346	.336	.309	.264	.205	.138	.073	.021
	4	.000	.000	.000	.002	.006	.015	.028	.049	.077	.113	.156	.206	.259	.312	.360	.396	.410	.392	.328	.204
	5	.000	.000	.000	.000	.000	.001	.002	.005	.010	.019	.031	.050	.078	.116	.168	.237	.328	.444	.590	.774
6	0	.941	.735	.531	.377	.262	.178	.118	.075	.047	.028	.016	.008	.004	.002	.001	.000	.000	.000	.000	.000
	1	.057	.232	.354	.399	.393	.356	.303	.244	.187	.136	.094	.061	.037	.020	.010	.004	.002	.000	.000	.000
	2	.001	.031	.098	.176	.246	.297	.324	.328	.311	.278	.234	.186	.138	.095	.060	.033	.015	.006	.001	.000
	3	.000	.002	.015	.042	.082	.132	.185	.236	.276	.303	.312	.303	.276	.236	.185	.132	.082	.042	.015	.002
	4	.000	.000	.001	.006	.015	.033	.060	.095	.138	.186	.234	.278	.311	.328	.324	.297	.246	.176	.098	.031
	5	.000	.000	.000	.000	.002	.004	.010	.020	.037	.061	.094	.136	.187	.244	.303	.356	.393	.399	.354	.232
	6	.000	.000	.000	.000	.000	.000	.001	.002	.004	.008	.016	.028	.047	.075	.118	.178	.262	.377	.531	.735
7	0	.932	.698	.478	.321	.210	.133	.082	.049	.028	.015	.008	.004	.002	.001	.000	.000	.000	.000	.000	.000
	1	.066	.257	.372	.396	.367	.311	.247	.185	.131	.087	.055	.032	.017	.008	.004	.001	.000	.000	.000	.000
	2	.002	.041	.124	.210	.275	.311	.318	.299	.261	.214	.164	.117	.077	.047	.025	.012	.004	.001	.000	.000
	3	.000	.004	.023	.062	.115	.173	.227	.268	.290	.292	.273	.239	.194	.144	.097	.058	.029	.011	.003	.000
	4	.000	.000	.003	.011	.029	.058	.097	.144	.194	.239	.273	.292	.290	.268	.227	.173	.115	.062	.023	.004
	5	.000	.000	.000	.001	.004	.012	.025	.047	.077	.117	.164	.214	.261	.299	.318	.311	.275	.210	.124	.041
	6	.000	.000	.000	.000	.000	.001	.004	.008	.017	.032	.055	.087	.131	.185	.247	.311	.367	.396	.372	.257
	7	.000	.000	.000	.000	.000	.000	.000	.001	.002	.004	.008	.015	.028	.049	.082	.133	.210	.321	.478	.698

Table 3 continued

n	r	.01	.05	.10	.15	.20	.25	.30	.35	.40	.45	.50	.55	.60	.65	.70	.75	.80	.85	.90	.95
8	0	.923	.663	.430	.272	.168	.100	.058	.032	.017	.008	.004	.002	.001	.000	.000	.000	.000	.000	.000	.000
	1	.075	.279	.383	.385	.336	.267	.198	.137	.090	.055	.031	.016	.008	.003	.001	.000	.000	.000	.000	.000
	2	.003	.051	.149	.238	.294	.311	.296	.259	.209	.157	.109	.070	.041	.022	.010	.004	.001	.000	.000	.000
	3	.000	.005	.033	.084	.147	.208	.254	.279	.279	.257	.219	.172	.124	.081	.047	.023	.009	.003	.000	.000
	4	.000	.000	.005	.018	.046	.087	.136	.188	.232	.263	.273	.263	.232	.188	.136	.087	.046	.018	.005	.000
	5	.000	.000	.000	.003	.009	.023	.047	.081	.124	.172	.219	.257	.279	.279	.254	.208	.147	.084	.033	.005
	6	.000	.000	.000	.000	.001	.004	.010	.022	.041	.070	.109	.157	.209	.259	.296	.311	.294	.238	.149	.051
	7	.000	.000	.000	.000	.000	.000	.001	.003	.008	.016	.031	.055	.090	.137	.198	.267	.336	.385	.383	.279
	8	.000	.000	.000	.000	.000	.000	.000	.000	.001	.002	.004	.008	.017	.032	.058	.100	.168	.272	.430	.663
9	0	.914	.630	.387	.232	.134	.075	.040	.021	.010	.005	.002	.001	.000	.000	.000	.000	.000	.000	.000	.000
	1	.083	.299	.387	.368	.302	.225	.156	.100	.060	.034	.018	.008	.004	.001	.000	.000	.000	.000	.000	.000
	2	.003	.063	.172	.260	.302	.300	.267	.216	.161	.111	.070	.041	.021	.010	.004	.001	.000	.000	.000	.000
	3	.000	.008	.045	.107	.176	.234	.267	.272	.251	.212	.164	.116	.074	.042	.021	.009	.003	.001	.000	.000
	4	.000	.001	.007	.028	.066	.117	.172	.219	.251	.260	.246	.213	.167	.118	.074	.039	.017	.005	.000	.000
	5	.000	.000	.001	.005	.017	.039	.074	.118	.167	.213	.246	.260	.251	.219	.172	.117	.066	.028	.007	.001
	6	.000	.000	.000	.001	.003	.009	.021	.042	.074	.116	.164	.212	.251	.272	.267	.234	.176	.107	.045	.008
	7	.000	.000	.000	.000	.000	.001	.004	.010	.021	.041	.070	.111	.161	.216	.267	.300	.302	.260	.172	.063
	8	.000	.000	.000	.000	.000	.000	.000	.001	.004	.008	.018	.034	.060	.100	.156	.225	.302	.368	.387	.299
	9	.000	.000	.000	.000	.000	.000	.000	.000	.000	.001	.002	.005	.010	.021	.040	.075	.134	.232	.387	.630
10	0	.904	.599	.349	.197	.107	.056	.028	.014	.006	.003	.001	.000	.000	.000	.000	.000	.000	.000	.000	.000
	1	.091	.315	.387	.347	.268	.188	.121	.072	.040	.021	.010	.004	.002	.000	.000	.000	.000	.000	.000	.000
	2	.004	.075	.194	.276	.302	.282	.233	.176	.121	.076	.044	.023	.011	.004	.001	.000	.000	.000	.000	.000
	3	.000	.010	.057	.130	.201	.250	.267	.252	.215	.166	.117	.075	.042	.021	.009	.003	.001	.000	.000	.000
	4	.000	.001	.011	.040	.088	.146	.200	.238	.251	.238	.205	.160	.111	.069	.037	.016	.006	.001	.000	.000
	5	.000	.000	.001	.008	.026	.058	.103	.154	.201	.234	.246	.234	.201	.154	.103	.058	.026	.008	.001	.000
	6	.000	.000	.000	.001	.006	.016	.037	.069	.111	.160	.205	.238	.251	.238	.200	.146	.088	.040	.011	.001
	7	.000	.000	.000	.000	.001	.003	.009	.021	.042	.075	.117	.166	.215	.252	.267	.250	.201	.130	.057	.010
	8	.000	.000	.000	.000	.000	.000	.001	.004	.011	.023	.044	.076	.121	.176	.233	.282	.302	.276	.194	.075
	9	.000	.000	.000	.000	.000	.000	.000	.000	.002	.004	.010	.021	.040	.072	.121	.188	.268	.347	.387	.315
	10	.000	.000	.000	.000	.000	.000	.000	.000	.000	.000	.001	.003	.006	.014	.028	.056	.107	.197	.349	.599
11	0	.895	.569	.314	.167	.086	.042	.020	.009	.004	.001	.000	.000	.000	.000	.000	.000	.000	.000	.000	.000
	1	.099	.329	.384	.325	.236	.155	.093	.052	.027	.013	.005	.002	.001	.000	.000	.000	.000	.000	.000	.000
	2	.005	.087	.213	.287	.295	.258	.200	.140	.089	.051	.027	.013	.005	.002	.001	.000	.000	.000	.000	.000
	3	.000	.014	.071	.152	.221	.258	.257	.225	.177	.126	.081	.046	.023	.010	.004	.001	.000	.000	.000	.000
	4	.000	.001	.016	.054	.111	.172	.220	.243	.236	.206	.161	.113	.070	.038	.017	.006	.002	.000	.000	.000
	5	.000	.000	.002	.013	.039	.080	.132	.183	.221	.236	.226	.193	.147	.099	.057	.027	.010	.002	.000	.000

p

Table 3 continued

										p											
n	r	.01	.05	.10	.15	.20	.25	.30	.35	.40	.45	.50	.55	.60	.65	.70	.75	.80	.85	.90	.95
11	6	.000	.000	.000	.002	.010	.027	.057	.099	.147	.193	.226	.236	.221	.183	.132	.080	.039	.013	.002	.000
	7	.000	.000	.000	.000	.002	.006	.017	.038	.070	.113	.161	.206	.236	.243	.220	.172	.111	.054	.016	.001
	8	.000	.000	.000	.000	.000	.001	.004	.010	.023	.046	.081	.126	.177	.225	.257	.258	.221	.152	.071	.014
	9	.000	.000	.000	.000	.000	.000	.001	.002	.005	.013	.027	.051	.089	.140	.200	.258	.295	.287	.213	.087
	10	.000	.000	.000	.000	.000	.000	.000	.000	.001	.002	.005	.013	.027	.052	.093	.155	.236	.325	.384	.329
	11	.000	.000	.000	.000	.000	.000	.000	.000	.000	.000	.000	.001	.004	.009	.020	.042	.086	.167	.314	.569
12	0	.886	.540	.282	.142	.069	.032	.014	.006	.002	.001	.000	.000	.000	.000	.000	.000	.000	.000	.000	.000
	1	.107	.341	.377	.301	.206	.127	.071	.037	.017	.008	.003	.001	.000	.000	.000	.000	.000	.000	.000	.000
	2	.006	.099	.230	.292	.283	.232	.168	.109	.064	.034	.016	.007	.002	.001	.000	.000	.000	.000	.000	.000
	3	.000	.017	.085	.172	.236	.258	.240	.195	.142	.092	.054	.028	.012	.005	.001	.000	.000	.000	.000	.000
	4	.000	.002	.021	.068	.133	.194	.231	.237	.213	.170	.121	.076	.042	.020	.008	.002	.001	.000	.000	.000
	5	.000	.000	.004	.019	.053	.103	.158	.204	.227	.223	.193	.149	.101	.059	.029	.011	.003	.001	.000	.000
	6	.000	.000	.000	.004	.016	.040	.079	.128	.177	.212	.226	.212	.177	.128	.079	.040	.016	.004	.000	.000
	7	.000	.000	.000	.001	.003	.011	.029	.059	.101	.149	.193	.223	.227	.204	.158	.103	.053	.019	.004	.000
	8	.000	.000	.000	.000	.001	.002	.008	.020	.042	.076	.121	.170	.213	.237	.231	.194	.133	.068	.021	.002
	9	.000	.000	.000	.000	.000	.000	.001	.005	.012	.028	.054	.092	.142	.195	.240	.258	.236	.172	.085	.017
	10	.000	.000	.000	.000	.000	.000	.000	.001	.002	.007	.016	.034	.064	.109	.168	.232	.283	.292	.230	.099
	11	.000	.000	.000	.000	.000	.000	.000	.000	.000	.001	.003	.008	.017	.037	.071	.127	.206	.301	.377	.341
	12	.000	.000	.000	.000	.000	.000	.000	.000	.000	.000	.000	.001	.002	.006	.014	.032	.069	.142	.282	.540
15	0	.860	.463	.206	.087	.035	.013	.005	.002	.000	.000	.000	.000	.000	.000	.000	.000	.000	.000	.000	.000
	1	.130	.366	.343	.231	.132	.067	.031	.013	.005	.002	.000	.000	.000	.000	.000	.000	.000	.000	.000	.000
	2	.009	.135	.267	.286	.231	.156	.092	.048	.022	.009	.003	.001	.000	.000	.000	.000	.000	.000	.000	.000
	3	.000	.031	.129	.218	.250	.225	.170	.111	.063	.032	.014	.005	.002	.000	.000	.000	.000	.000	.000	.000
	4	.000	.005	.043	.116	.188	.225	.219	.179	.127	.078	.042	.019	.007	.002	.001	.000	.000	.000	.000	.000
	5	.000	.001	.010	.045	.103	.165	.206	.212	.186	.140	.092	.051	.024	.010	.003	.001	.000	.000	.000	.000
	6	.000	.000	.002	.013	.043	.092	.147	.191	.207	.191	.153	.105	.061	.030	.012	.003	.001	.000	.000	.000
	7	.000	.000	.000	.003	.014	.039	.081	.132	.177	.201	.196	.165	.118	.071	.035	.013	.003	.001	.000	.000
	8	.000	.000	.000	.001	.003	.013	.035	.071	.118	.165	.196	.201	.177	.132	.081	.039	.014	.003	.000	.000
	9	.000	.000	.000	.000	.001	.003	.012	.030	.061	.105	.153	.191	.207	.191	.147	.092	.043	.013	.002	.000
	10	.000	.000	.000	.000	.000	.001	.003	.010	.024	.051	.092	.140	.186	.212	.206	.165	.103	.045	.010	.001
	11	.000	.000	.000	.000	.000	.000	.001	.002	.007	.019	.042	.078	.127	.179	.219	.225	.188	.116	.043	.005
	12	.000	.000	.000	.000	.000	.000	.000	.000	.002	.005	.014	.032	.063	.111	.170	.225	.250	.218	.129	.031
	13	.000	.000	.000	.000	.000	.000	.000	.000	.000	.001	.003	.009	.022	.048	.092	.156	.231	.286	.267	.135
	14	.000	.000	.000	.000	.000	.000	.000	.000	.000	.000	.000	.002	.005	.013	.031	.067	.132	.231	.343	.366
	15	.000	.000	.000	.000	.000	.000	.000	.000	.000	.000	.000	.000	.000	.002	.005	.013	.035	.087	.206	.463
16	0	.851	.440	.185	.074	.028	.010	.003	.001	.000	.000	.000	.000	.000	.000	.000	.000	.000	.000	.000	.000
	1	.138	.371	.329	.210	.113	.053	.023	.009	.003	.001	.000	.000	.000	.000	.000	.000	.000	.000	.000	.000

Table 3 continued

n	r	p																			
		.01	.05	.10	.15	.20	.25	.30	.35	.40	.45	.50	.55	.60	.65	.70	.75	.80	.85	.90	.95
16	2	.010	.146	.275	.277	.211	.134	.073	.035	.015	.006	.002	.001	.000	.000	.000	.000	.000	.000	.000	.000
	3	.000	.036	.142	.229	.246	.208	.146	.089	.047	.022	.009	.003	.001	.000	.000	.000	.000	.000	.000	.000
	4	.000	.006	.051	.131	.200	.225	.204	.155	.101	.057	.028	.011	.004	.001	.000	.000	.000	.000	.000	.000
	5	.000	.001	.014	.056	.120	.180	.210	.201	.162	.112	.067	.034	.014	.005	.001	.000	.000	.000	.000	.000
	6	.000	.000	.003	.018	.055	.110	.165	.198	.198	.168	.122	.075	.039	.017	.006	.001	.000	.000	.000	.000
	7	.000	.000	.000	.005	.020	.052	.101	.152	.189	.197	.175	.132	.084	.044	.019	.006	.001	.000	.000	.000
	8	.000	.000	.000	.001	.006	.020	.049	.092	.142	.181	.196	.181	.142	.092	.049	.020	.006	.001	.000	.000
	9	.000	.000	.000	.000	.001	.006	.019	.044	.084	.132	.175	.197	.189	.152	.101	.052	.020	.005	.000	.000
	10	.000	.000	.000	.000	.000	.001	.006	.017	.039	.075	.122	.168	.198	.198	.165	.110	.055	.018	.003	.000
	11	.000	.000	.000	.000	.000	.000	.001	.005	.014	.034	.067	.112	.162	.201	.210	.180	.120	.056	.014	.001
	12	.000	.000	.000	.000	.000	.000	.000	.001	.004	.011	.028	.057	.101	.155	.204	.225	.200	.131	.051	.006
	13	.000	.000	.000	.000	.000	.000	.000	.000	.001	.003	.009	.022	.047	.089	.146	.208	.246	.229	.142	.036
	14	.000	.000	.000	.000	.000	.000	.000	.000	.000	.001	.002	.006	.015	.035	.073	.134	.211	.277	.275	.146
	15	.000	.000	.000	.000	.000	.000	.000	.000	.000	.000	.000	.001	.003	.009	.023	.053	.113	.210	.329	.371
	16	.000	.000	.000	.000	.000	.000	.000	.000	.000	.000	.000	.000	.000	.001	.003	.010	.028	.074	.185	.440
20	0	.818	.358	.122	.039	.012	.003	.001	.000	.000	.000	.000	.000	.000	.000	.000	.000	.000	.000	.000	.000
	1	.165	.377	.270	.137	.058	.021	.007	.002	.000	.000	.000	.000	.000	.000	.000	.000	.000	.000	.000	.000
	2	.016	.189	.285	.229	.137	.067	.028	.010	.003	.001	.000	.000	.000	.000	.000	.000	.000	.000	.000	.000
	3	.001	.060	.190	.243	.205	.134	.072	.032	.012	.004	.001	.000	.000	.000	.000	.000	.000	.000	.000	.000
	4	.000	.013	.090	.182	.218	.190	.130	.074	.035	.014	.005	.001	.000	.000	.000	.000	.000	.000	.000	.000
	5	.000	.002	.032	.103	.175	.202	.179	.127	.075	.036	.015	.005	.001	.000	.000	.000	.000	.000	.000	.000
	6	.000	.000	.009	.045	.109	.169	.192	.171	.124	.075	.037	.015	.005	.001	.000	.000	.000	.000	.000	.000
	7	.000	.000	.002	.016	.055	.112	.164	.184	.166	.122	.074	.037	.015	.005	.001	.000	.000	.000	.000	.000
	8	.000	.000	.000	.005	.022	.061	.114	.161	.180	.162	.120	.073	.035	.014	.004	.001	.000	.000	.000	.000
	9	.000	.000	.000	.001	.007	.027	.065	.116	.160	.177	.160	.119	.071	.034	.012	.003	.000	.000	.000	.000
	10	.000	.000	.000	.000	.002	.010	.031	.069	.117	.159	.176	.159	.117	.069	.031	.010	.002	.000	.000	.000
	11	.000	.000	.000	.000	.000	.003	.012	.034	.071	.119	.160	.177	.160	.116	.065	.027	.007	.001	.000	.000
	12	.000	.000	.000	.000	.000	.001	.004	.014	.035	.073	.120	.162	.180	.161	.114	.061	.022	.005	.000	.000
	13	.000	.000	.000	.000	.000	.000	.001	.005	.015	.037	.074	.122	.166	.184	.164	.112	.055	.016	.002	.000
	14	.000	.000	.000	.000	.000	.000	.000	.001	.005	.015	.037	.075	.124	.171	.192	.169	.109	.045	.009	.000
	15	.000	.000	.000	.000	.000	.000	.000	.000	.001	.005	.015	.036	.075	.127	.179	.202	.175	.103	.032	.002
	16	.000	.000	.000	.000	.000	.000	.000	.000	.000	.001	.005	.014	.035	.074	.130	.190	.218	.182	.090	.013
	17	.000	.000	.000	.000	.000	.000	.000	.000	.000	.000	.001	.004	.012	.032	.072	.134	.205	.243	.190	.060
	18	.000	.000	.000	.000	.000	.000	.000	.000	.000	.000	.000	.001	.003	.010	.028	.067	.137	.229	.285	.189
	19	.000	.000	.000	.000	.000	.000	.000	.000	.000	.000	.000	.000	.000	.002	.007	.021	.058	.137	.270	.377
	20	.000	.000	.000	.000	.000	.000	.000	.000	.000	.000	.000	.000	.000	.000	.001	.003	.012	.039	.122	.358

Table 4 Areas of a Standard Normal Distribution
(a) Table of Areas from 0 to z

The table entries represent the area under the standard normal curve from 0 to the specified value of z.

z	.00	.01	.02	.03	.04	.05	.06	.07	.08	.09
0.0	.0000	.0040	.0080	.0120	.0160	.0199	.0239	.0279	.0319	.0359
0.1	.0398	.0438	.0478	.0517	.0557	.0596	.0636	.0675	.0714	.0753
0.2	.0793	.0832	.0871	.0910	.0948	.0987	.1026	.1064	.1103	.1141
0.3	.1179	.1217	.1255	.1293	.1331	.1368	.1406	.1443	.1480	.1517
0.4	.1554	.1591	.1628	.1664	.1700	.1736	.1772	.1808	.1844	.1879
0.5	.1915	.1950	.1985	.2019	.2054	.2088	.2123	.2157	.2190	.2224
0.6	.2257	.2291	.2324	.2357	.2389	.2422	.2454	.2486	.2517	.2549
0.7	.2580	.2611	.2642	.2673	.2704	.2734	.2764	.2794	.2823	.2852
0.8	.2881	.2910	.2939	.2967	.2995	.3023	.3051	.3078	.3106	.3133
0.9	.3159	.3186	.3212	.3238	.3264	.3289	.3315	.3340	.3365	.3389
1.0	.3413	.3438	.3461	.3485	.3508	.3531	.3554	.3577	.3599	.3621
1.1	.3643	.3665	.3686	.3708	.3729	.3749	.3770	.3790	.3810	.3830
1.2	.3849	.3869	.3888	.3907	.3925	.3944	.3962	.3980	.3997	.4015
1.3	.4032	.4049	.4066	.4082	.4099	.4115	.4131	.4147	.4162	.4177
1.4	.4192	.4207	.4222	.4236	.4251	.4265	.4279	.4292	.4306	.4319
1.5	.4332	.4345	.4357	.4370	.4382	.4394	.4406	.4418	.4429	.4441
1.6	.4452	.4463	.4474	.4484	.4495	.4505	.4515	.4525	.4535	.4545
1.7	.4554	.4564	.4573	.4582	.4591	.4599	.4608	.4616	.4625	.4633
1.8	.4641	.4649	.4656	.4664	.4671	.4678	.4686	.4693	.4699	.4706
1.9	.4713	.4719	.4726	.4732	.4738	.4744	.4750	.4756	.4761	.4767
2.0	.4772	.4778	.4783	.4788	.4793	.4798	.4803	.4808	.4812	.4817
2.1	.4821	.4826	.4830	.4834	.4838	.4842	.4846	.4850	.4854	.4857
2.2	.4861	.4864	.4868	.4871	.4875	.4878	.4881	.4884	.4887	.4890
2.3	.4893	.4896	.4898	.4901	.4904	.4906	.4909	.4911	.4913	.4916
2.4	.4918	.4920	.4922	.4925	.4927	.4929	.4931	.4932	.4934	.4936
2.5	.4938	.4940	.4941	.4943	.4945	.4946	.4948	.4949	.4951	.4952
2.6	.4953	.4955	.4956	.4957	.4959	.4960	.4961	.4962	.4963	.4964
2.7	.4965	.4966	.4967	.4968	.4969	.4970	.4971	.4972	.4973	.4974
2.8	.4974	.4975	.4976	.4977	.4977	.4978	.4979	.4979	.4980	.4981
2.9	.4981	.4982	.4982	.4983	.4984	.4984	.4985	.4985	.4986	.4986
3.0	.4987	.4987	.4987	.4988	.4988	.4989	.4989	.4989	.4990	.4990
3.1	.4990	.4991	.4991	.4991	.4992	.4992	.4992	.4992	.4993	.4993
3.2	.4993	.4993	.4994	.4994	.4994	.4994	.4994	.4995	.4995	.4995
3.3	.4995	.4995	.4995	.4996	.4996	.4996	.4996	.4996	.4996	.4997
3.4	.4997	.4997	.4997	.4997	.4997	.4997	.4997	.4997	.4997	.4998
3.5	.4998	.4998	.4998	.4998	.4998	.4998	.4998	.4998	.4998	.4998
3.6	.4998	.4998	.4998	.4999	.4999	.4999	.4999	.4999	.4999	.4999

For values of z greater than or equal to 3.70, use 0.4999 to approximate the shaded area under the standard normal curve.

Table 4 continued

(b) Confidence Interval
     Critical Values z_c

Level of Confidence c	Critical Value z_c
0.75, or 75%	1.15
0.80, or 80%	1.28
0.85, or 85%	1.44
0.90, or 90%	1.645
0.95, or 95%	1.96
0.98, or 98%	2.33
0.99, or 99%	2.58

(c) Hypothesis Testing
     Critical Values z_0

Level of Significance	$\alpha = 0.05$	$\alpha = 0.01$
Critical value z_0 for a left-tailed test	−1.645	−2.33
Critical value z_0 for a right-tailed test	1.645	2.33
Critical values $\pm z_0$ for a two-tailed test	±1.96	±2.58

Table 5   Student's *t* Distribution

Student's *t* values generated by Minitab Version 9.2

c is a confidence level:

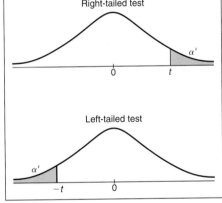

α' is the level of significance for a one-tailed test:

Right-tailed test

Left-tailed test

α'' is the level of significance for a two-tailed test:

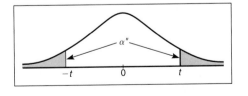

d.f.	c = 0.750 α' = 0.125 α'' = 0.250	0.800 0.100 0.200	0.850 0.075 0.150	0.900 0.050 0.100	0.950 0.025 0.050	0.980 0.010 0.020	0.990 0.005 0.010
1	2.414	3.078	4.165	6.314	12.706	31.821	63.657
2	1.604	1.886	2.282	2.920	4.303	6.965	9.925
3	1.423	1.638	1.924	2.353	3.182	4.541	5.841
4	1.344	1.533	1.778	2.132	2.776	3.747	4.604
5	1.301	1.476	1.699	2.015	2.571	3.365	4.032
6	1.273	1.440	1.650	1.943	2.447	3.143	3.707
7	1.254	1.415	1.617	1.895	2.365	2.998	3.499
8	1.240	1.397	1.592	1.860	2.306	2.896	3.355
9	1.230	1.383	1.574	1.833	2.262	2.821	3.250
10	1.221	1.372	1.559	1.812	2.228	2.764	3.169
11	1.214	1.363	1.548	1.796	2.201	2.718	3.106
12	1.209	1.356	1.538	1.782	2.179	2.681	3.055
13	1.204	1.350	1.530	1.771	2.160	2.650	3.012
14	1.200	1.345	1.523	1.761	2.145	2.624	2.977
15	1.197	1.341	1.517	1.753	2.131	2.602	2.947
16	1.194	1.337	1.512	1.746	2.120	2.583	2.921
17	1.191	1.333	1.508	1.740	2.110	2.567	2.898
18	1.189	1.330	1.504	1.734	2.101	2.552	2.878
19	1.187	1.328	1.500	1.729	2.093	2.539	2.861
20	1.185	1.325	1.497	1.725	2.086	2.528	2.845
21	1.183	1.323	1.494	1.721	2.080	2.518	2.831
22	1.182	1.321	1.492	1.717	2.074	2.508	2.819
23	1.180	1.319	1.489	1.714	2.069	2.500	2.807
24	1.179	1.318	1.487	1.711	2.064	2.492	2.797
25	1.178	1.316	1.485	1.708	2.060	2.485	2.787
26	1.177	1.315	1.483	1.706	2.056	2.479	2.779
27	1.176	1.314	1.482	1.703	2.052	2.473	2.771
28	1.175	1.313	1.480	1.701	2.048	2.467	2.763
29	1.174	1.311	1.479	1.699	2.045	2.462	2.756
30	1.173	1.310	1.477	1.697	2.042	2.457	2.750
35	1.170	1.306	1.472	1.690	2.030	2.438	2.724
40	1.167	1.303	1.468	1.684	2.021	2.423	2.704
45	1.165	1.301	1.465	1.679	2.014	2.412	2.690
50	1.164	1.299	1.462	1.676	2.009	2.403	2.678
55	1.163	1.297	1.460	1.673	2.004	2.396	2.668
60	1.162	1.296	1.458	1.671	2.000	2.390	2.660
90	1.158	1.291	1.452	1.662	1.987	2.369	2.632
120	1.156	1.289	1.449	1.658	1.980	2.358	2.617
∞	1.150	1.282	1.440	1.645	1.960	2.326	2.58

Table 6 The χ^2 Distribution

For d.f. ≥ 3

For d.f. = 1 or 2

d.f.\α	.995	.990	.975	.950	.900	.100	.050	.025	.010	.005
1	0.0^4393	0.0^3157	0.0^3982	0.0^2393	0.0158	2.71	3.84	5.02	6.63	7.88
2	0.0100	0.0201	0.0506	0.103	0.211	4.61	5.99	7.38	9.21	10.60
3	0.072	0.115	0.216	0.352	0.584	6.25	7.81	9.35	11.34	12.84
4	0.207	0.297	0.484	0.711	1.064	7.78	9.49	11.14	13.28	14.86
5	0.412	0.554	0.831	1.145	1.61	9.24	11.07	12.83	15.09	16.75
6	0.676	0.872	1.24	1.64	2.20	10.64	12.59	14.45	16.81	18.55
7	0.989	1.24	1.69	2.17	2.83	12.02	14.07	16.01	18.48	20.28
8	1.34	1.65	2.18	2.73	3.49	13.36	15.51	17.53	20.09	21.96
9	1.73	2.09	2.70	3.33	4.17	14.68	16.92	19.02	21.67	23.59
10	2.16	2.56	3.25	3.94	4.87	15.99	18.31	20.48	23.21	25.19
11	2.60	3.05	3.82	4.57	5.58	17.28	19.68	21.92	24.72	26.76
12	3.07	3.57	4.40	5.23	6.30	18.55	21.03	23.34	26.22	28.30
13	3.57	4.11	5.01	5.89	7.04	19.81	22.36	24.74	27.69	29.82
14	4.07	4.66	5.63	6.57	7.79	21.06	23.68	26.12	29.14	31.32
15	4.60	5.23	6.26	7.26	8.55	22.31	25.00	27.49	30.58	32.80
16	5.14	5.81	6.91	7.96	9.31	23.54	26.30	28.85	32.00	34.27
17	5.70	6.41	7.56	8.67	10.09	24.77	27.59	30.19	33.41	35.72
18	6.26	7.01	8.23	9.39	10.86	25.99	28.87	31.53	34.81	37.16
19	6.84	7.63	8.91	10.12	11.65	27.20	30.14	32.85	36.19	38.58
20	7.43	8.26	8.59	10.85	12.44	28.41	31.41	34.17	37.57	40.00
21	8.03	8.90	10.28	11.59	13.24	29.62	32.67	35.48	38.93	41.40
22	8.64	9.54	10.98	12.34	14.04	30.81	33.92	36.78	40.29	42.80
23	9.26	10.20	11.69	13.09	14.85	32.01	35.17	38.08	41.64	44.18
24	9.89	10.86	12.40	13.85	15.66	33.20	36.42	39.36	42.98	45.56
25	10.52	11.52	13.12	14.61	16.47	34.38	37.65	40.65	44.31	46.93
26	11.16	12.20	13.84	15.38	17.29	35.56	38.89	41.92	45.64	48.29
27	11.81	12.88	14.57	16.15	18.11	36.74	40.11	43.19	46.96	49.64
28	12.46	13.56	15.31	16.93	18.94	37.92	41.34	44.46	48.28	50.99
29	13.21	14.26	16.05	17.71	19.77	39.09	42.56	45.72	49.59	52.34
30	13.79	14.95	16.79	18.49	20.60	40.26	43.77	46.98	50.89	53.67
40	20.71	22.16	24.43	26.51	29.05	51.80	55.76	59.34	63.69	66.77
50	27.99	29.71	32.36	34.76	37.69	63.17	67.50	71.42	76.15	79.49
60	35.53	37.48	40.48	43.19	46.46	74.40	79.08	83.30	88.38	91.95
70	43.28	45.44	48.76	51.74	55.33	85.53	90.53	95.02	100.4	104.2
80	51.17	53.54	57.15	60.39	64.28	96.58	101.9	106.6	112.3	116.3
90	59.20	61.75	65.65	69.13	73.29	107.6	113.1	118.1	124.1	128.3
100	67.33	70.06	74.22	77.93	82.36	118.5	124.3	129.6	135.8	140.2

Source: From H. L. Herter, *Biometrika*, June 1964. Printed by permission of Biometrika Trustees.

Table 7   Critical Values of Pearson Product-Moment Correlation Coefficient, *r*

For a right-tailed test, use a positive *r* value:

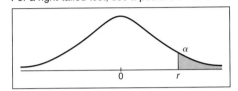

For a left-tailed test, use a negative *r* value:

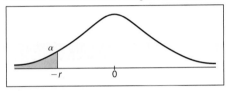

For a two-tailed test, use a positive *r* value and negative *r* value:

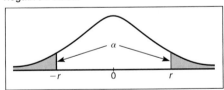

	$\alpha = 0.01$		$\alpha = 0.05$	
n	one tail	two tails	one tail	two tails
3	1.00	1.00	.99	1.00
4	.98	.99	.90	.95
5	.93	.96	.81	.88
6	.88	.92	.73	.81
7	.83	.87	.67	.75
8	.79	.83	.62	.71
9	.75	.80	.58	.67
10	.72	.76	.54	.63
11	.69	.73	.52	.60
12	.66	.71	.50	.58
13	.63	.68	.48	.53
14	.61	.66	.46	.53
15	.59	.64	.44	.51
16	.57	.61	.42	.50
17	.56	.61	.41	.48
18	.54	.59	.40	.47
19	.53	.58	.39	.46
20	.52	.56	.38	.44
21	.50	.55	.37	.43
22	.49	.54	.36	.42
23	.48	.53	.35	.41
24	.47	.52	.34	.40
25	.46	.51	.34	.40
26	.45	.50	.33	.39
27	.45	.49	.32	.38
28	.44	.48	.32	.37
29	.43	.47	.31	.37
30	.42	.46	.31	.36

Appendix II Data Files

The following data files may be used for class demonstrations or group projects. These files are on a data disk available from the publisher in formats for Minitab, Excel, or Texas Instruments TI-83. The TI-83 files are in ASCII format. In addition, the files are included as class demonstrations in the statistical software package ComputerStat (available free from Houghton Mifflin for adopters of this text).

The files contained in this Appendix are a subcollection of 35 data files available on the data disk.

Data: Weights of Professional Football Players

225	230	235	238	232	227
244	222	250	226	242	253
251	225	229	247	239	223
233	222	243	237	230	240
255	230	245	240	235	252
245	231	235	234	248	242
238	240	240	240	235	244
247	250	236	246	243	255
241	245				

DATA FILE 1

Weights of Professional Football Players

(a) Description: Random sample of weights (lb) of professional football players. (*Source: The Sports Encyclopedia, Pro Football.*)

(b) Recommended reading: Any of the following sections: 1.4, 2.1, 2.2, 2.3, 8.1, 9.2, 9.3.

(c) ComputerStat ➤ Main menu
   i. ➤ Descriptive Statistics ➤ Frequency Distributions and Grouped Data
   ii. ➤ Descriptive Statistics ➤ Averages, Variation, Box-and-Whisker Plots
   iii. ➤ Confidence Intervals ➤ Confidence Intervals for a Population Mean Mu
   iv. ➤ Hypothesis Testing ➤ Testing a Single Population Mean Mu

(d) Minitab portable worksheet name: Weights.mtp

(e) Excel workbook name: Weights.xls

(f) TI-83 data file name: Wts.txt

DATA FILE 2

Heights of Professional Basketball Players

(a) Description: Random sample of heights (ft) of professional basketball players. (*Source: All-Time Player Directory, The Official NBA Encyclopedia.*)

(b) Recommended reading: Any of the following sections: 1.4, 2.1, 2.2, 2.3, 8.1, 9.2, 9.3.

(c) ComputerStat ➤ Main menu
   i. ➤ Descriptive Statistics ➤ Frequency Distributions and Grouped Data
   ii. ➤ Descriptive Statistics ➤ Averages, Variation, Box-and-Whisker Plots
   iii. ➤ Confidence Intervals ➤ Confidence Intervals for a Population Mean Mu
   iv. ➤ Hypothesis Testing ➤ Testing a Single Population Mean Mu

(d) Minitab portable worksheet name: Heights.mtp

(e) Excel workbook name: Heights.xls

(f) TI-83 data file name: Hts.txt

Data: Heights of Professional Basketball Players

6.50	6.25	6.33	6.50	6.42	6.67
6.83	6.82	6.17	7.00	5.67	6.50
6.75	6.54	6.42	6.58	6.00	6.75
7.00	6.58	6.29	7.00	6.92	6.42
5.92	6.08	7.00	6.17	6.92	7.00
5.92	6.42	6.00	6.25	6.75	6.17
6.75	6.58	6.58	6.46	5.92	6.58
6.13	6.50	6.58	6.63	6.75	6.25
6.67	6.17	6.17	6.25	6.00	6.75
6.17	6.83	6.00	6.42	6.92	6.50
6.33	6.92	6.67	6.33	6.08	

Data: Miles per Gallon

30	27	22	25	24	25	24
15	35	35	33	52	49	10
27	18	20	23	24	25	30
24	24	24	18	20	25	27
24	32	29	27	24	27	26
25	24	28	33	30	13	13
21	28	37	35	32	33	29
31	28	28	25	29	31	

DATA FILE 3

Miles per Gallon Gasoline Consumption

(a) Description: Miles per gallon gasoline consumption (highway) for a random sample of 55 makes and models of passenger cars. (*Source:* Environmental Protection Agency.)

(b) Recommended reading: Any of the following sections: 1.4, 2.1, 2.2, 2.3, 8.1, 9.2, 9.3.

(c) ComputerStat ➤ Main menu
   i. ➤ Descriptive Statistics ➤ Frequency Distributions and Grouped Data
   ii. ➤ Descriptive Statistics ➤ Averages, Variation, Box-and-Whisker Plots
   iii. ➤ Confidence Intervals ➤ Confidence Intervals for a Population Mean Mu
   iv. ➤ Hypothesis Testing ➤ Testing a Single Population Mean Mu

(d) Minitab portable worksheet name: Mpgal.mtp

(e) Excel workbook name: Mpgal.xls

(f) TI-83 data file name: Mpg.txt

DATA FILE 4

Glucose Blood Tests

(a) Description: Random sample of glucose blood level (mg/100 mL) for women after 12-hour fast. (*Source: American Journal of Clinical Nutrition,* Vol. 19, pp. 345–351.)

(b) Recommended reading: Any of the following sections: 1.4, 2.1, 2.2, 2.3, 8.1, 9.2, 9.3.

(c) ComputerStat ➤ Main menu
   i. ➤ Descriptive Statistics ➤ Frequency Distributions and Grouped Data
   ii. ➤ Descriptive Statistics ➤ Averages, Variation, Box-and-Whisker Plots
   iii. ➤ Confidence Intervals ➤ Confidence Intervals for a Population Mean Mu
   iv. ➤ Hypothesis Testing ➤ Testing a Single Population Mean Mu

(d) Minitab portable worksheet name: Glucos.mtp

(e) Excel workbook name: Glucos.xls

(f) TI-83 data file name: Glu.txt

Data: Glucose Blood Tests

45	66	83	71	76	64	59
59	76	82	80	81	85	77
82	90	87	72	79	69	83
71	87	69	81	76	96	83
67	94	101	94	89	94	73
99	93	85	83	80	78	80
85	83	84	74	81	70	65
89	70	80	84	77	65	46
80	70	75	45	101	71	109
73	73	80	72	81	63	74

DATA FILE 5

Using "List Price" to Predict "Best Price" for a New GMC Pickup Truck (Linear Regression)

(a) Description: In the data pairs (x, y), x = list price (in $1000) for a GMC pickup truck and y = best price (in $1000) for a GMC pickup truck. (*Source: Consumers Digest.*)

(b) Recommended reading: Sections 3.1, 3.2, 3.3, 11.4, 11.5.

(c) ComputerStat ➤ Main menu ➤ Linear Regression and Correlation ➤ Linear Regression for Data Pairs (X, Y)

(d) Minitab portable worksheet name: Truck.mtp

(e) Excel workbook name: Truck.xls

(f) TI-83 data file name: Truck.txt

Data: (List Price, Best Price) in $1000 for a GMC Pickup Truck

(12.4, 11.2)	(14.3, 12.5)	(14.5, 12.7)	(14.9, 13.1)
(16.1, 14.1)	16.9, 14.8)	(16.5, 14.4)	(15.4, 13.4)
(17.0, 14.9)	(17.9, 15.6)	(18.8, 16.4)	(20.3, 17.7)
(22.4, 19.6)	(19.4, 16.9)	(15.5, 14.0)	(16.7, 14.6)
(17.3, 15.1)	(18.4, 16.1)	(19.2, 16.8)	(17.4, 15.2)
(19.5, 17.0)	(19.7, 17.2)	(21.2, 18.6)	

DATA FILE 6

Predicting Temperature Using Cricket Chirps (Linear Regression)

(a) Description: In the data pairs (x, y), x = chirps per second for the striped ground cricket and y = temperature in degrees Fahrenheit. (*Source: G. W. Pierce, The Song of Insects,* Harvard University Press.)

(b) Recommended reading: Sections 3.1, 3.2, 3.3, 11.4, 11.5.

(c) ComputerStat ➤ Main menu ➤ Linear Regression and Correlation ➤ Linear Regression for Data Pairs (X, Y)

(d) Minitab portable worksheet name: Cricket.mtp

(e) Excel workbook name: Cricket.xls

(f) TI-83 data file name: Crick.txt

Data: (Chirps/Sec, Temperature)

(20.0, 88.6)	(16.0, 71.6)	(19.8, 93.3)	(18.4, 84.3)
(17.1, 80.6)	(15.5, 75.2)	(14.7, 69.7)	(17.1, 82.0)
(15.4, 69.4)	(16.2, 83.3)	(15.0, 79.6)	(17.2, 82.6)
(16.0, 80.6)	(17.0, 83.5)	(14.4, 76.3)	

DATA FILE 7

Yield of Wheat at Rothamsted Experiment Station, England

(a) Description: Annual yield of wheat in tonnes (1 ton = 1.016 tonne) for an experimental plot of land at Rothamsted Experiment Station over a period of 30 consecutive years. (*Source: Rothamsted Experiment Station, U.K.*)

(b) Recommended reading: Section 6.1 (Control Charts).

(c) ComputerStat ➤ Main menu ➤ Probability Distributions and Central Limit Theorem ➤ Control Charts

(d) Minitab portable worksheet name: Wheat.mtp

(e) Excel workbook name: Wheat.xls

(f) TI-83 data file name: Wht.txt

Data: Yield of Wheat

1.73	1.66	1.36	1.19	2.66	2.14
2.25	2.25	2.36	2.82	2.61	2.51
2.61	2.75	3.49	3.22	2.37	2.52
3.43	3.47	3.20	2.72	3.02	3.03
2.36	2.83	2.76	2.07	1.63	3.02

DATA FILE 8

Salary for Male Compared to Female Assistant Professors (Paired Difference)

(a) Description: The data are paired by college. In the data pairs (A, B), A = average salary for males ($1000/yr) and B = average salary for for females ($1000/yr). (*Source: Academe, Bulletin of the American Association of University Professors.*)

(b) Recommended reading: Section 10.1.

(c) ComputerStat ➤ Main menu ➤ Hypothesis Testing ➤ $\mu_1 - \mu_2$ (Dependent Samples)

(d) Minitab portable worksheet name: Fsalary.mtp

(e) Excel workbook name: Fsalary.xls

(f) TI-83 data file name: Sal.txt

Data: Faculty Salary (Male, Female)

(34.5, 33.9)	(30.5, 31.2)	(35.1, 35.0)	(35.7, 34.2)
(31.5, 32.4)	(34.4, 34.1)	(32.1, 32.7)	(30.7, 29.9)
(33.7, 31.2)	(35.3, 35.5)	(30.7, 30.2)	(34.2, 34.8)
(39.6, 38.7)	(30.5, 30.0)	(33.8, 33.8)	(31.7, 32.4)
(32.8, 31.7)	(38.5, 38.9)	(40.5, 41.2)	(25.3, 25.5)
(28.6, 28.0)	(35.8, 35.1)		

DATA FILE 9

Unemployment by Education: College Graduates Compared to Only High School (Paired Difference)

(a) Description: The data are paired by year, and in the data pairs (A, B), A = percent unemployment for college graduates and B = percent unemployment for high school only. (*Source: Statistical Abstract of the United States.*)

(b) Recommended reading: Section 10.1.

(c) ComputerStat ➤ Main menu ➤ Hypothesis Testing ➤ $\mu_1 - \mu_2$ (Dependent Samples)

(d) Minitab portable worksheet name: Unempl.mtp

(e) Excel workbook name: Unempl.xls

(f) TI-83 data file name: Unem.txt

Data: Paired Difference, Percent Unemployment (College Grad, High School Only)

(2.8, 5.9)	(2.2, 4.9)	(2.2, 4.8)	(1.7, 5.4)
(2.3, 6.3)	(2.3, 6.9)	(2.4, 6.9)	(2.7, 7.2)
(3.5, 10.0)	(3.0, 8.5)	(1.9, 5.1)	(2.5, 6.9)

DATA FILE 10

Number of Traditional Hogans Compared to Modern Houses on the Navajo Indian Reservation (Paired Difference)

(a) Description: The data are paired by districts on the Navajo Reservation. A random sample of 8 districts was used. In the data pairs (A, B), A = the number of traditional Navajo hogans and B = the number of modern houses in a given district. (*Source: S. C. Jett and V. E. Spencer, Navajo Architecture, Forms, History, Distributions, University of Arizona Press.*)

(b) Recommended reading: Section 10.1.

(c) ComputerStat ➤ Main menu ➤ Hypothesis Testing ➤ $\mu_1 - \mu_2$ (Dependent Samples)

(d) Minitab portable worksheet name: Dwell.mtp

(e) Excel workbook name: Dwell.xls

(f) TI-83 data file name: Dwel.txt

Data: Paired Difference (Traditional Hogans, Modern Houses)

(13, 18)	(14, 16)	(46, 68)	(32, 9)
(15, 11)	(47, 28)	(17, 50)	(18, 50)

DATA FILE 11

Heights of Professional Football Players Compared to
Heights of Professional Basketball Players

(a) Description: The data represent heights in feet for 45
   randomly selected professional football players and 40
   randomly selected professional basketball players.
   (*Source: Sports Encyclopedia, Pro Football* and *Official NBA Basketball Encyclopedia.*)

(b) Recommended reading: Section 10.2.

(c) ComputerStat ➤ Main menu
   i. ➤ Confidence Intervals ➤ Confidence Intervals for
      $\mu_1 - \mu_2$ (Independent Samples)
   ii. ➤ Hypothesis Testing ➤ $\mu_1 - \mu_2$ (Independent
      Samples)

(d) Minitab portable worksheet name: Fhvbh.mtp

(e) Excel workbook name: Fhvbh.xls

(f) TI-83 data file name: Fvbh.txt

Data: Heights of Professional Football Players

6.33	6.50	6.50	6.25	6.50	6.33
6.25	6.17	6.42	6.33	6.42	6.58
6.08	6.58	6.50	6.42	6.25	6.67
5.91	6.00	5.83	6.00	5.83	5.08
6.75	5.83	6.17	5.75	6.00	5.75
6.50	5.83	5.91	5.67	6.00	6.08
6.17	6.58	6.50	6.25	6.33	5.25
6.67	6.50	5.83			

Data: Heights of Professional Basketball Players

6.08	6.58	6.25	6.58	6.25	5.92
7.00	6.41	6.75	6.25	6.00	6.92
6.83	6.58	6.41	6.67	6.67	5.75
6.25	6.25	6.50	6.00	6.92	6.25
6.42	6.58	6.58	6.08	6.75	6.50
6.83	6.08	6.92	6.00	6.33	6.50
6.58	6.83	6.50	6.58		

DATA FILE 12

Petal Length for Iris Virginica Compared to Petal Length for
Iris Setosa

(a) Description: The following data represent petal length
   (cm) for a random sample of 35 Iris Virginica and a
   random sample of 38 Iris Setosa. Source: Anderson,
   E., Bull. Amer. Iris Soc.

(b) Recommended reading: Section 10.2.

(c) ComputerStat ➤ Main menu
   i. ➤ Confidence Intervals ➤ Confidence Intervals for
      $\mu_1 - \mu_2$ (Independent Samples)
   ii. ➤ Hypothesis Testing ➤ $\mu_1 - \mu_2$ (Independent
      Samples)

(d) Minitab portable worksheet name: Petal.mtp

(e) Excel workbook name: Petal.xls

(f) TI-83 data file name: Petal.txt

Data: Petal Length (cm) of Iris Virginica

5.1	5.8	6.3	6.1	5.1	5.5
5.3	5.5	6.9	5.0	4.9	6.0
4.8	6.1	5.6	5.1	5.6	4.8
5.4	5.1	5.1	5.9	5.2	5.7
5.4	4.5	6.1	5.3	5.5	6.7
5.7	4.9	4.8	5.8	5.1	

Data: Petal Length (cm) of Iris Setosa

1.5	1.7	1.4	1.5	1.5	1.6
1.4	1.1	1.2	1.4	1.7	1.0
1.7	1.9	1.6	1.4	1.5	1.4
1.2	1.3	1.5	1.3	1.6	1.9
1.4	1.6	1.5	1.4	1.6	1.2
1.9	1.5	1.6	1.4	1.3	1.7
1.5	1.7				

DATA FILE 13

Weights of Professional Football Players Compared to
Weights of Professional Basketball Players

(a) Description: The data represent weights in pounds of
21 randomly selected professional football players and
19 randomly selected professional basketball players.
(*Source: Sports Encyclopedia, Pro Football and Official NBA Basketball Encyclopedia.*)

(b) Recommended reading: Section 10.3.

(c) ComputerStat ➤ Main menu
   i. ➤ Confidence Intervals ➤ Confidence Intervals for
   $\mu_1 - \mu_2$ (Independent Samples)
   ii. ➤ Hypothesis Testing ➤ $\mu_1 - \mu_2$ (Independent
   Samples)

(d) Minitab portable worksheet name: Fwvbw.mtp

(e) Excel workbook name: Fwvbw.xls

(f) TI-83 data file name: Fvbw.txt

Data: Weights of Professional Football Players

245	262	255	251	244	276
240	265	257	252	282	256
250	264	270	275	245	275
253	265	270			

Data: Weights of Professional Basketball Players

205	200	220	210	191	215
221	216	228	207	225	208
195	191	207	196	181	193
201					

DATA FILE 14

Number of Cases of Red Fox Rabies
in Two Regions of Germany

(a) Description: The data represent number of cases of red
fox rabies for a random sample of 16 areas in each of
two different regions of southern Germany. (*Source:*
B. Sayers, *Medical Informatics,* Vol. 2, pp. 11–34.)

(b) Recommended reading: Section 10.3.

(c) ComputerStat ➤ Main menu
   i. ➤ Confidence Intervals ➤ Confidence Intervals for
   $\mu_1 - \mu_2$ (Independent Samples)
   ii. ➤ Hypothesis Testing ➤ $\mu_1 - \mu_2$ (Independent
   Samples)

(d) Minitab portable worksheet name: Rabies.mtp

(e) Excel workbook name: Rabies.xls

(f) TI-83 data file name: Rab.txt

Data: Number of Cases of Rabies in Region 1

10	2	2	5	3	4
3	3	4	0	2	6
4	8	7	4		

Data: Number of Cases of Rabies in Region 2

1	1	2	1	3	9
2	2	4	5	4	2
2	0	0	2		

Answers and Key Steps to Odd-Numbered Problems

CHAPTER 1

Section 1.1

1. (a) Population: The responses to the question from all adults in the United States.
   (b) Sample: The responses from the 1261 adults.

3. (a) Population: The time interval between the arrival of an insurance payment check and the time it clears for *all* checks received by the company from the five-state region. (b) Sample: The time interval between the arrival of the check and the time the check clears for the 32 payment checks in the sample.

5. (a) ratio (b) interval (c) nominal (d) ordinal
   (e) ratio (f) ratio

7. (a) Census, since *all* Super Bowl games through Super Bowl XXXI were used to compute the average.
   (b) Experiment. The subjects were treated to determine level of pain tolerance. (c) Simulation. Computer images were used to model the relation between stride length and running efficiency. (d) Sampling. Gallup Chinese surveyed only a portion of the Chinese population.

9. Answers vary. (a) "Over the past few years" is likely to mean different time spans to people of different ages. "During the past 3 years" or "within the past 3 years" would be better. (b) The responses could well be different. If one considers the effects of other people running stop signs, the answer would likely be "yes." If a person recalls that he or she has inadvertently run stop signs, the answer is more likely to be "no." (c) Yes or no responses are likely to be unreliable. At certain times, such as for major sports events, the answer might be "yes." At other times, the same respondent could answer "no." Allowing a range of responses would probably produce more useful data.

Section 1.2

1. See text.

3. Because the largest number has two digits, use groups of two digits in the random-number table. Select a starting place, and proceed until you have obtained six different numbers from 01 to 99. For instance, starting in column 1, line 5 produces the sample 06, 34, 87, 69, 38, 90. Different starting places will produce different samples.

5. Use a random-number table to select four distinct numbers corresponding to people in your class.
   (a) Reasons may vary. For instance, the first four students may make a special effort to get to class on time. (b) Reasons may vary. For instance, four students who come in late might all be nursing students enrolled in an anatomy and physiology class that meets the hour before in a far-away building. They may be more motivated than other students to complete a degree requirement. (c) Reasons may vary. For instance, four students sitting in the back row might be less inclined to participate in class discussions. (d) Reasons may vary. For instance, the tallest students might all be male.

7. Since the largest number has two digits, read groups of two digits. Select a starting place in the random-number table, and proceed until you have six distinct numbers from 01 to 42. By starting in column 1 of row 11, we obtained the numbers 17, 13, 25, 29, 35, 06.

9. Assign even digits 0, 2, 4, 6, 8 to the outcome of heads and odd digits 1, 3, 5, 7, 9 to the outcome of tails. Select a starting position at random in the random-number table. If the first digit is even, record heads. If the first digit is odd, record tails. Continue in this manner until you have 25 outcomes recorded.

11. (a) simple random sample (b) cluster sample
    (c) convenience sample (d) systematic sample
    (e) stratified sample

13. (a) Yes, each time you roll a die, any of the outcomes 1 through 6 can occur, regardless of whether or not the outcome has occurred before. On the 4th roll, the outcome is 2. (b) 5/20 or 25%. (c) No, the outcomes occur at random.

Section 1.3

1. Highest Level of Education and Average Annual Household Income (in thousands of dollars)

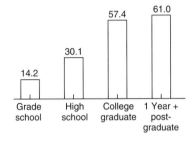

3. (a) Number of People Who Died in a Calendar Year from Listed Causes—Bar Graph

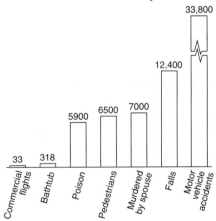

(b) Number of People Who Died in a Calendar Year from Listed Causes—Pareto Chart

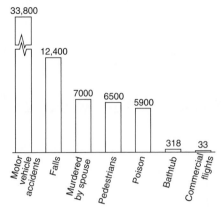

5. Where We Hide the Mess

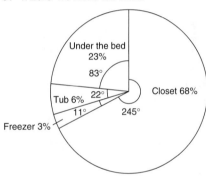

7. Percentage of Households with Phone Gadgets

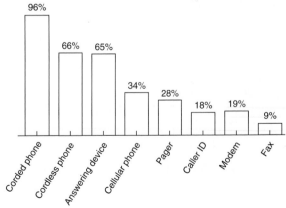

9. Elevation of Pyramid Lake Surface—Time Plot

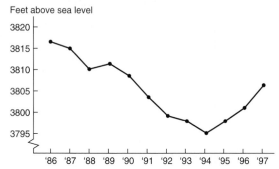

Section 1.4

1. (a) Class width = 4
   (b)

Class Limits	Boundaries	Midpoint	Frequency	Relative Frequency
29–32	28.5–32.5	30.5	1	0.03
33–36	32.5–36.5	34.5	6	0.19
37–40	36.5–40.5	38.5	12	0.39
41–44	40.5–44.5	42.5	7	0.23
45–48	44.5–48.5	46.5	4	0.13
49–52	48.5–52.5	50.5	1	0.03

(c–d) Record Maximum Temperature (°F)—
Histogram, Frequency Polygon, and Relative-
Frequency Histogram

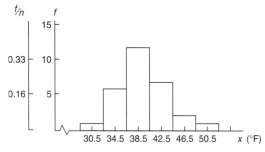

3. (a) Class width = 8

(b)

Class Limits	Boundaries	Midpoint	Frequency	Relative Frequency
62–69	61.5–69.5	65.5	2	0.06
70–77	69.5–77.5	73.5	15	0.48
78–85	77.5–85.5	81.5	8	0.26
86–93	85.5–93.5	89.5	1	0.03
94–101	93.5–101.5	97.5	2	0.06
102–109	101.5–109.5	105.5	3	0.10

(c–d) Attendance at Super Bowl—Histogram,
Frequency Polygon, Relative-Frequency Histogram

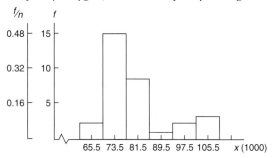

5. (a) Class width = 17
   (b) Nurses' Night Workload

Class Limits	Boundaries	Midpoint	Frequency	Relative Frequency
18–34	17.5–34.5	26	1	0.03
35–51	34.5–51.5	43	2	0.06
52–68	51.5–68.5	60	5	0.14
69–85	68.5–85.5	77	15	0.43
86–102	85.5–102.5	94	12	0.34

(c–d) Nurses' Night Workload—Histogram, Frequency Polygon, and Relative-Frequency Histogram

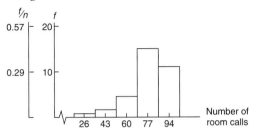

7. (a) Miami Dolphins seem heavier. Note that for the Dolphins, the class from 182 to 214 lb contains only 17 players compared to 27 players for the Chargers, while the next heavier class from 214 to 246 lb has 24 players for the Dolphins and only 15 for the Chargers. (b) The two additional Chargers seem to be lighter. Notice the Dolphins have only 3 players in the first two classes, while the Chargers have 5 players.

9. (a) Version 1 is skewed left; version 2 is uniform; version 3 is symmetrical; version 4 is bimodal; version 5 is skewed right. (b) Answers will vary.

11. (c) Class width = 0.40

Class Limits	Boundaries	Midpoint	Frequency
0.46–0.85	0.455–0.855	0.655	4
0.86–1.25	0.855–1.255	1.055	5
1.26–1.65	1.255–1.655	1.455	10
1.66–2.05	1.655–2.055	1.855	5
2.06–2.45	2.055–2.455	2.255	5
2.46–2.85	2.455–2.855	2.655	3

Tons of Wheat—Histogram

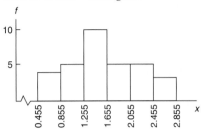

13. (a) One (b) 5/51 or 9.8% (c) Interval from 650 to 750

Section 1.5

1. (a) Walking Shoe Prices (7 years ago)

4	0 = $40
4	0 6
5	0 2 5 5 5 8 9 9
6	0 2 5 9
7	0 0 6
8	
9	
10	9
11	0

(b) Walking Shoe Prices (now)

4	0 = $40
4	0
5	
6	0 5 5 5 8 8
7	0 0 0 0 0 0 0 4 5 5
8	
9	0 5

Note: Leaves may be in any order unless instructions specify otherwise.

(c) In general, prices are higher now, probably due mostly to inflation. Note that seven years ago, two of the rated walking shoe models cost more than any recently rated ones.

3. Average Length of Hospital Stay

5	2 = 5.2 days
5	2 3 5 5 6 7
6	0 2 4 6 6 7 7 8 8 8 8 9 9
7	0 0 0 0 0 1 1 1 2 2 2 3 3 3 3 4 4 5 5 6 6 8
8	4 5 7
9	4 6 9
10	0 3
11	1

The distribution is skewed right.

5. (a) Minutes Beyond 2 Hours (1958–77)

0	9 = 9 minutes past 2 hours
0•	9
1*	0 3 3 4
1•	5 5 6 6 7 8 8 9
2*	0 0 2 2 3 3
2•	5

(b) Minutes Beyond 2 Hours (1978–97)

0	7 = 7 minutes past 2 hours
0•	7 7 8 8 8 8 9 9 9 9 9 9 9
1*	0 0 0 1 1 2 4

(c) In more recent years the winning times have been closer to 2 hours. In (a) there are 5 times under 2 hours 15 minutes. In (b) all 20 of the times are under 2 hours 15 minutes.

7. Milligrams of Tar per Cigarette

1	0 = 1.0 mg tar		11	4
1	0		11	4
2			12	0 4 8
3			13	7
4	1 5		14	1 5 9
5			15	0 1 2 8
6			16	0 6
7	3 8		17	0
8	0 6 8			
9	0			
10			29	8

Chapter 1 Review Problems

1. (a) Interval   (b) Ratio   (c) Nominal   (d) Ordinal
3. Problems with Tax Returns

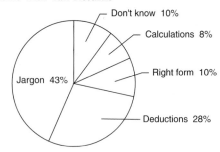

5. (a) Class width = 6.

Class Limits	Boundaries	Midpoint	Frequency	Relative Frequency
59–64	58.5–64.5	61.5	1	0.02
65–70	64.5–70.5	67.5	7	0.13
71–76	70.5–76.5	73.5	6	0.11
77–82	76.5–82.5	79.5	13	0.25
83–88	82.5–88.5	85.5	18	0.34
89–94	88.5–94.5	91.5	7	0.13
95–100	94.5–100.5	97.5	1	0.02

(b–c) Glucose Level—Histogram, Frequency Polygon, and Relative-Frequency Histogram

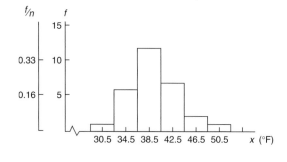

7. (a) 1240s had 40 data   (b) 75   (c) From 1203 to 1212. Little if any repairs or new construction.
9. Distribution of MVP Awards to Positions—Bar Graph

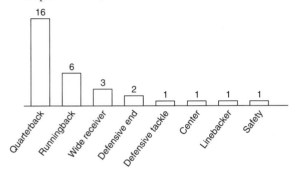

11. (a) Ages of Wealthy

3 | 4 = 34 years old

3	4
4	0 0 0 1 3 7 8 8 8
5	0 2 2 2 2 3 3 3 3 4 6 6 7 7 8 9
6	0 0 1 3 4 5 5 6 6 6 6 6 7 7 8 8
7	0 0 0 1 1 2 3 3 3 4 5 6 6 7 7 7 9
8	2 2 3 3 8
9	3

(b) Age Distribution of Billionaires—Histogram

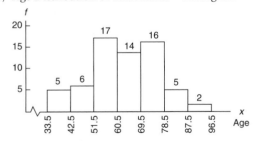

The histogram is somewhat symmetrical.

CHAPTER 2

Section 2.1

1. Mean = 156.33; median = 157; mode = 157. A gardener in Colorado should look at seed and plant descriptions to determine if the plant can thrive and mature in the designated number of frost-free days. The mean, median, and mode are all close. About half the locations have 157 or fewer frost-free days.
3. $\bar{x} \approx 167.3$ °F; median = 171 °F; mode = 178 °F.
5. (a) $\bar{x} \approx \$136.15$; median = \$66.5; mode = \$60.
   (b) 5% trimmed mean = \$121.28; yes, but still higher than the median.
   (c) Median. The low and high prices would be useful.
7. (a) Mean = 28.83 thousand dollars.
   (b) Median = 18.5 thousand dollars. The median best describes the salary of the majority of employees, since the mean is influenced by the high salaries of the president and vice president.
   (c) Mean = 17.3 thousand dollars; median = 17 thousand dollars.
   (d) Without the salaries for the two executives, the mean and the median are closer, and both reflect the salary of most of the other workers more accurately. The mean changed quite a bit, while the median did not, a difference that indicates that the mean is more sensitive to the absence or presence of extreme values.
9. (a) If the largest data value is *replaced* by a larger value, the mean will increase because the sum of the data values will increase, but the number of them will remain the same. The median will not change. The same value will still be in the eighth position when the data are ordered.
   (b) If the largest value is replaced by a value that is smaller (but still higher than the median), the mean will decrease because the sum of the data values will decrease. The median will not change. The same value will be in the eighth position in increasing order.
   (c) If the largest value is replaced by a value that is smaller than the median, the mean will decrease because the sum of the data values will decrease. The median also will decrease because the former value in the eighth position will move to the ninth position in increasing order. The median will be the new value in the eighth position.

Section 2.2

1. (a) Range = 54 deer/km²; sample mean $\bar{x} \approx 20.9$ deer/km²; sample variance $s^2 \approx 225.0$; sample standard deviation $s \approx 15.0$ deer/km².
 (b) $CV \approx 71.8\%$. Since the standard deviation is about 72% of the mean, there is considerable variation in the distribution of deer from one part of the park to another.
3. (a) Range = $160; $\bar{x}$ = $474; $s \approx$ $77.65
 (b) Range = $160; $\bar{x}$ = $642; $s \approx$ $62.61
 (c) Mean price of 32-inch sets is greater, but the standard deviation is less. For 27-inch sets, $CV \approx 16.4\%$ and for 32-inch sets $CV \approx 9.8\%$. There is less relative variation in the prices of the 32-inch sets.
5. (a) Range = 36; $\bar{x}$ = 33.5; s^2 = 236.7; $s \approx$ 15.39; $CV \approx 45.9\%$ (b) Range = 5; $\bar{x}$ = 9.5; s^2 = 3.5; $s \approx 1.87$; $CV \approx 19.7\%$ (c) Average cost for short flights is higher and more variable. Costs for longer flights are more consistent.
7. (a) 75% of the cycles should fall within 2 standard deviations of the mean: 6.67 to 15.35 years.
 (b) 93.8% of the cycles should fall within 4 standard deviations of the mean: 2.33 to 19.69 years.
9. (a) Range = 737; $\bar{x} \approx 566.9$.   (b) $s^2 \approx 71202$; $s \approx 266.8$.   (c) $CV \approx 47.1\%$.   (d) 33 to 1100.
11. (a) Midpoints are 65.5, 69.5, 73.5; $\bar{x} \approx 70.36$ years.
 (b) $s \approx 1.84$ years.
13. (a) For men, midpoints are 69.5 and 73.5; $\bar{x} \approx 70.46$ years; $s \approx 1.73$ years.   (b) For women, midpoints are 77 and 80; $\bar{x} \approx 77.84$ years; $s \approx 1.36$ years.
15. (a) Midpoints and frequencies are shown on the figure. $\bar{x} \approx 7.9$ hours; $s \approx 1.05$ hours; $CV \approx 13.29$.

Section 2.3

1. 82% or more of the scores were at or below Angela's score; 18% or fewer of the scores were above Angela's score.
3. No, the score 82 might have a percentile rank less than 70.
5. (a) 50% or fewer of the cardiovascular surgeons made more than $574,769.   (b) 10% or fewer of the cardiovascular surgeons made more than $887,057.
 (c) About 40% of the cardiovascular surgeons made between $574,765 and $887,057.
7. (a) Median = 6.2% increase; Q_1 = 6.0% increase; Q_3 = 6.7% increase; low value = 1.3% increase; high value = 10.2% increase. $IQR = 0.7\%$.

(b) Percentage Increase in Faculty Salaries in Oregon

6.7%
6.2%
6.0%

1.3%

9. Low = 3;
 Q_1 = 49;
 median = 59;
 Q_3 = 59;
 high = 99;
 IQR = 10

Cost per Minute for Cellular "Roam" Calls

Cents per minute

99

Q3 = median = 59
49

3

11. (a) Low value = 4;
 Q_1 = 61.5;
 median = 65.5;
 Q_3 = 71.5;
 high value = 80.

Students' Heights (inches)

Inches

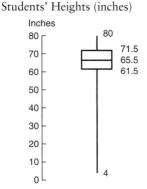

80

71.5
65.5
61.5

4

(b) $IQR = 10$.   (c) Lower limit = 46.5; upper limit
= 86.5.   (d) Yes, the value 4 is below the lower limit
and is probably an error. Our guess is that one of the
students is 4 feet tall and listed height in feet instead
of inches. There are no values above the upper limit.

Chapter 2 Review Problems

1. (a) $\bar{x} = 29$ years; median = 26.5 years; mode = 27
   years.   (b) Range = 33 years; $s = 11.06$ years
3. Suburban: low = 808; $Q_1 = 972$; median = 1081;
   $Q_3 = 1216$; high = 1292; $IQR = 244$.
   Urban: low = 1768; $Q_1 = 1968$; median = 2231.5;
   $Q_3 = 2674$; high = 2910; $IQR = 706$

Auto Insurance Premiums for Suburban and
Urban Customers (dollars)

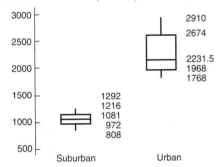

5. Mean weight = 156.25 lb.
7. (a) Low = 45; $Q_1 = 71$; median = 80; $Q_2 = 84$;
   High = 109. See figure at top of right column.
   (b)

Class Limits	Midpoint	Frequency
45–66	55.5	10
67–88	77.5	48
89–110	99.5	12

$\bar{x} \approx 78.1$; $s \approx 12.4$

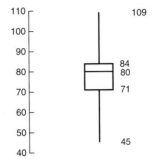

Section 3.1

1. Moderate or low linear correlation
3. High linear correlation
5. High linear correlation
7. (a) Ages and Average Weights of Shetland Ponies

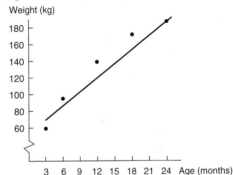

(b) Draw line you think best.
(c) High

9. (a, b) Magnitude (Richter Scale) and Depth (km) of Earthquakes

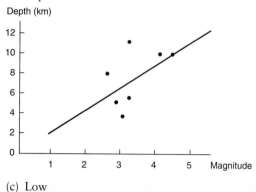

(c) Low

11. (a) Unit Length on *y* Same as That on *x*

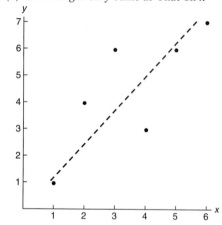

(b) Unit Length on *y* Twice That on *x*

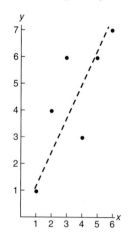

(c) Unit Length on *y* Half That on *x*

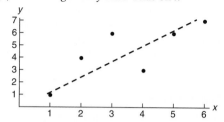

(d) The line in part b appears steeper than in part a, while the line in part c appears flatter than in part a. The slopes actually are all the same, but the lines look different because of the change in unit lengths on the *y* and *x* axes.

Section 3.2

Note: In this section and the next two, answers may vary slightly depending on how many significant digits are used throughout the calculations.

1. (a, c) Fouls and Basketball Wins

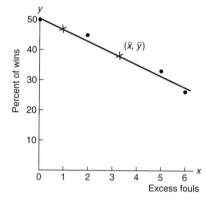

(b) $\bar{x} = 3.25$; $\bar{y} = 38.5$; $b \approx -3.934066$; $y = -3.934x + 51.29$
(d) 35.55%

3. (a) Weight of Cars and Gasoline Mileage

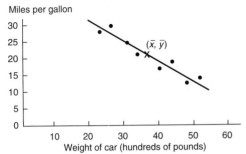

(b) $\bar{x} = 37.375$; $\bar{y} = 20.875$; $b = -0.6007$; $y = 43.3263 - 0.6007x$

(c) See figure of part a.

(d) 20.5

5. (a) Education and Income in Small Cities

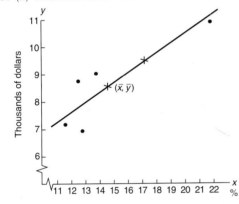

(b) $\bar{x} = 14.48$; $\bar{y} = 8.54$; $b \approx 0.31969$; $y = 0.320x + 3.91$ (c) See figure of part a. (d) 10.3 thousand

7. (a) Cultural Affiliation and Elevation of Archaeological Sites

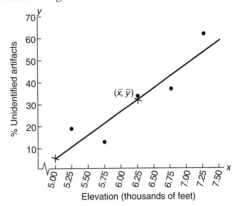

(b) $\bar{x} = 6.25$; $\bar{y} = 32.8$; $b \approx 22.0$; $y = 22.0x - 104.7$

(c) See figure of part a.

(d) 38.3

9. (a) (i) 820 (ii) 916 (iii) 96

(b) (i) 814.5 (ii) 897.5 (iii) 83

(c) (i) 34,100; 110,550 (ii) 114,700; 371,850

Section 3.3

1. (a) No

(b) Increase in population

3. (a) No

(b) Better medical treatment

5. (a) Daily Car Rental and Daily Room Rental (dollars)

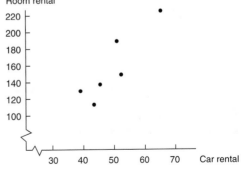

(b) r should be closest to 1.

(c) $r = 0.9089$; $r^2 = 0.8261$; 83% explained; 17% unexplained.

7. (a) Drivers' Ages and Fatal Accident Rate
Due to Speeding

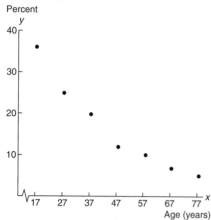

(b) Closest to -1.
(c) $r = -0.959$; $r^2 = 0.920$; 92% explained; 8% unexplained.

9. (a) Lowest Barometric Pressure and Maximum Wind Speed for Tropical Cyclones

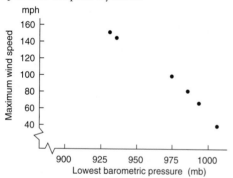

(b) Closest to -1
(c) $r = 0.9897$; $r^2 = 0.9796$; 98% explained; 2% unexplained

11. (a) $SS_{xy} = SS_{yx}$ (b) Same   (c) Same   (d) $r \approx 0.941$ in both cases; least-squares equation are not necessarily the same.

Chapter 3 Review Problems

1. (a) Age and Mortality Rate for Bighorn Sheep

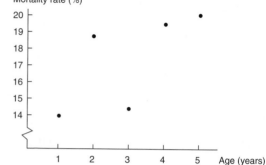

(b) $\bar{x} = 3$; $\bar{y} = 17.38$; $b = 1.27$; $y = 13.57 + 1.27x$.
(c) $r = 0.685$; $r^2 = 0.469$.

3. (a)

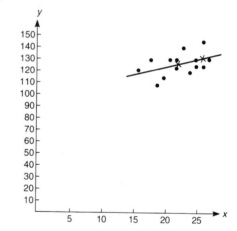

(b) $\bar{x} = 21.43$; $\bar{y} = 126.79$; $b = 1.285$; $y = 1.285x + 99.25$.
(c) See figure of part a.   (d) 124.95   (e) Positive
(f) $r = 0.47$; $r^2 = 0.221$.

5. (a)

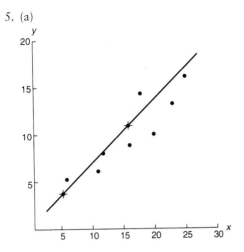

(b) $\overline{x} = 16.38$; $\overline{y} = 10.13$; $b = 0.554$; $y = 0.554x + 1.051$.   (c) See line in figure of part a.   (d) 9.36
(e) $r = 0.91$; $r^2 = 0.828$

CHAPTER 4

Section 4.1

1. See text.
3. b, since 4.1 is greater than 1; d, since −0.5 is less than 0; h, since 150% is greater than 100% or 1.
5. Answers vary. Probability as a relative frequency. One concern is whether the students in the class are more or less adept at wiggling their ears than people in the general population.
7. (a) $P(0) = 15/375$; $P(1) = 71/375$; $P(2) = 124/375$; $P(3) = 131/375$; $P(4) = 34/375$.   (b) Yes, the listed number of similar preferences form the sample space.
9. (a) P(best idea 6 A.M.–12 noon) $= 290/966 = 0.30$; P(best idea 12 noon–6 P.M.) $= 135/966 = 0.14$; P(best idea 6 P.M.–12 midnight) $= 319/966 = 0.33$; P(best idea from 12 midnight–6 A.M.) $= 222/966 = 0.23$.   (b) The probabilities add up to 1. They should add up to 1 provided that the intervals do not overlap and each inventor chose only one interval. The sample space is the set of four time intervals.

Section 4.2

1. (a) 20%; yes   (b) 40%; yes   (c) $100\% − 30\% = 70\%$
3. (a) 20% (no orange)   (b) 40%   (c) $100\% − 20\% = 80\%$
5. (a) Yes   (b) P(5 on green *and* 3 on red) $= P(5) \cdot P(3) = (1/6)(1/6) = 1/36 \approx 0.028$.   (c) P(3 on green *and* 5 on red) $= P(3) \cdot P(5) = (1/6)(1/6) = 1/36 \approx 0.028$.   (d) P((5 on green *and* 3 on red) *or* (3 on green *and* 5 on red)) $= (1/36) + (1/36) = 1/18 \approx 0.056$.
7. (a) P(sum of 6) $= P$(1 *and* 5) $+ P$(2 *and* 4) $+ P$(3 *and* 3) $+ P$(4 *and* 2) $+ P$(5 *and* 1) $= (1/36) + (1/36) + (1/36) + (1/36) + (1/36) = 5/36$.   (b) P(sum of 4) $= P$(1 *and* 3) $+ P$(2 *and* 2) $+ P$(3 *and* 1) $= (1/36) + (1/36) + (1/36) = 3/36$ or $1/12$.   (c) P(sum of 6 *or* sum of 4) $= P$(sum of 6) $+ P$(sum of 4) $= (5/36) + (3/36) = 8/36$ or $2/9$; yes.
9. (a) No, after the first draw the sample space becomes smaller and probabilities for events on the second draw change.   (b) P(ace on 1st *and* king on 2nd) $= P$(ace) $\cdot P$(king, *given* ace) $= (4/52)(4/51) = 4/663$.   (c) P(king on 1st *and* ace on 2nd) $= P$(king) $\cdot P$(ace, *given* king) $= (4/52)(4/51) = 4/663$.   (d) P(ace and king in either order) $= P$(ace on 1st *and* king on 2nd) $+ P$(king on 1st *and* ace on 2nd) $= (4/663) + (4/663) = 8/663$.
11. (a) Yes, replacement of the card restores the sample space and all probabilities for the second draw remain unchanged regardless of the outcome of the first card.   (b) P(ace on 1st *and* king on 2nd) $= P$(ace) $\cdot P$(king) $= (4/52)(4/52) = 1/169$.   (c) P(king on 1st *and* ace on 2nd) $= P$(king) $\cdot P$(ace) $= (4/52)(4/52) = 1/169$.   (d) P(ace and king in either order) $= P$(ace on 1st *and* king on 2nd) $+$ (king on 1st *and* ace on 2nd) $= (1/169) + (1/169) = 2/169$.
13. (a) P(6 years *or* older) $= P$(6–9) $+ P$(10–12) $+ P$(13 and over) $= 0.27 + 0.14 + 0.22 = 0.63$.   (b) P(12 years *or* younger) $= P$(2 and under) $+ P$(3–5) $+ P$(6–9) $+ P$(10–12) $= 0.15 + 0.22 + 0.27 + 0.14 = 0.78$.   (c) P(between 6 and 12) $= P$(6–9) $+ P$(10–12) $= 0.27 + 0.14 = 0.41$.   (d) P(between 3 and 9) $= P$(3–5) $+ P$(6–9) $= 0.22 + 0.27 = 0.49$. The category 13 and over contains far more ages than the group 10–12. It is not surprising that more toys are purchased for this group, since there are more children in this group.

15. The information from James Burke can be viewed as conditional probabilities. *P*(report lie, *given* person is lying) = 0.72 and *P*(report lie, *given* person is not lying) = 0.07.   (a) *P*(person is not lying) = 0.90; *P*(person is not lying *and* polygraph reports lie) = *P*(person is not lying) × *P*(reports lie, *given* person not lying) = (0.90)(0.07) = 0.063 or 6.3%.
(b) *P*(person is lying) = 0.10; *P*(person is lying *and* polygraph reports lie) = *P*(person is lying) × *P*(reports lie, *given* person is lying) = (0.10)(0.72) = 0.072 or 7.2%.   (c) *P*(person is not lying) = 0.5; *P*(person is lying) = 0.5; *P*(person is not lying *and* polygraph reports lie) = *P*(person is not lying) × *P*(reports lie, *given* person not lying) = (0.50)(0.07) = 0.035 or 3.5%. *P*(person is lying *and* polygraph reports lie) = *P*(person is lying) × *P*(reports lie, *given* person is lying) = (0.50)(0.72) = 0.36 or 36%.
(d) *P*(person is not lying) = 0.15; *P*(person is lying) = 0.85; *P*(person is not lying *and* polygraph reports lie) = *P*(person is not lying) × *P*(reports lie, *given* person is not lying) = (0.15)(0.07) = 0.0105 or 1.05%. *P*(person is lying *and* polygraph reports lie) = *P*(person is lying) × *P*(reports lie, *given* person is lying) = (0.85)(0.72) = 0.612 or 61.2%.

17. (a) 686/1160; 270/580; 416/580   (b) No
(c) 270/1160; 416/1160   (d) 474/1160; 310/580
(e) No   (f) 686/1160 + 580/1160 − 270/1160 = 996/1160

19. (a) 72/154   (b) 82/154   (c) 79/116   (d) 37/116
(e) 72/270   (f) 82/270

Section 4.3

1. (a) Outcomes for Tossing a Coin Three Times

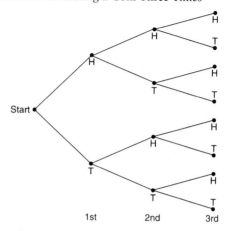

(b) 3

3. Choices for Three True/False Questions

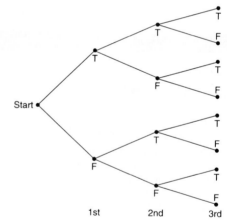

5. $4 \cdot 3 \cdot 2 \cdot 1 = 24$ ways
7. (a) $52 \cdot 52 = 2704$ (b) $4 \cdot 4 = 16$
   (c) $16/2704 = 0.006$
9. $P_{5,2} = (5!/3!) = 5 \cdot 4 = 20$
11. $P_{7,7} = (7!/0!) = 7! = 5040$
13. $C_{5,2} = (5!/(2!3!)) = 10$
15. $C_{7,7} = (7!/(7!0!)) = 1$
17. $15 \cdot 14 \cdot 13 = 2730$
19. $C_{15,5} = (15!/(5!10!)) = 3003$
21. (a) $C_{12,6} = (12!/(6!6!)) = 924$
    (b) $C_{7,6} = (7!/(6!1!)) = 7$ (c) $7/924 = 0.008$

Chapter 4 Review Problems

1. (a) $P(\text{under } 18) = 1.6/4.1 \approx 0.390$; $P(18–44) =$ $1.4/4.1 \approx 0.341$; $P(45–64) = 0.6/4.1 \approx 0.146$; $P(65$ or older$) = 0.5/4.1 \approx 0.122$ (b) $P(44 \text{ or younger}) =$ $1.4/4.1 + 1.6/4.1 \approx 0.732$.
3. (a) If the first card is replaced, the outcomes are independent. Replacing the first card restores the original sample space. If the first card is not replaced, the outcomes are not independent, because removing the first card changes the sample space.   (b) $P(\text{heart } and$ heart$) = (13/52)(13/52) = 0.063$.   (c) $P(\text{heart } and$ heart$) = (13/52)(12/51) = 0.059$.
5. (a) $P(\text{asked}) = 0.24$; $P(\text{received, } given \text{ asked}) = 0.45$; $P(\text{asked } and \text{ received}) = P(\text{asked}) \times P(\text{received, } given$ asked$) = (0.24)(0.45) = 0.108$.
7. No, the probabilities total to more than 100%, so the events are not mutually exclusive. Some companies will use several of the means of communication.
9. $C_{8,2} = (8!/(2!6!)) = (8 \cdot 7/2) = 28$.
11. $3 \cdot 2 \cdot 1 = 6$.
13. $4 \cdot 4 \cdot 4 \cdot 4 \cdot 4 = 1024$ choices; $P(\text{all correct}) =$ $1/1024 = 0.00098$.

CHAPTER 5

Section 5.1

1. (a) Discrete   (b) Continuous   (c) Continuous   (d) Discrete   (e) Continuous
3. (a) Yes   (b) No; probabilities total to more than 1.
5. (a) Yes, events are distinct and probabilities total 1
   (b) Income Distribution ($1000)

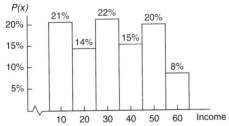

(c) 32.3   (d) 16.12

7. (a) Number of Fish Caught in a 6-Hour Period at Pyramid Lake, Nevada

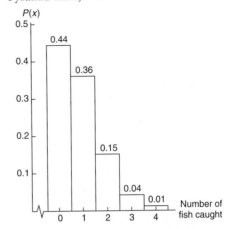

(b) 0.56   (c) 0.20   (d) 0.82   (e) 0.899
9. (a) 0.00756; $378   (b) $412.50; $448; 482.50; $517.50; $2238.5 total   (c) $2938.50   (d) $2761.50

Section 5.2

1. A trial is one flip of a fair quarter. Success = coin shows heads. Failure = coin shows tails. $n = 3; p = 0.5; q = 0.5$.   (a) $P(r = 3 \text{ heads}) = C_{3,3}p^3q^0 = 1(0.5)^3(0.5)^0 = 0.125$. To find this value in Table 3 of Appendix I, use the group in which $n = 3$, the column headed by $p = 0.5$, and the row headed by $r = 3$.   (b) $P(r = 2 \text{ heads}) = C_{3,2}p^2q^1 = 3(0.5)^2(0.5)^1 = 0.375$. To find this value in Table 3 of Appendix I, use the group in which $n = 3$, the column headed by $p = 0.5$, and the row headed by $r = 2$.   (c) $P(r \text{ is } 2$ or more$) = P(r = 2 \text{ heads}) + P(r = 3 \text{ heads}) = 0.375 + 0.125 = 0.500$.   (d) The probability of getting three tails when you toss a coin three times is the same as getting no or zero heads. Therefore, $P(3 \text{ tails}) = P(r = 0 \text{ heads}) = C_{3,0}p^0q^3 = 1(0.5)^0(0.5)^3 = 0.125$. To find this value in Table 3 of Appendix I, use the group in which $n = 3$, the column headed by $p = 0.5$, and the row headed by $r = 0$.
3. (a) A trial is a man's response to the question, "Would you marry the same woman again?" Success = a positive response. Failure = a negative response. $n = 10; p = 0.80; q = 0.20$. Using values in Table 3 of Appendix I, $P(r \text{ is at least } 7) = P(r = 7) + P(r = 8) + P(r = 9) + P(r = 10) = 0.201 + 0.302 + 0.268$

+ 0.107 = 0.878. $P(r$ is less than half of 10) = $P(r <$ 5) = $P(r = 0) + P(r = 1) = P(r = 2) + P(r = 3) + P(r = 4) = 0.000 + 0.000 + 0.000 + 0.001 + 0.006 = 0.007.   (b) A trial is a woman's response to the question, "Would you marry the same man again?" Success = a positive response. Failure = a negative response. $n = 10; p = 0.5; q = 0.5.$ Using values in Table 3 of Appendix I, $P(r$ is at least 7) = $P(r = 7) + P(r = 8) + P(r = 9) + P(r = 10) = 0.117 + 0.044 + 0.010 + 0.001 = 0.172.$ $P(r$ is less than half of 10) = $P(r < 5) = P(r = 0) + P(r = 1) + P(r = 2) + P(r = 3) + P(r = 4) = 0.001 + 0.010 + 0.044 + 0.117 + 0.205 = 0.377.$

5. A trial consists of a woman's response regarding her mother-in-law. Success = dislike. Failure = like. $n = 6; p = 0.90; q = 0.10.$ (a) $P(r = 6) = 0.531.$ (b) $P(r = 0) = 0.000$ (to 3 digits).   (c) $P(r \geq 4) = P(r = 4) + P(r = 5) + P(r = 6) = 0.098 + 0.354 + 0.531 = 0.983.$ (d) $P(r \leq 3) = 1 - P(r \geq 4) \approx 1 - 0.983 = 0.017$ or 0.016 directly from table.

7. A trial is taking a polygraph exam. Success = pass. Failure = fail. $n = 9; p = 0.85; q = 0.15.$ (a) $P(r = 9) = 0.232.$ (b) $P(r \geq 5) = P(r = 5) + P(r = 6) + P(r = 7) + P(r = 8) + P(r = 9) = 0.028 + 0.107 + 0.260 + 0.368 + 0.232 = 0.995.$ (c) $P(r \leq 4) = 1 - P(r \geq 5) \approx 1 - 0.995 = 0.005$ or 0.006 directly from table.   (d) $P(r = 0) = 0.000$ (to 3 digits).

9. A trial is catching and releasing a pike. Success = pike dies. Failure = pike lives. $n = 16; p = 0.05; q = 0.95.$ (a) $P(r = 0) = 0.440.$ (b) $P(r < 3) = 0.957.$ (c) $P(r = 0) = 0.440$ (all live is equivalent to none die).   (d) Change success to live; $p = 0.95; P(r > 14) = 0.811$

11. A trial consists of selecting the "best" coffee. Success = select Tasty Bean. Failure = not select Tasty Bean. $n = 4; p = 0.20; q = 0.80$ (a) $P(r = 4) = 0.002$ (b) $P(r = 0) = 0.410$ (c) $P(r \geq 3) = 0.028$

13. (a) $n = 10; p = 0.40; P(r = 0) = 0.006.$ (b) $n = 10; p = 0.40; P(r < 5) = 0.633.$ (c) $n = 10; p = 0.30; P(r \leq 2) = 0.382.$ (d) $n = 10; p = 0.30; P(r \geq 6) = 0.047.$

15. (a) 0.812515; yes, truncated at 5 digits   (b) 0.187486; 0.18749; yes, rounded to 5 digits.

17. (a) They are the same.   (b) They are the same.   (c) $r = 1$ (d) The one headed by $p = 0.80$

Section 5.3

1. (a) Binomial Distribution
The distribution is symmetrical.

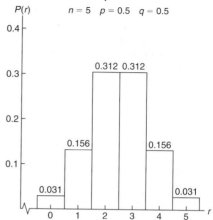

$P(r)$ $n = 5$ $p = 0.5$ $q = 0.5$

(b) Binomial Distribution
The distribution is skewed right.

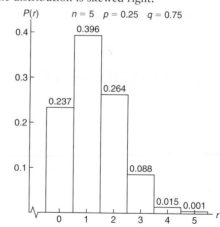

$P(r)$ $n = 5$ $p = 0.25$ $q = 0.75$

(c) Binomial Distribution
   The distribution is skewed left.

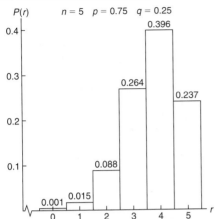

$P(r)$ $n = 5$ $p = 0.75$ $q = 0.25$

(d) The distributions are mirror images of one another. (e) The distribution would be skewed left for $p = 0.73$ because the more likely number of successes are to the right of the middle.

3. (a) Households with Children Under 2 That Buy Film

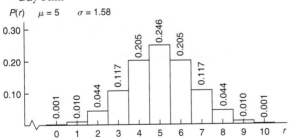

$P(r)$ $\mu = 8$ $\sigma = 1.26$

(b) Households with No Children Under 21 That Buy Film

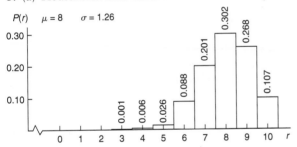

$P(r)$ $\mu = 5$ $\sigma = 1.58$

(c) Yes.

5. (a) Binomial Distribution for Number of Automobile Damage Claims by People Under Age 25

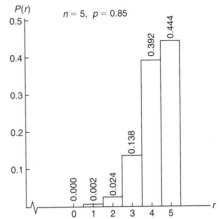

$P(r)$ $n = 5$, $p = 0.85$

(b) $\mu = 4.25$; $\sigma = 0.798$; expected number is about 4.
7. (a) Yes, symmetric; $\mu = 5.17$; Yes
   (b) Yes; yes; right; $\mu = 3.96$; left
   (c) 0.64; yes; left; $\mu = 7.04$; right.
   (d) They are the same respectively.
   (e) $\sigma = 1.59$; yes, because the values for p and q are exchanged.

Chapter 5 Review Problems

1. (a) Expected value $= \mu = 38$; $\sigma = 11.6$
   (b) Leases in Months

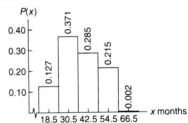

$P(x)$

3. (a) Claimant Under 25

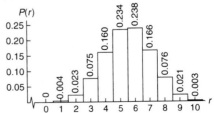

$P(r)$

(b) $P(r \geq 6) = 0.504$

(c) $\mu = 5.5$; $\sigma = 1.57$

5. (a) $P(r \geq 12) = 0.039$ (b) $P(r \leq 7) = 0.403$

(c) $\mu = 8$

7. (a)

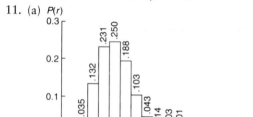

(b) $P(r \geq 9) = 0.244$, $P(r \geq 1) = 0.999$.

(c) The expected number is $\mu = 7.5$.

(d) $\sigma = 1.37$.

9. The expected number is $\mu = 102$.

11. (a)

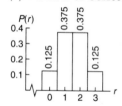

(b) The expected number is $\mu = 3$, $\sigma = 1.55$.

(c) $P(r \geq 3) = 0.602$, $P(r = 3) = 0.250$.

13. (a) Number of Correct Answers

P(r)
0.4 ┤ 0.375 0.375
0.3 ┤
0.2 ┤ 0.125   0.125
0.1 ┤
    └─────────────────── r
      0   1   2   3

(b) $\mu = 1.5$

(c) $\sigma = 0.866$

CHAPTER 6

Section 6.1

1. (a) No, it's skewed.

(b) No, it crosses the horizontal axis.

(c) No, it has three peaks.

(d) No, the curve is not smooth.

3. The figure on the left has the larger standard deviation. Its mean is $\mu = 10$. The mean of the figure on the right is $\mu = 4$.

5. (a) 50%   (b) 68.2%   (c) 99.7%

7. (a) 50%   (b) 50%   (c) 68.2%   (d) 95.4%

9. (a) From 1207 to 1279   (b) From 1171 to 1315

(c) From 1135 to 1351

11. (a) From 1.70mA to 4.50 mA

(b) From 0.25 mA to 6.05 mA

13. (a) Unusually high; type II; special tours, conventions, or events.

(b) Mixed; two type III on low side and one type I on high side; for low occupancy, last minute cancellations, nearby road construction, weather; for high occupancy, special events or conventions.

Section 6.2

1. (a) Robert, Jan, and Linda each scored above the mean.   (b) Joel scored on the mean.   (c) Susan and John scored below the mean.   (d) Robert, 172; Jan, 184; Susan, 110; Joel, 150; John, 134; Linda, 182.

3. (a) $-4.00 < z < 4.00$ (b) $z < -1.6$ (c) $1.00 < z$

(d) $81.75°F < x$ (e) $x < 63.5°F$ (f) $64°F < x < 81.25°F$

5. (a) $-1.77 < z$ (b) $z < 1.61$ (c) $-1.45 < z < 1.45$

(d) $3706 < x < 5907$ (e) $x < 5615$ (f) $6000 < x$

(g) A population of 2800 deer corresponds to a z value of -2.58. Data values this far below the mean occur less than 2.5% of the time. This would be an unusually low number. The population 6300 corresponds to a z value of 3.06. Fall deer populations are practically never that large. Such a population would be considered an unusually high population.

7. (a) $-1.00 < z$ (b) $z < -2.00$ (c) $-2.67 < z < 2.33$ (d) $x < 4.4$ (e) $5.2 < x$ (f) $4.1 < x < 4.5$

(g) A red blood cell count of 5.9 or higher corresponds to a standard z score of 3.67. Practically no data values occur this far above the mean. Such a count would be considered unusually high for a healthy female.

9. 0.4993

11. 0.4778

13. 0.8953

15. 0.3471

17. 0.0306

19. 0.5000

21. 0.0643

23. 0.0934

25. 0.8888

27. 0.6736

29. 0.4474

31. 0.2939

33. 0.6704

35. 0.3226

37. 0.8808

39. 0.5000

41. 0.0885

43. 0.4483

45. 0.8849

47. 0.8849

Section 6.3

1. $P(3 \leq x \leq 6) = P(-0.50 \leq z \leq 1.00) = 0.5328$
3. $P(50 \leq x \leq 70) = P(0.67 \leq z \leq 2.00) = 0.2286$
5. $P(8 \leq x \leq 12) = P(-2.19 \leq z \leq -0.94) = 0.1593$
7. $P(x \geq 30) = P(z \geq 2.94) = 0.0016$
9. $P(x \geq 90) = P(z \geq -0.67) = 0.7486$
11. 1.645
13. −1.41
15. −1.555
17. 1.41
19. ±2.33
21. (a) $P(x > 60) = P(z > -1) = 0.8413$
    (b) $P(x < 110) = P(z < 1) = 0.8413$
    (c) $P(60 \leq x \leq 110) = P(-1.00 \leq z \leq 1.00) = 2(0.3413) = 0.6826$
    (d) $P(x > 140) = P(z > 2.20) = 0.0139$
23. (a) $P(x > 675) = P(z > 1.75) = 0.0401$
    (b) $P(x < 450) = P(z < -0.50) = 0.3085$
    (c) $P(450 \leq x \leq 675) = P(-0.50 \leq z \leq 1.75) = 0.6514$
    (d) $P(x > 28) = P(z > 1.67) = 0.0475$
    (e) $P(x > 12) = P(z > -1.00) = 0.8413$ (f) $P(12 \leq x \leq 28) = P(-1.00 \leq z \leq 1.67) = 0.7938$
25. (a) $P(x < 3.0 \text{ mm}) = P(z < -2.33) = 0.0099$
    (b) $P(x > 7.0 \text{ mm}) = P(z > 2.11) = 0.0174$
    (c) $P(3.0 \text{ mm} < x < 7.0 \text{ mm}) = P(-2.33 < z < 2.11) = 0.9727$
27. (a) $P(x < 36 \text{ mo}) = P(z < -1.13) = 0.1292$. They will replace 13% of their batteries.
    (b) $P(z < z_0) = 10\%$ for $z_0 = -1.28$; $x = -1.28(8) + 45 = 34.76$; Guarantee the batteries for 35 months.
29. (a) $\sigma \approx 12$ beats/min
    (b) $P(x < 25) = P(z < -1.75) = 0.0401$
    (c) $P(x > 60) = P(z > 1.17) = 0.1210$
    (d) $P(25 \leq x \leq 60) = P(-1.75 \leq z \leq 1.17) = 0.8389$ (e) $P(z \leq z_0) = 0.90$ for $z = 1.28$; $x = 1.28(12) + 46 = 61.36$ beats/min. A heart rate of 61 beats/min corresponds to the 90% cutoff point of the distribution.
31. (a) red, black, blue   (b) blue, red
    (c) red, blue

Section 6.4

Note: Answers may differ slightly depending on how many digits are carried in the computation of the standard deviation and the computation of z.

1. (a) $P(r \geq 50) = P(x \geq 49.5) = P(z \geq -27.53) \approx 1$ or almost certain.
   (b) $P(r \geq 50) = P(x \geq 49.5) = P(z \geq 7.78) \approx 0$ or almost impossible for a random sample.
3. (a) $P(r \geq 50) = P(x \geq 49.5) = P(z \geq 3.04) = 0.0012$
   (b) $P(r \leq 30) = P(x \leq 30.5) = P(z \leq -0.46) = 0.3228$
   (c) $P(30 \leq r \leq 50) = P(29.5 \leq x \leq 50.5) = P(-0.65 \leq z \leq 3.23) = 0.7416$
   (d) $n = 300$; $p = 0.11$; $q = 0.89$; yes; the normal approximation to the binomial is appropriate.
5. (a) $P(r \geq 15) = P(x \geq 14.5) = P(z \geq -2.35) = 0.9906$.   (b) $P(r \geq 30) = P(x \geq 29.5) = P(z \geq 0.62) = 0.2676$.   (c) $P(25 \leq r \leq 35) + P(24.5 \leq x \leq 35.5) = P(-0.37 \leq z \leq 1.81) = 0.6092$.   (d) $P(r > 40) = P(r \geq 41) = P(x \geq 40.5) = P(z \geq 2.80) = 0.0026$.
7. (a) $P(r \geq 47) = P(x \geq 46.5) = P(z \geq -1.94) = 0.9738$.   (b) $P(r \leq 58) = P(x \leq 58.5) = P(z \leq 1.75) = 0.9599$. In parts c and d, let r be the number of products that succeed, and use $p = 1 - 0.80 = 0.20$.   (c) $P(r \geq 15) = P(x \geq 14.5) = P(z \geq 0.40) = 0.3446$.
   (d) $P(r < 10) = P(r \leq 9) = P(x \leq 9.5) = P(z \leq -1.14) = 0.1271$.
9. (a) $P(r > 280) = P(r \geq 281) = P(x > 280.5) = P(z \geq -2.16) = 0.9846$.   (b) $P(r \geq 320) = P(x \geq 319.5) = P(z \geq 1.95) = 0.0256$.   (c) $P(280 \leq r \leq 320) = P(279.5 \leq x \leq 320.5) = P(-2.26 \leq z \leq 2.05) = 0.9679$.   (d) $n = 430$; $p = 0.70$; $q = 0.30$; np and nq are both greater than 5. These conditions mean that the normal approximation to the binomial is appropriate.
11. (a) $P(r \geq 540) = P(x \geq 539.5) = P(z \geq 3.81) \approx 0.0001$.   (b) $P(r \leq 500) = P(x \leq 500.5) = P(z \leq 1.11) = 0.8665$.   (c) $P(485 \leq r \leq 525) = P(484.5 \leq x \leq 525.5) = P(0 \leq z \leq 2.84) = 0.4977$.

Chapter 6 Review Problems

1. (a) 0.4599   (b) 0.4015   (c) 0.0384   (d) 0.0104
   (e) 0.0250   (f) 0.8413
3. (a) 0.9821   (b) 0.3156   (c) 0.2977
5. 1.645
7. $z = \pm 1.96$
9. (a) 0.89   (b) 0   (c) 0.2514
11. (a) 0.0166   (b) 0.975
13. (a) 0.9772   (b) 17.3 hr.
15. (a) From \$1.81 to \$3.51 is a 68% range of errors.
    (b) From \$0.96 to \$4.36 is a 95% range of errors.
    (c) Almost all errors are from \$0.11 to \$5.21.
17. (a) 0.8665   (b) 0.7330

CHAPTER 7

Section 7.1

1. A set of measurements or counts either existing or conceptual. For example, the population of all ages of all people in Colorado; the population of weights of all students in your school; the population count of all antelope in Wyoming.
3. A numerical descriptive measure of a population, such as μ, the population mean; σ, the population standard deviation; σ^2, the population variance.
5. A statistical inference is a conclusion about the value of a population parameter. We will do both estimation and testing.
7. They help us visualize the sampling distribution by using tables and graphs that approximately represent the sampling distribution.
9. We studied the sampling distribution of mean trout lengths based on samples of size 5. Other such sampling distributions abound.

Section 7.2

Note: Answers may differ slightly depending on the number of digits carried in the standard deviation.

1. (a) $\mu_{\bar{x}} = 15$; $\sigma_{\bar{x}} = 2.0$; $P(15 \leq \bar{x} \leq 17) = P(0 \leq z \leq 1.00) = 0.3413$.   (b) $\mu_{\bar{x}} = 15$; $\sigma_{\bar{x}} = 1.75$; $P(15 \leq \bar{x} \leq 17) = P(0 \leq z \leq 1.14) = 0.3729$.   (c) The standard deviation is smaller in part b because of the larger sample size. Therefore, the distribution about $\mu_{\bar{x}}$ is narrower in part b.
3. (a) No; the sample size is only 9 and so is too small.
   (b) Yes; the $\bar{x}$ distribution also will be normal with $\mu_{\bar{x}} = 25$; $\sigma_{\bar{x}} = 3.5/3$; $P(23 \leq \bar{x} \leq 26) = P(-1.71 \leq z \leq 0.86) = 0.7615$.
5. (a) $P(x < 74.5) = P(z < -0.63) = 0.2643$.   (b) $P(\bar{x} < 74.5) = P(z < -2.79) = 0.0026$.   (c) No. If the weight of only one car were less than 74.5 tons, we cannot conclude that the loader is out of adjustment. If the mean weight for a sample of 20 cars were less than 74.5 tons, we would suspect that the loader is malfunctioning. As we see in part b, the probability of this happening is very low if the loader is correctly adjusted.
7. (a) $P(x < 40) = P(z < -1.80) = 0.0359$.   (b) Since the x distribution is approximately normal, the $\bar{x}$ distribution is approximately normal with mean 85 and standard deviation 17.678. $P(\bar{x} < 40) = P(z < -2.55) = 0.0054$.   (c) $P(\bar{x} < 40) = P(z < -3.12) = 0.0009$.   (d) $P(\bar{x} < 40) = P(z < -4.02) = 0.0001$.
   (e) Yes; If the average value based on five tests were less than 40, the patient is almost certain to have excess insulin.
9. (a) $P(x < 54) = P(z < -1.27) = 0.1020$.   (b) The expected number undernourished is 2200(0.1020), or about 224.   (c) $P(\bar{x} \leq 60) = P(z \leq -2.99) = 0.0014$.   (d) $P(\bar{x} < 64.2) = P(z < 1.20) = 0.8849$. Since the sample average is above the mean, it is quite unlikely that the doe population is undernourished.
11. (a) 30 or more   (b) No

Chapter 7 Review Problems

1. (a) A normal distribution.
   (b) The mean μ of the x distribution.
   (c) $\sigma/\sqrt{n}$, where σ is the standard deviation of the x distribution.
   (d) They will both be approximately normal with the same mean, but the standard deviations will be $\sigma/\sqrt{50}$ and $\sigma/\sqrt{100}$ respectively.
3. (a) $P(x \geq 40) = P(z \geq 0.71) = 0.2389$.
   (b) $P(\bar{x} \geq 40) = P(z \geq 2.14) = 0.0162$.
5. $P(98 \leq \bar{x} \leq 102) = P(-1.33 \leq z \leq 1.33) = 0.8164$

CHAPTER 8

Section 8.1

1. 11.3 mg/liter to 12.5 mg/liter
3. 15,342 cars to 16,658 cars
5. (a) 1.10 sec to 1.30 sec   (b) 560.87 Hz to 657.13 Hz
7. (a) The mean and standard deviation round to the values given.   (b) Using the rounded values of part a, the 75% interval is from 34.19 thousand to 37.81 thousand.   (c) Yes. 30 thousand dollars is below the mean annual profit per employee. We can say with 75% confidence that the mean lies between 34.19 thousand and 37.81 thousand.   (d) Yes. 40 thousand is above the mean annual profit per employee. (e) 33.41 thousand to 38.59 thousand. We can say with 90% confidence that the mean lies between 33.4 thousand and 38.6 thousand dollars. 30 thousand is below the mean and 40 thousand is above the mean.
9. (a) The mean and standard deviation round to the values given.   (b) Using the rounded values for the mean and standard deviation given in part a, the interval is from 143.8 to 149.2.   (c) Using the rounded values for the mean and standard deviation given in part a, the interval is from 143.0 to 150.0.   (d) Using the rounded values for the mean and standard deviation given in part a, the interval is from 141.0 to 152.0. (e) The lengths increase as c increases because the values of z_c increase as c increases. If we want to be more certain that μ is in the interval based on the given sample, we have to make the interval wider.
11. (a) 2.3 to 2.7   (b) 13.9 to 16.5   (c) 23.5 to 27.9 (d) The intervals got longer as s increased. This is to be expected, since the endpoints are $\pm z_c s/\sqrt{n}$ from the sample mean and z_c and n were fixed. The length of each interval is $2z_c s/\sqrt{n}$.

Section 8.2

1. 2.110
3. 1.721
5. (a) Use calculator   (b) $120.84 to $175.82
7. (a) Use calculator   (b) 74.7 lb to 107.3 lb
9. (a) Use calculator   (b) $10.84 to $13.86
11. (a) They differ in total length, location of box, median, and whisker length. Different samples contain different data values. The sample means $\bar{x}$ differ as well as the sample standard deviations.

(b) The length of the intervals range from 2.234 to 2936 units. Notice that the standard deviations of the samples vary, and so the error of estimate E varies. However, all the intervals contain the population mean $\mu = 68$. Note that if we generated more samples and confidence intervals, some intervals might not contain $\mu = 68$.

Section 8.3

1. (a) $\hat{p} = 39/62 = 0.6290$.   (b) 0.51 to 0.75. If this experiment were repeated many times, about 95% of the intervals would contain p.   (c) Both np and nq are greater than 5. If either is less than 5, the normal curve will not necessarily give a good approximation to the binomial.
3. (a) $\hat{p} = 1619/5222 = 0.3100$.   (b) 0.29 to 0.33. If we repeat the survey with many different samples of 5222 dwellings, about 99% of the intervals will contain p.   (c) Both np and nq are greater than 5. If either is less than 5, the normal curve will not necessarily give a good approximation to the binomial.
5. (a) $\hat{p} = 0.5420$.   (b) 0.53 to 0.56.   (c) Yes. np and nq are greater than 5.
7. (a) $\hat{p} = 0.0304$.   (b) 0.02 to 0.05.   (c) Yes. np and nq are greater than 5.
9. (a) 0.239 to 0.293
   (b) 0.257 to 0.275
   (c) The confidence interval based on a larger sample is shorter or narrower.
11. (a) $\hat{p} = 0.8603$ (b) 0.83 to 0.89   (c) A recent study shows that 86% of women shoppers remained loyal to their favorite supermarket last year. The margin of error was 2.5 percentage points.
13. 11% to 17%

Section 8.4

1. Estimate a mean. Use 75 plots.
3. Estimate a mean; 120 total or 64 more
5. Estimate a mean; 117 or 34 more
7. Estimate a mean; 385 or 218 more
9. (a) Estimate a proportion; 385   (b) 186
11. (a) Estimate a proportion; 666   (b) 659
13. (a) Estimate a proportion; 1068   (b) 1016

Chapter 8 Review Problems

1. See text.
3. Interval for mean, large sample; 176.91 to 180.49.
5. Interval for mean, small sample
   (a) Use calculator.   (b) 64.1 to 84.3.
7. Interval for proportion; 0.50 to 0.54
9. Interval for proportion
   (a) $\hat{p} = 0.4072$ (b) 0.333 to 0.482
11. (a) Mean and standard deviation round to results shown.
    (b) Confidence interval for a mean with small sample. Using the rounded values for the mean and standard deviation given in part a, the interval is from 8.3 to 10.8.

CHAPTER 9

Section 9.1

1. See text.
3. No, if we fail to reject the null hypothesis, we have not proven it beyond all doubt. We have failed only to find sufficient evidence to reject it.
5. (a) $H_0: \mu = 60$ kg.   (b) $H_1: \mu < 60$ kg.   (c) $H_1: \mu > 60$ kg.   (d) $H_1: \mu \neq 60$ kg.   (e) For part b the critical region is on the left. For part c the critical region is on the right. For part d the critical region is on both sides of the mean.
7. (a) $H_0: \mu = 288$ lb.   (b) Higher: $H_1: \mu > 288$ lb; lower: $H_1: \mu < 288$ lb; different: $H_1: \mu \neq 288$ lb. For higher the critical region is the right tail; for lower the critical region is the left tail; for different the critical region is both tails.

Section 9.2

1. $H_0: \mu = 16.4$ feet; $H_1: \mu < 16.4$ feet; left-tailed; normal distribution; $z_0 = -2.33$; sample $z = -2.44$; Reject H_0; The storm is lessening; Results statistically significant.
3. $H_0: \mu = 31.8$ calls/day; $H_1: \mu \neq 31.8$ calls/day; two-tailed; normal distribution; $z_0 = \pm 2.58$; sample $z = -2.45$; Do not reject H_0; There is not enough evidence that the mean number of messages has decreased; Results not statistically significant.

5. $H_0: \mu = \$4.75$; $H_1: \mu > \$4.75$; right-tailed; normal distribution; $z_0 = 2.33$; sample $z = 3.14$; Reject H_0; There is evidence that her average tip is more than \$4.75; Results statistically significant.
7. $H_0: \mu = 10.2$ seconds; $H_1: \mu < 10.2$ seconds; left-tailed; normal distribution; $z_0 = -1.645$; sample $z = -1.52$; Do not reject H_0. There is not enough evidence to conclude that the mean acceleration time is less; Results not statistically significant.
9. $H_0: \mu = 19.0$; $H_1: \mu > 19.0$; right-tailed; normal distribution; $z_0 = 2.33$; sample $z = 1.19$; Do not reject H_0. There is not sufficient evidence to conclude that the average oxygen capacity has increased; Results not statistically significant.
11. $H_0: \mu = 7.4$; $H_1: \mu \neq 7.4$; two-tailed; normal distribution; $z_0 = \pm 1.96$; sample $z = 6.70$; Reject H_0; The drug has changed the pH of the blood; Results are statistically significant.
13. Essay or class discussion.

Section 9.3

1. Sample $z = 2.78$; P value $= 0.0027$; The data are significant at the 1% level.
3. Sample $z = -1.18$; P value $= 2(0.1190) = 0.238$; No, not significant at the 5% level.
5. $H_0: \mu = 1.75$; $H_1: \mu > 1.75$; sample $z = 3.02$; P value $= 0.0013$; Yes, the data are significant at the 1% level.
7. $H_0: \mu = 19$ inches; $H_1: \mu < 19$ inches; sample $z = -0.80$; P value $= 0.2119$; No, the data are not significant at the 5% level. We cannot conclude that the average length of trout caught in Pyramid Lake is different from 19 inches.
9. $H_0: \mu = \$15.35$; $H_1: \mu < 15.35$; sample $z = -3.29$; P value $= 0.0005$; Yes, the data are significant at the 5% level. There is evidence that college students' average daily ownership expenses are less than the national average.

Section 9.4

1. $t_0 = -1.860$
3. $t_0 = \pm 2.807$
5. $t_0 = \pm 2.201$
7. (a) Rounded answers are used in part b.
   (b) $H_0: \mu = 4.8$; $H_1: \mu < 4.8$; left-tailed; $t_0 = -2.015$; sample $t = -4.06$; P value < 0.005; Reject H_0; At the 5% significance level, this patient's average red blood count is less than 4.8.
9. (a) Rounded answers are used in part b.
   (b) $H_0: \mu = 16.5$ days; $H_1: \mu \neq 16.5$ days; two-tailed; $t_0 = \pm 2.110$; sample $t = 2.565$; $0.020 < P$ value < 0.050; Reject H_0; At the 5% level the mean incubation time is different from 16.5 days.
11. (a) Rounded answers are used in part b.
    (b) $H_0: \mu = 1300$; $H_1: \mu \neq 1300$; two-tailed; $t_0 = \pm 3.250$; sample $t = -2.71$; $0.020 < P$ value < 0.050; Do not reject H_0 at the 1% level of significance. There is not enough evidence to conclude that the population mean of tree ring dates is different from 1300.
13. (a) Rounded answers used in part b.
    (b) $H_0: \mu = 6.6$; $H_1: \mu < 6.6$; Left-tailed; $t_0 = -1.895$; sample $t = -1.414$; $0.100 < P$ value < 0.125; Do not reject H_0. The population average annual profits as a percentage of assets is not less than 6.6.

Section 9.5

1. $H_0: p = 0.70$; $H_1: p \neq 0.70$; two-tailed; normal; $z_0 = \pm 2.58$; sample $\hat{p} = 0.75$ with $z = 0.62$; P value $= 2(0.2676) = 0.5352$; Do not reject H_0; The population proportion of such arrests is not significantly different from 0.70.
3. $H_0: p = 0.77$; $H_1: p < 0.77$; left-tailed; normal; $z_0 = -2.33$; sample $\hat{p} = 0.5556$ with $z = -2.65$; P value $= 0.004$; Reject H_0; The population proportion of driver fatalities related to alcohol is less than 77% at the 1% significance level.
5. $H_0: p = 0.50$; $H_1: p < 0.50$; left-tailed; normal; $z_0 = -2.33$; sample $\hat{p} = 0.2941$; $z = -2.40$; P value $= 0.0082$; Reject H_0. The population proportion of female wolves is less than 50% at the 1% significance level.

7. $H_0: p = 0.261$; $H_1: p \neq 0.261$; two-tailed; normal; $z_0 = \pm 2.58$; sample $\hat{p} = 0.1924$ with $z = -2.78$; P value $= 2(0.0027) = 0.0054$; Reject H_0. The population proportion of this type of five-syllable sequence is significantly different from that of Plato's *Republic*.
9. $H_0: p = 0.47$; $H_1: p > 0.47$; right-tailed; normal; $z_0 = 2.33$; sample $\hat{p} = 0.4871$ with $z = 1.09$; P value $= 0.1379$; Do not reject H_0. The population loyalty of Chevrolet owners is not significantly greater than 47%.
11. $H_0: p = 0.65$; $H_1: p \neq 0.65$; two-tailed test; $\pm z_0 = \pm 1.96$; z value corresponding to $\hat{p} = 60/78$ is $z = 2.21$; P value 0.0272. Since the sample test statistic falls in the critical region, we reject H_0. It seems that the proportion of office workers who decided to take their current job because of open communication with their supervisor is different from 65%.
13. $H_0: p = 0.82$; $H_1: p \neq 0.82$; two-tailed; normal; $z_0 = \pm 2.58$; sample $\hat{p} = 0.7671$ with $z = -1.18$; P value $= 2(0.1190) = 0.2380$; Do not reject H_0. The population proportion of extroverts among college leaders is not different from 82% at the 1% significance level.

Chapter 9 Review Problems

1. Single mean, large sample; $H_0: \mu = 11.3$; $H_1: \mu \neq 11.3$; two-tailed test; $z_0 = \pm 1.96$; z value corresponding to $\overline{x}$ is -3.00; P value ≈ 0.0026; reject H_0. The average number of miles driven in Chicago is different from the national average.
3. Single mean, small sample; $H_0: \mu = 10.8$; $H_1: \mu < 10.8$; left-tailed test; $d.f. = 19$; $t_0 = -2.539$; t value corresponding to $\overline{x}$ is $t = 2.385$; $0.025 > P$ value > 0.010; Do not reject H_0. There is not sufficient evidence at the 1% level of significance to conclude that a new patient waits less than 10.8 days for an appointment at Rockwood Clinic.
5. Single mean, large sample; $H_0: \mu = 88.9$; $H_1: \mu > 88.9$; right-tailed test; $z_0 = 1.645$; z value corresponding to $\overline{x}$ is $z = 1.01$; P value $= 0.1562$; Do not reject H_0. At the 5% level of significance there is not enough evidence to conclude that the average sales value per employee of the chain is greater than the industry apparel-store average.

7. Single proportion; H_0: $p = 0.846$; H_1: $p < 0.846$; left-tailed test; $z_0 = -2.33$; z value corresponding to $\hat{p} = 0.72$ is $z = -2.47$; P value $= 0.0068$; Reject H_0. At the 1% level of significance there is enough evidence to conclude that a smaller proportion of the older group believe in Mr./Ms. Right.

9. Single proportion; H_0: $p = 0.36$; H_1: $p < 0.36$; left-tailed test; $z_0 = -1.645$; z value corresponding to $\hat{p} = 33/120$ is $z = -1.94$; P value $= 0.0262$; Reject H_0. At the 5% level of significance we can conclude that the proportion of private sector employees holding bachelor or higher degrees is less than that in the federal civilian sector.

11. (a) The mean and standard deviation round to the given values. (b) Single mean, small sample; H_0: $\mu = 48$; H_1: $\mu < 48$; left-tailed test; $d.f. = 9$; $t_0 = -1.833$; t value of sample statistic is $t = -0.525$; $0.125 < P$ value; fail to reject H_0. There is not enough evidence to conclude that the average fuel injection system lasts less than 48 months.

CHAPTER 10

Section 10.1

1. H_0: $\mu_d = 0$; H_1: $\mu_d \neq 0$; two-tailed; t distribution with $d.f. = 7$; $t_0 = \pm2.365$; $\overline{d} = 2.25$; $s_d = 7.78$; sample $t = 0.818$; P value > 0.25; Do not reject H_0. There is not a significant difference between the population mean percentage increase in corporate revenue and the population mean percentage increase of CEO salary.

3. H_0: $\mu_d = 0$; H_1: $\mu_d > 0$; right-tailed; t distribution with $d.f. = 4$; $t_0 = 3.747$; $\overline{d} = 12.6$; $s_d = 22.66$; sample $t = 1.243$; P value > 0.125; Fail to reject H_0; At the 1% level of significance we do not conclude that average peak wind gusts are higher in January than they are in April.

5. H_0: $\mu_d = 0$; H_1: $\mu_d \neq 0$; two-tailed; t distribution with $d.f. = 11$; $t_0 = \pm2.201$; $\overline{d} = -1.15$; $s_d = 1.633$; sample $t = -2.440$; $0.020 < P$ value < 0.050; Reject H_0. We have enough evidence to conclude at the 5% level that the average temperature in Buffalo is different from the average temperature during this period in Grand Rapids.

7. H_0: $\mu_d = 0$; H_1: $\mu_d > 0$; right-tailed; t distribution with $d.f. = 7$; $t_0 = 1.895$; $\overline{d} = 6$; $s_d = 21.5$; sample $t = 0.789$; P value > 0.125; Do not reject H_0; We do not have enough evidence to conclude that the average number of inhabited houses is greater than the average number of hogans on the Navaho Reservation.

9. H_0: $\mu_d = 0$; H_1: $\mu_d \neq 0$; two-tailed; t distribution with $d.f. = 4$; $t_0 = \pm2.776$; $\overline{d} = 1.0$; $s_d = 5.24$; sample $t = 0.427$; P value > 0.250. Do not reject H_0. There is not enough evidence to conclude that there is a difference in the average number of service ware shards in subarea 1 compared to subarea 2.

11. H_0: $\mu_d = 0$; H_1: $\mu_d < 0$; left-tailed; t distribution with $d.f. = 5$; $t_0 = -2.015$; $\overline{d} = -10.5$; $s_d = 5.17$; sample $t = -4.97$; P value < 0.005; Reject H_0; The population mean heart rate after the test is higher than that before the test.

13. H_0: $\mu_d = 0$; H_1: $\mu_d > 0$; right-tailed; t distribution with $d.f. = 8$; $t_0 = 1.860$; $\overline{d} = 2.0$; $s_d = 4.5$; sample $t = 1.33$; $0.10 < P$ value < 0.125; Do not reject H_0; The population mean score on the last round is not significantly higher than the population mean score on the first round.

Section 10.2

1. H_0: $\mu_1 = \mu_2$; H_1: $\mu_1 > \mu_2$; right-tailed test with $z_0 = 2.33$. The z value corresponding to the sample test statistic $\overline{x}_1 - \overline{x}_2 = 0.7$ is $z = 4.22$. P value ≈ 0.0001. We reject H_0 and conclude that, on average, 10-year-old children have more REM sleep than do 35-year-old adults.

3. H_0: $\mu_1 = \mu_2$; H_1: $\mu_1 > \mu_2$; right-tailed test with $z_0 = 2.33$. The z value corresponding to the sample test statistic $\overline{x}_1 - \overline{x}_2 = 13$ is $z = 3.50$. P value ≈ 0.0002. We reject H_0 and conclude that the mean pollution index for Englewood is less than that for Denver.

5. H_0: $\mu_1 = \mu_2$; H_1: $\mu_1 \neq \mu_2$; two-tailed test with $\pm z_0 = 1.96$. The z value corresponding to the sample test statistic $\overline{x}_1 - \overline{x}_2 = -1.4$ is $z = -0.11$. P value $= 0.9124$. Since the sample test statistic does not fall in the critical region, we do not reject H_0. There is virtually no evidence of any difference in the average scores of the two groups.

7. (a) $z = -3.718$ (rounded) (b) $H_1: \mu_1 - \mu_2 \neq 0$; For a two-tailed test, the P value is "$P(Z < z)$ two-tail $= 0.0002$" (rounded). Since the P value is <0.05, we reject H_0 at the 5% level of significance. Based on our sample evidence, the football player average height differs from the basketball player average height.

9. (a) $z_c = 2.58$; $E = 7.51$; interval from 2.93 to 17.95. (b) The interval contains only positive values, so we can conclude at the 99% confidence level that the mothers' mean score is higher than that for fathers.

11. (a) $z_c = 1.96$; $E = 7.9689$; the interval is from -28.46 to -12.52. (b) The confidence inteval contains values that are all negative. At the 95% confidence level, we conclude that $\mu_1 < \mu_2$.

13. Based on the same data, a 989% confidence interval is longer than a 95% confidence interval. Therefore, if the 95% confidence interval has both positive and negative values, so will the 99% confidence interval. However, for the same data, a 90% confidence interval is shorter than a 95% confidence interval. The 90% confidence interval might contain only positive or only negative values even if the 95% interval contains both.

Section 10.3

1. (a) The means and standard deviations round to the results given.
(b) $H_0: \mu_1 = \mu_2$; $H_1: \mu_1 \neq \mu_2$; two-tailed test; $d.f. = 29$; $t_0 = \pm 2.045$; $s = 2.6389$. Sample test statistic $\bar{x}_1 - \bar{x}_2 = 0.82$ with $t = 0.865$; $0.25 < P$ value. Since the sample test statistic does not fall in the critical region, we do not reject H_0. There is not enough evidence to conclude that the average number of cases for fox rabies is different in the two regions.

3. (a) The means and standard deviations round to the results given. (b) $H_0: \mu_1 = \mu_2$; $H_1: \mu_1 \neq \mu_2$; two-tailed; $d.f. = 12$; $t_0 = \pm 2.179$; $s = 3.6745$; sample test statistics $\bar{x}_1 - \bar{x}_2 = 0.15$ with $t = 0.0764$; P value > 0.25; Do not reject H_0. The data do not indicate that the population mean time lost for hot tempers is different from that lost due to disputes.

5. (a) The means and standard deviations round to the results given. (b) $H_0: \mu_1 = \mu_2$; $H_1: \mu_1 < \mu_2$; left-tailed tests; $d.f. = 20$; $t_0 = -1.725$; $s = 2.592$. Sample test statistic $\bar{x}_1 - \bar{x}_2 = -3.6$ with $t = -3.244$. P

value < 0.005. Since the sample test statistic falls in the critical region, we reject H_0 and conclude that the average water temperature has increased.

7. (a) Use calculator to check that the means and standard deviations round to the values given.
(b) $s = 0.9573$; interval from 2.07 to 4.13.
(c) The interval consists of all positive numbers. At the 99% confidence level it appears that electronics companies have a higher profit as a percentage of revenue than food and drug companies.

9. (a) $\bar{x}_1 - \bar{x}_2 = 0.52$; $s = 3.3563$; interval from -1.29 to 2.33. (b) $\bar{x}_1 - \bar{x}_3 = 1.96$; $s = 3.4214$; interval from 0.11 to 3.81. (c) $\bar{x}_2 - \bar{x}_3 = 1.44$; $s = 3.6805$; interval from -0.55 to 3.43. (d) At the 85% confidence level we can say that the mean index of self-esteem based on competence is greater than the mean index of self-esteem based on physical attractiveness. We cannot conclude that there is a difference in mean index of self-esteem based on competence and that based on social acceptance. We cannot conclude that there is a difference in the mean indices based on social acceptance and physical attractiveness.

11. (a) Different signs; at the 95% confidence level, there does not appear to be a difference in family size.
(b) Since the P value for the test is 0.47, we see that we fail to reject H_0 at both the 1% and 5% levels of significance. There is not sufficient evidence that the family sizes differ. This result is consistent with the conclusion of part (a).

Section 10.4

1. $H_0: p_1 = p_2$; $H_1: p_1 \neq p_2$; two-tailed; $z_0 = \pm 2.58$; $\hat{p} = 0.0676$; sample test statistic $\hat{p}_1 - \hat{p}_2 = 0.0237$ with $z = 0.79$; P value $= 0.4296$; Do not reject H_0; The data do not indicate that the population proportions are different.

3. Let $p_1 =$ proportion who did not attend college and who believe in extraterrestrials and $p_2 =$ proportion who did attend college and who believe in extraterrestrials; $H_0: p_1 = p_2$; $H_1: p_1 < p_2$; left-tailed test; $z_0 = -2.33$; $\hat{p} = 0.42$. Sample test statistic $\hat{p}_1 - \hat{p}_2 = -0.10$ with $z = -1.43$. P value $= 0.0764$. Since the sample test statistic does not fall in the critical region, we do not reject H_0. There is not enough evidence to conclude that the first proportion is less than the second.

5. H_0: $p_1 = p_2$; H_1: $p_1 \neq p_2$; two-tailed; $z_0 = \pm 1.96$; $\hat{p}$ = 0.0234; sample test statistics $\hat{p}_1 - \hat{p}_2 = -0.0020$ with $z = -0.18$; P value = 0.8572; Do not reject H_0; The data do not indicate that the population proportions differ.

7. (a) $z_c = 2.58$; $\hat{\sigma} = 0.0232$; $E = 0.0599$; the interval is from 0.67 to 0.79.
   (b) The confidence interval contains values that are all positive, so we can be 99% sure that $p_1 > p_2$.

9. (a) $\hat{p}_1 = 0.3095$; $\hat{p}_2 = 0.1184$; $\hat{\sigma} = 0.0413$; 0.08 to 0.30   (b) Greater proportion of hogans occurs at Fort Defiance.

11. (a) All positive; Group I placement appears to have a higher hatch proportion.
   (b) The p value to 3 digits after the decimal is 0.000. Since the P value is less than 0.01, we reject H_0 at the 1% level of significance. The evidence indicates that the hatch proportion is higher for Group I placement. This result is consistent with the conclusion in part a.

Chapter 10 Review Problems

1. (a) Independent; normal; H_0: $\mu_1 = \mu_2$; H_1: $\mu_1 \neq \mu_2$; $z_0 = \pm 1.96$; $\bar{x}_1 - \bar{x}_2 = -1.06$ corresponds to $z = -4.94$; P value ≈ 0.0002; Reject H_0. There appears to be a difference in average ratings.
   (b) -1.48 to -0.64

3. Dependent, student's t; $\bar{d} = 0.775$; $s_d = 1.0539$; H_0: $\mu_d = 0$; H_1: $\mu_d > 0$; right-tailed test; $d.f. = 7$; $t_0 = 1.895$. The t value corresponding to the sample test statistic is $t = 2.080$; $0.025 < P$ value < 0.05. Reject H_0. At the 5% level of significance it seems that rats receiving larger rewards perform better on the ladder climb.

5. (a) Difference of proportions; normal; H_0: $p_1 = p_2$; H_1: $p_1 \neq p_2$; $z_0 = \pm 2.58$; $\hat{p}_1 - \hat{p}_2 = -0.0421$ corresponds to $z = -0.83$; P value = 0.4066; Do not reject H_0. Evidence does not show a difference in response accuracy. (b) -0.1721 to 0.088

7. (a) $s = 1.910$; interval from -1.59 to 0.79 min;
   (b) Some positive and some negative; There does not appear to be a difference in average waiting times at the 90% confidence level.

9. Difference of means; large independent samples; H_0: $\mu_1 = \mu_2$; H_1: $\mu_1 \neq \mu_2$; two-tailed; $z_0 = \pm 1.96$; $z = 1.82$ for $\bar{x}_1 - \bar{x}_2 = 0.3$; P value = 0.0688; Fail to reject H_0. The data do not indicate a difference in the population mean lengths of the two types of projectile points.

11. (a) Reject H_0; P value < 0.01
   (b) Reject H_0; P value < 0.05

CHAPTER 11

Section 11.1

1. H_0: Meyers-Briggs preference and profession are independent; H_1: Meyers-Briggs preference and profession are not independent; $\chi^2 = 43.5562$; $d.f. = 2$; $\chi^2_{0.01} = 9.21$; the sample statistic falls in the critical region; Reject H_0. Meyers-Briggs preference and profession are not independent.

3. H_0: Site type and pottery type are independent; H_1: Site type and pottery type are not independent; $\chi^2 = 0.5552$; $d.f. = 4$; $\chi^2_{0.05} = 9.49$; the sample statistic falls outside the critical region; Do not reject H_0; There is not sufficient evidence to conclude that site type and pottery type are not independent.

5. H_0: Age distribution and location are independent; H_1: Age and location are not independent; $\chi^2 = 0.6710$; $d.f. = 4$; $\chi^2_{0.05} = 9.49$; the sample statistic falls outside the critical region; Do not reject H_0; Age distribution and location appear to be independent.

7. H_0: Ages of young adults and movie preferences are independent; H_1: Ages of young adults and movie preferences are not independent; $\chi^2 = 3.6230$; $d.f. = 4$; $\chi^2_{0.05} = 9.49$; The sample statistic falls outside the critical region; Do not reject H_0; Age of young adults and movie preference appear to be independent.

9. (a) H_0: Ticket sales and type of billing are independent; H_1: Ticket sales and type of billing are not independent
   (b) 7.52
   (c) 1.868
   (d) Since the P value = 0.60 is greater than $\alpha = 0.05$, we do not reject H_0. It appears that type of billing and ticket sales are independent.

Section 11.2

1. H_0: the distributions are the same; H_1: the distributions are different; $\chi^2 = 11.788$; $d.f. = 3$; $\chi^2_{0.05} = 7.81$; Reject H_0; The distributions are different.

3. H_0: the distributions are the same; H_1: the distributions are different; $\chi^2 = 0.1984$; $d.f. = 4$; $\chi^2_{0.01} = 13.28$; Do not reject H_0; There is no evidence that the distributions are different.

5. (a) Essay (b) H_0: the distributions of temperatures fit a normal distribution; H_1: the distribution is not normal; $\chi^2 = 1.7076$; $d.f. = 5$; $\chi^2_{0.01} = 15.09$; Do not reject H_0.

7. H_0: the distributions are the same; H_1: the distributions are different; $\chi^2 = 9.333$; $d.f. = 3$; $\chi^2_{0.05} = 7.81$; Reject H_0; The fish distribution has changed.

9. H_0: the distributions are the same; H_1: the distributions are different; $\chi^2 = 13.70$; $\chi^2_{0.01} = 15.09$; Do not reject H_0; The distributions are the same.

Section 11.3

1. H_0: $\sigma^2 = 42.3$; H_1: $\sigma^2 > 42.3$; right-tailed; $d.f. = 22$; $\chi^2_{0.05} = 33.92$; $\chi^2 = 23.98$; Do not reject H_0.

3. H_0: $\sigma^2 = 136.2$; H_1: $\sigma^2 < 136.2$; left-tailed; $d.f. = 7$; $\chi^2_{0.99} = 1.24$; $\chi^2 = 5.92$; Do not reject H_0. The variance has not decreased.

5. H_0: $\sigma^2 = 9$; H_1: $\sigma^2 < 9$; left-tailed; $d.f. = 23$; $\chi^2_{0.95} = 13.09$; $\chi^2 = 9.23$; Reject H_0. The new typhoid shot has a smaller variance of protection times.

7. H_0: $\sigma^2 = 0.15$; H_1: $\sigma^2 > 0.15$; right-tailed; $d.f. = 60$; $\chi^2_{0.01} = 88.38$; $\chi^2 = 108$; Reject H_0; The variance is greater than 0.15 mm^2. All fan blades should be replaced.

Section 11.4

1. (c) $S_e = 2.1103$ (d) 35.55%
   (e) 31.06 to 40.04%

3. (c) $S_e = 2.2361$ (d) 20.5 miles per gallon
   (e) 17.1 to 23.9 miles per gallon

5. (c) $S_e = 0.9248$ (d) 10.3 thousand
   (e) 8.4 to 12.2 thousand

7. (c) $S_e = 8.996$ (d) 38.3
   (e) 24 to 52 (Interval will vary according to rounding)

9. (a) Yes. The pattern of residuals appears randomly scattered around the horizontal line at 0.
   (b) No. There do not appear to be any outliers.

Section 11.5

1. H_0: $\rho = 0$; H_1: $\rho > 0$. Critical value is 0.54. Because the observed sample statistic $r = 0.384$ does not fall in the critical region, we fail to reject H_0. At the 5% level, r is not significant. There is not enough evidence to conclude that there is a positive correlation between the time a movie has been running and its total revenue.

3. H_0: $\rho = 0$; H_1: $\rho > 0$. Critical value is 0.67. The sample statistic $r = 0.918$ falls in the critical region, so it is significant. We reject H_0 and conclude that there is a positive correlation between percentage change in wages and percentage change in consumer prices.

5. H_0: $\rho = 0$; H_1: $\rho < 0$. Critical value is -0.54. The sample statistic $r = -0.766$ falls in the critical region, so it is significant. We reject H_0 and conclude that there is a negative correlation between the number who had 4 or more years of college and the death rate.

7. H_0: $\rho = 0$; H_1: $\rho > 0$. Critical value is 0.66. The sample statistic $r = 0.945$ falls in the critical region, so it is significant. We reject H_0 and conclude that there is a positive correlation between elevation and water content of snowpack.

9. H_0: $\rho = 0$; H_1: $\rho > 0$; Critical value is 0.66; The sample statistic falls in the critical region so it is significant. Reject H_0. There appears to be a positive correlation between the tar content and the amount of carbon monoxide given off by a burning cigarette.

Chapter 11 Review Problems

1. H_0: time and score are independent; H_1: time and score are not independent; $\chi^2 = 3.92$; $\chi^2_{0.01} = 11.34$. Do not reject H_0.

3. H_0: the distributions are the same; H_1: the distributions are different; $\chi^2 = 33.93$; $\chi^2_{0.01} = 13.28$. Reject H_0.

5. H_0: $\sigma^2 = 15$; H_1: $\sigma^2 \neq 15$; $\chi^2 = 20.02$; Critical values are 35.48 and 10.28. (*Hint:* look up $\chi^2_{0.975}$ and $\chi^2_{0.025}$ so that 2.5% of the area is in each tail.) Do not reject H_0.

7. (a) 5.49 thousand (b) $S_e = 0.5368$; 4.45 to 6.53
   (c) H_0: $\rho = 0$; H_1: $\rho > 0$; Critical value = 0.81. Reject H_0; $r = 0.875$ is significant.

Index

FREQUENTLY USED FORMULAS

n = sample size N = population size f = frequency

Chapter 1

Class Width $= \dfrac{\text{high} - \text{low}}{\text{number classes}}$ (increase to next integer)

Class Midpoint $= \dfrac{\text{upper limit} + \text{lower limit}}{2}$

Lower boundary = lower boundary of previous class + class width

Chapter 2

Sample mean $\bar{x} = \dfrac{\Sigma x}{n}$

Population mean $\mu = \dfrac{\Sigma x}{N}$

Range = largest data value − smallest data value

Sample standard deviation $s = \sqrt{\dfrac{\Sigma(x - \bar{x})^2}{n - 1}}$

Computation formula $s = \sqrt{\dfrac{SS_x}{n - 1}}$ where

$SS_x = \Sigma x^2 - \dfrac{(\Sigma x)^2}{n}$

Population standard deviation $\sigma = \sqrt{\dfrac{\Sigma(x - \mu)^2}{N}}$

Sample variance s^2

Population variance σ^2

Sample Coefficient of Variation $CV = \dfrac{s}{\bar{x}} \cdot 100$

Sample mean for grouped data $\bar{x} = \dfrac{\Sigma xf}{n}$

Sample standard deviation for grouped data

$s = \sqrt{\dfrac{\Sigma(x - \bar{x})^2 f}{n - 1}}$

Chapter 3

Regression and Correlation

In all these formulas

$SS_x = \Sigma x^2 - \dfrac{(\Sigma x)^2}{n}$

$SS_y = \Sigma y^2 - \dfrac{(\Sigma y)^2}{n}$

$SS_{xy} = \Sigma xy - \dfrac{(\Sigma x)(\Sigma y)}{n}$

Least squares line $y = a + bx$ where $b = \dfrac{SS_{xy}}{SS_x}$ and

$a = \bar{y} - b\bar{x}$

Pearson product-moment correlation coefficient

$r = \dfrac{SS_{xy}}{\sqrt{SS_x SS_y}}$

Coefficient of determination $= r^2$

Chapter 4

Probability of the complement of event A
$P(not\ A) = 1 - P(A)$

Multiplication rule for independent events
$P(A\ and\ B) = P(A) \cdot P(B)$

General multiplication rules
$P(A\ and\ B) = P(A) \cdot P(B,\ given\ A)$
$P(A\ and\ B) = P(B) \cdot P(A,\ given\ B)$

Addition rule for mutually exclusive events
$P(A\ or\ B) = P(A) + P(B)$

General addition rule
$P(A\ or\ B) = P(A) + P(B) - P(A\ and\ B)$

Permutation rule $P_{n,r} = \dfrac{n!}{(n - r)!}$

Combination rule $C_{n,r} = \dfrac{n!}{r!(n - r)!}$

Chapter 5

Mean of a discrete probability distribution $\mu = \Sigma x P(x)$

Standard deviation of a discrete probability distribution
$\sigma = \sqrt{\Sigma(x - \mu)^2 P(x)}$

For Binomial Distributions

r = number of successes; p = probability of success; $q = 1 - p$

Binomial probability distribution $P(r) = \dfrac{n!}{r!(n - r)!} p^r q^{n-r}$

Mean $\mu = np$

Standard deviation $\sigma = \sqrt{npq}$

Chapter 6

Raw score $x = z\sigma + \mu$

Standard score $z = \dfrac{x - \mu}{\sigma}$

Chapter 7

Mean of $\bar{x}$ distribution $\mu_{\bar{x}} = \mu$

Standard deviation of $\bar{x}$ distribution $\sigma_{\bar{x}} = \dfrac{\sigma}{\sqrt{n}}$

Standard score for $\bar{x}$ $\quad z = \dfrac{\bar{x} - \mu}{\sigma/\sqrt{n}}$

Chapter 8

Confidence Interval

for μ(when $n \geq 30$)

$$\bar{x} - z_c \frac{\sigma}{\sqrt{n}} < \mu < \bar{x} + z_c \frac{\sigma}{\sqrt{n}}$$

for μ(when $n < 30$)

$$d.f. = n - 1$$

$$\bar{x} - t_c \frac{s}{\sqrt{n}} < \mu < \bar{x} + t_c \frac{s}{\sqrt{n}}$$

for p(when $np > 5$ and $nq > 5$)

$$\hat{p} - z_c \sqrt{\frac{\hat{p}(1 - \hat{p})}{n}} < p < \hat{p} + z_c \sqrt{\frac{\hat{p}(1 - \hat{p})}{n}} \text{ where } \hat{p} = r/n$$

Sample Size for Estimating

means $n = \left(\dfrac{z_c \sigma}{E}\right)^2$

proportions

$n = p(1 - p)\left(\dfrac{z_c}{E}\right)^2$ with preliminary estimate for p

$n = \dfrac{1}{4}\left(\dfrac{z_c}{E}\right)^2$ without preliminary estimate for p

Chapter 9

Sample Test Statistics for Tests of Hypotheses

for μ(when $n \geq 30$) $\quad z = \dfrac{\bar{x} - \mu}{\sigma/\sqrt{n}}$

for μ(when $n < 30$); $t = \dfrac{\bar{x} - \mu}{s/\sqrt{n}}$ with $d.f. = n - 1$

for p $\quad z = \dfrac{\hat{p} - p}{\sqrt{pq/n}}$ where $q = 1 - p$

Chapter 10

Sample Test Statistics for Tests of Hypothesis

for paired differences d $\quad t = \dfrac{\bar{d} - \mu_d}{s_d/\sqrt{n}}$ with $d.f. = n - 1$

difference of means large sample

$$z = \frac{(\bar{x}_1 - \bar{x}_2) - (\mu_1 - \mu_2)}{\sqrt{\dfrac{\sigma_1^2}{n_1} + \dfrac{\sigma_2^2}{n_2}}}$$

difference of means small sample with $\sigma_1 \approx \sigma_2$;
$$d.f. = n_1 + n_2 - 2$$

$$t = \frac{(\bar{x}_1 - \bar{x}_2) - (\mu_1 - \mu_2)}{s\sqrt{\dfrac{1}{n_1} + \dfrac{1}{n_2}}}$$

where $s = \sqrt{\dfrac{(n_1 - 1)s_1^2 + (n_2 - 1)s_2^2}{n_1 + n_2 - 2}}$

difference of proportions

$$z = \frac{\hat{p}_1 - \hat{p}_2}{\sqrt{\dfrac{\bar{p}\bar{q}}{n_1} + \dfrac{\bar{p}\bar{q}}{n_2}}} \text{ where } \bar{p} = \frac{r_1 + r_2}{n_1 + n_2}; \ \bar{q} = 1 - \bar{p};$$

$$\hat{p}_1 = r_1/n_1; \hat{p}_2 = r_2/n_2$$

Confidence Intervals

for difference of means (when $n_1 \geq 30$ and $n_2 \geq 30$)

$$(\bar{x}_1 - \bar{x}_2) - z_c \sqrt{\frac{\sigma_1^2}{n_1} + \frac{\sigma_2^2}{n_2}} < \mu_1 - \mu_2 < (\bar{x}_1 - \bar{x}_2)$$
$$+ z_c \sqrt{\frac{\sigma_1^2}{n_1} + \frac{\sigma_2^2}{n_2}}$$

for difference of means ($n_1 < 30$ and/or $n_2 < 30$ and $\sigma_1 \approx \sigma_2$
$$d.f. = n_1 + n_2 - 2$$

$$(\bar{x}_1 - \bar{x}_2) - t_c s \sqrt{\frac{1}{n_1} + \frac{1}{n_2}} < \mu_1 - \mu_2 < (\bar{x}_1 - \bar{x}_2)$$
$$+ t_c s \sqrt{\frac{1}{n_1} + \frac{1}{n_2}}$$

where $s = \sqrt{\dfrac{(n_1 - 1)s_1^2 + (n_2 - 1)s_2^2}{n_1 + n_2 - 2}}$

for difference of proportions

where $\hat{p}_1 = r_1/n_1; \hat{p}_2 = r_2/n_2; \hat{q}_1 = 1 - \hat{p}_1; \hat{q}_2 = 1 - \hat{p}_2$

$$(\hat{p}_1 - \hat{p}_2) - z_c\sqrt{\frac{\hat{p}_1\hat{q}_1}{n_1} + \frac{\hat{p}_2\hat{q}_2}{n_2}} < p_1 - p_2 < (\hat{p}_1 - \hat{p}_2)$$
$$+ z_c\sqrt{\frac{\hat{p}_1\hat{q}_1}{n_1} + \frac{\hat{p}_2\hat{q}_2}{n_2}}$$

Chapter 11

$$\chi^2 = \sum \frac{(O - E)^2}{E} \text{ where } E = \frac{\text{(row total)(column total)}}{\text{sample size}}$$

Tests of Independence $d.f. = (R - 1)(C - 1)$

Goodness of fit $d.f. = (\text{number of entries}) - 1$

Sample test statistic for H_0: $\sigma^2 = k$; $d.f. = n - 1$
$$\chi^2 = \frac{(n - 1)s^2}{\sigma^2}$$

Linear Regression

Standard error of estimate $S_e = \sqrt{\dfrac{SS_y - bSS_{xy}}{n - 2}}$

where $b = \dfrac{SS_{xy}}{SS_x}$

Confidence interval for y

$y_p - E < y < y_p + E$ where y_p is the predicted y value
for x and
$$E = t_c S_e\sqrt{1 + \frac{1}{n} + \frac{(x - \bar{x})^2}{SS_x}} \text{ with } d.f. = n - 2$$